Y0-BRL-922

SERIES ON SEMICONDUCTOR
SCIENCE AND TECHNOLOGY

Series Editors

SERIES ON SEMICONDUCTOR SCIENCE AND TECHNOLOGY

---

1. M. Jaros: *Physics and applications of semiconductor microstructures*
2. V.N. Dobrovolsky and V.G. Litovchenko: *Surface electronic transport phenomena in semiconductors*
3. M.J. Kelly: *Low-dimensional semiconductors*
4. P.K. Basu: *Theory of optical processes in semiconductors*
5. N. Balkan: *Hot electrons in semiconductors*
6. B. Gil: *Group III nitride semiconductor compounds: physics and applications*
7. M. Sugawara: *Plasma etching*
8. M. Balkanski and R.F. Wallis: *Semiconductor physics and applications*
9. B. Gil: *Low-dimensional nitride semiconductors*
10. L.J. Challis: *Electron–phonon interaction in low-dimensional structures*
11. V. Ustinov, A. Zhukov, A. Egorov, N. Maleev: *Quantum dot lasers*
12. H. Spieler: *Semiconductor detector systems*
13. S. Maekawa: *Concepts in spin electronics*

Concepts in Spin Electronics

Edited by

Sadamichi Maekawa

Institute for Materials Research,
Tohoku University, Japan

OXFORD
UNIVERSITY PRESS

Great Clarendon Street, Oxford OX2 6DP

Oxford University Press is a department of the University of Oxford.
It furthers the University's objective of excellence in research, scholarship,
and education by publishing worldwide in

Oxford New York

Auckland Cape Town Dar es Salaam Hong Kong Karachi
Kuala Lumpur Madrid Melbourne Mexico City Nairobi
New Delhi Shanghai Taipei Toronto

With offices in

Argentina Austria Brazil Chile Czech Republic France Greece
Guatemala Hungary Italy Japan Poland Portugal Singapore
South Korea Switzerland Thailand Turkey Ukraine Vietnam

Oxford is a registered trade mark of Oxford University Press
in the UK and in certain other countries

Published in the United States
by Oxford University Press Inc., New York

First published 2006

British Library Cataloguing in Publication Data
Data available

Library of Congress Cataloging in Publication Data
Data available

Printed in Great Britain
on acid-free paper by
Biddles Ltd., King's Lynn

ISBN 0–19–856821–5 978–0–19–856821–6

1 3 5 7 9 10 8 6 4 2

Preface

Nowadays information technology is based on semiconductor and ferromagnetic materials. Information processing and computation are performed using electron charge by semiconductor transistors and integrated circuits, but on the other hand the information is stored on magnetic high-density hard disks by electron spins. Recently, a new branch of physics and nanotechnology, called magnetoelectronics, spintronics, or spin electronics, has emerged, which aims to simultaneously exploit both the charge and the spin of electrons in the same device and describes the new physics raised. One of its tasks is to merge the processing and storage of data in the same basic building blocks of integrated circuits, but a broader goal is to develop new functionality that does not exist separately in a ferromagnet or a semiconductor.

Research in magnetic materials has long been characterized by unusually rapid transitions to technology. A prominent example is the discovery in 1988 of one of the first spin electronics effects, namely the giant magnetoresistance (GMR) effect in magnetic layered structures, which has already found market application in read heads in computer hard disk drives and also in magnetic sensors. Recently new technology based on the tunneling magnetoresistance (TMR) of magnetic tunnel junctions as magnetic random access memory (MRAM) is emerging into the electronic memory market. It is to be expected that future progress in spin electronics will lead to similarly rapid applications, in particular once the merging of semiconductor and magnetic technologies is achieved.

The aim of this book is to present new directions in the development of spin electronics, both the basic physics and technology in recent years, which will become the foundation of future technology. In the first part we give an introduction to ferromagnetic semiconductors: recent developments, new effects and devices. Further it will demonstrate how a spin current can be created, maintained, measured, and manipulated by light or an electric field, in several types of devices.

One very interesting and promising group of such devices, which allow us to control and manipulate a single spin, is ultrasmall systems called quantum dots (QDs), where Coulomb interaction (Coulomb blockade) plays an important role. In quantum dots due to the control of a single electron charge, the possibility of manipulating of a single spin is opened up, which can be important for quantum computing. On the level of a few spins, the new physics related to exchange interaction, spin blockade, Larmor precession, electron spin resonance (ESR), the Kondo effect, and hyperfine interactions with nuclear spins is raised. Also combining ferromagnetic materials with QDs opens up the new possibility of control and manipulation of a QD single spin by direct exchange interactions and construction of ferromagnetic single-electron transistors (F-SET).

Recent study of spin-dependent transport in hybrid structures involving a combination of ferromagnetic (both metallic and semiconducting) and normal or superconducting materials is reviewed. The interplay between the different types of interactions and correlations present in each produces a host of interesting spin-dependent effects, many of which have direct potentials for applications.

A very promising new effect and technology of spin current induced magnetization switching in magnetic nanostructures are discussed, together with potential applications. Another interesting field closely related with the miniaturization of magnetic systems is nanoscopic magnetism, where the cross-over between Stoner magnetism of the bulk magnetism to Hund's rules in molecular systems using tunneling spectroscopy can be studied.

In summary, spin electronics and spin optoelectronics promise to lead to a growing collection of novel devices and circuits that possibly can be integrated into high-performance chips to perform complex functions, where the key element will be integration of complex magnetic materials with mainstream semiconductor technology.

April 2005

Sadamichi Maekawa
(On behalf of the authors)

Contents

7 Theory of spin-transfer torque and domain wall motion in magnetic nanostructures 293

S. E. Barnes and S. Maekawa

Contributors

J. Barnaś
Department of Physics,
Adam Mickiewicz University,
61-614 Poznań,
Poland
Institute of Molecular Physics,
Polish Academy of Sciences,
60-179 Poznań,
Poland
barnas@main.amu.edu.pl

S. E. Barnes
Department of Physics,
University of Miami,
Coral Gables, Florida 33124,
U.S.A.
barnes@physics.miami.edu

R. A. Buhrman
Laboratory of Atomic and Solid State Physics,
Cornell University,
Ithaca, New York 14853,
U.S.A.
rab8@cornell.edu

J. Fabian
Institute of Theoretical Physics,
University Regensburg,
93040 Regensburg,
Germany
jaroslav.fabian@physik.uni-regensburg.de

H. Imamura
Graduate School of Engineering,
Tohoku University,
Sendai 980-8579,
Japan
hima@cmt.is.tohoku.ac.jp

X. Jiang
IBM Almaden Research Center,
650 Harry Road, K11-D2, San Jose,
CA95150-6099,
U.S.A.
xinjiang@us.ibm.com

S. Maekawa
Institute for Materials Research,
Tohoku University,
Sendai 980-8577,
Japan
CREST, Japan Science and Technology Agency,
Kawaguchi 332-0012,
Japan
maekawa@imr.tohoku.ac.jp

J. Martinek
Institut für Theoretische Festkörperphysik
Universität Karlsruhe,
76128 Karlsruhe,
Germany
Institute for Materials Research,
Tohoku University,
Sendai 980-8577,
Japan
Institute of Molecular Physics,
Polish Academy of Sciences,
60-179 Poznań,
Poland
martinek@ifmpan.poznan.pl

H. Munekata
Imaging Science and Engineering

Laboratory,
Tokyo Institute of Technology,
Yokohama, Kanagawa 226-8502,
Japan
hiro@isl.titech.ac.jp

K. Ono
Department of Applied Physics,
University of Tokyo,
Hongo, Bunkyo-ku, Tokyo 113-0033,
Japan
k-ono@riken.jp

S. S. P. Parkin
IBM Almaden Research Center,
650 Harry Road, K11-D2, San Jose,
CA95150-6099,
U.S.A.
parkin@almaden.ibm.com

D. C. Ralph
Laboratory of Atomic and Solid State Physics,
Cornell University,
Ithaca, New York 14853,
U.S.A.
ralph@ccmr.cornell.edu

S. Sasaki
NTT Basic Research Laboratories,
NTT Corporation,
Japan
satoshi@nttbrl.jp

M. Stopa
ERATO-SORST Quantum Spin Information Project,
JST, Atsugi-shi, Kanagawa 243-0198,
Japan
stopa@deas.harvard.edu

S. Takahashi
Institute for Materials Research,
Tohoku University,
Sendai 980-8577,
Japan
takahasi@imr.tohoku.ac.jp

S. Tarucha
Department of Applied Physics,
University of Tokyo,
Hongo, Bunkyo-ku, Tokyo 113-0033,
Japan
tarucha@ap.t.u-tokyo.ac.jp

I. Žutić
Center for Computational Materials Science,
Naval Research Laboratory, Washington, D.C. 20375,
U.S.A.
Department of Physics,
SUNY at Buffalo,
Buffalo, NY 14260-1500,
U.S.A.
Department of Physics,
Condensed Matter Theory Center,
University of Maryland,
College Park, MD 20742-4111,
U.S.A.
zigor@buffalo.edu

1 Optical phenomena in magnetic semiconductors

H. Munekata

1.1 Introduction

An electronic system having itinerant character with moderate carrier concentrations (10^{18}–10^{21} cm^{-3}) is an interesting system in that it allows us to study novel aspects of the interaction between charges and spins, with light and an electric field/current as tools to manipulate the interaction and to detect changes a consequence of the manipulation. Among various materials systems, semiconductor-based nanostructures are especially interesting, since findings obtained from the structures can be examined from both the fundamental and device-application points of view. The study of optical access to the spin degree of freedom can be found in early work on ferromagnetic compound semiconductors of multi-elements [1, 2], in which controlling magnetism via the direct optical excitation of local states (such as intra-atomic transition and defect states) was the primal interest. Lately, when II-VI-based paramagnetic semiconductors, named diluted magnetic semiconductors or semimagnetic semiconductors, became available in the laboratory [3], the band-to-band optical excitation was found possible to enhance magnetization through the effective magnetic field [4,5] and the formation of magnetic polarons [6]. This led us to the understanding of the spin exchange interaction between carrier spins and local spins, together with the possibility of optical access to the spin system through the delocalized states. In this chapter, we are concerned with the *manipulation of ferromagnetism* by the band-to-band optical excitation. The materials of choice are another class of magnetic semiconductor alloy based on III-V compounds. This class of semiconductor materials, as first demonstrated successfully by the molecular beam epitaxy of (In,Mn)As [7], contain a large number of magnetic ions (10^{20}–10^{21} cm^{-3}) in the host crystals despite of its low equilibrium solubility limit. The work carried out for more than a decade has revealed that various transition metal elements could be incorporated beyond the solubility limit in every III-V host crystal by properly choosing the epitaxial growth conditions. Furthermore, for the mid- and narrow gap III-V hosts, it became clear that Mn ions that occupy the group III sublattice sites are electrically and magnetically active and form the spin-selective, acceptor-like states near the top of the valence band. This spin-selective character causes both Mn ions and valence bands to be spin polarized when electrons (or holes) are shared between the two subsystems (hole-mediated ferromagnetism), making it possible to open ways to manipulate spins by changing characteristic parameters of carrier spins by external stimulation such as optical excitation.

Overall reviews of III-V-based magnetic semiconductors, hereafter called III-V-based magnetic alloy semiconductors (III-V-MAS), and hole-mediated ferromagnetism are already available [6,7]. Therefore, we do not take the redundant step of reviewing every aspect of the physical properties of III-V-based MAS, but rather focus on the essence of hole-mediated ferromagnetism together with a brief description of the historical development, and focus on the optical properties of the III-V-MAS. We then move on to the optical manipulation of ferromagnetically coupled spins by the energy and angular momentum of light. This is followed by the discussions of utilizing spin-dependent optical transitions and spin-dependent carrier transport to fabricate spin-dependent optical devices, e.g., polarized-light detectors and emitters, with specific applications in mind.

1.2 Optical properties of III-V-based MAS

1.2.1 *Brief history*

The realization of III-V-MAS was first reported in 1989 with paramagnetic n-type (In,Mn)As epitaxial layers prepared on GaAs(001) substrates by molecular beam epitaxy (MBE) [5]. They found that setting the substrate temperature $T_s(\approx 200\,^\circ\mathrm{C})$ during epitaxial growth far below the value used for conventional epitaxial growth ($T_s \approx$ 400–450 °C) resulted in the suppression of the second phase (MnAs, $GaMn_3$) in the regime beyond the equilibrium solubility limit. Consequently, the incorporation of Mn ions up to $x = 0.2$ in the form of $In_{1-x}Mn_xAs$ was realized. On the other hand, they were puzzled by the fact that the electrical conduction was n-type. In 1991, activation of holes from incorporated Mn ions was found possible for the (In,Mn)As layers prepared at $T_s \approx 300$°C [8], which has resulted in the establishment of hole-mediated ferromagnetism in p-(In,Mn)As in 1992 [9,10], and in the strained p-(In,Mn)As/(Al,Ga)Sb heterostructure in 1993 [11]. The study of ferromagnetic, electronic, and optical properties of the heterostructures revealed that those properties are correlated to each other and can be controlled by the band edge profile and strain, the representative parameters in band-gap engineering. The advantage coming from the marriage of band-gap engineering and ferromagnetism became even clearer in the study of ferromagnetic p-type (Ga,Mn)As which became available in 1996 [12,13] born after the study of GaAs:MnAs [14]. Since (Ga,Mn)As, the study of ferromagnetism in semiconductor-based materials became one of the subjects of intensive interest in solid state physics and spintronics. Theoretical treatments of carrier-mediated ferromagnetism and experimental investigations toward room-temperature ferromagnetism have been pursued by many laboratories from various points of views. Since the first demonstration of optical inducement of ferromagnetic order in 1997 [15], manipulation of strongly coupled spins by external stimulations, in particularly by light and an electric field, became an important research subject in semiconductor spintronics. Experimental demonstrations were initially carried out primarily in (In,Mn)As/(Al,Ga)Sb heterostructures [15–18], and are now taking place in the

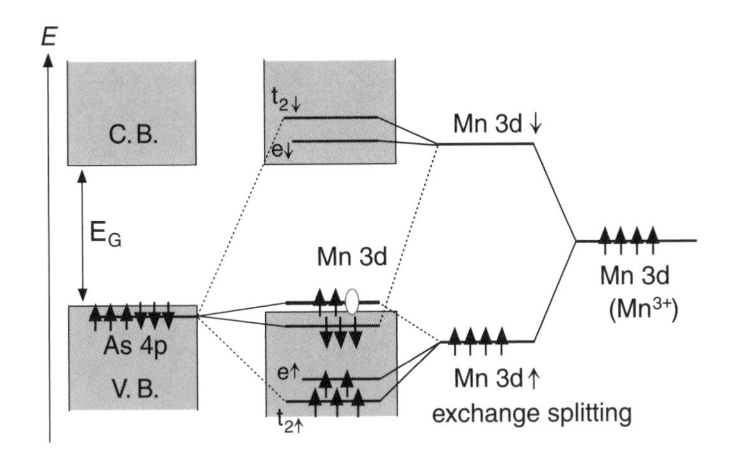

FIG. 1.1. Schematic illustration of Mn d orbital in the tetrahedral site with four As ions. The initial state of a Mn ion is regarded as Mn^{3+} as a consequence of transferring one electron to As atoms. (By courtesy of J. Okabayashi.)

technologically more important ferromagnetic semiconductor (Ga,Mn)As and other exotic material systems.

1.2.2 *Hole-mediated ferromagnetism*

It is very important to grab the concept of hole-mediated ferromagnetism and understand the influence of this phenomenon on the semiconductor band structure. We do so by describing qualitatively hole-mediated ferromagnetism in terms of the conventional band structure picture used in semiconductor physics.

(1) When Mn ($4s^2 3d^5$) is introduced substitutionally in the group III sublattice sites of a III-V matrix, their orbitals are hybridized with those of the host crystal, yielding two sets of states with different characters: one kind is the split-off states near the top of the valence band and another kind is the localized states deep inside the valence band, as shown schematically in Fig. 1.1. There is one hole in the former states, having acceptor-like character [24, 25], whereas the later states preserve the $3d^5$ character, exhibiting the local magnetic moment of $S = 5/2$. As a whole, in the dilute limit, the ground state of Mn in the III-V host is "neutral Mn plus one hole" [26].

(2) When the split-off state E_A captures an electron from the valence band by, say, thermal excitation, the relative spin alignment between the captured electron and the resultant hole in the valence band is antiparallel, whereas, between the captured electron and the local spins, the alignment is parallel. Therefore, the relative spin alignment is antiparallel between the hole and the local spins, as shown schematically in Fig. 1.2(a).

(3) When the Mn concentration is increased and reaches a value high enough to realize the degenerate condition in the valence band, the itinerant char-

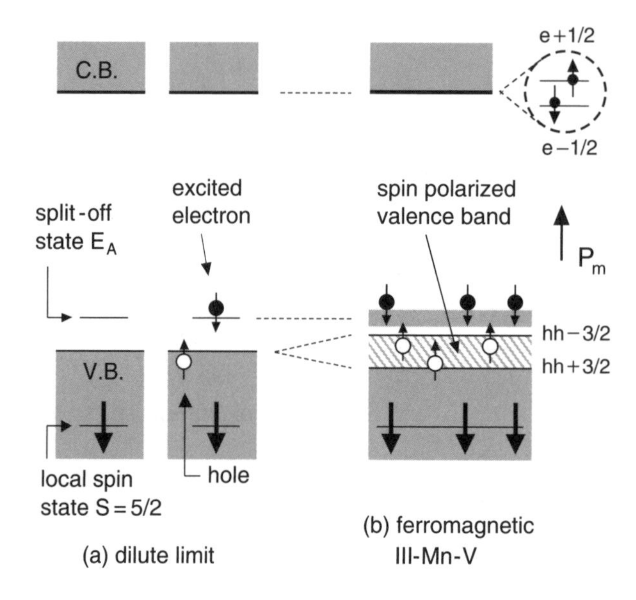

FIG. 1.2. Schematic band structures of a III-V compound semiconductor doped with the transition metal Mn for (a) low and (b) very high concentrations. The energy separation between the top of the valence band and E_A is 110 meV in the dilute limit for GaAs [24, 25]. In the III-V-MAS, the itinerant character of holes is strong enough to sustain the spin-split valence band and ferromagnetic coupling among Mn local spins. Note that the axis of the magnetic moment P_m (magnetic polarization) is pointing upward, whereas the axis of the angular momentum of an electron spin points downward. Spin splitting of electrons in the conduction band is also shown in the area surrounded by the dotted circle.

acter of the hole manifests itself as hole-mediated ferromagnetism since holes have to be shared among numbers of Mn ions and, in order to do so, antiparallel alignment has to be retained between Mn local spins and hole spins, as shown schematically in Fig. 1.2(b). Simultaneously, holes are spontaneously spin polarized, being antiparallel to the local spins. In terms of the band expression based on electrons, the valence band is spin-split with the hh−3/2 state being closer to the impurity state. Resonant states may be formed in relation with the itinerant process [27]. The strength of ferromagnetic coupling can be viewed as the reduction of the total energy in the carrier system when they are populated in the spin-split band [28, 29]. In this description, Curie temperature T_c can be expressed as $T_c \propto x' \times p^{1/3}$, where x' and p are the effective Mn content out of the chemical content x in the form of $III_{1-x}Mn_xV$ and the hole concentration.

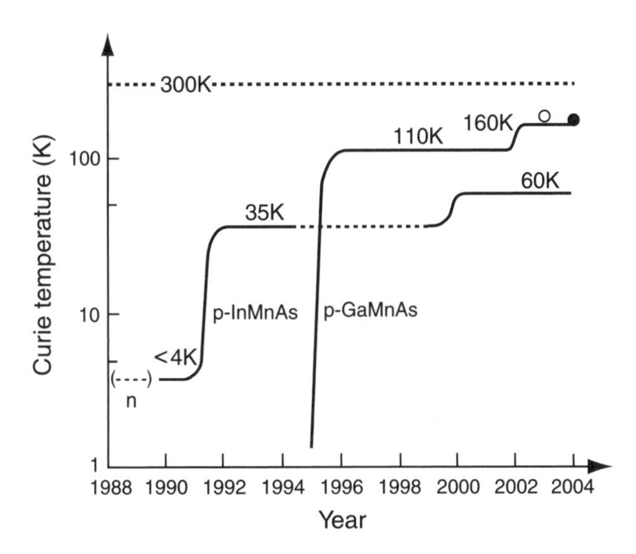

FIG. 1.3. Chronological evolution of the Curie temperature T_c of ferromagnetic III-V-MAS. The open circle depicts the T_c value obtained for modulation doped structures [37], whereas the closed circle represents that achieved for bulk (Ga,Mn)As [33].

(4) When the Mn concentration is low ($\lesssim 10^{18}\,\mathrm{cm}^{-3}$) and the system is in the insulator (semiconductor) regime, holes are localized in the Mn split-off state at low temperatures so that there is no particular long-range alignment among Mn local spins. At high enough temperatures, at which holes are activated, the energy gain caused by the hole-mediated ferromagnetic order is inferred to be substantially smaller than the thermal energy. As a whole, in the insulator regime, paramagnetism would be the dominant magnetic behavior. Note that, in the II-VI based diluted magnetic semiconductors, the split-off states are fully occupied by electrons, and for this reason the path for the spin exchange with itinerant electrons is completely blocked.

On the basis of the pictures given above, we are able to infer that one of the important points to realize hole-mediated ferromagnetism is the energy level of the split-off state E_A with respect to the top of the valence band. This state should be close enough to generate sufficient numbers of holes in the valence band. The energy separation between the top of the valence band and E_A depends on the extent of hybridization between the Mn d orbital and the host sp^3 orbital, and therefore being dependent on the host semiconductor. Among various combinations of transition metal elements and host III-V semiconductors, it is (Ga,Mn)As and other III-Mn-V having the band gap smaller than that of GaAs (except for InP) in which hole-mediated ferromagnetism has been clearly observed. Direct evidence of the effective-mass type holes had not been found for a long time because of the low carrier mobility (1–100 $cm^2/V\,sec$). However, the presence of effective-mass-type holes was finally established in 2004 through the observation of cyclotron resonance absorption at ultrahigh magnetic fields in ferromagnetic p-(In,Mn)As [30].

Figure 1.3 summarizes the evolution of the Curie temperature T_c of III-V-MAS based on hole-mediated ferromagentism. At the time of writing this chapter, the highest T_c is 55–60 K with $x = 0.07$ for (In,Mn)As [31], and 160–170 K with $x = 0.07$–0.09 [32, 33] for (Ga,Mn)As. T_c of (In,Ga,Mn)As is 100–130 K with relatively high Mn contents [34, 35]. Hole concentrations are in the range of $p = 10^{20}$–$10^{21}\,cm^{-3}$. In addition, with Hall effect measurements, the electrochemical capacitance method has been proven to be another useful technique to estimate the p value [36]. For delta-doped Mn ($x = 0.3\,ML$) with two-dimensional holes ($p = 2 \times 10^{12}\,cm^{-3}$), $T_c = 172\,K$ has been reported [37]. It is also worth noting that ferromagnetic behavior up to around 200 K for GaMnP:C [38], as well as $T_c = 600$–$900\,K$ [39, 40] for GaMnN and GaCrN, has been reported. The origin of very high T_c values in nitrides and other wide-gap systems [6] is not well understood at present. This is primarily because of their highly resistive characteristics, which makes it difficult to simply apply the framework of hole-mediated ferromagnetism. In contrast to Mn in GaAs, the Mn level in GaN appears to be deep, being around the middle of the gap [41,42]. The question arises whether the hopping conduction through the Mn impurity band results in ferromagnetism, since the exchange interaction may become short range reflecting the localized nature of the mid-gap states.

For mid- and narrow-gap III-Vs, the factor that limits T_c at present is the saturation of numbers of substitutional Mn ions. Instead of being incorporated substitutionally, the number of Mn ions incorporated interstitially as double donors [43] increases at the relatively high Mn contents for which a high Curie temperature is expected. This problem, being reminiscent of the charge compensation problem which often encountered in wide-gap semiconductors, can be recognized as an important research subject in the field of epitaxial growth.

1.2.3 *Optical properties*

As one can infer from the schematic spin-polarized band structures shown in Fig. 1.2, the influence of the selection rule for the angular momentum ($\Delta J = \pm 1$) on the optical transition becomes important, giving rise to circular dichroism and optical rotary power of magnetic origin, the so-called magneto-optical (MO) effects. In fact, a large magneto-optical effects with ferromagnetic characteristics has been established for (In,Mn)As as well as (Ga,Mn)As so far. One such example is shown, respectively, in Figs 1.4(a) and (b) for the Faraday ellipticity (η_F) and rotation (θ_F) spectra of a ferromagnetic $Ga_{0.96}Mn_{0.04}As/Al_{0.39}Ga_{0.61}As/GaAs(001)$ sample. The hysteresis loop of the Faraday ellipticity is also shown in the inset. The Kerr ellipticity (η_K) and rotation (θ_K) spectrum obtained from ferromagnetic (In,Mn)As samples will be discussed and shown later in Fig. 1.6 for comparison. For (Ga,Mn)As, a large enhancement in both η and θ spectra is

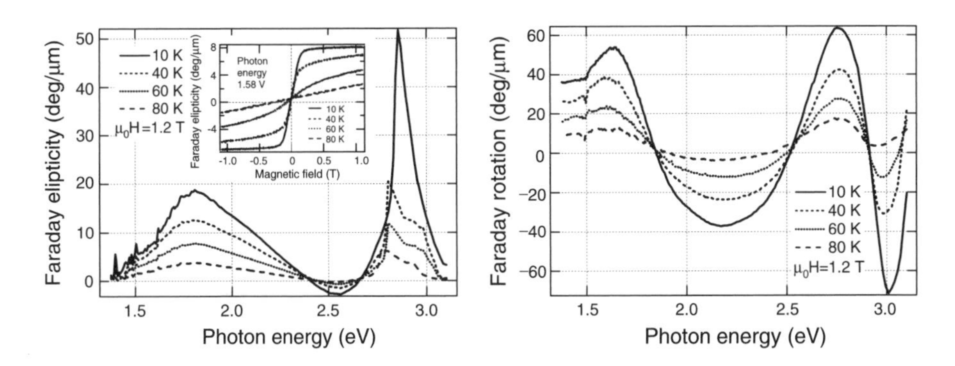

FIG. 1.4. (a) Faraday ellipticity spectra of a (Ga,Mn)As ($x = 0.04$) sample with an external field of $\mu_0 H_\perp = 1.2\,\mathrm{T}$ at various temperatures, together with hysteresis curves taken at $h\nu = 1.59\,\mathrm{eV}$, and (b) Faraday rotation spectra. A GaAs(001) substrate was removed by chemical selective wet etching. (Data obtained in Munekata Lab.)

noticeable at around $h\nu \sim 1.5\,\mathrm{eV}$ as well as $h\nu \sim 3.0\,\mathrm{eV}$. The spectra indicate that the incorporation of Mn ions of this level of quantity successfully dissolve the spin degeneracy of the energy bands through ferromagnetism without severely destroying the entire band structure of the host semiconductor GaAs. On the basis of this understanding, the circular dichroism bands at around $h\nu \sim 1.5\,\mathrm{eV}$ and $h\nu \sim 3.0\,\mathrm{eV}$ can probably be assigned due to the E_0 (the band gap; $\Gamma_{8\mathrm{v}} \rightarrow \Gamma_{6\mathrm{c}}$) and the E_1 transitions ($\Lambda_{4,5\mathrm{v}} \rightarrow \Lambda_{6\mathrm{c}}$). Magneto-optical effects with similar characteristic spectra were noticeable in earlier studies from several groups in the transmission [44,45] and reflection [46] configurations. It is worth noting that the ellipticity signal appears in the energy region which is slightly less than the band gap of GaAs. One would infer the reduced band gap of ferromagnetic (Ga,Mn)As with respect to GaAs due to spontaneous spin splitting. However, the presence of this low-energy absorption tail above the Curie temperature ($T_c \sim 60\,\mathrm{K}$) suggests that this part of the optical transition is rather attributed to the band tailing induced by the alloy fluctuation and the Mn impurity band. Nevertheless, it is important to keep in mind that the band tail and the delocalized band can both also be spin polarized through hole-mediated ferromagnetism.

It is very important to consider the sign of the Faraday ellipticity in relation with the spin-dependent band structure. Let us start by stating the definition of the ellipticity in the experiments by which the spectra shown in Fig. 1.4 were obtained. Firstly, we set the direction of the magnetic polarization, as well as the magnetic flux, parallel to the propagation direction of light. Viewing from the light source (Fig. 1.5), we define the clockwise rotation of an electric field plane toward the direction of the propagation of light as the "*positive*" rotation (the right circular polarization σ^+) [47]. On the other hand, the clockwise rotation of the transmitted light is detected as the counter-clockwise rotation, namely the

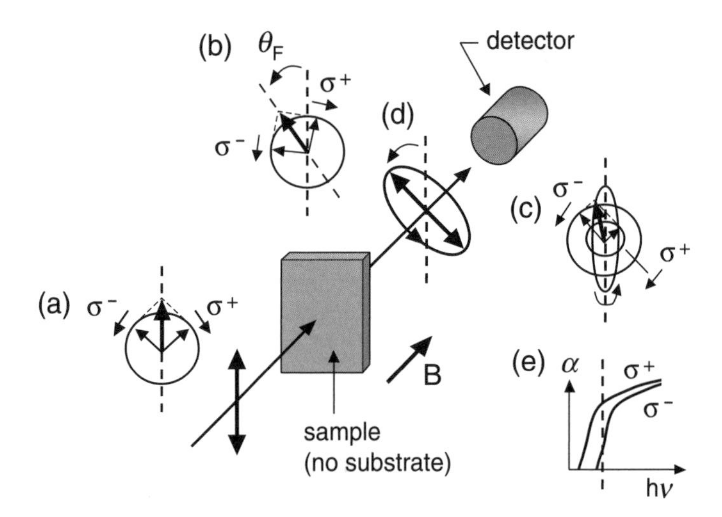

FIG. 1.5. Right σ^+ (left σ^-) circular polarization is defined as the polarization whose electrical plane rotates clockwise (counter-clockwise) toward the direction of light propagation. The linearly polarized light is the superposition of the σ^+ and σ^- light, as shown in panel (a). The difference in the phase velocity between the σ^+ and σ^- light in the sample gives rise to rotation of the electrical plane as shown in panel (b), whereas the difference in the amplitude between the σ^+ and σ^- light in the sample yields the ellipticity (magnetic circular dichroism) as depicted in panel (c). These two effects change the linearly polarized incident light into light having Faraday rotation and ellipticity when it is transmitted through the sample, as shown in panel (d). In the case of (d), the signs of rotation and ellipticity are both negative. The negative ellipticity indicates the relatively larger absorption of the σ^+ light with respect to σ^-, as shown in panel (e). Note, in some of magneto-optical spectrometers, the sign of the rotation and ellipticity is defined by viewing the transmitted light signal from the detector. In this case, the sign of both the ellipticity and rotation are reversed, as was the case for Figs 1.4(a) and (b). The sign of the magnetic flux is defined to be positive when it is parallel to the propagation direction of light.

negative rotation, viewing from the detector. For the data shown in Figs 1.4(a) and (b), the sign is defined in this fashion. Therefore, the positive Faraday ellipticity signal obtained by experiment is actually negative (counter-clockwise), indicating that the absorption of the σ^+ light (α^+) is stronger than that of the σ^- light (α^-), as shown schematically in Fig. 1.5 (c). Similarly, the positive Faraday rotation signal near the band-gap energy is actually the negative rotation (Fig 1.5b). In terms of the optical transition, α^+ is attributed to the interband transition involving $\Delta J = +1$ and is given by the joint density of states consisting of the $-3/2$ valence band and the $-1/2$ conduction band, whereas α^- is due to the transition with $\Delta J = -1$ and is given by the joint density of states consisting of the $+3/2$ valence and the $+1/2$ conduction bands. This situation can be illustrated schematically in terms of polarization-dependent absorption spectra as shown schematically in Fig. 1.5(e) in which the onset energy of the interband transition with $\Delta J = +1$ is lower than that of the transition with $\Delta J = -1$. In conclusion, the negative (positive) sign of the Faraday ellipticity defined by the light source (detector) is an indication that, in the ferromagnetic (Ga,Mn)As, the $|-3/2\rangle$ state is lifted up with respect to the $|+3/2\rangle$ state in the heavy-hole valence band. This picture is consistent with the schematic band structure (Fig 1.2b) deduced independently by considering the itinerant process of electrons between the Mn split-off state and the heavy-hole valence band.

This simplified conclusion holds well in the real situation in which a large number of holes are present in the ferromagnetic *p*-type (Ga,Mn)As. Under this condition, the optical transition does not take place at the center of the Brillouin zone. Nevertheless, the joint density of states for the α^+ transition is always larger than that for the α^- transition, and the sign of the ellipticity remains negative in principle. However, as pointed out in Ref. [45], the spontaneous spin splitting below the Curie temperature results in the occurrence of an additional component whose sign of ellipticity is opposite to the main signals in the photon energy region between 1.5 and 1.75 eV. This can be seen as the evolution of the concave feature appearing in the ellipticity spectra below the Curie temperature (Fig. 1.4a).

That the $-3/2$ valence band is higher in energy than the $+3/2$ valence band in the ferromagnetic state is reminiscent of the Zeeman splitting in the valence band of (II,Mn)VI-based diluted magnetic semiconductors in which the sign of the spin exchange splitting $N_0\beta$ is defined to be negative [48]. Following this notation, we are able to state that the sign of the spontaneous Zeeman splitting in the valence band is negative in III-V based ferromagnetic semiconductors. The absolute value of $N_0\beta$ determined from the Faraday rotation measurement of a ferromagnetic (Ga,Mn)As ($x = 0.04$) is around 1 eV [44] which is comparable to the value deduced from the analysis of the x-ray photoemission spectrum for (Ga,Mn)As with Mn contents of $x = 0.07$ [49]. The interpretation of the ellipticity data taken with the reflection geometry has also reached the same conclusion [46].

Similarly, as shown in Fig. 1.6, a large MO effect has been observed in ferromagnetic (In,Mn)As [50–52] with characteristic structures around the critical points of the host InAs. Broad and strong magnetic circular dichroism (MCD) signals appear at photon energies that correspond to the E_1, $E_1 + \Delta_1$, E_0', and E_2 critical points of InAs. These facts are clearly evidence of spin-polarized bands due to hole-mediated ferromagnetism. Its $N_0\beta$ value, however, has not yet been determined by the optical method. The likely value for this system is around -1 eV, which has been deduced from the analysis of cyclotron resonance absorption of both *n*-type paramagnetic [53] and *p*-type ferromagnetic samples

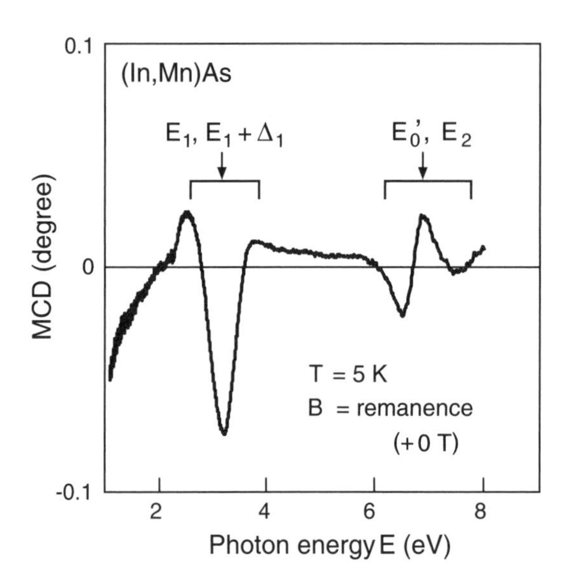

FIG. 1.6. Kerr ellipticity spectra of (In,Mn)As at an external field of $\mu_0 H_\perp = 1\,\mathrm{T}$ (after Ref. [52]).

[30], together with the $N_0\alpha$ (spin exchange splitting of the conduction band) of $+0.5\,\mathrm{eV}$. While the enhanced MCD signals have also been observed in the wide-band gap III-V-MAS, (Ga,Mn)N [54] and (In,Ga,Mn)N [55], the magnetic field dependence of the signal around the fundamental absorption edge reveals that the energy states associated with this transition are not spontaneously spin-polarized. MCD signals in the photon energy of about half of the band gap energy are also noticeable [55], which is most likely due to the state associated with the incorporated Mn.

Another important aspect of a ferromagnetic semiconductor is the shift of the fundamental absorption edge as a consequence of alloying with Mn or the formation of the spin polarized band, or both. As shown in Fig. 1.7 for (In,Mn)As [56], however, the absorption edges lose their sharpness by adding Mn, indicating a severe Ulback-type tailing [57]. This tailing can be understood qualitatively in terms of alloy fluctuation, formation of the Mn impurity band near the valence band top, plus the Burstein–Moss effect due to a large number of holes (10^{20}–$10^{21}\,\mathrm{cm}^{-3}$). Because of all these, the determination of the energy shift of magnetic origin is not well resolved in both (In,Mn)As and (Ga,Mn)As up to now. In (Ga,Mn)As, the presence of the Mn impurity band was found by infrared photoconductivity measurements [58, 59]. An approach with ellipsometry will be a strong tool to elucidate the influence of the incorporation of a large amount of Mn on the host band structure, as reported in (Ga,Fe)As [60] and most recently in (Ga,Mn)As [61].

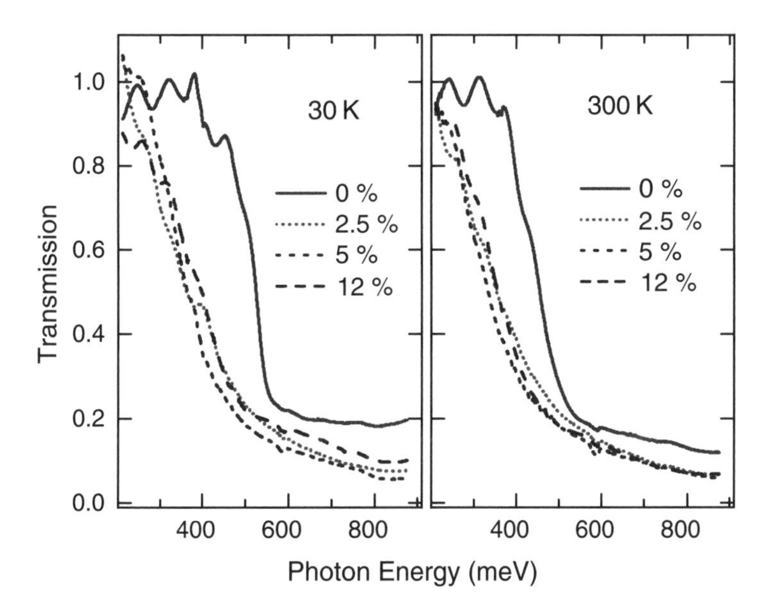

FIG. 1.7. Transmission spectra of paramagnetic, n-type (In,Mn)As epilayers with different Mn content x in the form of $In_{1-x}Mn_xAs$ at 30 and 300 K (after Ref. [56]). Epilayers are grown on GaAs(100) substrates.

Presumably because of the presence of large amounts of non-radiative recombination centers, studies on luminescence of III-V-MAS have been scarce. In nitrides [55,62–64], photoluminescence near the band edge were reported, but the magnitude of optical polarization remained unclear. Suppression of non-radiative processes, which probably shares common problems with the quest of room temperature ferromagnetism, is one of the remaining big challenges in III-V-MAS.

1.3 Photo-induced ferromagnetism

1.3.1 *Effect of charge injection I: photo-induced ferromagnetism*

Studies of magnetic phenomena induced by light in condensed matter have long been a subject of interest in both fundamental and applied research. For magnetic insulators and semiconductors, this subject has its root in the study of optical excitation of magnetic semiconductors of multi-elements [1, 2]. The same research interest can also be found in organic-based magnets [65]. The work in paramagnetic II-VI diluted magnetic semiconductors [4–6] and III-V-based magnetic alloy semiconductors (MAS) [20–22] has shed light on the importance of the spin exchange interaction between carrier spins and local moments for the manipulation of both weakly and strongly coupled spins via light. For metal systems, the transfer of quasi-momentum from photons to the free electron system at the metal surface has been demonstrated [66, 67]. In all of these experiments, the photonic process can be seen clearly at cryogenic temperatures. However, in

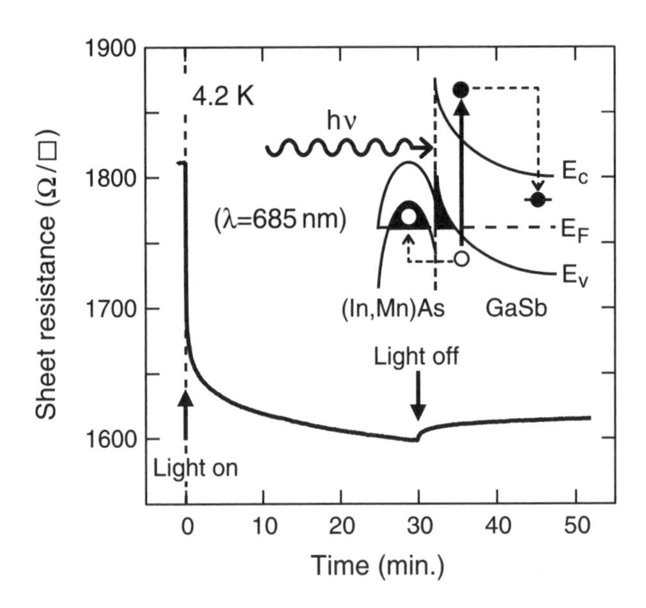

FIG. 1.8. Persistent photoconductivity in the (In,Mn)As/GaSb heterostructure, and the schematic band-edge profile across the heterostructure (after Ref. [69]).

general, one has to carry out very careful experiments to distinguish the photonic process from the effects associated with light-induced heating.

The first demonstration of photo-induced effects in III-V-MAS was carried out in the (In,Mn)As/GaSb heterostructure [20]. Properly adjusting the ratio between uncompensated and compensated Mn ions by controlling the growth condition during molecular beam epitaxy, the hole concentration was tuned up slightly below the value necessary for induced ferromagnetism [68]. When the heterostructure is irradiated with light, the light passes through a very thin (In,Mn)As layer ($E_g = 0.4\,\text{eV}$) and is absorbed predominantly in the relatively thick GaSb layer ($E_g = 0.8\,\text{eV}$), yielding electrons and holes in the *non-magnetic* GaSb layer. Here, the important contrivance is the type II band alignment. As shown schematically in Fig. 1.8, photogenerated holes are transferred into the *magnetic* (In,Mn)As layer by the electric field across the interface, and are stored in the (In,Mn)As layer. This is the optical charge injection and accumulation. An increase in hole numbers enhances electrical conductivity as well as the strength of the carrier-mediated ferromagnetic interaction. This results in persistent photoconductivity (Fig. 1.8) and an increase in magnetization as shown in Fig. 1.9.

In this experiment, a change in magnetization was measured in the dark after light illumination, by which the effect coming from the light-induced heating could be excluded. This is why magnetization before and after the light illumination could be compared quantitatively, making it possible to confirm a magnetic

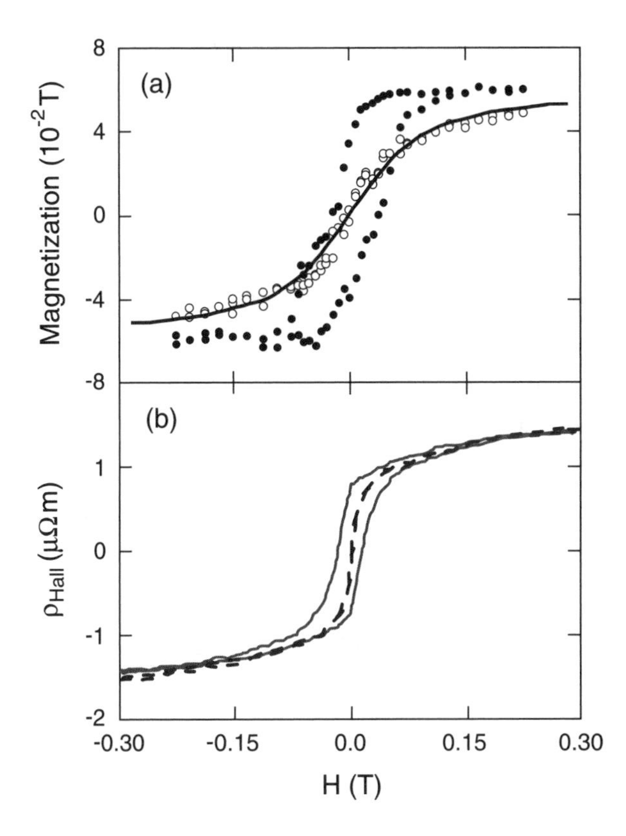

FIG. 1.9. Magnetization (a) and Hall resistance (b) data before and after light illumination. Data were taken at 4 K. In (a), open and close circles are the data taken before and after the illumination, respectively. In (b), broken and continuous lines are the data taken before and after the illumination, respectively (after Ref. [15]).

phase transition induced by optical charge injection. The value of the saturation magnetization after the optical injection indicates that about 80% of Mn ions in the (In,Mn)As ($x = 0.07$) epilayers contribute to the occurrence of ferromagnetism. The magnetization curve measured before the illumination showed that the sample was close to the ferromagnetic transition condition. The enhanced magnetization eventually vanishes at around 40 K, whereas persistent photoconductivity is noticeable up to around 175 K. Taking the vanishing temperature to be roughly comparable to the Curie temperature, the number of holes accumulated by the light illumination is inferred to be $10^{20}\,\text{cm}^{-3}$ or higher. The experiment in which a change in magnetization was detected electrically by the anomalous Hall effect also demonstrated optically induced ferromagnetism as shown in Fig. 1.9(b). These data indicate that magnetization and carrier transport are strongly correlated. The difference in the shape of hysteresis loops between Figs. 1.9(a) and (b) presumably comes from inhomogeneity of the sample wafer. An increase in hole numbers after light illumination can be seen as slight reduction in the Hall resistance in the saturation region. These results have established experimentally that ferromagnetism in III-V-MAS is hole mediated.

The data shown in Fig. 1.9 were taken after the sample was illuminated with the white light source whose wavelengths extend between 0.8 and 1.5 μm (0.83–1.55 eV). Other experiments carried out by using semiconductor lasers for which the excitation wavelength is monochromatic revealed the discrepancy in wavelength dependence between persistent photoconductivity and enhanced magnetization [69]. In the wavelength region of 700 nm–1.55 μm, a change in magnetization became smaller with increasing the wavelength and vanished at the wavelength of 1.55 μm ($h\nu = 0.8\,\mathrm{eV}$), whereas photoconductivity did not. One likely explanation for this discrepancy is the presence of a hole accumulation region in the GaSb side of the (In,Mn)As/GaSb interface, as shown schematically in the inset of Fig. 1.8 by the black triangular area. Similar photo-induced enhancement in magnetization has also been observed in (In,Mn)(As,Sb)/InSb heterostructures [70], and more recently in the modulation doped p-type GaAs/AlGaAs:Be having a sheet of Mn delta-doping in the two-dimensional hole gas [71]. These experiments have confirmed the universality of the optical charge injection and enhanced magnetization in III-V-MAS. Placing a gate electrode on top of the p-(In,Mn)As/AlSb heterostructure, one also achieved gate-induced switching between ferromagnetic and paramagnetic states [21], the adiabatic way to manipulate ferromagnetism. Photo-induced changes in magnetization in the paramagnetic regime have also been pursued by using (In,Mn)As/GaSb heterostructures. The induced change appeared to be significantly smaller compared with that in the ferromagnetic regime [72].

1.3.2 *Effect of charge injection II: optical control of coercive force*

Another important aspect of carrier-mediated magnetism is that the alteration of carrier numbers causes a significant influence on magnetic characteristics. One such example is the reduction of coercive force by light illumination [22]. In Fig. 1.10, we show one of the experimental results that demonstrates the reduction of coercive force by light irradiation. Magnetization hysteresis curves before and after light illumination were obtained electrically from magnetoresistance R_{Sheet} and Hall resistance R_{Hall} data, assuming the side-jump scattering mechanism [73]. As detailed in the previous section, the holes primarily generated in the GaSb layer are transferred to the ferromagnetic (In,Mn)As layer and stored in this layer. The experimental data clearly show that the coercive force is reduced after light illumination. In this experiment, the coercive force was reduced to about 45% at 35 K. The reduced coercive force lasts even after turning the light off, reflecting hole storage in the (In,Mn)As layer.

Figure 1.11 shows the relation between the number of extra holes Δp and the coercive force H_c. Here, Δp is estimated from the magnitude of persistent photoconductivity $\Delta\sigma = e\mu_p\Delta p$ which is depicted in the inset. The number of

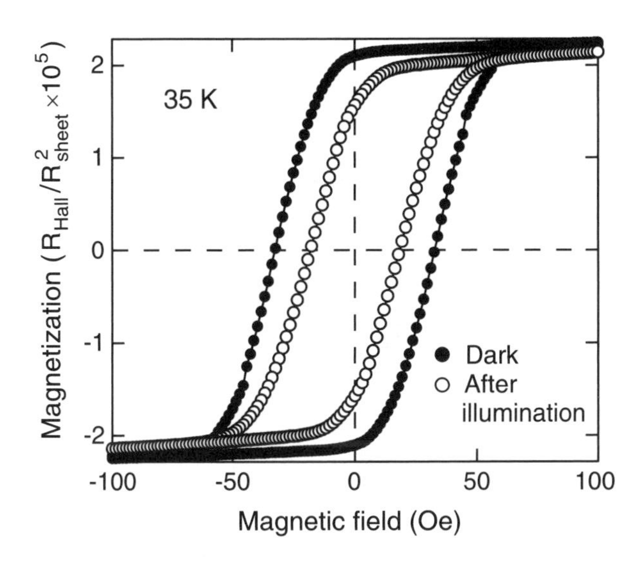

FIG. 1.10. Photo-induced change of coercive force observed in a ferromagnetic (In,Mn)As/GaSb heterostructure with Mn contents $x = 0.07$ (after Ref. [22]). The excitation wavelength and power are $\lambda = 685\,\text{nm}$ and $P = 6\,\text{mW/cm}^2$ at the (In,Mn)As surface, respectively. The sample shows perpendicular magnetic anisotropy with Curie temperature $T_c \sim 40\,\text{K}$.

holes that were accumulated in the (In,Mn)As layers during the light illumination was 10^{17}–$10^{18}\,\text{cm}^{-3}$, being about 3% of the background hole concentration. On the other hand, ΔH_c at 4.2 K is about 1.5%. The difference in the magnitude of reduction in coercive force between high (35 K) and low (4.2 K) temperature indicates the contribution of thermal activation in the process of magnetization reversal.

In general, there are two different mechanisms in the process of magnetization reversal; firstly, reversal through the domain wall motion, and, secondly, that involving the rotation of magnetic objects. In the domain wall model, magnetization reversal occurs through the displacement of magnetic domain walls [74], whereas, in the particle model, the reversal takes place through the rotation of the aggregates of single-domain ferromagnetic particles [75, 76]. When the behavior shown in Fig. 1.10 was found, the observed phenomenon was explained in terms of the domain wall model which is the mechanism applicable for conventional ferromagnets. However, as shown in Fig. 1.12, it was found lately that the dependence of the direction of an applied magnetic field on the coercive force favors the particle model. The likelihood of this picture comes from the fact that unsaturated ferromagnetic *p*-InMnAs consists of large magnetic clusters composed of 500 Mn ions and holes [11]. A sheet of fully magnetized layer may be decomposed into single-domain-type magnetic clusters in the event of magnetization reversal.

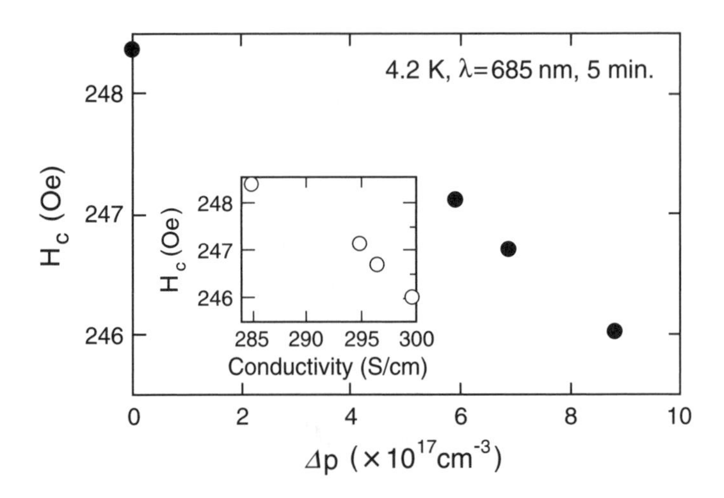

FIG. 1.11. Dependence of the number of photogenerated extra holes on the coercive force H_c. In the inset, H_c is plotted as a function of conductivity. The initial hole concentration in the dark is inferred to be about $10^{20}\,\mathrm{cm}^{-3}$. (By courtesy of A. Oiwa.)

In the particle model, a single particle, being a magnetic cluster in the present case, is flipped at the coercive force given by the difference in the applied magnetic field H and the interaction field $A \cdot p \cdot M_p$ where A, p and M_p are a proportional constant, the volumetric packing factor of the particles, and the magnetization of a particle, respectively. When the sample is illuminated with light, the ferromagnetic exchange coupling between Mn moments is increased due to an increase in carrier concentration. Carrier localization is also relaxed. These mechanisms give rise to an enlargement of magnetic clusters and thus an increase in the volumetric packing factor ($p' > p$). Consequently, the interaction field acting on the reversed clusters becomes $A \cdot p' \cdot M_p$, which is larger than that before the light illumination. This mechanism will in turn reduces the coercive force. Further studies on the effects of light illumination in (In,Mn)As have revealed that the change in magnetic anisotropy induced by the change in the number of holes may be another important mechanism which results in a change in magnetic behavior [77]. Nevertheless, the data shown in Figs 1.10 and 1.11 suggest the feasibility of magnetization reversal with external stimulations such as light illumination and an external electric field, without changing the external magnetic field. Indeed, electrical-field-assisted magnetization reversal was demonstrated after the light-illumination experiments by using gate-controlled (In,Mn)As devices [78]. Being opposite from the trend shown in Fig. 1.11, a gradual increase in coercive force was observed on increasing the negative gate bias.

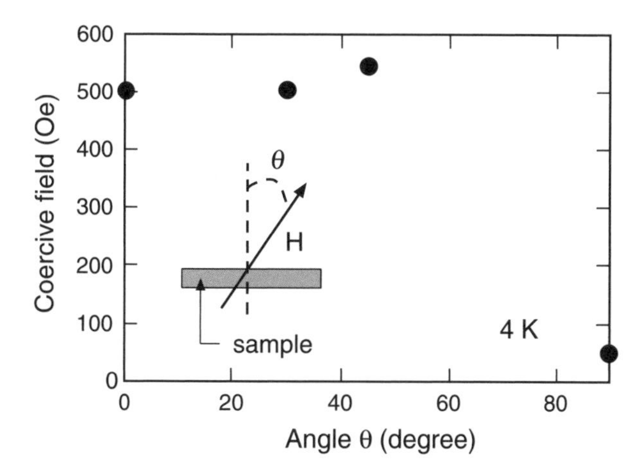

FIG. 1.12. Dependence of the direction of an external magnetic field on the coercive force of the ferromagnetic *p*-(In,Mn)As sample used for the experiment. (By courtesy of A. Oiwa.) The data resemble the coherent rotation model of single-domain particles.

1.4 Photo-induced magnetization rotation effect of spin injection

As discussed in Section 1.2, ferromagnetism in (In,Mn)As and (Ga,Mn)As is realized by the fact that holes are shared by the Mn acceptor states (hole-mediated ferromagnetism). Hole-spin and Mn-spin subsystems are coupled by the *p*-*d* exchange interaction, and both subsystems are spin polarized. Alteration of either of these two subsystems is expected to cause significant influence on the counter-subsystem, giving ways to manipulate spins by external stimulations. In the previous section, we have seen the cause and effect in such a magnetically coupled system in the case of the optical injection of hole-charges. In this section, we turn our attention to the effect brought by the optical injection of hole spins. The injection can be achieved by adequate interband optical excitation, and the most fundamental question in this situation is to what extent the injected carrier spins can influence the ferromagnetically coupled local spin subsystem or the coupled spin system.

In zincblende semiconductors, illumination with circularly polarized light generates non-equilibrium spin polarized electrons and holes in the conduction and valence bands, respectively [79] (Fig. 1.13). The magnitude of polarization is 50% in principle. Pioneering work was carried out in the 1980s for paramagnetic II-VI-based diluted magnetic semiconductors (II-VI-DMS) (Hg,Mn)Te [4, 5] and (Cd,Mn)Te [6, 80, 81]. It has been shown through the band-to-band excitation by circularly polarized light that the orientation of local spins can be changed and causes photo-induced magnetization. The observed changes in magnetization have been explained in terms of the effective field by spin-polarized electrons for (Hg,Mn)Te, and magnetic polaron formation for (Cd,Mn)Te. Dynamical studies

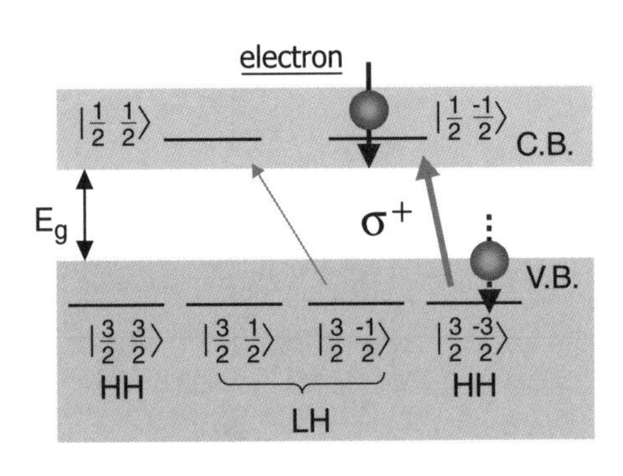

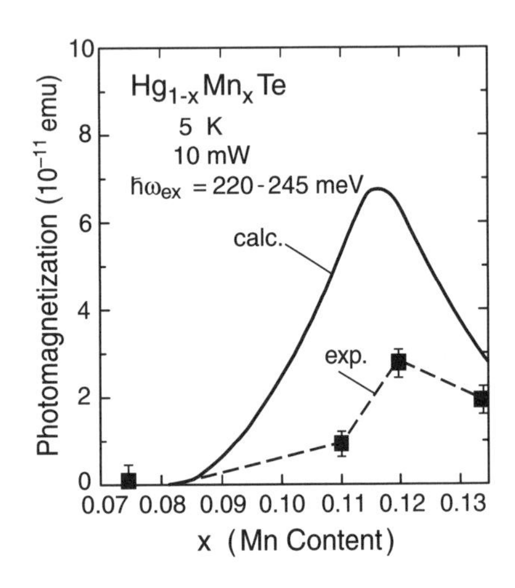

FIG. 1.13. (Left) schematic illustration of interband optical transition with $\Delta J = +1$. The transition rate from the HH state to the conduction band (C.B.) state is three times larger than that from the LH state to the C.B. state. (Right) magnitude of photomagnetization as a function of Mn content x in $Hg_{1-x}Mn_xTe$ (after Ref. [4, 5]).

on polaron formation have been carried out in II-VI-DMS by time-resolved luminescence, magnetization and Faraday rotation measurements [6,80–83]. Through these studies, researchers have recognized that one of the most interesting questions is to ask whether the energy and angular momentum given by the optical excitation to an electron system are first absorbed by the lattice system and then partially given to the spin system, or can they be directly transferred to the spin system without making a detour via the lattice system. This question also stands for III-V-MAS, even more so, since, in contrast to the paramagnetic II-VI-DMS, III-V-MAS inherently has a direct communication path (via the spin-selective acceptor states) between the electron and spin systems. Another motivation, which comes from the device application point of view, is to develop knowledge for the manipulation of magnetization without the use of an external magnetic field.

The experiment to find answers to these questions was first carried out by the cw-excitation of ferromagnetic p-(Ga,Mn)As with circularly polarized light without an external magnetic field [84]. The experimental configuration and detection principle are shown schematically in Fig. 1.14. The sample used for the experiment was a ferromagnetic, p-(Ga,Mn)As single layer grown on a GaAs(100) substrate by molecular beam epitaxy at the substrate temperature of 250 °C. The magnetization axis of the sample lies in the sample plane due to the compressive strain induced by a slight lattice mismatch between (Ga,Mn)As and GaAs [8,9,14–17]. Reflecting the nature of hole-mediated ferromagnetism (Fig. 1.2), the axis of hole spin, as well as that of magnetization, lies in the (100) plane in the

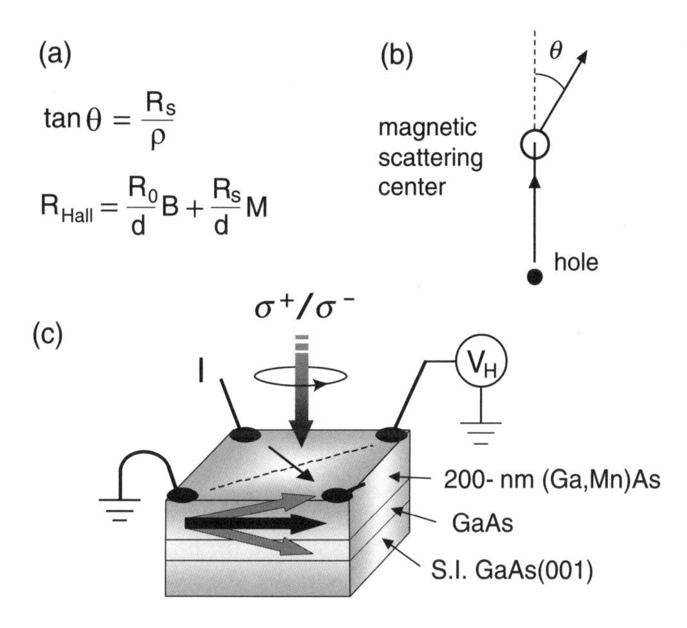

FIG. 1.14. Hall resistance in a (Ga,Mn)As layer of low Mn content (a), (b), together with schematic illustration of experimental setup (c). The Mn content and Curie temperature of a (Ga,Mn)As epitaxial layer is $x = 0.011$ and $T_c \sim 30\,\mathrm{K}$ [84]. The Hole concentration is inferred to be around $10^{20}\,\mathrm{cm}^{-3}$ or lower, referring to the electrical transport data of representative (Ga,Mn)As with similar Mn content [86]. The range of excitation wavelength was from 685 nm to 1064 nm. The positive magnetic flux is defined as the flux which is parallel to the propagation direction of light (the Faraday configuration). This gives rise to the observation of a positive Hall voltage V_{H} for a p-type sample.

dark. This situation does not yield a Hall electromotive force of both normal and anomalous origin without external magnetic fields. When the circularly polarized light of clockwise rotation (σ^+) with appropriate photon energy is illuminated normal to the sample plane, spin-polarized electrons and heavy holes with their spin axes being antiparallel and parallel, respectively, to the direction of light propagation are generated. Note that the spin axes of photogenerated carriers are orthogonal to those of hole spins and Mn spins existing in the (Ga,Mn)As layer. The effect induced by the optical spin injection was detected electronically by the Hall effect using the van der Pauw configuration. For example, inducement of the perpendicular magnetization component is expected to be picked up through the anomalous Hall effect [85]. Photoconductivity was also measured to monitor the amount of photogenerated carriers.

Figure 1.15 shows, from the top panel, the change in Hall resistance, sheet resistance, and the magnified Hall resistance data around the vicinity of the zero Hall resistance. No external magnetic field was applied to the sample. Reflecting the in-plane magnetic anisotropy, there is no noticeable remanent Hall component along the direction normal to the sample plane at zero magnetic field (inset of Fig. 1.15). When circularly polarized light is illuminated, the R_{Hall} exhibits a non-zero value whose sign is altered with the direction of chirality of the circularly polarized light. The sample resistance R_{sheet} reduces because of photoconductivity whose magnitude does not depend on the chirality of the excitation light. As seen in the lowest panel having a magnified vertical scale, R_{Hall} does not return to its original value when the excitation light is turned off, leaving a small residual component whose sign is the same as that during the optical excitation. The magnitude of the residual component is estimated to be less than 0.1% of the saturation magnetization. As a whole, the observed results indicate that perpendicular magnetization is induced by illumination with circularly polarized light. Let us now check several points to understand this phenomenon in more detail.

Judging from the sign of the Hall voltage, the induced perpendicular component is pointing downward (from the top surface to the bottom substrate) when the chirality of the incident light is σ^+. On the other hand, as shown schematically in Fig. 1.13, the excitation with the σ^+ light yields electrons and holes whose spin axes are respectively antiparallel and parallel to the direction of light propagation. Putting these two facts together, one obtains the relation that the direction of induced magnetization is parallel to the spin axis of photogenerated holes. This is qualitatively consistent with the relation shown schematically in Fig. 1.2 for hole-mediated ferromagnetism. Assuming that the induced magnetic component can be summed up with the already existing in-plane magnetization, we are able to regard the observed phenomenon as the rotation of magnetization as a consequence of optical excitation with circular polarized light. It should be noted that the observed Hall voltage is not due to the circular photogalvanic effect reported earlier for *p*-doped (100)-oriented GaAs multiple quantum well structures. This effect is supposed to appear only along $[1\bar{1}0]$ in zincblende crystals [87], whereas in our experiments, the electromotive force is observed along the $\langle 100 \rangle$ crystallographic direction.

The magnitude of the induced-magnetization δM reaches about 15% (2 mT) of the saturation magnetization at the excitation power of $600\,\mathrm{mW/cm^2}$. This δM value corresponds to $N_{\mathrm{Mn}} = 4 \times 10^{19}\,\mathrm{cm^{-3}}$ ($S_{\mathrm{Mn}} = 5/2$) in terms of the number of Mn ions, which is 10^4–10^5 times larger than that reported in the (Hg,Mn)Te system (Fig. 1.13, right). In the case of paramagnetic (Hg,Mn)Te, the magnitude of induced magnetization was explained successfully in terms of the effective magnetic field H_{ph} given by the spin-polarized photogenerated carriers; electrons in the case of (Hg,Mn)Te! If we simply adopt this mechanism for ferromagnetic (Ga,Mn)As, H_{ph} of 20–40 mT is required to realize the observed change in magnetization (2 mT) according to the M-H curve (right panel in Fig. 1.15). This value is far from the estimated value of $H_{\mathrm{ph}} \sim 10^{-5}$–$10^{-3}\,\mathrm{mT}$ based on Eq. (15) in Ref. [4, 5] with the numbers of photogenerated carriers of $n = 10^{11}$–

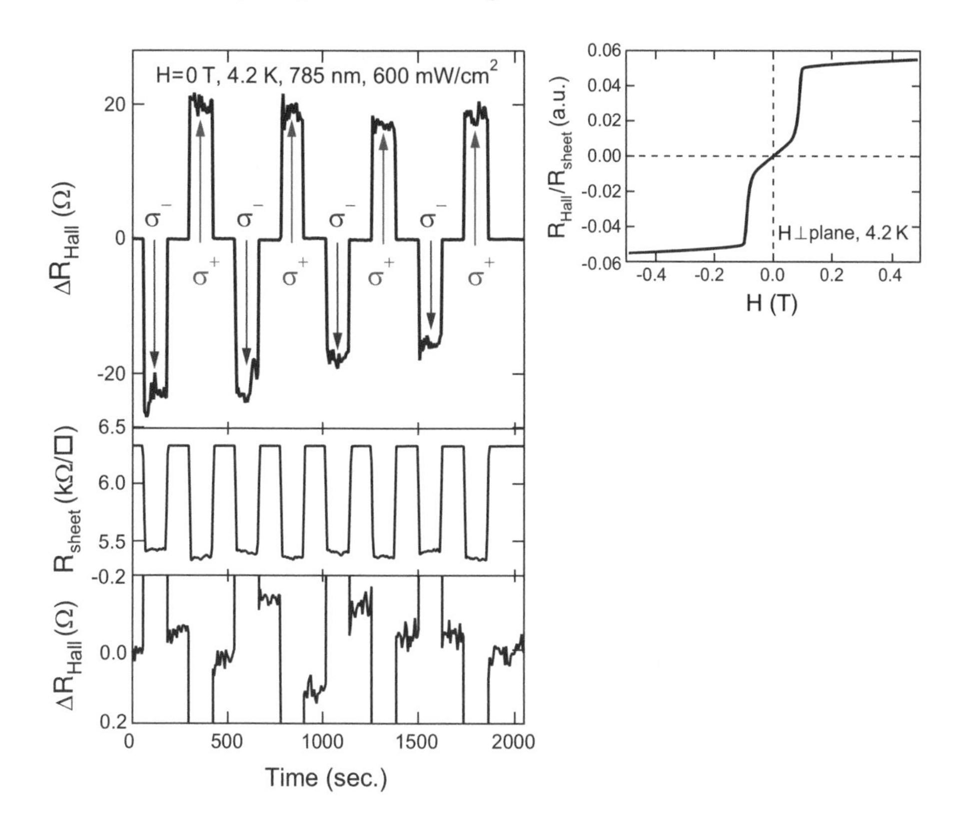

FIG. 1.15. Changes in Hall resistance $R_{\rm Hall}$ (upper and bottom) and sheet resistance $R_{\rm sheet}$ (middle) at 4.2 K for a 20-nm $\rm Ga_{0.989}Mn_{0.011}As$/GaAs(100) sample with in-plane magnetic anisotropy under light illumination with different chirality of incident light [84]. No external magnetic field was applied. Wavelength and power of the excitation light were 785 nm and 600 mW/cm^2, respectively. The inset shows the magnetization hysteresis curve extracted from the AHE measurements.

$10^{12}\,\rm cm^{-3}$ with $N_0\alpha = 0.5\,\rm eV$, and $p = 10^{11}$–$10^{13}\,\rm cm^{-3}$ with $N_0\beta = -1.2\,\rm eV$. It becomes clear that the effect of optical spin injection in ferromagnetic III-V-MAS is quantitatively different from that in pramagnetic II-VI-DMS. In other words, the observed results suggest that the conservation of angular momentum is broken between excitation light and resultant δM. Some non-linear mechanism should be introduced, and one possible mechanism to account for this problem may be the dynamic polarization via the *direct* spin-flip scattering between effective mass carriers (electronic system) and localized magnetic moments (spin system). This mechanism was considered to be less important in II-VI-DMS. In ferromagnetic III-V semiconductors, however, this mechanism may become important since the Mn spin system and carrier charge/spin system are strongly

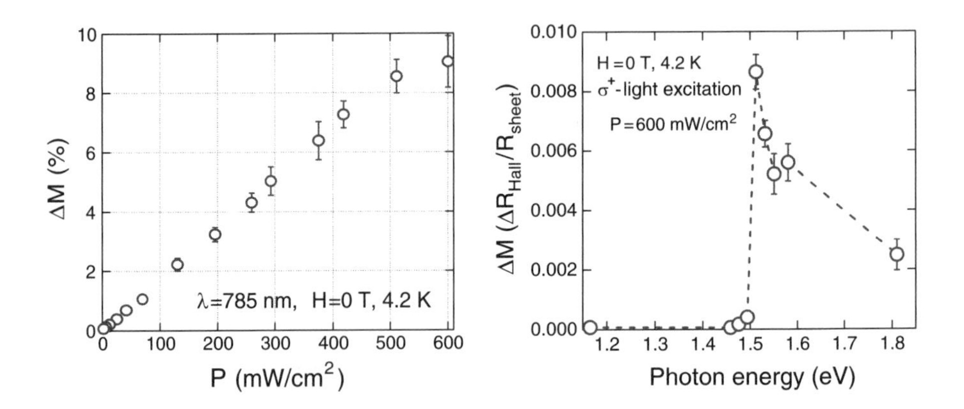

FIG. 1.16. Dependencies of (a) excitation power P and (b) excitation photon energy on photo-induced magnetization δM at 4.2 K for a $Ga_{0.989}Mn_{0.011}As/GaAs$ sample. σ^+ polarized light was used for the excitation. No external magnetic field was applied to the sample (after Ref. [84]).

coupled to each other. We extend our discussion of this problem in Section 1.5, incorporating the examination of the influence of ferromagnetic domains on the observed light-induced magnetization rotation.

The dependencies of the excitation power P and excitation photon energy on the photo-induced magnetization δM is shown in Figs 1.16(a) and (b), respectively, for excitation with σ^+ polarized light. The former experiment was taken at excitation photon energy of 1.51 eV whereas the latter experiment was carried out at $P = 600$ mW/cm^2. An almost linear increase is observed with increasing the P value up to around 500 mW/cm^2 beyond which the change in δM per unit excitation power tends to be reduced because of the contribution of light-induced heating. The absence of the threshold power for the onset of δM suggests that the observed light-induced magnetization rotation does not involve the storage of energy which is necessary in critical phenomena. Turning to the excitation spectrum, the δM value is negligibly small when the excitation photon energy is smaller than the band-gap energy E_g ($\sim$ 1.5 eV) of (Ga,Mn)As which is eventually very close to that of GaAs at $x \sim 0.01$. When the photon energy is increased and becomes higher than the E_g value, δM increases steeply and shows a maximum value at around 1.51 eV. The δM decreases sharply with further increasing the photon energy. The excitation spectrum clearly indicates that the interband excitation is responsible for the observed change in magnetization. On the other hand, the sharp feature around the band edge suggests either the loss of spin polarization for the optical excitation which is off from the center of the Brillouin zone or the contribution of the split-off states formed near the top of the valence band.

Another important clue indicating that the observed light induced effect is of magnetic origin is the memorization effect which we have already seen in the lowest panel of Fig. 1.15. $R_{\rm Hall}$ does not return to its original value in the dark after the excitation light is turned off, yielding a small amount of residual resistance whose sign is the same as that during the excitation. The observed memorization effect does not occur when the experiment is done under the application of a sufficiently strong perpendicular magnetic field at which magnetization almost saturates [88]. Under this condition, the sample is magnetically nearly homogeneous in the dark so that the collectively rotated Mn moments are forced to return back to the original homogeneous state when the light is turned off. Based on this observation, we reach the conclusion that the memorization effect is the remnant of the photo-induced perpendicular magnetization component that is stabilized presumably in the form of small perpendicular ferromagnetic domains imbedded in the in-plane ferromagnetic domains. The origin of nucleation sites is not well understood at the present stage.

On the basis of the observed experimental results, one would be able to expect the rotation or reversal of the magnetization by the electrical injection of spin-polarized carriers in III-V-MAS. Current-induced switching of magnetization, a method that may lead us to the elimination of an external magnetic field to control magnetization orientation, is an important research subject for both fundamental and application points of view in the field of magnetism in magnetic materials. Experimentally, in magnetic metal multilayer structures [89, 90], a large amount of spin-polarized carriers of the order of 10^7 A/cm^2 or higher, have been found necessary for the switching. By using the non-linear effect found in III-V-MAS, the electrical control of magnetization orientation may be realized with a spin-polarized current that is substantially lower than that in metallic systems. This problem is discussed in Section 1.6.

1.5 Spin dynamics

The discovery of magnetization rotation induced by optical excitation with circularly polarized light has addressed various interesting questions that were not clearly recognized in the study of paramagnetic II-VI-DMS. In particular, the magnitude of the relatively large perpendicular magnetization component δM cannot be explained by the generation of an effective magnetic field due to the photogenerated spin-polarized electrons and holes. Some non-linear mechanism should be introduced to account for the observed phenomenon. The role of ferromagnetic domains should also be clarified. Regardless of these questions, we should pursue the microscopic picture as to whether the conservation of angular momentum is broken between excitation light and resultant δM or not. We now look into the dynamic behavior of the optical excitation with all these questions in mind.

The dynamics of photo-induced perpendicular magnetization was studied by carefully comparing the temporal profile of the photo-induced polar Kerr rotation with that of the photo-induced reflectivity change [91]. The polar Kerr rotation signal consists of the perpendicular spin components originating from the optically injected spin polarized carriers and Mn spins whose orientation is

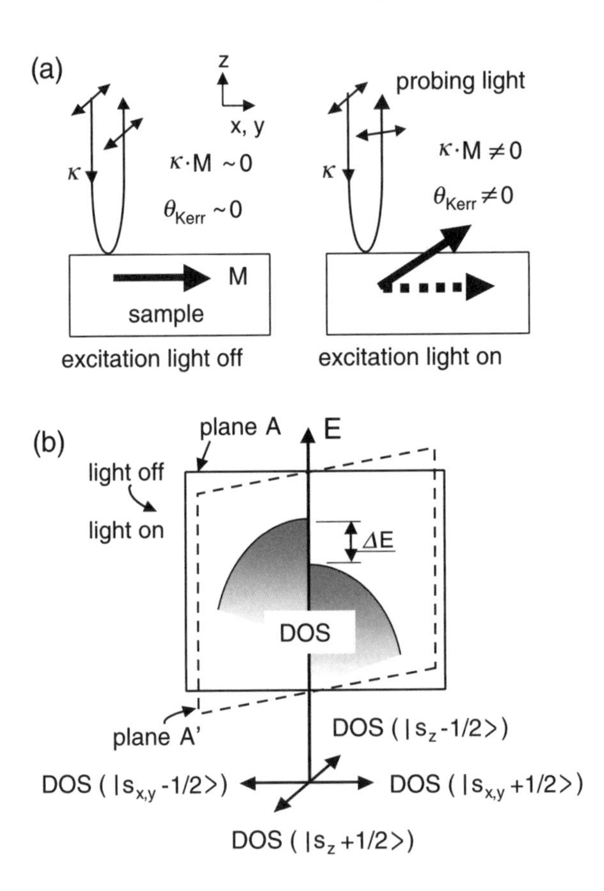

FIG. 1.17. Schematic illustration of the relation between the polar Kerr effect and the orientation of magnetization with and without an excitation light pulse. k, M, and θ_{Kerr} represent light propagation vector, magnetization, and angle of polar Kerr rotation, respectively. (b) Simplified illustration of density-of-states (DOS) of the spin-polarized valence band incorporating the spin quantum axis. Spins pointing towards the (x-y) and z directions are depicted by two horizontal axes which are orthogonal to each other. The DOS in the spin-plane A shows that carrier spins are aligned in the (x-y) plane (after Ref. [91]).

changed by hole spins. Therefore, the contribution of these components should be evaluated rigorously to study the spin dynamics of Mn spins. For this reason, the data obtained from the temporal polar Kerr rotation were carefully compared with the temporal change in reflectivity which gives information as to the lifetime of photogenerated carriers.

The cause and effect between the optical excitation and the resultant magneto-optical Kerr rotation is as follows. Before the excitation without an external magnetic field, the axis of magnetization of magnetic domains is in the sample plane, either in the virgin state or after full saturation, due to the in-plane magnetic anisotropy caused by the lattice mismatch between a (Ga,Mn)As layer and a substrate [8, 9, 14–17]. Reflecting the hole-mediated ferromagnetism, the spin-split density of states (DOS) of the valence band is expressed by two half-parabolas in the spin plane 'A' whose spin quantum axes are along $|S_{x,y}; \pm 1/2\rangle$ (Fig. 1.17b). The magnitude of the splitting is given by the exchange-split energy ΔE. Here, we neglect the orbital angular momentum L of the holes for simplicity. Linearly polarized probing light reflected normal from the sample surface only exhibits a small longitudinal/transverse Kerr rotation. Through the band-to-band excitation by the circularly polarized light impinged normal to the sample plane, spin-polarized electrons and heavy holes are selectively generated [79] with spin quantum axes perpendicular to the sample plane $|S_z; \pm 1/2\rangle$. This is followed by the transfer of angular momentum and excitation energy from the photogenerated carriers to the host electron system (the plane A) and to the Mn spin system, resulting in the rotation of the *spin plane* from A to A'. Temporal polar Kerr rotation signals will be observed during the rotation and recovery processes of the *spin plane*, in which the magnitude and response speed depend on the mechanism of the transfer of angular momentum and excitation energy. Participation of magnetic domains can also be elucidated since dynamics of domains are usually accompanied by characteristic dinging [92]. Photogenerated carrier spins themselves also give an additional polar Kerr rotation signal whose contribution should be separated from that due to magnetization rotation. This is the reason why time-resolved Kerr rotation (TRKR) data should be carefully compared with photo-induced temporal changes in reflectivity ($\Delta R/R$) at the temperatures below and above the Curie temperature T_c. Actual experiments were carried out by the pump-and-probe technique using a cw mode-locked Ti:Sapphire laser with the same pump and probe photon energy. The change in the rotation angle was measured by a polarization-sensitive optical bridge technique with an accuracy of 10^{-4} deg. As to photo-induced reflectivity change, polarization of both pump and probe pulses was in the linear regime.

The temporal profiles of light-induced Kerr rotation obtained by the above mentioned technique are summarized in Fig. 1.18 for two different ferromagnetic (Ga,Mn)As samples. As shown in the upper part of Fig. 1.18, the Kerr rotation signal rises very abruptly within the pulse width of 170 fsec and decays within 100 psec. The profile is symmetric with respect to the chirality of a circular polarized pump pulse, which is qualitatively consistent with the experimental data obtained by cw excitation [84]. In the lower part of the figure, the temporal profile of Kerr rotation is plotted in a semilogarithmic scale and compared with that of the reflectivity change. With this procedure, we are able to clearly notice the presence of two different components which follow two exponential curves $A_1 \exp(-t/\tau_1)$ and $A_1 \exp(-t/\tau_2)$ with different time constants $\tau_1 = 13 \pm 1$ ps and $\tau_2 = 54 \pm 1$ ps, respectively. This indicates that two independent excitations/relaxations take place by optical pumping. The τ_1 value coincides with

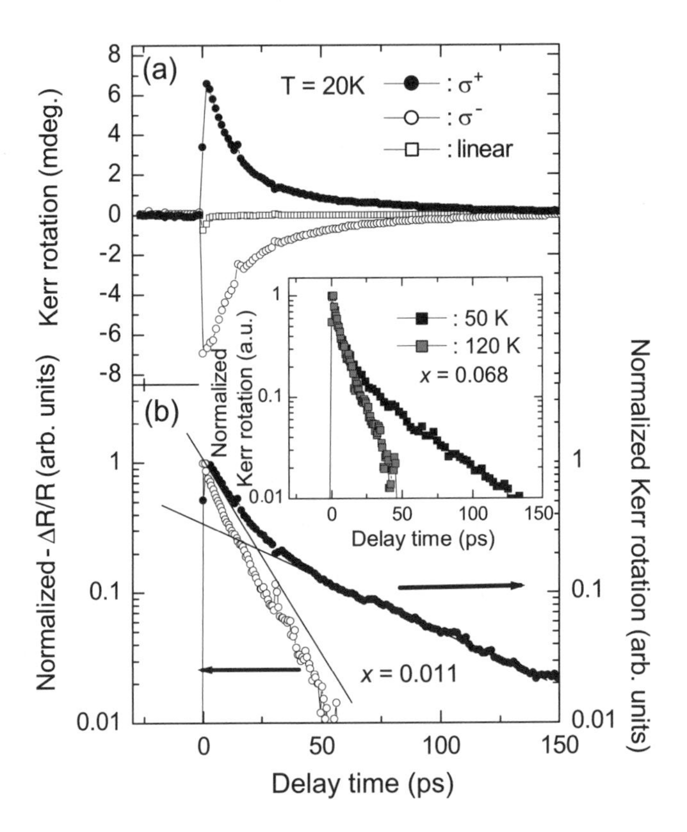

FIG. 1.18. Temporal profiles of TRKR together with the temporal change in reflectivity for the $x = 0.011$ sample ($T_c = 35\,\mathrm{K}$) in the ferromagnetic (20 K) regime (after Ref. [91]). The inset shows TRKR data for the $x = 0.068$ sample ($T_c = 90\,\mathrm{K}$). In both samples, the second component appears below their Curie temperature (Ref. [85]). The photon energy is 1.579 eV for both pump and probe pulses. The duration and repetition of pulses were 120 fs and 78 MHz, respectively. The intensity of the circularly polarized excitation pulse was 3×10^{12} photons/pulse/cm^2, whereas that of the linearly polarized probe pulse was fixed at 1×10^{11} photons/pulse/cm^2. Hole concentration is inferred to be $10^{20}\,\mathrm{cm}^{-3}$ or lower for the $x = 0.011$ sample and around $10^{20}\,\mathrm{cm}^{-3}$ for the $x = 0.068$ sample, referring to the existing carrier transport data for the representative samples [86].

that of the single-exponential decay profile of the reflectivity which is governed by the trapping of photogenerated electron spins [93–95]. The second component represented by the relatively long decay time constant τ_2 can be attributed to the behavior of magnetic origin, because it develops at the Curie temperature or lower, as shown in Fig. 1.19. This component is inferred to be the perpendicular component induced by the illumination of circularly polarized light. It is very

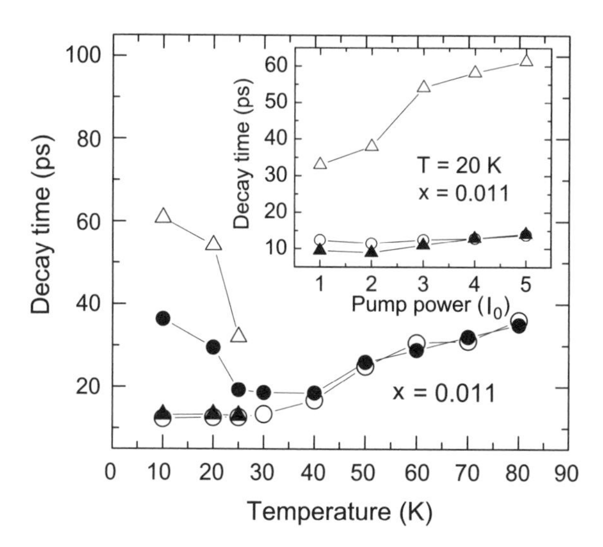

FIG. 1.19. Various decay time constants extracted from TRKR and $\Delta R/R$ profiles plotted at different temperatures for the $x = 0.011$ sample. TRKR data were analyzed by both single-exponential and bi-exponential fits, whereas $\Delta R/R$ data were analyzed by the single-exponential fit. The inset shows the dependence of the decay time constant on the excitation power (after Ref. [91]).

striking that inducement of the perpendicular magnetization component occurs very fast, being comparable to the pumping pulse width (170 fsec). It should also be noted that the fast trapping process of spin-polarized electrons suggests that the lifetime of photogenerated holes is rather long.

As shown in the inset, similar behavior can be observed in the sample with different Mn contents $x = 0.068$ ($p \simeq 10^{20}\,\mathrm{cm}^{-3}$, $T_c = 90\,\mathrm{K}$). This fact suggests that the critical sample conditions, such as critical x and p, are not required for the inducement of perpendicular magnetization. Another important insight is that the trapping of carrier spins is not responsible for the observed magnetization rotation. In studies of the traditional ferromagnetic semiconductors [1, 2], elongation of the spin lifetime by trapping centers was discussed as one of the effective ways to utilize the angular momentum of light. The data shown in Fig. 1.18 suggest that the trapping process is remote because of the ultrafast rising behavior.

There are two different pictures to account for the inducement and relaxation of the perpendicular magnetization component. The first picture is the description based on the effective magnetic field that works on the ferromagnetically coupled Mn spins. In this picture, which was used when the effect of circularly polarized light was reported for the first time [84, 91], the inducement of the perpendicular magnetization component has been viewed as a consequence of rotation of magnetization due to the impulsive effective magnetic field caused by the spin polarized holes through the *p-d* exchange interaction. The relaxation of rotated magnetization was viewed as the transverse spin relaxation by the built-in effective field originating from the in-plane magnetic anisotropy [91]. The lack of precessional motion during the recovery process was explained in terms of a strong damping in this picture. The absence of the oscillatory behavior also suggests that the contribution of signals coming from the formation and relaxation of out-of-plane magnetic domains is negligibly small.

The second picture has became clearer very recently when the optical inducement of the effective magnetic field in the sample plane and subsequent precessional motion of ferromagnetically coupled Mn spins was found [96]. In this picture, the inducement of the perpendicular magnetization component δM is regarded as the optical excitation of electron-spin coupled system. In other words, δM is not explained in terms of magnetization rotation due to the effective magnetic field, but is described as the excitation involving the dynamic polarization via the direct spin-flip scattering between carriers and localized magnetic moments. During this event, a hole-Mn spin complex whose overall spin axis is off from the in-plane direction is formed. A non-linear mechanism involving the breakdown of the conservation of angular momentum between excitation light and resultant δM may be anticipated in this mechanism [83]. The relaxation of this excitation is the relaxation of angular momentum and energy into the *cold* (equilibrium) hole-Mn system (plane A in Fig. 1.17). This picture does not require the effective magnetic field to recover the equilibrium, and explains fairly well the absence of the oscillatory behavior in the TRKR data. From the data shown in Fig. 1.18, the angle of Kerr rotation at time zero is found to be 3×10^{-3} deg., from which the canting angle of magnetization with respect to the sample plane is estimated to be 0.3 deg. On the other hand, the canting angle of the spin axis of the valence band is estimated to be $3 \times 10^{16}/10^{20} = 3 \times 10^{-4}$ deg. These analyses lead us to infer that about one hundred Mn spins are excited by the injection of one hole spin on average.

While the elucidation of the optical excitation with circular polarized light is still an on-going research subject, it is clear that this subject contains rich physics that were not accessible by ferromagnetic metals and paramagnetic II-VI-based diluted magnetic semiconductors. The study of spin dynamics with an external magnetic field will further shed light on this interesting subject. The relaxation of the induced perpendicular magnetization is found to become faster by the application of an in-plane magnetic field, and it eventually vanishes at the field around $\mu_0 H = 1.5\,\mathrm{T}$, as shown in Fig. 1.20. It has also been found that the response time of the spin system becomes rather slow when (Ga,Mn)As is excited by a stronger excitation pulse under a perpendicular magnetic field [97]. This behavior has been explained in terms of the demagnetization and its recovery via phonons in a half-metal in which spin-flip event is suppressed because the density-of-states of only one spin polarity is available [97]. Modification of the electronic structure with a heterojunction and low-dimensional structures would

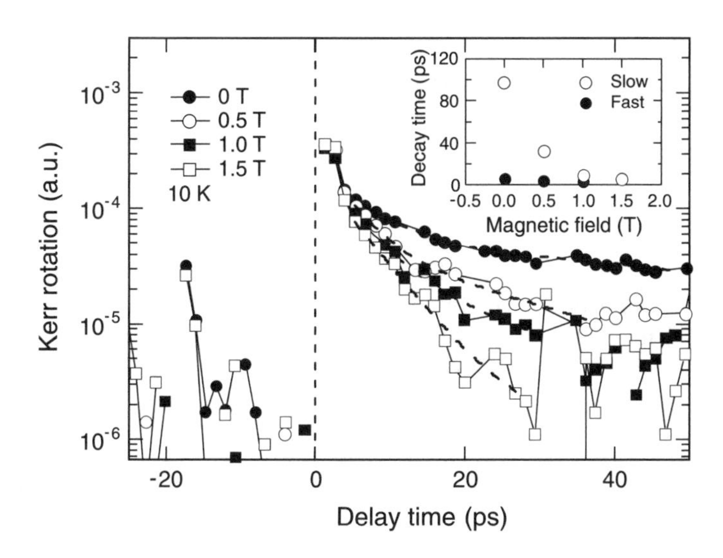

FIG. 1.20. TRKR data obtained from the $x = 0.011$ sample under different external magnetic fields. The field was applied in the sample plane, along the hard magnetic axis of [110]. (By courtesy of A. Oiwa.)

also be another interesting approach to study the spin dynamics of hole-mediated ferromagnetic systems [23, 98].

1.6 Possible applications

Spin-related phenomena in semiconductor-based materials and structures will find places for applications when the phenomena can be viewed either as novel functionality that cannot be realized by magnetic metals or semiconductors. The system of low carrier concentration and small magnetization are the two major differences which distinguish magnetic semiconductors from magnetic metals. The advantage of low carrier concentration is the ease in electrical manipulation and the accessibility to optical excitation across the band gap, whereas the small magnetization reduces the energy required to manipulate the magnetization. In the dilute limit, or even without magnetic impurities, local magnetic moments and free carriers manifest the feature of spins with which novel functionality can also be expected. The drawback of the relatively low numbers of carrier spins and local spins is the relatively weak spin exchange interaction between the two, as exemplified by the relatively low Curie temperature of hole-mediated ferromagnetism (Section 1.2.2). However, room temperature ferromagnetism seems feasible if one can overcome the problem associated with controlling the occupation site in which magnetic ions are incorporated. Hybrid structures consisting of semiconductors and ferromagnetic semiconductors/metals may also become one of the solutions to circumvent this problem. Up to now, various interesting electronic and opto-electronic devices have been proposed and tested in various laboratories and universities with ferromagnetic semiconductors, hybrid structures, and non-magnetic semiconductor nanostructures; for example, electrically switchable ferromagnets [78], unipolar and bipolar spin-diodes and transistors [99, 100], spin-MOSFETs [101], ultrafast all-optical switches [102], spin-LED [103], spin photodetectors (spin-PD) [104], and a laser integrated in one body with an optical isolator [105]. Quantum manipulation of spins in semiconductor-based structures has also been proposed and has been under investigation experimentally [106]. In the following, three examples associated with the application of ferromagnetic semiconductors and spin-dependent optical transitions are given. These are (1) magnetization reversal in a (Ga,Mn)As thin film by electrical spin injection, (2) circularly polarized light emitters, and (3) circularly polarized light detectors.

1.6.1 *Magnetization reversal by electrical spin injection*

Current-induced switching of magnetization, the method that may lead us to the elimination of an external magnetic field to control magnetization orientation, is one of the important and interesting research subjects from both fundamental and application points of view. In magnetic metal multilayer structures [89, 90], a large number of spin-polarized carriers of the order of 10^7 A/cm^2 or higher, has been found necessary for switching the magnetization. Taking advantage of the small magnetization of III-V-MAS, one may be able to realize magnetization reversal by a spin-polarized current that is substantially lower than that needed for metallic systems. Furthermore, on the basis of the non-linear effect found in optically induced perpendicular magnetization, one would be able to further reduce the amount of current needed to reverse the magnetization.

As can be found in other chapters in this book, magnetization reversal by electrical spin injection has been studied primarily in the metallic trilayer structure consisting of ferromagnet metal (FM-I)/non-magnetic metal/ferromagnet metal (FM-II) in which the spin injection from one FM layer to another results in the change in the relative configuration of the in-plane magnetization between parallel and antiparallel. The spin transfer torque brought about by the injected spin-polarized carriers is the mechanism for the magnetization reversal [107]. The investigation using a III-V-based ferromagnetic semiconductor has been carried out in the (Ga,Mn)As/AlAs/(Ga,Mn)As trilayer structure [108] as well as the (Ga,Mn)As/GaAs/(Ga,Mn)As trilayer structure [109]. In both cases, the non-magnetic layer works as a tunneling barrier rather than an electrically conductive region, which, by itself, addresses a new problem of whether magnetization reversal by the tunneling process is possible or not.

In both experiments, a change in relative magnetization configuration induced by the spin-polarized current was detected by the change in device resistance with the current-perpendicular-to-plane (CPP) configuration. Here in this section, we will review the experiment done by using the (Ga,Mn)As/AlAs/(Ga,Mn)As trilayer structure prepared by molecular beam epitaxy (MBE) on Zn-doped p^+-GaAs (100) substrates. As seen in Fig. 1.21, the thicknesses and Mn contents of

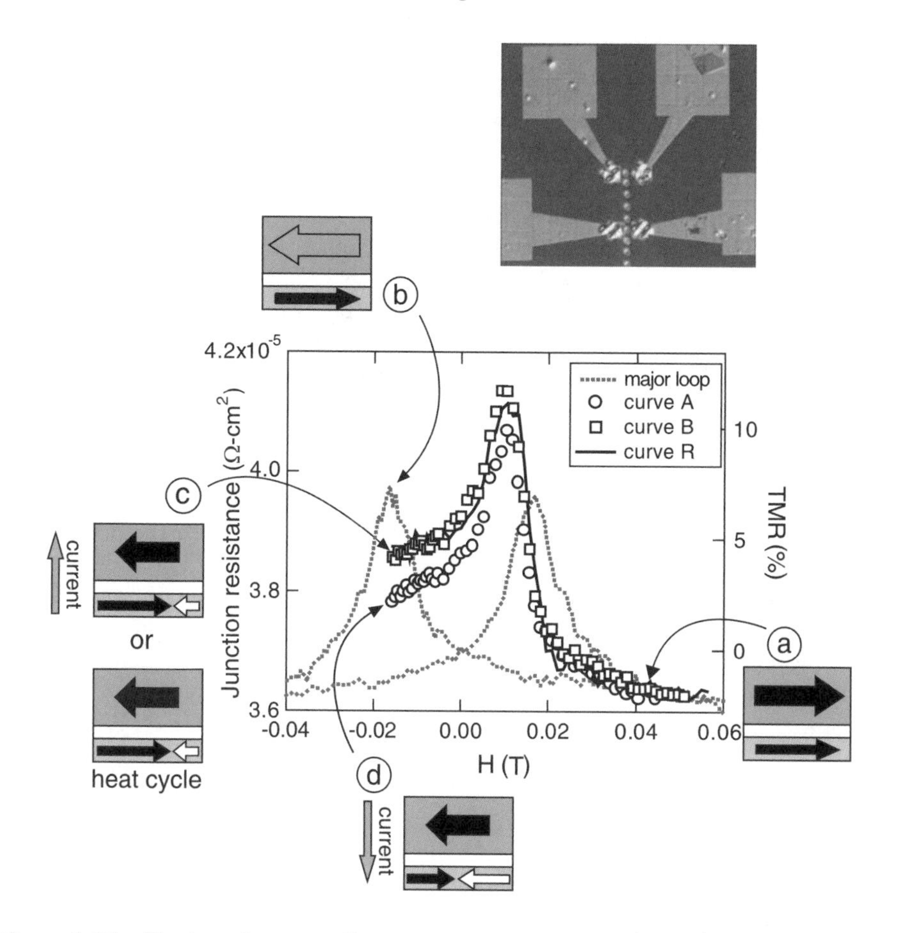

FIG. 1.21. Hysteretic tunneling magnetoresistance (TMR) curves of the (Ga,Mn)As/AlAs/(Ga,Mn)As trilayer tunneling diode processed in the mesa structures of $1\,\mu$m in the diameter [108] (after Physica E **25**, 160 (2004)). Different TMR curves are obtained depending on the magnitude and direction of the current. Arrows in the (Ga,Mn)As layers indicate the direction of magnetization. The upper figure shows a top view image of the 1-μm device taken by scanning electron microscopy.

top and bottom (Ga,Mn)As layers were 30 and 8 nm, and $x = 0.04$ and 0.05, respectively, in order to distinguish each layer magnetically by the difference in the coercive force. The thickness of the AlAs layer is 1.8 nm. Magnetization of the thinner (Ga,Mn)As is expected to be reversed with a smaller amount of spin-injected carriers as compared with that needed for the thicker layer. Micron-size mesa structures were used for the experiment. A top-view image of the device taken by scanning electron microscopy is shown in the upper part of Fig. 1.21.

Experiments were carried out as follows at 10 K. A high magnetic field was first applied to yield a parallel configuration (point a in Fig. 1.21). Then, the field was swept down to $-20\,\mathrm{mT}$ at which the antiparallel configuration was realized, and the device was in the high-resistance state (point b). This was followed by the application of positive or negative DC current for 1 sec with a current density of $1 \times 10^6\,\mathrm{A/cm^2}$. The resistance is supposed to decrease if the effect of spin-induced magnetization reversal is sufficiently large. Finally, magnetoresistance curves were measured to investigate the relative magnetic configuration in more detail.

Figure 1.21 shows the magnetoresistance (MR) curves obtained by the experiments. The reference experiment was carried out by heat-cycling the sample between 10 and 20 K without current. It is clearly seen that the resistance obtained after the application of negative current coincides with that produced by the reference experiment (point c). Both magnetoresistance curves (curve B and R) also coincide fairly well to each other. On the other hand, the resistance obtained after the positive current (point d) shows further reduction with respect to the value at the point c, which is believed to be due to the occurrence of magnetization reversal induced by electrical spin injection. The difference in resistance between the point c and d corresponds to the magnetization reversal of about 25% with respect to the complete antiparallel configuration. Magnetization reversal presumably took place in the form of magnetic domains.

Detail analysis of the *I-V* characteristics reveals that about 90% of the current is attributed to the leak current which is not spin polarized. Within the limit of this analysis, partial magnetization reversal is achieved with the spin-polarized tunneling current of the order of $10^5\,\mathrm{A/cm^2}$. The study using the (Ga,Mn)As/GaAs/(Ga,Mn)As trilayer structure has also concluded that magnetization reversal with the current of the order of $10^5\,\mathrm{A/cm^2}$ is possible. The current of $10^5\,\mathrm{A/cm^2}$ is obviously smaller than that in the metallic system. This value can probably be understood in the framework of the spin transfer mechanism [107]. The non-linear effect, which is supposed to reduce the current, has not been observed in the electrical spin injection experiment up to now.

1.6.2 *Circularly polarized light emitters and detector*

A combination of spin-dependent optical transition and spin-dependent carrier transport will produce unique optical devices that generate and detect the optical polarization. Let us consider circular dichroism (CD) measurements, one of the conventional optical characterization methods to analyze physical and chemical properties of various kinds of materials. In general, the CD measurement requires an instrument that modulates the chirality of the circular polarization; for example, a Polaroid-film rotator and a photo-elastic modulator. If the polarization of the light source can be modulated directly without using the external device, the CD measurement can be simplified significantly. Fast modification of polarization in the light source, when it is paired with a polarization detector whose detection speed is comparable to the modulation frequency, will give an opportunity to not only enhance the performance of the CD measurement in

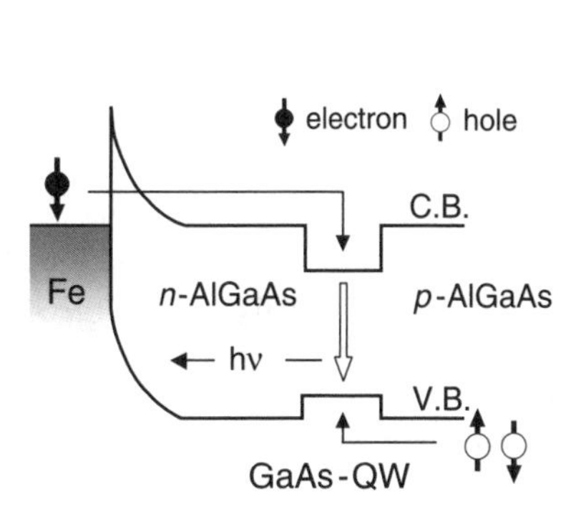

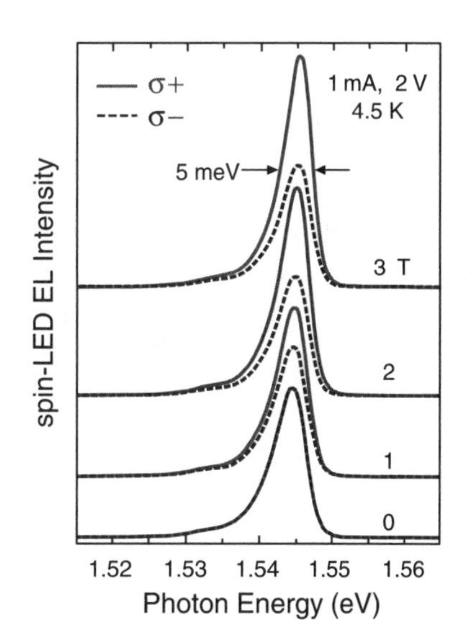

FIG. 1.22. (Left) conduction and valence band edge profiles of the spin-LED consisting of Fe/AlGaAs/GaAs-QW/AlGaAs, and (Right) electroluminescence spectra of the spin-LED under different magnetic fields (after Ref. [103]). Here, an external perpendicular magnetic field was applied in order to rotate the magnetization of a ferromagnet electrode from in-plane to out-of-plane to realize the Faraday configuration in which spin axis and optical orientation is either parallel or antiparallel.

the fast time domain, but also yield new applications for laboratory/industry uses, in the areas of chemistry, physics, and biology, as well as pharmacy, image recognition, and optical data communication. Devices and systems that deal not only with circularly polarized light but also with s- and p-linearly polarized light will even expand the frontier of novel applications. At present, these devices are *the devices of the future*, and the research to propose and test their working principles has just started.

Towards the realization of a circularly polarized light source having the functionality of switching the chirality of the polarization, a light-emitting diode combined with an electrode in which carriers are spin-polarized (the so-called spin-LED) is a candidate semiconductor device. As shown schematically in Fig. 1.22, spin-polarized carriers, electrons in this case, are injected from a ferromagnet electrode into an n-type region of a non-magnetic semiconductor. These electrons are transported to the optically active region, a quantum well structure in this case, and recombine radiatively there with holes supplied from the p-type region. Because of the optical selection rule (Fig. 1.13), the emission is accompanied with circular polarization; the σ^+ light for the transition between the states $|-1/2\rangle_{\mathrm{CB}}$ and $|-3/2\rangle_{\mathrm{VB}}$ and the σ^- light for the transition between the states $|+1/2\rangle_{\mathrm{CB}}$ and $|+3/2\rangle_{\mathrm{VB}}$. In the quantum well, degeneracy between the light-

and the heavy-hole states is broken, so that the transition associated with the light-hole states $| \pm 1/2\rangle_{\rm VB}$ is not energetically favorable. Circular polarization of 100% is theoretically expected in the spin-LED imbedded with the quantum well if carriers in the well are 100% spin polarized.

The spin-LED, when it was invented for the first time [110,111], aimed at the demonstration of spin injection into a non-magnetic semiconductor layer from either a ferromagnetic or paramagnetic semiconductor. The quantum wells were regarded as an optical detector that could analyze the magnitude of spin polarization of injected carriers at the point of recombination. These works have been extended to the study of spin injection from a ferromagnetic metal electrode into non-magnetic semiconductor structures [103], through which researchers have found that efficient spin injection is possible across the metal and semiconductor interface through the tunneling barrier, as shown in the left panel of Fig. 1.22. Here, the magnitude of circular polarization is about 30% at 4.5 K. The polarization value decreases with increasing the temperature, in the range 5–10% at 100–250 K. This fact indicates the presence of a temperature-dependent spin relaxation process in the spin injection at the metal/semiconductor interface, in the spin transport process across the n-type region, as well as in the quantum well during the energy relaxation before the recombination [112]. The switching of polarization can be realized by reversing the magnetization direction of the ferromagnetic electrode. Study of other switching schemes is desired to pursue the fast modification of circular polarization.

There are presently two different working principles for the photodetector that has the functionality of converting the polarization of light directly into an electrical signal (spin photodetector, spin-PD). The first method is to utilize the perpendicular magnetization component induced by the illumination of circularly polarized light which was discussed in Sections 1.4 and 1.5. The data shown in Figs 1.15 and 1.18 already represent the feasibility of direct and fast conversion of optical polarization into an electrical signal. This functionality comes directly from the behavior of hole-mediated ferromagnetism in III-V-MAS, so that the maximum working temperature of the detector is presumably limited by the Curie temperature which is still less than room temperature but is way higher than liquid nitrogen temperature. Efforts for device fabrication are strongly desired.

Another interesting approach is to utilize the spin-photovoltaic effect [113, 114] which is based on the difference in the Zeeman splitting of the conduction (or valence) band between the n- and p-type regions (see the discussion in Chapter 2). This gives rise to the difference in the diffusion potential between the up- and down-spin carriers across the p-n junction and thus the difference in the diffusive current between up- and down-spin across the p-n junction, as shown schematically in Fig. 1.23. One way to realize this situation is to use a pair of semiconductors whose lattice constants are close to each other, yet the g-factors are different from each other. Application of an external magnetic field breaks the spin degeneracy across the junction. When the circularly polarized light is

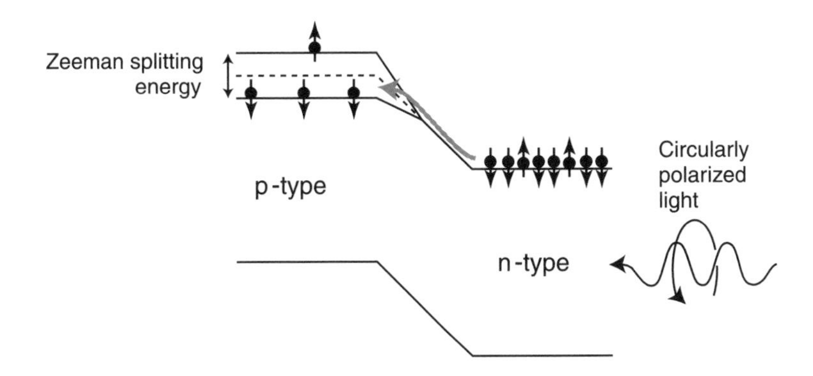

FIG. 1.23. Schematic illustration of the band profile of the p-n junction with external magnetic fields. Reflecting the difference in g-factors for conduction bands between p and n regions, the diffusion potential splits into spin-up and spin-down levels by the application of magnetic fields.

irradiated on an n-type region, spin polarized carriers are generated in this region and diffuse toward the junction at which the diffusion potential is spin dependent. For one particular spin orientation, the barrier height is low, whereas it is high for the opposite spin orientation. This gives rise to circularly polarization-dependent photovoltage and/or photocurrent since the spin orientation of photogenerated carriers depends on the chirality of the polarization.

Very recently, a photodiode incorporating the p-n heterojunction with a pair of semiconductors whose g-factors are nearly zero and non-zero for n- and p-type layers, respectively, was fabricated by molecular beam epitaxy and tested in the laboratory [104]. It consisted of n-$Al_{0.12}Ga_{0.88}As$ having $g_{CB} = 0$ and p-$In_{0.3}Ga_{0.7}As$ having $g_{CB} = -1.9$. Structure optimization, taking into account the lattice mismatch, associated critical thickness, the efficiency of light absorption, the width of the depletion region, as well as the spin diffusion length [115], resulted in the thicknesses and room-temperature carrier concentrations of p-type $In_{0.3}Ga_{0.7}As$ and n-type $Al_{0.12}Ga_{0.88}As$ layers being $d = 15$ nm and $p = 5\times10^{18}$ cm^{-3}, and $d = 1$ nm and $n = 3\times10^{17}$ cm^{-3}, respectively. The photodiode consisted of a mesa structure having an ohmic contact electrode with an optical access window of 240 μm in diameter. The magnetic-field dependence of the current-voltage (I-V) characteristics at 4 K has exhibited an increase in current flow, from which the reduction in the diffusion potential of 0.3 meV was estimated for the magnetic field of 3 tesla. This value was comparable to the value calculated on the basis of the Zeeman splitting with $\Delta g_{CB} = -1.9$.

The behavior of the I-V curve under illumination ($\lambda = 785$ nm) with circular polarized light was investigated, in which a change in the open-circuit voltage V_{oc} was observed. Figure 1.24 shows the V_{oc} signal at 4 K measured with changing the circular polarization between σ^+ and σ^-. At the field of +3 T, the V_{oc} with the σ^+ polarization is larger than that with the σ^- polarization, whereas, at

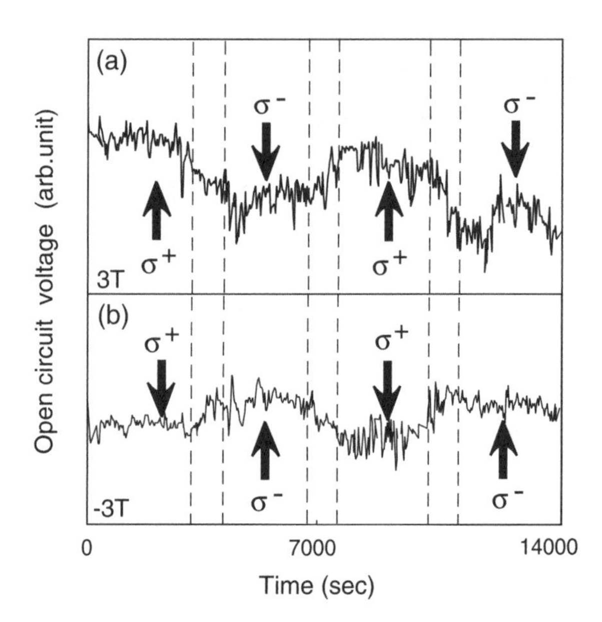

FIG. 1.24. Circular polarization dependence of photo-induced open-circuit voltage of the n-$Al_{0.12}Ga_{0.88}As$/p-$In_{0.3}Ga_{0.7}As$ sample (after Ref. [104]). The data were taken under cw-illumination. Circular polarization was changed by manually rotating a $\lambda/4$ plate. The open-circuit voltage V_{oc} was detected by using the lock-in technique with the optical chopping frequency of $f = 70\,\mathrm{Hz}$. The wavelength and power of the illumination was $\lambda = 785\,\mathrm{nm}$ and $P = 15\,\mathrm{mW}$.

$-3\,\mathrm{T}$, the dependence of chirality was reversed. The observed data, although qualitative at present, are consistent with the photovoltaic effect and are very encouraging experimental results. Detailed studies concerning the wavelength and temperature dependencies are desired. Polarization-dependent behavior can probably be enhanced by using semiconductors with larger g-factors, e.g., III-Sb-based alloys, or ferromagnetic semiconductors [116].

Acknowledgements
The author devotes this chapter with deep thanks to Dr. A. Oiwa and other members of the Munekata laboratory.

References

[1] V. F. Kovalenko and E. L. Nagaev, Sov. Phys. Usp. **29**, 297 (1986).
[2] E. L. Nagaev, Phys. Status Solidi B **145**, 11 (1988).
[3] Semiconductor and Semimetals, Vol. 25, *Diluted Magnetic Semiconductors*, edited by J. Frudyna and J. Kossute (Academic Press, 1988).

[4] H. Krenn, W. Zawadzki, and G. Bauer, Phys. Rev. Lett. **55**, 1510 (1985).
[5] H. Krenn, K. Kaltenegger, T. Dietl, J. Spalek, and G. Bauer, Phys. Rev. B **39**, 10918 (1989).
[6] D. D. Awschalom, J. Warnock, and S. von Molnár, Phys. Rev. Lett. **58**, 812 (1987).
[7] H. Munekata, H. Ohno, S. von Molnar, A. Segmuller, L. L. Chang, and L. Esaki, Phys. Rev. Lett. **63**, 1849 (1989).
[8] F. Matsukura, H. Ohno, and T. Dietl, "III-V Ferromagnetic Semiconductors" in *Handbook of Magnetic Materials*, edited by K. H. Buschow (Elsevier Science, 2003).
[9] T. Dietl and H. Ohno, MRS Bulletin **28**, 714 (2003).
[10] H. Munekata, H. Ohno, R.R. Ruf, R. J. Gambino, and L. L. Chang, J. Cryst. Growth **111**, 1011 (1991).
[11] H. Ohno, H. Munekata, T. Penney, S. von Molnár, and L. L. Chang, Phys. Rev. Lett. **68**, 2664 (1992).
[12] H. Munekata, T. Penney, and L. L, Chang, Surf. Sci. **267**, 342 (1992).
[13] H. Munekata, A. Zaslavsky, P. Fumagalli, and R. J. Gambino, Appl. Phys. Lett. **63**, 2929 (1993).
[14] H. Ohno, A. Shen, F. Matsukura, A. Oiwa, A. Endo, S. Katsumoto, and Y. Iye, Appl. Phys. Lett. **69**, 363 (1996).
[15] H. Ohno, Science **200**, 110 (1998).
[16] H. Ohno, J. Magn. Magn. Mater. **200**, 110 (1999).
[17] H. Ohno and F. Matsukura, Solid State Commun. **117**, 179 (2001).
[18] T. Hayashi, M. Tanaka, T. Nishinaga, H. Shimada, H. Tsuchiya and Y. Otuka, J. Cryst. Growth **175/176**, 1063 (1997).
[19] J. De Boeck, R. Oesterholt, A. Van Esch, H. Bender, C. Bruynseraede, C. Van Hoof, and G. Borghs, Appl. Phys. Lett. **68**, 2744 (1996).
[20] S. Koshihara, A. Oiwa, M. Hirasawa, S. Katsumoto, Y. Iye, C. Urano, H. Takagi, and H. Munekata, Phys. Rev. Lett. **78**, 4617 (1997).
[21] H. Ohno, D. Chiba, F. Matsukura, T. Omiya, E. Abe, T. Dietl, Y. Ohno, and K. Ohtani, Nature **408**, 944 (2000).
[22] A. Oiwa, T. Slupinski and H. Munekata, Appl. Phys. Lett. **78**, 518 (2001).
[23] J. Wang, G. A. Khodaparast, J. Kono, T. Slupinski, A. Oiwa, and H. Munekata, J. Superconductivity: Incorporating Novel Magnetism **16**, 373 (2003).
[24] M. Ilegems, R. Dingle, and L. W. Rupp, Jr., J. Appl. Phys. **46**, 3059 (1975).
[25] J. S. Blakemore, W. J. Brown, Jr., M. L. Stass, and D. A. Woodbury, J. Appl. Phys. **44**, 3352 (1973).
[26] J. Schneider, U. Kaufmann, W. Wilkening, M. Baeumler, and F. Köhl, Phys. Rev. Lett. **59**, 240 (1987).
[27] J. Inoue, S. Nonoyama, and H. Itoh, Phys. Rev. Lett. **86**, 4610 (2000).
[28] T. Dietl, H. Ohno, F. Matsukura, J. Cibert, D. Ferrand, Science, **287**, 1019 (2000).
[29] T. Dietl, H. Ohno, and F. Matsukura, Phys. Rev. B **63**, 195205 (2001).
[30] Y. H. Matsuda, G. A. Kodaparast, M. A. Zudov, J. Kono, Y. Sun, F. V.

Kyrychenko, G. D. Sanders, C. J. Stanton, N. Miura, S. Ikeda, Y. Hashimoto, and S. Katsumoto, and H. Munekata, Phys. Rev. B **70**, 195211 (2004).
[31] T. Slupinski, A. Oiwa, S. Yanagi and H. Munekata, J. Crystal Growth **237-239**, 1326 (2002).
[32] D. Chiba, K. Takamura, F. Matsukura, and H. Ohno, Appl. Phys. Lett. **82**, 3020 (2003).
[33] C. T. Foxon, R. P. Campion, K. W. Edmonds, L. Zhao, K. Wang, N.R. S. Farley, C.R. Staddon, and B. L. Gallagher, J. Mater. Sci. Mater. Electronics **15**, 727 (2004).
[34] T. Slupinski, H. Munekata, and A. Oiwa, Appl. Phys. Lett. **80**, 1592 (2002).
[35] S. Ohya, H. Kobayashi, and M. Tanaka, Appl. Phys. Lett. **83**, 2175 (2003)
[36] R. Moriya and H. Munekata, J. Appl. Phys. **93**, 4603 (2003).
[37] A. M. Nazmul, S. Sugahara, and M. Tanaka, Phys. Rev. B **67**, 241308 (2003); very recently, $T_c = 190 \sim 250\,\mathrm{K}$ with $x = 0.5\,\mathrm{ML}$ has also been reported from the same group.
[38] M. E. Overberg, B. P. Gila, C.R. Abernathy, S. J. Pearton, N. A. Theodoropoulou, K. T. McCarthy, and S. B. Arnason, Appl. Phys. Lett. **79**, 3128 (2001).
[39] S. Sonoda, S. Shimizu, T. Sasaki, Y. Yamamoto, and H. Hori, J. Cryst. Growth **237-239**, 1358 (2002).
[40] H. Hashimoto, Y. K. Zhou, M. Kanaura, and H. Asahi, Solid State Commun. **122**, 37 (2002).
[41] T. Graf, M. Gjukic, M. S. Brandt, M. Stuzmann, and O. Ambacher, Appl. Phys. Lett. **81**, 5159 (2002).
[42] P. Mahadevan and A. Zunger, Appl. Phys. Lett. **85**, 2860 (2004).
[43] K. M. Yu, W. Walukiewicz, I. Kuryliszyn, X. Liu, Y. Sasaki, and J. K. Furdyna, Phys. Rev. B **65**, 201303 (2002).
[44] T. Kuroiwa, T. Yasuda, F. Matsukura, A. Shen, Y. Ohno, Y. Segawa, and H. Ohno, Electron. Lett. **34**, 190 (1998).
[45] B. Beschoten, P. A. Crowell, I. Malajovich, D. D. Awschalom, F. Matsukura, A. Shen, and H. Ohno, Phys. Rev. Lett. **83**, 3073 (1999).
[46] K. Ando, T. Hayashi, M. Tanaka, and A. Twardowski, J. Appl. Phys. **83**, 6548 (1998).
[47] *IEEE Standard Dictionary of Electrical and Electronics Terms* (IEEE, New York and Wiley-Interscience, 3rd Edition, Ed. in chief by F. Jay), p. 483 and p. 789.
[48] Semiconductor and Semimetals, Vol. 25, *Diluted Magnetic Semiconductors*, edited by J. Frudyna and J. Kossute (Academic Press, 1988), p. 294.
[49] J. Okabayashi, A. Kimura, O. Rader, T. Mizokawa, A. Fujimori, T. Hayashi, and M. Tanaka, Phys. Rev. B **58**, R2486 (1998).
[50] P. Fumagalli, H. Munekata, and R. J. Gambino, IEEE Transactions on Magnetics **29**, 3411 (1993).
[51] P. Fumagalli and H. Munekata, Phys. Rev. B **53**, 15045 (1996).
[52] K. Ando and H. Munekata, J. Magn. Magn. Mater. **272**, 2004 (2004).

[53] M. Z. Zudov, J. Kono, Y. H. Matsuda, T. Ikeda, N. Miura, H. Munekata, G. D. Sanders, Y. Sun, and C. J. Stanton, Phys. Rev. B **66**, 161307(R) (2002).
[54] K. Ando, Appl. Phys. Lett. **82**, 100 (2003).
[55] T. Kondo, J. Hayafuji, A. Oiwa, and H. Munekata, Jpn. J. Appl. Phys. **43**, L851 (2004).
[56] G. D. Sanders, Y. Sun, F. V. Kyrychenko, C. J. Stanton, G. A. Khodaparast, M. A. Zudov, J. Kono, Y. H. Matsuda, N. Miura, and H. Munekata, Phys. Rev. B **68**, 165205 (2003).
[57] N. F. Mott and E. A. Davis, *Electronic Processes in Non-Crystalline Materials*, 2nd. ed. (Clarendon Press, Oxford), p. 273.
[58] E. J. Singley, R. Kawakami, D. D. Awschalom, and D. N. Basov, Phys. Rev. Lett. **89**, 097203 (2002).
[59] E. J. Singley, K. S. Burch, R. Kawakami, J. Stephens, D. D. Awschalom, and D. N. Basov, Phys. Rev. B **68**, 165204 (2003).
[60] H. Lee, T. D. Kang, Y. J. Park, H. Y. Cho, R. Moriya, and H. Munekata, J. Korean Phys. Soci. **42**, 441 (2003).
[61] K. S. Burch, J. Stephens, R. K. Kawakami, D. D. Awschalom, and D. N. Basov, Phys.Rev. B **70**, 205208 (2004).
[62] M. Hashimoto, Y. K. Zhou, H. Tampo, M. Kanamura, and H. Asahi, J. Cryst. Growth **252**, 499 (2003).
[63] Y. K. Zhou, M. Hashimoto, M. Kanamura, and H. Asahi, J. Supercond. Incorp. Novel Mag. **16**, 37 (2003).
[64] M. Hashimoto, H. Tanaka, R. Asano, S. Hasegawa, and H. Asahi, Appl. Phys. Lett. **84**, 4191 (2004).
[65] J. Epstein, MRS Bulletin **28**, 492 (2003).
[66] V. L. Gurevich, R. Laiho, and A. V. Lashkul, Phys. Rev. Lett. **69**, 180 (1992).
[67] V. V. Afonin, V. L. Gurevich, and R. Laiho, Phys. Rev. B **52**, 2090 (1995).
[68] H. Munekata, T. Abe, S. Koshihara, A. Oiwa, M. Hirasawa, S. Katsumoto, Y. Iye, C. Urano, and H. Takagi, J. Appl. Phys. **81**, 4862 (1997).
[69] H. Munekata, A. Oiwa and T. Slupinski, Physica E **13**, 516 (2002).
[70] Y. K. Zhou, H. Asahi, S. Okumura, M. Kanamura, J. Asakura, K. Asami, M. Nakajima, H. Harima, and S. Gonda, J. Cryst. Growth **227/228**, 614 (2001).
[71] A. M. Nazmul, S. Kobayashi, S. Sugahara, and M. Tanaka, Jpn. J. Appl. Phys. **43**, L233 (2004).
[72] H. Munekata and S. Koshihara, Superlattices and Microstructures **25**, 251 (1999).
[73] A. Oiwa, A. Endo, S. Katsumoto, Y. Iye, H. Ohno, H. Munekata, Phys. Rev. B **59**, 5826 (1999).
[74] S. Chikazumi, in *Physics of Magnetism* (English version assisted by S. H. Charap), (Kreiger Publishing, Florida, 1978), p. 260.
[75] L. Néel, Compt. Rend. **224**, 1550 (1947).
[76] D. F. Eldridge, J. Appl. Phys. **32**, 247S (1961).
[77] X. Liu, W. L. Lim, L. V. Titova, T. Wojtowicz, M. Kutrowski, K. J. Yee, M. Dobrowolska, J. K. Furdyna, S. J. Potashnik, M. B. Stone, P. Schiffer, I.

Vurgaftman, and J.R. Meyer, Physica E **20**, 370 (2004).
[78] D. Chiba, M. Yamanouch, F. Matsukura, H. Ohno, Science **301**, 943 (2003).
[79] M. I. D'yakonov and V. I. Perel', in *Optical Orientation*, eds. F. Meier and B. P. Zakharchenya (North-Holland, Amsterdam, 1984), p. 11.
[80] D. D. Awschalom, J.-M. Halbout, S. von Molnár, T. Siegrist, and F. Holtzberg, Phys. Rev. Lett. **55**, 1128 (1985).
[81] D. D. Awschalom, M.R. Freeman, N. Samarth, H. Luo, and J. K. Frudyna, Phys. Rev. Lett. **66**, 1212 (1991).
[82] G. Mackh, M. Hilpert, D. R. Yakovlev, W. Ossau, H. Heinke, T. Litz, F. Fischer, A. Waag, and G. Landwehr, Phys. Rev. B **50**, 14069 (1994).
[83] T. Dietl, P. Peyla, W. Grieshaber, and Y. Merle d'Aubigné, Phys. Rev. Lett. **74**, 474 (1995).
[84] A. Oiwa, Y. Mitsumori, R. Moriya, T. Slupinski, and H. Munekata, Phys. Rev. Lett. **88**, 137202 (2002).
[85] T.R. McGuire, R. J. Gambino, and R. C. O'Handley, in *The Hall Effect and Its Application*, edited by C. L. Chien and C. R. Westgate (Plenum, New York, 1980), p. 147.
[86] F. Matsukura, H. Ohno, A. Shen, and Y. Sugawara, Phys. Rev. B **57**, R2037 (1998).
[87] S. D. Ganichev, H. Ketterl, W. Prettl, E. L. Ivchenko, and L. E. Vorobjev, Appl. Phys. Lett. **77**, 3146 (2000).
[88] A. Oiwa, R. Moriya, Y. Mitsumori, T. Slupinski, H. Munekata, J. Superconductivity: Incorporating Novel Magnetism **16**, 439 (2003).
[89] K. Bussmann, G. A. Prinz, S-F. Cheng, and D. Wang, Appl. Phys. Lett. **75**, 2476 (1999)
[90] J. A. Katine, F. J. Albert, R. A. Buhman, E. B. Myers, and D. C. Ralph, Phys. Rev. Lett. **84**, 3149 (2000).
[91] Y. Mitsumori, A. Oiwa, T. Slupinski, H. Muraki, Y. Kashimura, F. Minami, and H. Munekata, Phys. Rev. B **69**, 033203 (2004).
[92] S. Chikazumi, in *Physics of Magnetism* (English version assisted by S. H. Charap), (Kreiger Publishing, Florida, 1978), p. 338.
[93] S. Gupta, M. Y. Frankel, J. A. Valdmanis, J. F. Whitaker, G. A. Mourou, F. W. Smith, and A. R. Calawa, Appl. Phys. Lett. **59**, 3276 (1991).
[94] A. J. Lochtefeld, M. R. Melloch, J. C. P. Chang, and E. S. Harmon, Appl. Phys. Lett. **69**, 1465 (1996).
[95] M. Haiml, U. Siegner, F. Morier-Genoud, U. Keller, M. Luysberg, P. Specht, and E. R. Weber, Appl. Phys. Lett. **74**, 1269 (1999).
[96] A. Oiwa, H. Takechi, H. Munekata, J. Supercond. Incorp. Novel Magnetism, **18**, 9 (2005).
[97] E. Kojima, R. Shimano, Y. Hashimoto, S. Katsumoto, Y. Iye, and M. Kuwata-Gonokami, Phys. Rev. B **68**, 193203 (2003).
[98] Y. Kashimura, R. Moriya, A. Oiwa, and H. Munekata, Physica E **21**, 987 (2004).
[99] M. E, Flatté and G. Vignale, Appl. Phys. Lett. **78**, 1273 (2001).

[100] J. Fabian, I. Žutić, and S. Das Sarma, Phys. Rev. B **66**, 165301 (2002).
[101] S. Sugahara and M. Tanaka, Appl. Phys. Lett. **84**, 2307 (2004).
[102] Y. Nishikawa, A. Tackeuchi, and S. Muto, Appl. Phys. Lett. **66**, 839 (1995).
[103] B. T. Jonker, Proceedings of the IEEE **91**, 727 (2003).
[104] J. Hayafuji, T. Kondo, and H. Munekata, Inst. Phys. Conf. Ser. No. **184**, 127 (2005).
[105] H. Shimizu and Y. Nakano, Jpn. J. Appl. Phys. **43**, L1561 (2004).
[106] See e.g., T. D. Ladd, J.R. Goldman, F. Yamaguchi, Y. Yamamoto, E. Abe, and K. Itoh, Phys. Rev. Lett. **89**, 017901 (2002).
[107] J. C. Slonczewski, J. Magn. Magn. Mater. **159**, L1 (1996).
[108] R. Moriya, K. Hamaya, A. Oiwa, and H. Munekata, Jpn. J. Appl. Phys. **43**, L825 (2004).
[109] D. Chiba, Y. Sato, T. Kita, F. Matsukura, and H. Ohno, Phys. Rev. Lett. **93**, 216602 (2004).
[110] R. Fiederling, M. Kelm, G. Reuscher, W. Ossau, G. Schmidt, A. Waag, and L. W. Molencamp, Nature **402**, 787 (1999).
[111] Y. Ohno, D. K. Young, B. Beschoten, F. Matsukura, H. Ohno, and D. D. Awschalom, Nature **402**, 790 (1999).
[112] A. Tackeuchi, T. Kuroda, S. Muto, Y. Nishikawa, and O. Wada, Jpn. J. Appl. Phys. **38**, 4680 (1999).
[113] I. Žutić, J. Fabian, and S. Das Sarma, Appl. Phys. Lett. **79**, 1558 (2001).
[114] I. Žutić, J. Fabian, and S. Das Sarma, Phys. Rev. B **64**, 121201 (2001).
[115] J. M. Kikkawa and D. D. Awshalom, Phys. Rev. Lett. **80**, 4313 (1998).
[116] I. Žutić, J. Fabian, and S. Das Sarma, Phys. Rev. Lett. **88**, 66603 (2002).

2 Bipolar spintronics

Igor Žutić and Jaroslav Fabian

2.1 Preliminaries

2.1.1 *Introduction*

Spintronics, or spin electronics, involves the study of active control and manipulation of spin degrees of freedom in solid-state systems [1]. Conventionally, the term spin stands for either the spin of a single electron, which can be detected by its magnetic moment, or the average spin of an ensemble of electrons, manifested by magnetization. The control of spin is then a control of either the population and the phase of the spin of an ensemble of particles, or a coherent spin manipulation of a single- or a few-spin system.

Bipolar spintronics, typically realized in systems which include semiconductors, is an emerging subfield of spintronics in which carriers of both polarities (electrons and holes) are important. In contrast to unipolar spintronics, characteristic for metallic systems, there are large deviations from local charge neutrality and intrinsic non-linearities in the current-voltage characteristics, which are important even at small applied bias. Together with the ease of manipulating the minority charge carriers (electrons and holes), these distinguishing features of bipolar spintronics are suitable for realizing active devices which could amplify signals and provide an additional degree of control not available in charge-based electronics. Spin-polarized bipolar transport can be thought of as a generalization of its unipolar counterpart. Specifically, spin-polarized unipolar transport can then be obtained as a limiting case by setting the electron-hole recombination rate to zero and considering only one type of carrier (either electrons or holes).

In the first section we present some background material that covers several important findings of unipolar spintronics which are later contrasted with the situation relevant for the bipolar case. Spintronic studies typically rely on the lifting of spin degeneracy in various physical properties. The different behavior for "spin up" and "spin down" is not limited to ferromagnetic or paramagnetic materials (in an applied magnetic field) but can also be realized even in nonmagnetic materials with the aid of transport, optical, and resonance methods to generate non-equilibrium spin polarization. We describe two such methods: optical spin orientation and spin injection, which are further discussed throughout this book. Spin transport differs from charge transport in that spin is a nonconserved quantity in solids due to spin-orbit and hyperfine coupling. We focus on the mechanisms for spin relaxation in semiconductors through which spin relaxes towards equilibrium. Such mechanisms usually involve spin-orbit coupling to provide the spin-dependent potential, in combination with momentum scattering providing a randomizing force. Typical time scales for spin relaxation in electronic systems are measured in nanoseconds, while the range is from pico- to microseconds.

In the second section we develop a theory of bipolar spin-polarized transport based on drift-diffusion and Poisson equations. We examine the case of spin-polarized and magnetic *p-n* junctions to illustrate several novel effects. We then consider three terminal transistor structures. In particular, we briefly describe a unipolar spin field effect transistor and discuss in detail our proposal for a magnetic bipolar transistor. We conclude this section by presenting possible future directions.

2.1.2 *Concept of spin polarization*

Spin polarization not only of electrons, but also of holes, nuclei, and excitations can be conveniently defined as

$$P_X = X_s/X, \tag{2.1}$$

the ratio of the difference $X_s = X_\lambda - X_{-\lambda}$ and the sum $X = X_\lambda + X_{-\lambda}$, of the spin-resolved λ components for a particular quantity X. To avoid ambiguity as to what precisely is meant by spin polarization both the choice of the spin-resolved components and the relevant physical quantity X need to be specified. Conventionally, λ is taken to be $\uparrow$ or $+$ (numerical value $+1$) for spin up, $\downarrow$ or $-$ (numerical value -1) for spin down, with respect to the chosen axis of quantization. For example, the quantization axis can be chosen along the spin angular momentum, applied magnetic field, magnetization, or direction of light propagation. For a free electron, the spin angular momentum and magnetic moment are in opposite directions, and what precisely is denoted by "spin up" varies in the literature [2].

In ferromagnetic metals it is customary to refer to $\uparrow$ ($\downarrow$) as carriers with magnetic moment parallel (antiparallel) to the magnetization or, equivalently, as carriers with majority (minority) spin [4]. In semiconductors the terms majority and minority usually refer to relative populations of the carriers while $\uparrow$ or $+$ and $\downarrow$ or $-$ correspond to the quantum numbers m_j with respect to the z-axis taken along the direction of light propagation or along the applied magnetic field. It is important to emphasize that both the magnitude and the sign of the spin polarization in Eq. (2.1) depends of the choice of X, relevant to the detection technique employed, say optical vs. transport and bulk vs. surface measurements [2, 5]. Even in the same homogeneous material the measured P_X can vary for different X, and it is crucial to identify which physical quantity–charge current, carrier density, conductivity, or the density of states–is being measured experimentally.

The interest in highly spin-polarized materials can be illustrated with the example of tunneling magnetoresistance (TMR) in a magnetic tunnel junction

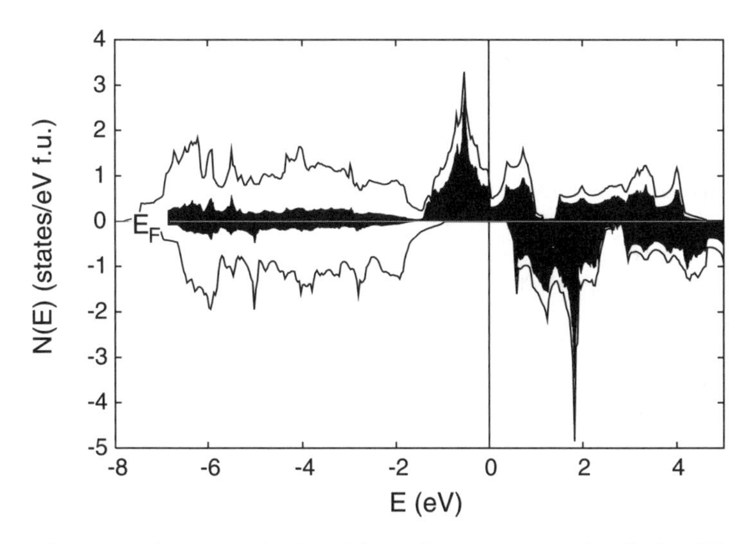

FIG. 2.1. Density of states calculated from first principles for CrO_2. The shaded area is the Cr d-states contribution. The Fermi level is at 0 where only the spin majority electrons have non-vanishing density of states. From [3].

(MTJ) consisting of two ferromagnets (F1, F2) separated by an insulator (I). The tunneling magnetoresistance can be defined as

$$\mathrm{TMR} = \frac{R_{\uparrow\downarrow} - R_{\uparrow\uparrow}}{R_{\uparrow\uparrow}}, \tag{2.2}$$

where $\uparrow\uparrow$ and $\uparrow\downarrow$ denote the parallel and antiparallel orientation of magnetizations in regions F1 and F2 (a detailed discussion of TMR is given in [6–8]). The difference between $R_{\uparrow\downarrow}$ and $R_{\uparrow\uparrow}$ is usually referred to as the spin-valve effect [9]. A simplified relation between TMR and the spin-polarization of the density of states [$P_N = (N_\uparrow - N_\downarrow)/(N_\uparrow + N_\downarrow)$, from Eq. (2.1)] can be obtained from Jullière's model [10]

$$\mathrm{TMR} = \frac{2P_{N1}P_{N2}}{1 - P_{N1}P_{N2}}, \tag{2.3}$$

suggesting that the highly polarized materials could yield desirably large TMR values. Materials which have a completely polarized density of states at the Fermi level ($P_N = \pm 1$) are also known as half-metals (or half-metallic ferromagnets). Near the Fermi level they behave as metals only for one spin. The density of states vanishes completely for the other spin [11], as illustrated in Fig. 2.1. Idealized half-metallicity, at $T = 0$ and in bulk samples, has been predicted in some oxides (such as, CrO_2, Fe_3O_4, $La_{0.7}Sr_{0.3}MnO_3$, and $SrFeMoO_6$), Heusller alloys (such as NiMnSb and Ni_2MnGa), zincblende ferromagnets (such as MnAs and CrAs) and ferromagnetic semiconductors (see also Chapter 1 [12–14]. However, experiments often show substantially lower values of spin polarization than what is expected for half-metals. A particular technique, where such a reduction of measured spin polarization has been extensively explored, uses transport properties in ferromagnet/superconductor junctions and the process of Andreev reflection [15–21]. This topic is further discussed in Chapter 9. The absence of predicted half-metalicity at finite temperatures was attributed to the presence of magnons and phonons, while spin-orbit effects, surfaces, interfaces, various inhomogeneities and defects could reduce a complete spin polarization even at zero temperature [22]. Interestingly, a deviation from the idealized situation of T=0 and bulk material can also yield an *increase* of effective spin polarization. Important experiments, discussed further in Chapter 6, have shown that TMR can be increased several times by merely replacing a non-magnetic tunnel barrier (aluminum oxide was replaced by magnesium oxide) [23]. Even with conventional ferromagnetic electrodes (CoFe) extraordinarily large values of TMR ($>$ 200% at room temperature) can be achieved. These considerations show that an effective spin polarization used to interpret TMR values from Eq. (2.3) is not an intrinsic bulk property of the F region.

Ferromagnetism in semiconductors, discussed in detail in Chapter 1, has some intriguing implications on the concept of spin polarization. The first direct measurement of spin polarization in these materials was realized in a superconducting junction with (Ga,Mn)As [24], following a similar theoretical proposal [25] that the spin polarization of a ferromagnetic semiconductor will strongly influence the process of Andreev reflection and low bias conductance in such junctions. There is a wide range of ferromagnetic semiconductors, including for example, II-VI (Zn,Cr)Te [26] and chalcopyrite materials [14,27,28], in which room-temperature ferromagnetism has been reported. However, the carrier-mediated nature of ferromagnetism, which allows novel means to manipulate equilibrium spin polarization, has most extensively been studied in the (III,Mn)V family. A change in the number of carriers which can be realized optically (illumination by laser) [29–31] or electrically (by applying a gate voltage) [31–33] could turn the ferromagnetism on and off and provide a net carrier spin polarization.

2.1.3 *Optical spin orientation*

The generation of non-equilibrium spin polarization can be realized by optical methods known as optical orientation or optical pumping. This technique, frequently used in semiconductors, is derived from optical pumping proposed by Kastler [34] in which optical irradiation changes the relative populations within the Zeeman and the hyperfine levels of the ground states of atoms. In optical orientation, the angular momentum of absorbed circularly polarized light is transferred to the medium. Electron orbital momenta are directly oriented by light and through the spin-orbit interaction electron spins become polarized. The demonstration of optical orientation in semiconductors was first realized in silicon by Lampel [35]. Polarized carriers in a direct band-gap semiconductor can be simply detected by observing circularly polarized light created by recombination of electrons and holes. However, an indirect band gap in Si precludes this ap-

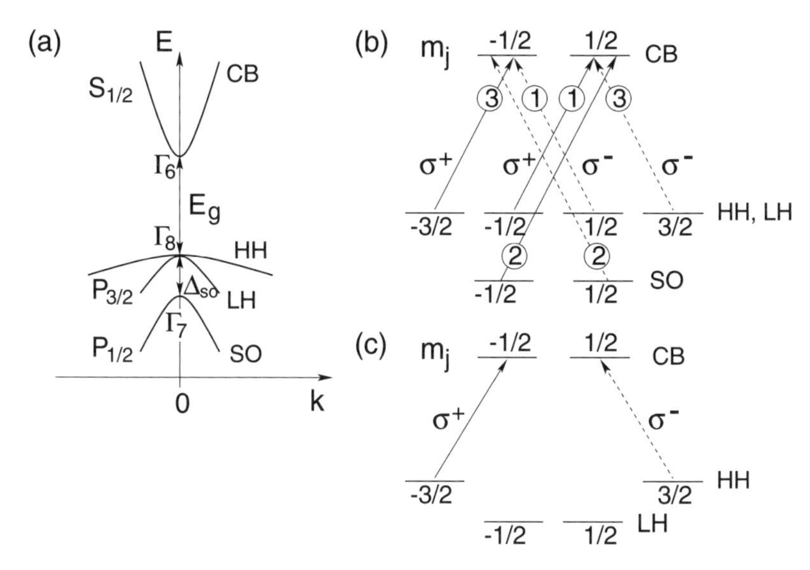

FIG. 2.2. Interband transitions in GaAs: (a) schematic band structure of GaAs near the center of the Brillouin zone (Γ point), where E_g is the band gap and Δ_{so} is the spin-orbit splitting; CB, conduction band; HH, valence heavy hole; LH, light hole; SO, spin-orbit-split-off subbands; $\Gamma_{6,7,8}$ are the corresponding symmetries at the $k = 0$ point; (b) Selection rules for interband transitions between m_j sublevels for circularly polarized light σ^+ and σ^- (positive and negative helicity). The circled numbers denote the relative transition intensities that apply for both excitations (depicted by the arrows) and radiative recombinations; (c) Interband transitions in a quantum well where the HH and LH degeneracy is lifted by quantum confinement.

proach and optical orientation of electrons was inferred using nuclear magnetic resonance since through hyperfine interaction polarized carriers also induce polarization of nuclei.

In a semiconductor the photo-excited spin-polarized electrons and holes exist for a time τ before they recombine. If a fraction of the carriers' initial orientation survives longer than the recombination time, that is, if $\tau < \tau_s$, where τ_s is the spin relaxation time (see Section 2.1.5), the luminescence (recombination radiation) will be partially polarized. By measuring the circular polarization of the luminescence it is possible to study the spin dynamics of the non-equilibrium carriers in semiconductors and to extract such useful quantities as the spin orientation, the recombination time, or the spin relaxation time of the carriers [36, 37].

We illustrate the basic principles of optical orientation by the example of GaAs which is representative of a large class of III-V and II-VI zincblende semiconductors (such as, AlAs, AlP, CdS, CdSe, GaP, GaSb, HgS, HgSe, InAs, InSb, ZnO, and ZnS). The band structure is depicted in Fig. 2.2(a). The band gap is $E_g = 1.52$ eV at $T = 0$ K, while the spin split-off band is separated from

Table 2.1 Angular and spin part of the wave function at Γ.

Symmetry	$\|J, m_j\rangle$	Wave function
Γ_6	$\|1/2, 1/2\rangle$	$\|S\uparrow\rangle$
	$\|1/2, -1/2\rangle$	$\|S\downarrow\rangle$
Γ_7	$\|1/2, 1/2\rangle$	$\|-(1/3)^{1/2}[(X+iY)\downarrow -Z\uparrow]\rangle$
	$\|1/2, -1/2\rangle$	$\|(1/3)^{1/2}[(X-iY)\uparrow +Z\downarrow]\rangle$
Γ_8	$\|3/2, 3/2\rangle$	$\|(1/2)^{1/2}(X+iY)\uparrow\rangle$
	$\|3/2, 1/2\rangle$	$\|(1/6)^{1/2}[(X+iY)\downarrow +2Z\uparrow]\rangle$
	$\|3/2, -1/2\rangle$	$\|-(1/6)^{1/2}[(X-iY)\uparrow -2Z\downarrow]\rangle$
	$\|3/2, -3/2\rangle$	$\|(1/2)^{1/2}(X-iY)\downarrow\rangle$

the light and heavy hole bands by $\Delta_{so} = 0.34$ eV. We denote the Bloch states according to the total angular momentum J and its projection onto the positive z axis m_j: $|J, m_j\rangle$. Expressing the wave functions with the symmetry of s, p_x, p_y, and p_z orbitals as $|S\rangle$, $|X\rangle$, $|Y\rangle$, and $|Z\rangle$, respectively, the band wave functions can be written as listed in Table 2.1 (see also [38]).

To obtain the excitation (or recombination) probabilities consider photons arriving in the z direction. Let $\sigma^{\pm}$ represent the helicity of the exciting light (for outgoing light in the $-z$ direction the helicities are reversed). When we represent the dipole operator corresponding to the $\sigma^{\pm}$ optical transitions as $\propto (X \pm iY) \propto Y_1^{\pm 1}$, where Y_l^m is a spherical harmonic, it follows from Table 2.1 that

$$\frac{|\langle 1/2, -1/2|Y_1^1|3/2, -3/2\rangle|^2}{|\langle 1/2, 1/2|Y_1^1|3/2, -1/2\rangle|^2} = 3 \tag{2.4}$$

for the relative intensity of the σ^+ transition between the heavy ($|m_j = 3/2|$) and the light ($|m_j = 1/2|$) hole subbands and the conduction band. Other transitions are analogous. The relative transition rates are indicated in Fig. 2.2(b). The same selection rules apply to the optical orientation of shallow impurities [39].

We consider here only a spin polarization of the excited electrons which depends on the photon energy $\hbar\omega$. Although holes are initially polarized too, they lose spin orientation very fast, on the time scale of the momentum relaxation time. This can be illustrated on the example of GaAs where a characteristic spin relaxation time for electrons is $\sim$ ns and for holes $\sim$ 100 fs [40]. For $\hbar\omega$ between E_g and $E_g + \Delta_{so}$, only the light and heavy hole subbands contribute. Denoting by n_+ and n_- the density of electrons polarized parallel ($m_j = 1/2$) and antiparallel ($m_j = -1/2$) to the direction of light propagation, we define the spin polarization as (see Section 2.1.2)

$$P_n = (n_+ - n_-)/(n_+ + n_-). \tag{2.5}$$

A simple reversal in the polarization of the illuminating light (from positive to negative helicity) also reverses the sign of the electron density polarization. For our example of the zincblende structure,

$$P_n = (1 - 3)/(3 + 1) = -1/2 \tag{2.6}$$

is the spin polarization at the moment of photo-excitation. The spin is oriented against the direction of light propagation, since there are more transitions from the heavy hole than from the light-hole subbands. The circular polarization of the luminescence is defined as

$$P_{\text{circ}} = (I^+ - I^-)/(I^+ + I^-), \tag{2.7}$$

where $I^\pm$ is the radiation intensity for the helicity $\sigma^\pm$. The polarization of the σ^+ photoluminescence is then

$$P_{\text{circ}} = \frac{(n_+ + 3n_-) - (3n_+ + n_-)}{(n_+ + 3n_-) + (3n_+ + n_-)} = -\frac{P_n}{2} = \frac{1}{4}. \tag{2.8}$$

If the excitation involves transitions from the spin split-off subband, that is, if $\hbar\omega \gg E_g + \Delta_{so}$, the electrons will not be spin polarized ($P_n = P_{\text{circ}} = 0$), indicating the importance of spin-orbit coupling for spin orientation. On the other hand, Fig. 2.2(c) suggests that a removal of the heavy/light hole degeneracy can substantially increase P_n up to the limit of complete spin polarization (for excitations involving only transitions from the heavy-hole subband). Such an increase in P_n and P_{circ} can be realized, for example, due to confinement in quantum well heterostructures or due to strain.

The relation between P_{circ} and P_n, for example, given by Eq. (2.8) in the bulk sample or by $|P_{circ}| = |P_n|$ in a quantum well (see Fig. 2.2), has been successfully employed for spin detection. A class of structures known as spin light-emitting diodes (LEDs) are now widely used to detect electrically injected spin in semiconductors [41–45]. Similar to an ordinary LED [46], electrons and holes recombine (in a quantum well or a *p-n* junction) and produce electroluminescence. However, in a spin LED, as a consequence of radiative recombination of spin-polarized carriers, the emitted light is circularly polarized and could be used to trace back the degree of polarization of carrier density upon injection into a semiconductor. Additional discussion of spin detection using spin LEDs is given in Chapter 6.

2.1.4 *Spin injection in metallic F/N junctions*

Electrical spin injection is an example of a transport method for generating non-equilibrium spin. We focus here on a simple case of a single metallic F/N junction in the regime of linear response appropriate for a single type of carrier (electrons or holes), while more general geometries for studying spin injection

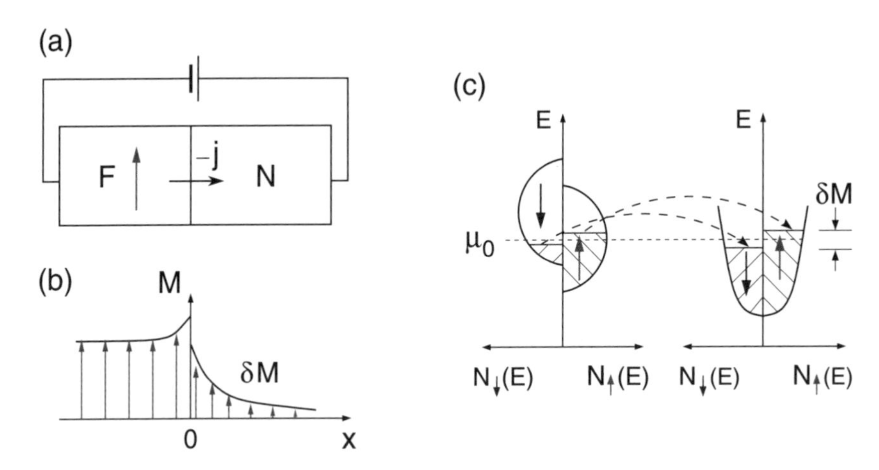

FIG. 2.3. Pedagogical illustration of the concept of electrical spin injection from a ferromagnet (F) into a normal metal (N). Electrons flow from F to N: (a) schematic device geometry; (b) magnetization M as a function of position. Non-equilibrium magnetization δM (spin accumulation) is injected into a normal metal; (c) contribution of different spin-resolved densities of states to charge and spin transport across the F/N interface. Unequal filled levels in the density of states depict spin-resolved electrochemical potentials different from the equilibrium value μ_0. From [1].

[47] will be discussed in Chapter 8. It is important to note that in the context of spin injection in semiconductors, the underlying assumptions of the local charge neutrality and linear response are often violated. We will revisit the implications of such violation in Section 2.2 where we consider the case of spin injection in which both electrons and holes could contribute to the transport.

When a charge current flows across the F/N junction (Fig. 2.3) spin-polarized carriers in a ferromagnet contribute to the net current of magnetization entering the non-magnetic region and lead to non-equilibrium magnetization δM, depicted in Fig. 2.3(b), with the spatial extent given by the spin diffusion length. Magnetization can be decomposed into an equilibrium part and a non-equilibrium part $M = M_0 + \delta M$, where δM is also equivalent to a *non-equilibrium* spin accumulation, first measured in metals by Johnson and Silsbee [48]. In the steady state δM is realized as a balance between spins added by the magnetization current and spins removed by spin relaxation.

Here we follow the approach of [49, 50] using the notation introduced in [1]. We show that the main findings: expressions for current spin polarization and the non-equilibrium resistance due to spin injection across F/N junction can also be obtained from a simple equivalent resistor scheme proposed by Petukhov [51]. Let us consider a steady-state flow of electrons along the x direction in a three-dimensional (3D) geometry consisting of a metallic ferromagnet (region $x < 0$) and a paramagnetic metal or a degenerate semiconductor (region $x > 0$). The

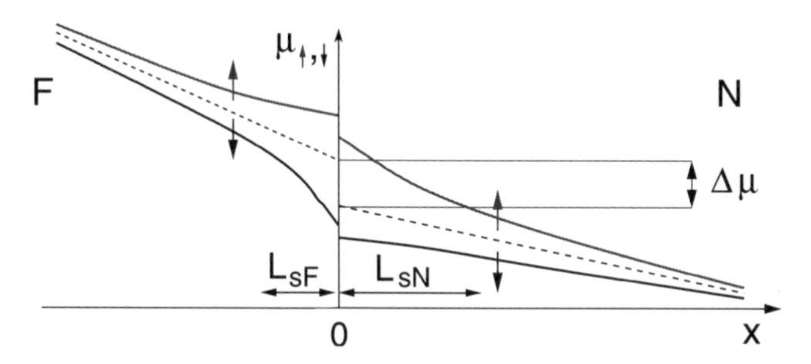

FIG. 2.4. Spatial variation of the electrochemical potentials near a spin-selective resistive interface at an F/N junction. At the interface $x = 0$ both the spin-resolved electrochemical potentials (μ_λ, $\lambda = \uparrow, \downarrow$, denoted with solid lines) and the average electrochemical potential (μ_F, μ_N, dashed lines) are discontinuous. The spin diffusion lengths L_{sF} and L_{sN} characterize the decay of $\mu_s = \mu_\uparrow - \mu_\downarrow$ (or equivalently the decay of spin accumulation and the non-equilibrium magnetization) away from the interface and into the bulk F and N regions, respectively. From [1].

two regions, F and N, form a contact at $x = 0$, as depicted in Fig. 2.4. These are the contact resistance r_c and the two characteristic resistances, r_N and r_F, each given by the ratio of the spin diffusion length and the effective bulk conductivity in the corresponding region. Two limiting cases correspond to the transparent limit, where $r_c \to 0$, and the low-transmission limit, where $r_c \gg r_N, r_F$.

Spin-resolved quantities are labeled by $\lambda = 1$ or $\uparrow$ for spin up, $\lambda = -1$ or $\downarrow$ for spin down along the chosen quantization axis. For a free electron, the spin angular momentum and magnetic moment are in opposite directions, and what precisely is denoted by "spin up" varies in the literature [2]. Conventionally, in metallic systems [52], spin up refers to carriers with majority spin. This means that the spin (angular momentum) of such carriers is antiparallel to the magnetization. Spin-resolved charge current (density) in a diffusive regime can be expressed as

$$j_\lambda = \sigma_\lambda \nabla \mu_\lambda, \tag{2.9}$$

where σ_λ is the conductivity and the electrochemical potential is

$$\mu_\lambda = (qD_\lambda/\sigma_\lambda)\delta n_\lambda - \phi, \tag{2.10}$$

with q the proton charge, D_λ the diffusion coefficient, $\delta n_\lambda = n_\lambda - n_{\lambda 0}$ the change of electron density from the equilibrium value for spin λ, and ϕ the electric potential [we note that often in the literature the same quantity μ_λ is also referred to as the (spin resolved) chemical potential, see also Section 2.2]. More generally, for a noncollinear magnetization, addressed in other chapters in the context of spin-transfer torque and current magnetization reversal, j_λ becomes a second-rank tensor [53].

In the steady state the continuity equation is

$$\nabla j_\lambda = \lambda q \left[\frac{\delta n_\lambda}{\tau_{\lambda-\lambda}} - \frac{\delta n_{-\lambda}}{\tau_{-\lambda\lambda}} \right], \tag{2.11}$$

and $\tau_{\lambda\lambda'}$ is the average time for flipping a λ-spin to λ'-spin. For a degenerate conductor the Einstein relation is

$$\sigma_\lambda = q^2 N_\lambda D_\lambda, \tag{2.12}$$

where $\sigma = \sigma_\uparrow + \sigma_\downarrow$ and $N = N_\uparrow + N_\downarrow$ is the density of states. Using detailed balance $N_\uparrow/\tau_{\uparrow\downarrow} = N_\downarrow/\tau_{\downarrow\uparrow}$ [54] together with Eqs (2.10) and (2.12), the continuity equation can be expressed as

$$\nabla j_\lambda = \lambda q^2 \frac{N_\uparrow N_\downarrow}{N_\uparrow + N_\downarrow} \frac{\mu_\lambda - \mu_{-\lambda}}{\tau_s}, \tag{2.13}$$

where $\tau_s = \tau_{\uparrow\downarrow}\tau_{\downarrow\uparrow}/(\tau_{\uparrow\downarrow} + \tau_{\downarrow\uparrow})$ is the spin relaxation time. Equation (2.13) implies the conservation of charge current $j = j_\uparrow + j_\downarrow = const.$, while the spin counterpart, the difference of the spin-polarized currents $j_s = j_\uparrow - j_\downarrow$ is position dependent. Other "spin quantities," X_s, unless explicitly defined, are analogously expressed with the corresponding (spin) polarization given by $P_X = X_s/X$. For example, the current polarization $P_j = j_s/j$, generally different from the density polarization $P_n = (n_\uparrow - n_\downarrow)/n$, is related to the conductivity polarization P_σ as

$$P_j = 2(\sigma_\uparrow\sigma_\downarrow/\sigma)\nabla\mu_s/j + P_\sigma \tag{2.14}$$

where $\mu_s = \mu_\uparrow - \mu_\downarrow$. In terms of the average electrochemical potential $\mu = (\mu_\uparrow + \mu_\downarrow)/2$, P_σ further satisfies

$$\nabla\mu = -P_\sigma \nabla\mu_s/2 + j/\sigma. \tag{2.15}$$

From Eqs (2.10) and (2.13) it follows that μ_s satisfies the diffusion equation [54–57]

$$\nabla^2\mu_s = \mu_s/L_s^2, \tag{2.16}$$

where the spin diffusion length is $L_s = (\overline{D}\tau_s)^{1/2}$ with the spin averaged diffusion coefficient $\overline{D} = (\sigma_\downarrow D_\uparrow + \sigma_\uparrow D_\downarrow)/\sigma = N(N_\downarrow/D_\uparrow + N_\uparrow/D_\downarrow)^{-1}$. Using Eq. (2.10) and the local charge quasineutrality $\delta n_\uparrow + \delta n_\downarrow = 0$ shows that μ_s is proportional to the non-equilibrium (or excess) spin density $\delta s = \delta n_\uparrow - \delta n_\downarrow$ ($s = s_0 + \delta s = n_\uparrow - n_\downarrow$)

$$\mu_s = \frac{1}{2q} \frac{N_\uparrow + N_\downarrow}{N_\uparrow N_\downarrow} \delta s. \tag{2.17}$$

Correspondingly, μ_s is often referred to as the (non-equilibrium) *spin accumulation.*

The preceding equations are simplified for the N region by noting that $\sigma_\lambda = \sigma/2$, $\sigma_s = 0$, and $D_\lambda = \overline{D}$. Quantities pertaining to a particular region are denoted by the index F or N.

Some care is needed to establish the appropriate boundary conditions at the F/N interface. In the absence of spin-flip scattering at the F/N interface (which can arise, for example, due to spin-orbit coupling or magnetic impurities) the spin current is continuous and thus $P_{jF}(0^-) = P_{jN}(0^+) \equiv P_j$ (omitting $x = 0^\pm$ for brevity, and superscripts $\pm$ in other quantities).

Unless the F/N contact is highly transparent, μ_λ is discontinuous across the interface and the boundary condition is

$$j_\lambda(0) = \Sigma_\lambda[\mu_{\lambda N}(0) - \mu_{\lambda F}(0)], \tag{2.18}$$

where

$$\Sigma = \Sigma_\uparrow + \Sigma_\downarrow \tag{2.19}$$

is the contact conductivity. For a free-electron model $\Sigma_\uparrow \neq \Sigma_\downarrow$ can be simply inferred from the effect of the exchange energy, which would yield spin-dependent Fermi wave vectors and transmission coefficients. A microscopic determination of the corresponding contact resistance (see Eq. 2.22) is complicated by the influence of disorder, surface roughness, and different scattering mechanisms and is usually obtained from model calculations [58, 59]. From Eqs (2.18) and (2.19) it follows that

$$\mu_{sN}(0) - \mu_{sF}(0) = 2r_c(P_j - P_\Sigma)j, \tag{2.20}$$

$$\mu_N(0) - \mu_F(0) = r_c(1 - P_\Sigma P_j)j, \tag{2.21}$$

where the effective contact resistance is

$$r_c = \Sigma/4\Sigma_\uparrow\Sigma_\downarrow. \tag{2.22}$$

The decay of μ_s, away from the interface, is characterized by the corresponding spin diffusion length

$$\mu_{sF} = \mu_{sF}(0)e^{x/L_{sF}}, \quad \mu_{sN} = \mu_{sN}(0)e^{-x/L_{sN}}. \tag{2.23}$$

A non-zero value for $\mu_{sN}(0)$ implies the existence of non-equilibrium magnetization δM in the N region (for non-interacting electrons $q\mu_s = \mu_B \delta M/\chi$, where χ is the magnetic susceptibility). Such a δM, as a result of electrical spin injection, was proposed by Aronov and Pikus [60] and first measured in metals by Johnson and Silsbee[48].

By applying Eq. (2.14), separately, to the F and N regions, one can obtain the amplitude of spin accumulation in terms of the current and density of states spin polarization and the effective resistances r_F and r_N,

$$\mu_{sF}(0) = 2r_F\left[P_j - P_{\sigma F}\right]j, \quad \mu_{sN}(0) = -2r_N P_j j, \tag{2.24}$$

where

$$r_N = L_{sN}/\sigma_N, \quad r_F = L_{sF}\sigma_F/(4\sigma_{\uparrow F}\sigma_{\downarrow F}). \tag{2.25}$$

From Eqs (2.24) and (2.20) the current polarization can be obtained as

$$P_j = [r_c P_\Sigma + r_F P_{\sigma F}]/r_{FN}, \tag{2.26}$$

where $r_{FN} = r_F + r_c + r_N$ is the effective equilibrium resistance of the F/N junction. It is important to emphasize that a measured highly polarized current, representing an efficient spin injection, does not itself imply a large spin accumulation or a large density polarization, typically measured by optical techniques. In contrast to the derivation of P_j from Eq. (2.26), determining P_n requires using Poisson's equation or a condition of the local charge quasineutrality.

An alternative approach to study spin injection, based on irreversible thermodynamics, was developed by Johnson and Silsbee [61] who were the first to obtain an expression equivalent to Eq. (2.26). The results obtained here can also be illustrated using a simple resistor scheme. Petukhov has shown [51] that Eqs (2.26) and (2.28) (given below) could be obtained by considering an equivalent circuit scheme with two resistors $\tilde{R}_\uparrow$, $\tilde{R}_\downarrow$ connected in parallel. Each of these resistors contains three contributions from the F and the N regions and the contact between them, where

$$\tilde{R}_\lambda = L_{sF}/\sigma_{\lambda F} + 1/\Sigma_\lambda + 2L_{sN}/\sigma_N, \tag{2.27}$$

and $\tilde{R}_\uparrow + \tilde{R}_\downarrow = 4r_{FN}$.

For such a resistor scheme, by noting that $j_\uparrow \tilde{R}_\uparrow = j_\downarrow \tilde{R}_\downarrow$, Eq. (2.26) is obtained as $P_j = -P_{\tilde{R}} \equiv -(\tilde{R}_\uparrow - \tilde{R}_\downarrow)/(\tilde{R}_\uparrow + \tilde{R}_\downarrow)$. δR in Eq. (2.28) is then obtained as the difference between the total resistance of the non-equilibrium spin-accumulation region of length $L_{sF} + L_{sN}$ [given by the equivalent resistance $\tilde{R}_\uparrow \tilde{R}_\downarrow/(\tilde{R}_\uparrow + \tilde{R}_\downarrow)$] and the equilibrium resistance for the same region, $L_{sF}/\sigma_F + L_{sN}/\sigma_N$.

By examining either Eq. (2.26) or Fig. 2.5 we can both infer some possible limitations and deduce several experimental strategies for effective spin injection, i.e. to increase P_j into semiconductors. For a perfect Ohmic contact $r_c = 0$, the typical resistance mismatch $r_F \ll r_N$ (where F is a metallic ferromagnet) implies inefficient spin injection with $P_j \approx r_F/r_N \ll 1$, referred to as the *conductivity mismatch* problem in [57]. Even in the absence of resistive contacts, effective spin injection into a semiconductor can be achieved if the resistance mismatch is reduced by using for spin injectors either a magnetic semiconductor or a highly spin-polarized ferromagnet.[1] As can be seen from Eq. (2.26) the spin-selective resistive contact $r_c \gg r_F, r_N$ (such as a tunnel or Schottky contact) would contribute to effective spin injection with $P_j \approx P_\Sigma$ being dominated by the effect r_c and not the ratio r_F/r_N.[2] This limit is also instructive to illustrate the principle of spin filtering [63, 64]. In a spin-discriminating transport

[1] From Eq. (2.25) a half-metallic ferromagnet implies a large r_F.

[2] A similar result was stated previously in [62].

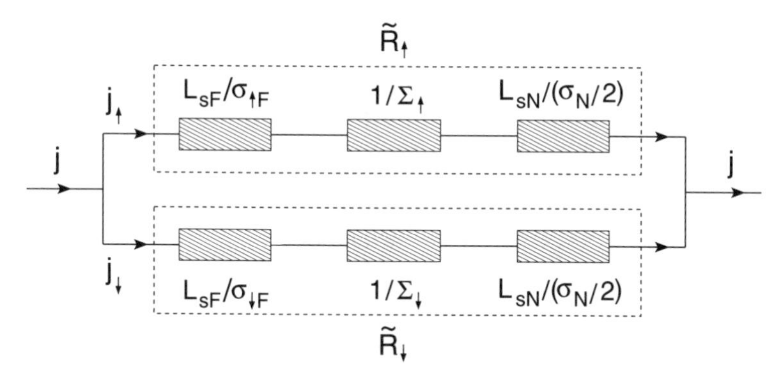

FIG. 2.5. Equivalent resistor scheme illustrating spin injection across an F/N junction in a linear regime.

process the resulting degree of spin polarization is changed. Consequently the effect of spin filtering, similar to spin injection, leads to the generation of (non-equilibrium) spin polarization. For example, at low temperature EuS and EuSe can act as spin-selective barriers. In the extreme case, initially spin-unpolarized carriers (say, injected from a non-magnetic material) via spin-filtering could attain complete polarization. For a strong spin-filtering contact $P_\Sigma > P_{\sigma F}$, the sign of the spin accumulation is reversed in the F and N regions, near the interface (recall Eq. 2.20), in contrast to the behavior sketched in Fig. 2.4, where $\mu_{sF,N} > 0$.

The spin injection process alters the potential drop across the F/N interface because differences of spin-dependent electrochemical potentials on either side of the interface generate an effective resistance δR. By integrating Eq. (2.15) for the N and F regions, separately, it follows that $Rj = \mu_N(0) - \mu_F(0) + P_{\sigma F}\mu_{sF}(0)/2$, where R is the junction resistance. Using Eqs (2.21), (2.25), and (2.26) allows us to express $R = R_0 + \delta R$, where $R_0 = 1/\Sigma$ ($R_0 = r_c$ if $\Sigma_\uparrow = \Sigma_\downarrow$) is the equilibrium resistance, in the absence of spin injection, and

$$\delta R = \left[r_N \left(r_F P_{\sigma F}^2 + r_c P_\Sigma^2 \right) + r_F r_c (P_{\sigma F} - P_\Sigma)^2 \right] / r_{FN}, \tag{2.28}$$

where $\delta R > 0$ is the non-equilibrium resistance. In the previous analysis, spin-flip processes at an F/N interface have been neglected. However, such processes could be increasingly important at higher temperatures and would require a generalization to the usual experimental analysis for spin injection in metallic systems [65].

2.1.5 *Spin relaxation in semiconductors*

Spin relaxation is a process in which spin relaxes towards equilibrium. The corresponding spin relaxation time is denoted as T_1. This time is also called longitudinal since it is usually defined for the spin population along an applied magnetic field. Spin dephasing is then a process in which the transverse spin randomizes and reaches equilibrium. The corresponding time T_2 is thus also called the transverse time. During spin relaxation energy needs to be exchanged with the environment (to bring the spin population into equilibrium characterized by the magnetic field), in spin dephasing energy is preserved. It turns out that for small magnetic fields conduction electron spin relaxation and dephasing in cubic systems is the same process, and $T_1 = T_2 = \tau_s$ [66]. This is due to motional narrowing: fluctuating microscopic magnetic fields that cause spin dephasing change on the order of the correlation time τ_c. Typically for conduction electrons $\gamma B_0 \ll 1/\tau_c$, where γ is the gyromagnetic ratio, meaning that parallel spins will have enough time to precess fully over the dephasing fields, which then cause equally dephasing and decoherence. As the applied magnetic field increases, the precession of longitudinal spin is inhibited.

Spin relaxation times T_1 and T_2 are conventionally defined by the Bloch equations:

$$\partial S_x/\partial t = \gamma(\mathbf{S} \times \mathbf{B})_x - \frac{S_x}{T_2}, \tag{2.29}$$

$$\partial S_y/\partial t = \gamma(\mathbf{S} \times \mathbf{B})_y - \frac{S_y}{T_2}, \tag{2.30}$$

$$\partial S_z/\partial t = \gamma(\mathbf{S} \times \mathbf{B})_z - \frac{S_z - S_z^0}{T_1}, \tag{2.31}$$

where S_i is the ith component of the average spin of the electron statistical ensemble, $\mathbf{B}$ is the applied magnetic field containing a static $B_0\mathbf{z}$ component and an oscillating transverse component; S_z^0 is the equilibrium spin in the $\mathbf{z}$ direction. Microscopic expressions for T_1 and T_2 can be obtained from model Hamiltonians and scattering interactions, by deriving effective equations of the above Bloch form.

We introduce relevant mechanisms of spin relaxation in semiconductors. We discuss in detail the Dyakonov–Perel mechanism, and use bulk n-GaAs as a nice illustration of the rich physics involved. Spin relaxation in quantum dots and quantum wells is discussed in Chapters 3 and 6.

2.1.5.1 *Mechanisms of spin relaxation* There are four principal mechanisms of spin relaxation in semiconductors.

(i) In the Elliott–Yafet mechanism [67,68] the spin relaxes by momentum scattering. Bloch states in the presence of spin-orbit coupling are an admixture of spin up and spin down Pauli states. These states can be oriented in a way to give the average magnetic moment parallel and antiparallel to a magnetic field or a chosen spin polarization axis. Scattering, which connects different momentum states, now includes a spin flip. Typically an electron undergoes a thousand to a million scattering events before its spin is flipped. The more the electron scatters, the faster its spin relaxes.

(ii) The Dyakonov–Perel spin relaxation [69] operates in systems with no inversion symmetry (such as zincblende structures). In these systems the

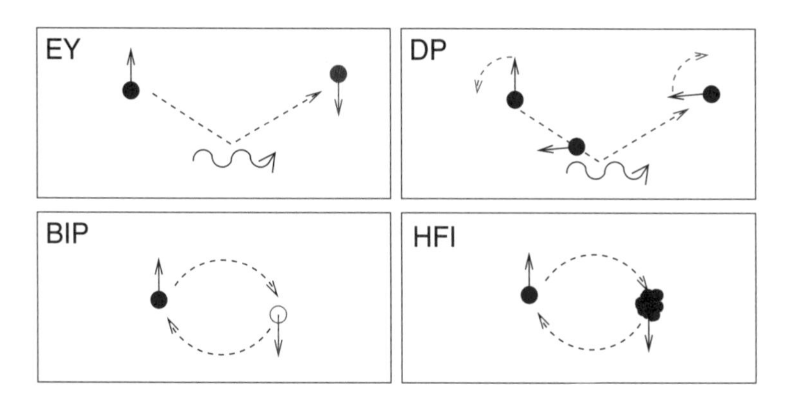

FIG. 2.6. Spin relaxation mechanisms. In the Elliott–Yafet (EY) mechanism electrons have a small probability (say 1 per 10^5) of being scattered (here illustrated on a phonon) with a spin flip. In the Dyakonov–Perel (DP) mechanism the electron spins precess in between scatterings. Scattering events change the precession direction. The Bir–Aronov–Pikus (BIP) mechanism applies for p-doped semiconductors. Here electron spin is exchanged with the spin of holes (preserving the total spin), while the hole spin then soon relaxes. Finally, the mechanism based on the hyperfine interaction (HFI) is dominant for confined electrons on donors or in quantum dots. The electron spin is exchanged with that of nuclei. Adapted from [72].

momentum spin doublet is split into spin up and spin down singlets. The energy difference is proportional to the spin-orbit coupling. This is equivalent of having a momentum-dependent magnetic field. Spins precess in such a field, then scatter to precess along a different field. Scattering acts to randomize the spin precession and leads to spin dephasing. We will describe this mechanism below in detail.

(iii) In p-doped semiconductors electron spin relaxation can be due to electron-hole exchange coupling – the so called Bir–Aronov–Pikus mechanism [70]. Electron spin is exchanged with the spin of holes. While the total spin is preserved, hole spin decays much faster, providing a randomizing environment for electron holes.

(iv) Finally, for states localized on donors or in quantum dots, hyperfine coupling [71] is a dominant mechanism of spin decoherence. The four mechanisms are illustrated in Fig. 2.6.

2.1.5.2 *Dyakonov–Perel mechanism* The mechanism of Dyakonov and Perel uses the following spin Hamiltonian

$$H' = \frac{1}{2}\hbar\Omega_{\mathbf{k}} \cdot \sigma, \tag{2.32}$$

where $\Omega_{\mathbf{k}}$ is the precession angle vector simulating the **k**-dependent spin splitting of the conduction band in zincblende systems. In bulk GaAs, for example, as first found by Dresselhaus [73],

$$\Omega = \alpha\hbar^2 \left(2m_c^3 E_g\right)^{-1/2} \kappa, \tag{2.33}$$

where

$$\kappa = [k_x(k_y^2 - k_z^2), ky(k_z^2 - k_x^2), k_z(k_x^2 - k_y^2)]. \tag{2.34}$$

The parameter α is about 0.07 for GaAs. Similar spin splittings appear in two-dimensional systems where quantization of κ (in which k are treated as operators) along the confining axis leads to terms linear in k. For example, in quantum wells oriented along [001],

$$\kappa = k_n^2(-k_x, k_y, 0), \tag{2.35}$$

where k_n^2 is the expectation value of $-i\nabla$ along [001]. In addition to this so-called Dresselhaus term, a term resulting from structure inversion asymmetry arises, as first observed by Bychkov and Rashba [74,75]. This term has a form independent of the orientation of the 2D system:

$$\Omega = 2\alpha_{\mathrm{BR}}(\mathbf{k} \times \mathbf{n}), \tag{2.36}$$

where α_{BR} is the Bychkov–Rashba parameter which depends on spin-orbit coupling as well as the degree of the confinement asymmetry, and **n** is a unit vector along the confining axis. Typically α_{BR} is of the order of 10^{-11} eV m. Various cases of growth orientation are discussed in [1].

We will now give the formula for calculating the spin relaxation time τ_s in the Dyakonov–Perel mechanism for an arbitrary function $\Omega_{\mathbf{k}}$:

$$1/\tau_{s,ii} = \gamma_l^{-1}\tau_p(\overline{\Omega^2} - \overline{\Omega_i^2}), \quad 1/\tau_{s,i\neq j} = -\gamma_l^{-1}\tau_p\overline{\Omega_i\Omega_j}. \tag{2.37}$$

The spin relaxation rate $1/\tau_s$ is in general a tensor. The averaging denoted by the overline is over all directions of **k** on the energy surface. The factor $\gamma_l = \tau_p/\tilde{\tau}_l$, where τ_p is the spin-conserving momentum relaxation time and

$$1/\tilde{\tau}_l = \int_{-1}^{1} W(\theta)\,[1 - P_l(\cos\theta)]\,d\cos\theta, \tag{2.38}$$

is the effective randomization time for Ω. In the above W is the isotropic scattering form factor and P_l is the lth Legendre polynomial. The power of l is given by the power of k in Ω [$l = 3$ for the Dresselhaus bulk term, Eq. (2.33)]. The derivation of Eq. (2.37) can be found in [76].

For cubic systems $1/\tau_s$ is a scalar. Calculation for bulk GaAs, using Ω of Eq. (2.33), gives the following spin relaxation rate for electrons at energy $E_{\mathbf{k}}$:

$$1/\tau_s(E_{\mathbf{k}}) = \frac{32}{105}\gamma_3^{-1}\tau_p(E_{\mathbf{k}})\alpha^2\frac{E_{\mathbf{k}}^3}{\hbar^2 E_g}. \tag{2.39}$$

The numerical value of γ_3 depends on the scattering mechanism. For example, impurity scattering gives $\gamma_4 \approx 6$. In this case also $\tau_p \sim T^{3/2}$, so that the spin relaxation time decreases as $\tau_s \sim T^{-3/2}$. When the electron density is in the non-degenerate regime, the averaging over the thermal distribution (denoted by angular brackets) of $E_{\mathbf{k}}$ gives

$$1/\tau_s = Q\tau_m \alpha^2 \frac{(k_{\mathrm{B}}T)^3}{\hbar^2 E_g}, \tag{2.40}$$

where the thermal momentum relaxation time is

$$\tau_m = \langle \tau_p(E_{\mathbf{k}})E_{\mathbf{k}}\rangle / \langle E_{\mathbf{k}}\rangle, \tag{2.41}$$

and the coefficient

$$Q = \frac{16}{35}\gamma_3^{-1}(\nu + 7/2)(\nu + 5/2), \tag{2.42}$$

and ν is the coefficient in the power law $\tau_p \sim E_{\mathbf{k}}^{\nu}$. Ionized impurity scattering gives $Q \approx 1.5$, in which case also $\tau_s \sim T^{-9/2}$, a rather steep decrease.

The peculiar feature of the DP mechanism is the fact that $1/\tau_s \sim \tau_p$, meaning that more momentum scattering translates into less spin dephasing. The opposite holds for the Elliott–Yafet mechanism. The reason for the peculiar behavior is motional narrowing, which is the physics behind DP relaxation. Consider an electron spin precessing angle $\delta\phi = \Omega\tau_p$ in time τ_p. This is as far as the spin can precess along a specific Ω. After that time, the angle and magnitude of Ω change due to momentum scattering. The phase than goes through a random walk. After time t the average phase will be zero, but the rms $\phi = \delta\phi(t/\tau_p)^{1/2}$. The spin will be dephased when $\phi \approx 1$, which happens at time $t = \tau_s = 1/(\Omega)^2\tau_p$, which is roughly Eq. (2.37). It is interesting to point out that the spin diffusion length $L_s = (D\tau_s)^{1/2}$ will not depend on τ_p, since $D \sim \tau_p$ and $\tau_s \sim 1/\tau_p$.

2.1.5.3 *Bulk semiconductors* Measured electron spin relaxation times in semiconductors range from 100 ns to 1 ps. Holes typically lose their spin orientation on time scales of τ_p, since their spin-orbit coupling is much larger. We will illustrate the experimental and theoretical understanding of spin relaxation in semiconductors on the example of GaAs, the most studied case. Figure 2.7 is a nice compilation of measured low-temperature τ_s as a function of doping. The measured values range over four orders of magnitude. At low temperatures GaAs undergoes a metal to insulator transition (MIT) at about $2\times10^{16}\,\mathrm{cm}^{-3}$. At larger dopings, where GaAs is metallic and electrons are delocalized in the conduction band, τ_s decreases as N_d^2, consistent with the DP mechanism induced by ionized impurity scattering.[3] One can see that spin relaxation in reasonably conductive samples is rather large.

Much has been said about the regime around MIT where $\tau_s \sim 100$ ns [78]. Unfortunately, this happens only for a small window of doping levels where electrons are bound on donors and conductivity is through the impurity band. A

[3]Consider that $E_{\mathrm{F}} \sim N_d^{2/3}$ and $1/\tau_p \sim N_d/E_{\mathrm{F}}^{3/2}$ (the Brooks–Herring formula). Then Eq. (2.37) gives $\tau_s \sim N_d^2$

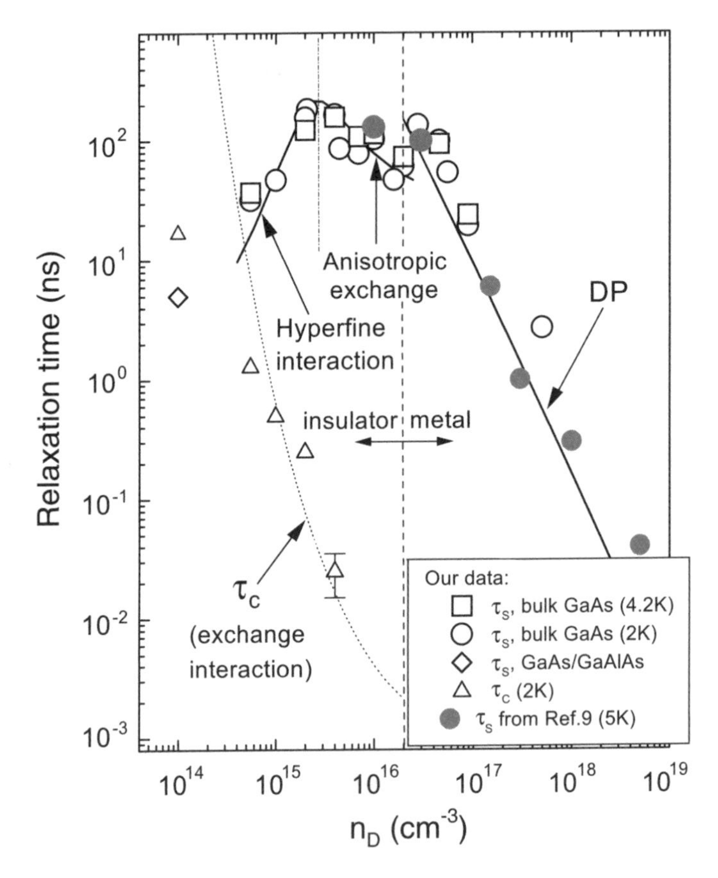

FIG. 2.7. Spin relaxation in bulk GaAs. Symbols are measurements and solid lines are theoretical estimates. Empty symbols come from [77], filled circles from [78]. The vertical dashed line indicates the metal-to-insulator transition at $N_{dc} = 2 \times 10^{16}\,\mathrm{cm}^{-3}$. The dotted line is the correlation time τ_c due to exchange coupling. Adapted from [77].

new mechanism of spin relaxation due to so called anisotropic exchange (of the type $\mathbf{S}_1 \times \mathbf{S}_2$) was suggested by Kavokin [79] to explain this regime. Finally, at still smaller densities the electron hopping is strongly suppressed and hyperfine coupling with Ga and As nuclei causes spin relaxation (HFI mechanism). The figure also shows the extracted exchange interaction time in this regime (obtained by observing spin relaxation suppression in a longitudinal magnetic field). It was suggested by Dzhioev [80] that τ_c is a measure of single-spin decoherence time. Then it is clear that single spin would lose spin polarization much faster than the ensemble spin. It is the single spin electron time which is useful for quantum computation. At higher temperatures also electrons in the low doped samples become itinerant and the DP mechanism becomes relevant there too.

A recent report [81], however, finds an apparent breakdown of the DP mechanism, Eq. (2.37), for high mobility samples at $T = 77\,\mathrm{K}$, and suggests that our understanding of this mechanism is still not yet satisfactory.

2.2 Bipolar spin-polarized transport and applications

2.2.1 *Spin-polarized drift-diffusion equations*

In the absence of any spin polarization, equations which aim to describe spin-polarized bipolar transport need to recover a description of charge transport. We recall that conventional charge transport in semiconductors is often accompanied with large deviations from local charge neutrality (for example, due to material inhomogeneities, interfaces, and surfaces) and Poisson's equation needs to be explicitly included. If we consider (generally inhomogeneous) doping with density of N_a ionized acceptors and N_d donors we can then write

$$\nabla \cdot (\epsilon \nabla \phi) = q(n - p + N_a - N_d), \tag{2.43}$$

were n, p (electron and hole densities) also depend on the electrostatic potential ϕ and the permittivity can be spatially dependent. In contrast to the metallic regime considered in Section 2.1.4, even equilibrium carrier density can have large spatial variations which can be routinely tailored by the appropriate choice of the doping profile $[N_d(x) - N_a(x)]$. Furthermore, charge transport in semiconductors can display strong non-linearities, for example, the exponential-like current-voltage dependence of a diode [82].

Returning to the case of spin-polarized transport in semiconductors, we formulate a drift-diffusion model which will generalize the considerations of Eqs (2.9)-(2.13) to include both electrons and holes [83–85]. We recall that from Eqs (2.9) and (2.10) the spin-resolved current has a drift part (proportional to the electric field, i.e., $\propto \nabla\phi$) and a diffusive part ($\propto \nabla n_\lambda$), which we want to extend to also capture the effects of band bending, band offsets, various material inhomogeneities, and the presence of two type of charge carriers.

To introduce the notation and terminology, which is a direct generalization of what is conventionally used in semiconductor physics [82], we first give an expression for quasi-equilibrium carrier densities. For non-degenerate doping levels (Boltzmann statistics) the spin-resolved components are

$$n_\lambda = \frac{N_c}{2} e^{-[E_{c\lambda} - \mu_{n\lambda}]/k_B T}, \quad p_\lambda = \frac{N_v}{2} e^{-[\mu_{p\lambda} - E_{v\lambda}]/k_B T}, \tag{2.44}$$

where subscripts "c" and "v" label quantities which pertain to the conduction and valence bands. For example, $N_{c,v} = 2(2\pi m^*_{c,v} k_B T/h^2)^{3/2}$ are effective densities of states with the corresponding effective masses $m^*_{c,v}$ and k_B is the Boltzmann constant. We consider a general case where the spin splitting of the conduction and valence bands, expressed as $2q\zeta_c$ and $2q\zeta_v$ can be spatially inhomogeneous [84]. Splitting of carrier bands (Zeeman or exchange) can arise due to

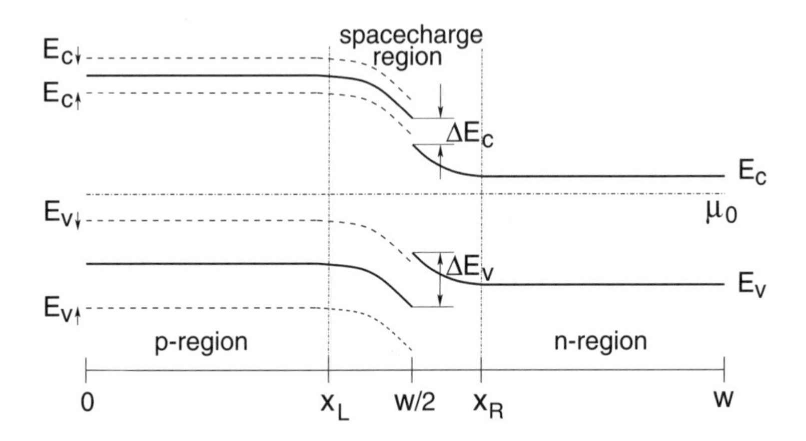

FIG. 2.8. Band-energy schemes for a magnetic heterojunction. In equilibrium the chemical potential μ_0 is constant. Conductance and valence-band edges (E_c and E_v) are spin split in the magnetic p-region, while in the non-magnetic n-region there is no spin splitting. For a sharp doping profile, at $x = w/2$, there are generally discontinuities in the conduction and valence bands (ΔE_c and ΔE_v) and in other quantities, such as effective mass, permittivity, and diffusion coefficient.

doping with magnetic impurities (see Chapter 1) and/or applied magnetic field. The spin-λ conduction band edge (see Fig. 2.8)

$$E_{c\lambda} = E_{c0} - q\phi - \lambda q\zeta_c \tag{2.45}$$

differs from the corresponding non-magnetic bulk value E_{c0} due to the electrostatic potential ϕ and the spin splitting $\lambda q\zeta_c$. The discontinuity of the conduction band edge is denoted by ΔE_c. In the non-equilibrium state the chemical potential for λ-electrons is $\mu_{n\lambda}$ and generally differs from the corresponding quantity for holes. While $\mu_{n\lambda}$ has an analogous role as the electrochemical potential in Eqs (2.9) and (2.10), following conventional semiconductor terminology, we refer to it here as the chemical potential, which is also known as the quasi-Fermi level. An analogous notation holds for p_λ in Eq. (2.44) where, for example, $E_{v\lambda} = E_{v0} - q\phi - \lambda q\zeta_v$.

By assuming the drift-diffusion dominated transport across a heterojunction, the spin-resolved charge current densities can be expressed as [86]

$$\mathbf{j}_{n\lambda} = \bar{\mu}_{n\lambda} n_\lambda \nabla E_{c\lambda} + qD_{n\lambda} N_c \nabla(n_\lambda/N_c), \tag{2.46}$$

$$\mathbf{j}_{p\lambda} = \bar{\mu}_{p\lambda} p_\lambda \nabla E_{v\lambda} - qD_{p\lambda} N_v \nabla(p_\lambda/N_v), \tag{2.47}$$

where $\bar{\mu}$ and D are the mobility and diffusion coefficients (we use the symbol $\bar{\mu}$ to distinguish it from the chemical potential μ). In non-degenerate semiconductors $\bar{\mu}$ and D are related by Einstein's relation

$$\bar{\mu}_{n,p\lambda} = qD_{n,p\lambda}/k_{\mathrm{B}}T, \tag{2.48}$$

which differs from the metallic (completely degenerate) case given by Eq. (2.12).

With two type of carriers the continuity equations are more complex than those in metallic systems (see Section 2.1.4). After including additional terms for recombination of electrons and holes as well as photo-excitation of electron-hole pairs, we can write these equations as

$$-\frac{\partial n_\lambda}{\partial t} + \nabla \cdot \frac{\mathbf{j}_{n\lambda}}{q} = + r_\lambda(n_\lambda p_\lambda - n_{\lambda 0}p_{\lambda 0}) + \frac{n_\lambda - n_{-\lambda} - \lambda \tilde{s}_n}{2\tau_{sn}} - G_\lambda, \tag{2.49}$$

$$+\frac{\partial p_\lambda}{\partial t} + \nabla \cdot \frac{\mathbf{j}_{p\lambda}}{q} = - r_\lambda(n_\lambda p_\lambda - n_{\lambda 0}p_{\lambda 0}) - \frac{p_\lambda - p_{-\lambda} - \lambda \tilde{s}_p}{2\tau_{sp}} + G_\lambda. \tag{2.50}$$

Generation and recombination of electrons and holes of spin λ can be characterized by the rate coefficient r_λ, the spin relaxation time for electrons and holes is denoted by $\tau_{sn,p}$ and the photo-excitation rate G_λ represents the effects of electron-hole pair generation and optical orientation (when $G_\uparrow \neq G_\downarrow$, see Section 2.1.3). Spin relaxation equilibrates carrier spin while preserving the non-equilibrium carrier density and for non-degenerate semiconductors $\tilde{s}_n = nP_{n0}$, where from Eq. (2.44) the equilibrium polarization of electron density is

$$P_{n0} = \tanh(q\zeta_c/k_{\mathrm{B}}T), \tag{2.51}$$

and an analogous expression holds for holes and $\tilde{s}_p$.

The system of drift-diffusion equations (Poisson and continuity equations) can be self-consistently solved numerically [83, 84, 87] and under simplifying assumptions (similar to the case of charge transport) analytically [85, 88]. Heterojunctions, such as the one sketched in Fig. 2.8, can be thought of as building blocks of bipolar spintronics. To obtain a self-consistent solution in such a geometry, only the boundary conditions at $x = 0$ and $x = w$ need to be specified. On the other hand, for an analytical solution we also need to specify the matching conditions at x_L and x_R, the two edges of the space charge region (or depletion region), in which there is a large deviation from the local charge neutrality, accompanied by band bending and a strong built-in electric field.

We illustrate how the matching conditions for spin and carrier density can be applied within the small-bias or low-injection approximation, widely used to obtain analytical results for charge transport [82]. In this case non-equilibrium carrier densities are small compared to the density of majority carriers in the corresponding semiconductor region. For materials such as GaAs a small bias approximation gives good agreement with the full self-consistent solution up to approximately 1 V [84, 87]. To simplify our notation, we consider a model where only electrons are spin polarized ($p_\uparrow = p_\downarrow = p/2$), while it is straightforward to also include spin-polarized holes [85, 86]. Outside the depletion charge region material parameters (such as, N_a, N_d, N_c, N_v, $\bar{\mu}$, and D) are taken to be constant. The voltage drop is confined to the depletion region which is highly resistive and depleted of carriers. In thermal equilibrium ($\mu_{n\lambda} = \mu_{p\lambda} = \mu_0$) the built-in voltage V_{bi} can be simply evaluated from Eq. (2.44) as

$$V_{bi} = \phi_{0R} - \phi_{0L}, \tag{2.52}$$

while the applied bias V (taken to be positive for forward bias) can be expressed as

$$V = -(\delta\phi_R - \delta\phi_L), \tag{2.53}$$

implying that the total junction potential between $x = 0$ and $x = w$ is $V - V_{bi}$. Outside of the depletion region the system of the drift-diffusion equations reduces to only diffusion equations for spin density and the density of minority carriers, while the density of majority carriers is simply given by the density of donors and acceptors [84, 85].

From Eq. (2.44) we rewrite the electron density by separating various quantities into equilibrium and non-equilibrium parts as

$$n_\lambda = n_{\lambda 0} \exp[(q\delta\phi + \delta\mu_{n\lambda})/k_{\mathrm{B}}T], \tag{2.54}$$

and the electron carrier and spin density (for simplicity we omit subscript "n" when writing $s = n_\uparrow - n_\downarrow$) can be expressed as [85]

$$n = e^{(\delta\phi + \delta\mu_+)/k_{\mathrm{B}}T} \left[n_0 \cosh\left(\frac{q\mu_-}{k_{\mathrm{B}}T}\right) + s_0 \sinh\left(\frac{q\mu_-}{k_{\mathrm{B}}T}\right) \right], \tag{2.55}$$

$$s = e^{(\delta\phi + \delta\mu_+)/k_{\mathrm{B}}T} \left[n_0 \sinh\left(\frac{q\mu_-}{k_{\mathrm{B}}T}\right) + s_0 \cosh\left(\frac{q\mu_-}{k_{\mathrm{B}}T}\right) \right], \tag{2.56}$$

where $\mu_\pm \equiv (\mu_{n\uparrow} \pm \mu_{n\downarrow})/2$, and the polarization of electron density is

$$P_n = \frac{\tanh(q\mu_-/k_{\mathrm{B}}T) + P_{n0}}{1 + P_{n0} \tanh(q\mu_-/k_{\mathrm{B}}T)}. \tag{2.57}$$

If we assume that the spin-resolved chemical potentials are constant for $x_L \le x \le x_R$ (which means that the depletion region is sufficiently narrow so that the spin relaxation and carrier recombination can be neglected there) it follows from Eq. (2.57) and $\tanh(q\mu_-/k_{\mathrm{B}}T) \equiv const.$ that

$$P_n^L = \frac{P_{n0}^L[1 - (P_{n0}^R)^2] + \delta P_n^R (1 - P_{n0}^L P_{n0}^R)}{1 - (P_{n0}^R)^2 + \delta P_n^R (P_{n0}^L - P_{n0}^R)}, \tag{2.58}$$

where L (left) and R (right) label the edges of the space-charge (depletion) region of a p-n junction. Correspondingly, δP_n^R represents the non-equilibrium electron polarization, evaluated at R, arising from a spin source. For a homogeneous equilibrium magnetization ($P_{n0}^L = P_{n0}^R$), $\delta P_n^L = \delta P_n^R$; the non-equilibrium spin polarization is the same across the depletion region. Equation (2.58) demonstrates that only *non-equilibrium* spin, already present in the bulk region, can be transferred through the depletion region at small biases [83–85].

Our assumption of constant spin-resolved chemical potentials is a generalization of a conventional model for charge transport in which both μ_n and μ_p are assumed to be constant across the depletion region [82]. From Eqs (2.53), (2.55), and (2.56) we can obtain minority carrier and spin densities at $x = x_L$

$$n_L = n_{0L} e^{qV/k_B T} \left[1 + \delta P_n^R \frac{P_{n0}^L - P_{n0}^R}{1 - (P_{n0}^R)^2} \right], \tag{2.59}$$

$$s_L = s_{0L} e^{qV/k_B T} \left[1 + \frac{\delta P_n^R}{P_{n0}^L} \frac{1 - P_{n0}^L P_{n0}^R}{1 - (P_{n0}^R)^2} \right], \tag{2.60}$$

which in the absence of non-equilibrium spin ($\delta P_n^R = 0$) reduce to the well-known Shockley relation for the minority carrier density at the depletion region [89]

$$n_L = n_{0L} e^{qV/k_B T}, \tag{2.61}$$

and an analogous formula holds for spin

$$s_L = s_{0L} e^{qV/k_B T}. \tag{2.62}$$

2.2.2 *Spin-polarized p-n junctions*

On one hand, spin-polarized p-n junctions can be thought of as a generalization of conventional (non-magnetic) p-n junctions in which only a non-equilibrium spin is introduced. In the absence of equilibrium spin (i.e., equilibrium magnetization vanishes identically, or $\zeta_c = \zeta_v = 0$), various well-studied charge properties in conventional p-n junctions also apply to spin-polarized p-n junctions. On the other hand, spin-polarized p-n junctions can also be viewed as a generalization of homogeneously doped semiconductors where the optical orientation (see Section 2.1.3) was traditionally used to generate the non-equilibrium spin. Here our goal will be twofold. We will show how, within the framework of spin-polarized transport equations for p-n junctions, one can recover results previously known for optical orientation in either homogeneously p- or n-doped semiconductors, as well as discuss how the presence of inhomogeneous doping, and the resulting built-in field can lead to novel effects.

To derive the steady-state expressions for the spin polarization due to optical orientation (see Section 2.1.3), we simplify our considerations from Section 2.2.1 and examine a homogeneously doped non-magnetic semiconductor with unpolarized holes. Having spin-unpolarized holes is an accurate approximation in materials such as GaAs (see Section 2.1.3); the spin-relaxation time for holes can be several orders of magnitude shorter than the corresponding time for electrons [40]. We can then adopt the continuity Eqs (2.50) and (2.51) such that holes would recombine with the electrons of either spin. In the steady state, the balance between direct electron-hole recombination and optical pair creation can be expressed from the sum of Eq. (2.50) for $\lambda = \uparrow$ and $\downarrow$ as

$$r(np - n_0 p_0) = G, \tag{2.63}$$

where $r = 2r_\uparrow = 2r_\downarrow$ is the total generation-recombination rate coefficient, and $G = G_\uparrow + G_\downarrow$ is the total electron-hole photo-excitation rate. Similarly, from the difference of Eq. (2.50) for $\lambda = \uparrow$ and $\downarrow$ the balance between spin relaxation and spin generation is expressed by

$$rsp + s/\tau_s = P_n(t = 0)G, \tag{2.64}$$

where $P_n(t = 0)$ is the spin polarization at the moment of photo-excitation, given by Eq. (2.5). The first term in Eq. (2.64) describes the disappearance of the spin density due to carrier recombination, while the second term describes the intrinsic spin relaxation. From Eqs (2.63) and (2.64) we obtain the steady-state electron polarization as [83]

$$P_n = P_n(t = 0) \frac{1 - n_0 p_0 / np}{1 + 1/\tau_s rp}. \tag{2.65}$$

In a p-doped sample $p \approx p_0$, $n \gg n_0$, and Eq. (2.65) gives

$$P_n = P_n(t = 0)/(1 + \tau_n/\tau_s), \tag{2.66}$$

where $\tau_n = 1/rp_0$ is the electron lifetime. After the illumination is switched off, the electron spin density, or equivalently the non-equilibrium magnetization, will decrease exponentially with the inverse time constant [39]

$$1/T_s = 1/\tau_n + 1/\tau_s. \tag{2.67}$$

The steady-state polarization is independent of the illumination intensity, being reduced from the initial spin polarization $P_n(t = 0)$.[4] The polarization of the photoluminescence is $P_{\text{circ}} = P_n(t = 0)P_n$ [39].

For spin pumping in an n-doped sample, where $n \approx n_0$ and $p \gg p_0$, Eqs (2.63) and (2.65) give [92]

$$P_n = P_n(t = 0)/(1 + n_0/G\tau_s). \tag{2.68}$$

In contrast to the previous case, the hole lifetime (not to be confused with the momentum relaxation time) $\tau_p = 1/rn_0$ has no effect on P_n. However, P_n

[4] The effect of a finite length for the light absorption on P_n is discussed in [90]. The absorption length α^{-1} is typically a micron for GaAs. It varies with frequency roughly as $\alpha(\hbar\omega) \propto (\hbar\omega - E_g)^{1/2}$ [91].

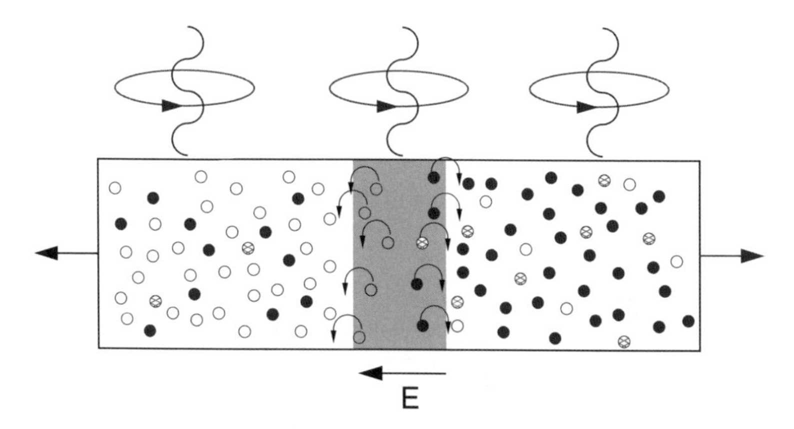

FIG. 2.9. Spin-polarized solar battery. The top figure shows a configuration where circularly-polarized light creates electron-hole pairs at the p-region (left). Holes are denoted as empty circles while spin up and spin down electrons are represented by filled and patterned circles, respectively. Spin-polarized electrons diffuse towards the depletion region where they are swept by the built-in field E to the n-side. Spin-polarization is pumped into the majority region. A uniform illumination is assumed throughout the sample, giving rise to spin-polarized current. Adapted from [87].

depends on the photo-excitation intensity G, as expected for a pumping process. The effective carrier lifetime is $\tau_J = n_0/G$, where J represents the intensity of the illuminating light. If it is comparable to or shorter than τ_s, spin pumping is very effective. Spin pumping works because the photo-excited spin-polarized electrons do not need to recombine with holes. There are plenty of unpolarized electrons in the conduction band available for recombination. The spin is thus pumped in to the electron system.

We next consider our proposal for a spin-polarized solar cell which could be a source of spin electromotive force (EMF) to both generate spin-polarized currents at no applied bias as well as to provide an open circuit voltage [87]. There is also a wide range of other structures which have been recently suggested as a source of spin EMF [93–96], often referred to as spin(-polarized) pumps, cells, or batteries.

We focus on a particular realization based on a spin-polarized p-n junction which combines (see Fig. 2.9) two key ingredients: (1) non-equilibrium spin produced by optical orientation and (2) a built-in field which separates electron-hole pairs created by illumination. Additionally, we show that in such a structure spin polarizations of current P_j and of electron density P_n have very different behavior.

Consider a GaAs-based sample at room temperature, of length w (extending on the x-axis from $x = 0$ to $12\ \mu$m), doped with $N_a = 3 \times 10^{15}\,\text{cm}^{-3}$ acceptors on the left and with $N_d = 5 \times 10^{15}\,\text{cm}^{-3}$ donors on the right [the doping profile, $N_d(x) - N_a(x)$, is shown in Fig. 2.10]. The intrinsic carrier concentration is

$n_i = 1.8 \times 10^6\,\mathrm{cm}^{-3}$. For an undoped semiconductor $n_0 = p_0 = n_i$ and can be expressed from Eq. (2.44) as $n_i^2 = N_c N_v \exp(-E_g/k_\mathrm{B}T)$. The electron (hole) mobility and diffusion coefficients are 4000 (400) $\mathrm{cm}^2\cdot\,\mathrm{V}^{-1}\cdot\mathrm{s}^{-1}$ and 103.6 (10.36) $\mathrm{cm}^2\cdot\mathrm{s}^{-1}$. The total recombination rate is taken to be $r = (1/3) \times 10^{-5}\,\mathrm{cm}^3\cdot\mathrm{s}^{-1}$, giving an electron lifetime in the p-region of $\tau_n = 1/rN_a = 0.1$ ns, and a hole lifetime in the n-region of $\tau_p = 1/rN_d = 0.06$ ns. The spin relaxation time (which is the spin lifetime in the n-region) is $\tau_s = 0.2$ ns. In the p-region electron spin decays on the time scale of (recall Eq. 2.67) $T_s = \tau_s\tau_n/(\tau_s + \tau_n) \approx 0.067$ ns. The minority diffusion lengths are $L_n = (D_n\tau_n)^{0.5} \approx 1\,\mu\mathrm{m}$ for electrons in the p-region, and $L_p = (D_p\tau_p)^{0.5} \approx 0.25$ for holes in the n-region. The spin decays on the length scale of $L_{sp} = (D_n T_s)^{0.5} \approx 0.8\,\mu\mathrm{m}$ in the p- and $L_{sn} = (D_n\tau_s)^{0.5} \approx 1.4\,\mu\mathrm{m}$ in the n-region. At no applied voltage, the depletion region formed around $x_d = w/2 = 6\ \mu\mathrm{m}$ has a width of $d \approx 0.9\,\mu\mathrm{m}$, of $d_p = (5/8)d$ in the p-side and $d_n = (3/8)d$ in the n-side.

Let the sample be uniformly illuminated with circularly polarized light with photon energy higher than the band gap (bipolar photogeneration). The pair generation rate is chosen to be $G = 3 \times 10^{23}\,\mathrm{cm}^{-3}\cdot\mathrm{s}^{-1}$ (which corresponds to concentrated solar light of intensity about $1\,\mathrm{W}\cdot\mathrm{cm}^{-2}\cdot\mathrm{s}^{-1}$), so that in the bulk of the p-side there are $\delta n = G\tau_n \approx 3 \times 10^{13}\,\mathrm{cm}^{-3}$ non-equilibrium electrons and holes; in the n-side the density is $\delta p = G\tau_p = 1.8 \times 10^{13}\,\mathrm{cm}^{-3}$. Recall from Section 2.1.3 that spin polarization at the moment of creation is $P_n(t=0) = G_s/G = 0.5$, where $G_s = G_\uparrow - G_\downarrow$ is the difference in the generation rates for spin up and down electrons. For homogeneous doping, the spin density in the p-side would be $s_p = G_s T_s \approx 1 \times 10^{13}\,\mathrm{cm}^{-3}$, while in the n-side $s_n = G_s\tau_s \approx 3\times10^{13}\,\mathrm{cm}^{-3}$. The physical situation and the geometry are illustrated in Fig. 2.9.

We solve numerically the drift-diffusion equations for inhomogeneously doped spin-polarized semiconductors to obtain electron n, hole p, and spin s densities, as well as charge j and spin $j_s = j_\uparrow - j_\downarrow$ current densities. We consider ideal Ohmic contacts attached at both ends of the sample, providing infinite carrier and spin recombination velocities (so that both non-equilibrium carrier densities and spin density vanish at $x = 0$ and $x = w$).

Calculated spatial profiles of carrier and spin densities, as well as carrier and current polarizations $P_n = n/s$ and $P_j = j_s/j$, are shown in Fig. 2.10. There is no applied voltage V, but the illumination produces a reverse photo current $j_\mathrm{photo} = -eG(L_n + L_p + d)$ [46]. The behavior of carrier densities is the same as in the unpolarized case. The spin density essentially follows the non-equilibrium electronic density in the p-side, sharply decreases in the depletion region, and then rapidly increasing to a value larger than the normal excitation value in the n-side, s_n. We interpret this as a result of spin pumping through the minority channel [83]: electron spin excited within the distance L_{ps} from the depletion region, as well as generated inside that region, is swept into the n-side by the built-in field, thus pumping spin polarization into the n-region. In the rest of the n-region, spin density decreases, until it reaches zero at the right boundary. The carrier spin polarization P_n is reasonably high in the p-side, but diminishes

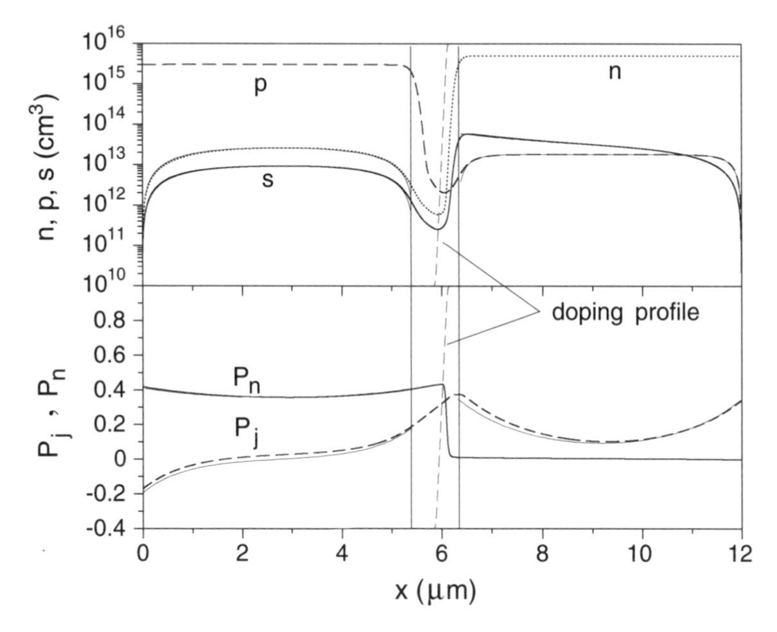

FIG. 2.10. Calculated spatial profiles of (top) carrier densities n and p, spin density s, and (bottom) carrier and current spin polarizations P_n and P_j. The thin dashed lines show the doping profile $N_d(x) - N_a(x)$ (not to scale), and the two vertical lines at $x_L \approx 5.4$ and $x_R \approx 6.3$ indicate the depletion region boundaries. The thin lines accompanying the numerical curves are analytical results for an ideal spin-polarized solar cell (if not visible, they overlap with the numerical results). From [87].

in the n-side. Current polarization, however, remains quite large throughout the sample. It changes sign in the p-region, and since $j(V = 0) = j_{\text{photo}} < 0$ is a constant, P_j shows the negative profile of spin current, and has a symmetric shape in the n-region, being much larger than P_n. These findings can be confirmed analytically (see Fig. 2.10) using a low-bias approximation and Shockley relation for the minority carrier density at the edge of the depletion region [87].

Several trends, shown in Fig. 2.10, are helpful to illustrate important differences with the unipolar transport from Section 2.1.4. For example, carrier density in semiconductors can have strong spatial variation and, within a depletion region, there is a large deviation from local charge neutrality. An applied bias would change the width of the depletion region and give rise to non-linear I-V characteristics. While in metallic F/N junctions P_j was calculated to illustrate the degree of spin injection efficiency, we see from Fig. 2.10 that P_j and P_n can have a qualitatively different behavior. It is useful to recall from the discussion of spin LEDs in Section 2.1.3 that spin injection in semiconductors is typically characterized by measuring P_n rather than P_j.

Previous discussion of spin injection in F/N metallic junctions gave an intuitive picture of non-equilibrium magnetization (spin accumulation) in the non-magnetic region. The spatial profile of spin accumulation sketched in Fig. 2.3(b) and Eq. (2.23) suggest that the non-equilibrium magnetization need to monotonically decay away from the point of spin injection (at F/N interface) to the interior of a non-magnetic material. However, we have shown that spin accumulation can even *increase* away from the point of spin injection, as a consequence of inhomogeneous doping in non-magnetic semiconductors. This behavior can be illustrated in the geometry slightly modified from the one shown in Fig. 2.9 in which optical spin pumping is applied only at the left end of a p-n junction (where spin-polarized electrons are minority carriers in the p-region). Away from the point of spin injection, following the increase in n from p- to the n-region, there will also be an increase in s which we termed spin amplification [83]. Similar behavior was also confirmed in [97].

2.2.3 *Magnetic p-n junctions*

For potential spintronic applications (see Section 2.2.4.3), as well as to demonstrate novel effects due to spin-polarized bipolar transport, it is desirable to have large carrier spin-subband splitting (see Fig. 2.8). In the absence of a magnetic field, such a splitting can be realized using ferromagnetic semiconductors (see Chapter 1), while in an applied magnetic field one could utilize large effective g-factors either due to magnetic impurities (for example, $|g| \approx 500$ at $T < 1\,\mathrm{K}$) or due to spin-orbit coupling in narrow band-gap semiconductors ($|g| \approx 50$ at room temperature in InSb), as discussed in [1]. Selective doping with magnetic impurities and/or the application of an inhomogeneous magnetic field could be used to realize a desirable, spatially inhomogeneous, spin splitting.

Although practical magnetic p-n junctions are still to be fabricated and the effects discussed here are currently being experimentally examined [98], magnetic p-n junctions have already been demonstrated. Indeed, Wen *et al.* [99] were perhaps the first to show that a ferromagnetic p-n junction, based on the ferromagnetic semiconductor $CdCr_2Se_4$ doped with Ag acceptors and In donors, could act as a diode. Photovoltaic diodes were also fabricated using (Hg,Mn)Te magnetic semiconductor [100]. However more extensive work on magnetic p-n junction has begun after the discovery of (III,Mn)V ferromagnetic semiconductors, discussed in Chapter 1. Heavily doped p-(Ga,Mn)As/n-GaAs junctions were fabricated [101–105] to demonstrate tunneling interband spin injection. Recently, Tsui *et al.* [106] have shown that the current in p-CoMnGe/n-Ge magnetic heterojunction diodes can indeed be controlled by a magnetic field.

We discuss several properties of magnetic p-n junctions which rely on the interplay of the carrier spin-subband splitting (implying that there is a finite equilibrium spin polarization of carrier density) and the non-equilibrium spin induced, for example, by optical or electrical means (recall Sections 2.1.3 and 2.1.4). We also focus here on the diffusive regime while a magnetic diode in a ballistic regime was recently discussed in [107]. For simplicity, we look at a

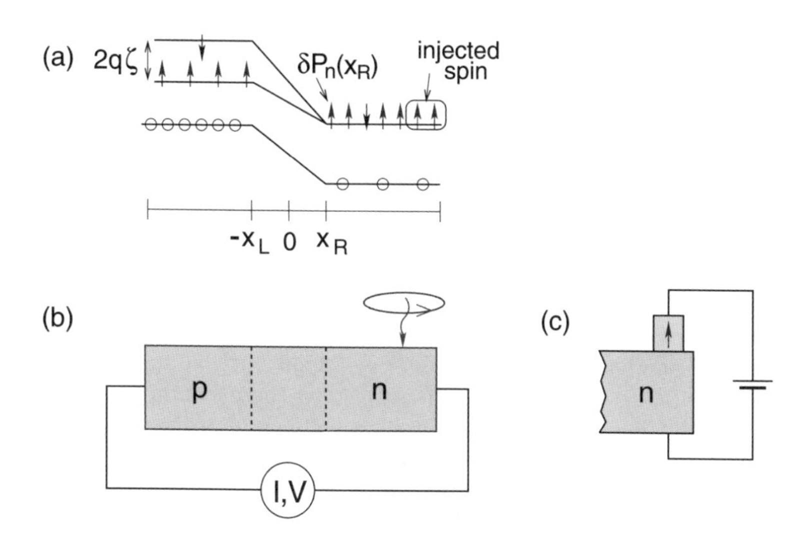

FIG. 2.11. Scheme of a magnetic p-n junction. (a) Band-energy diagram with spin-polarized electrons (arrows) and unpolarized holes (circles). The spin splitting $2q\zeta$, the non-equilibrium spin polarization at the depletion region edge $\delta P_n(x_R)$, and the region where the spin is injected are depicted. (b) Circuit geometry corresponding to panel (a). Using circularly polarized light (photo-excited electron-hole pairs absorb the angular momentum carried by incident photons), non-equilibrium spin is injected transversely in the non-magnetic n region and the circuit loop for I-V characteristics is indicated. Panel (c) indicates an alternative scheme to electrically inject spin into the n region. Adapted from [88].

particular case where the band-offsets (see Fig. 2.8) are negligible and the spin polarization of holes can be neglected and in both the notation for the carrier spin-splitting $2q\zeta$ and for the spin density s we can omit the index n. Simple scheme of such a magnetic p-n junction is given in Fig. 2.11.

From Eqs (2.44) and (2.45) we can rewrite the product of equilibrium densities as

$$n_0 p_0 = n_i^2 \cosh(q\zeta/k_{\mathrm{B}}T), \tag{2.69}$$

where n_i is the intrinsic (non-magnetic) carrier density [82] and we notice that the density of minority carriers in the p-region will depend on the spin splitting $n_0(\zeta) = n_0(\zeta = 0)\cosh(q\zeta/k_{\mathrm{B}}T)$. Similar to the theory of charge transport in non-magnetic junctions [89] the total charge current can be expressed as the sum of minority carrier currents at the deletion edges $j = j_{nL} + j_{pR}$ with

$$j_{nL} \propto \delta n_L, \quad j_{pR} \propto \delta p_R, \tag{2.70}$$

where δn_L is given by Eq. (2.59) with $P_{n0}^R = 0$, $\delta p_R = p_0[\exp(qV/k_B T) - 1]$, and V is the applied bias (positive for forward bias). Equation (2.69) implies that in the regime of large spin splitting, $q\zeta > k_B T$, the density of minority electrons changes exponentially with B ($\propto \zeta$) and can give rise to exponentially large magnetoresistance [84]. In the absence of an external spin source, a geometry depicted in Fig. 2.11(a) and (b), can also be used to illustrate the prediction of spin extraction [84], a process opposite to spin injection. Spin splitting in the p-region provides spin-dependent barriers for electron transport across the depletion region. With large forward applied bias and the generation of non-equilibrium carrier density there can be a significant spin extraction from the non-magnetic n-region into the magnetic p-region with spin densities having opposite signs in these two regions (for $s_{0L} > 0$ there is a spin accumulation $\delta s_R < 0$). These findings, obtained from a self-consistent numerical solution of drift-diffusion equations [84], can also be confirmed analytically, within the small bias approximation [85]. Similar spin extraction was recently observed experimentally in MnAs/GaAs junctions [108] and theoretical implications due to tunneling from non-magnetic semiconductors into metallic ferromagnets were considered [109].

The interplay between the P_{n0} (recall Eq. 2.51) in the p-region, and the non-equilibrium spin source of polarization δP_n in the n-region, at the edge of the depletion region, determines the I-V characteristics of the diodes. The dependence of the electric current j on $q\zeta$ and δP_n was obtained by both numerical and analytical methods. Numerical calculations [84] were performed by self-consistently solving for the system of drift-diffusion equations and analytical results [85, 88] were obtained using a small-bias approximation (see Section 2.2.1).

To illustrate the I-V characteristics of a magnetic p-n junction, consider the small-bias limit in the configuration of Fig. 2.11. The electron contribution to the total electric current can be expressed from Eqs (2.59) and (2.70) as [84, 85]

$$j_{nL} \sim n_0(\zeta) \left[e^{qV/k_B T} (1 + \delta P_n P_{n0}) - 1 \right]. \tag{2.71}$$

Equation (2.71) generalizes the Silsbee–Johnson spin-charge coupling [48, 110], originally proposed for ferromagnet/paramagnet metal interfaces, to the case of magnetic p-n junctions. The advantage of the spin-charge coupling in p-n junctions, as opposed to metals or degenerate systems, is the non-linear voltage dependence of the non-equilibrium carrier and spin densities [84, 85], allowing for the exponential enhancement of the effect with increasing V. Equation (2.71) can be understood qualitatively from Fig. 2.11. In equilibrium, $\delta P_n = 0$ and $V = 0$, no current flows through the depletion region, as the electron currents from both sides of the junction balance out. The balance is disturbed either by applying bias or by selectively populating different spin states, making the flow of one spin species greater than that of the other. In the latter case, the effective barriers for crossing of electrons from the n to the p side is different for spin up and down electrons (see Fig. 2.11). Current can flow even at $V = 0$

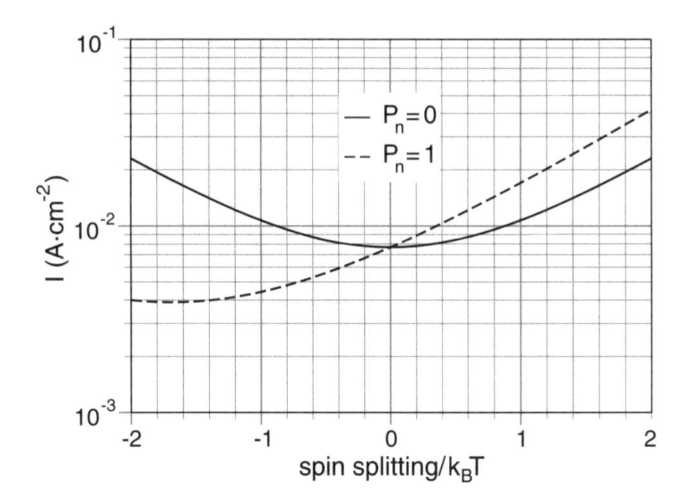

FIG. 2.12. Giant magnetoresistance (GMR) effect in magnetic diodes. Current/spin-splitting characteristics (I-ζ) are calculated self-consistently for $V = 0.8\,\mathrm{V}$ for the diode from Fig. 2.11. Spin splitting $2q\zeta$ on the p-side is normalized to k_BT. The solid curve corresponds to a switched-off spin source. The current is symmetric in ζ. With spin source on (the extreme case of 100% spin polarization injected into the n-region is shown), the current is a strongly asymmetric function of ζ, displaying large GMR, shown by the dashed curve. Material parameters of GaAs were applied. Adapted from [84].

when $\delta P_n \neq 0$. This is an example of the spin-voltaic effect (a spin analog of the photovoltaic effect), in which non-equilibrium spin causes an EMF [84, 111]. In addition, the direction of the zero-bias current is controlled by the relative sign of P_{n0} and δP_n. Experimental efforts to detect the spin-voltaic effect are discussed in Chapter 1. We will revisit the implications of the spin-voltaic effect in three-terminal structures, discussed in Section 2.2.4.3.

Magnetic p-n junctions can display an interesting giant magnetoresistance (GMR)-like effect, which follows from Eq. (2.71) [84]. The current depends strongly on the relative orientation of the non-equilibrium spin and the equilibrium magnetization. Figure 2.12 plots j, which also includes the contribution from holes, as a function of $2q\zeta/k_BT$ for both the unpolarized, $\delta P_n = 0$, and fully polarized, $\delta P_n = 1$, n-region. In the first case j is a symmetric function of ζ, increasing exponentially with increasing ζ due to the increase in the equilibrium minority carrier density $n_0(\zeta)$. In unipolar systems, where transport is due to the majority carriers, such a modulation of the current is not likely, as the majority carrier density is fixed by the density of dopants.

If $\delta P_n \neq 0$, the current will depend on the sign of $P_{n0} \cdot \delta P_n$. For parallel non-equilibrium (in the n-region) and equilibrium spins (in the p-region), most electrons cross the depletion region through the lower barrier (see Fig. 2.11), increasing the current. In the opposite case of antiparallel relative orientation, electrons experience a larger barrier and the current is inhibited. This is demonstrated in Fig. 2.12 by the strong asymmetry in j. The corresponding GMR ratio, the difference between j for parallel and antiparallel orientations, can also be calculated analytically from Eq. (2.71) as $2|\delta P_n P_{n0}|/(1-|\delta P_n P_{n0}|)$ [85]. If, for example, $|P_{n0}| = |\delta P_n| = 0.5$, the relative change is 66%. The GMR effect should be useful for measuring the spin relaxation rate of bulk semiconductors [88], as well as for detecting non-equilibrium spin in the non-magnetic region of the p-n junction.

2.2.4 *Spin transistors*

We describe two types of spin transistors: Datta–Das spin field effect transistor and magnetic bipolar transistor. These two represent two different views on spintronic devices. The Datta–Das transistor is based on individual electron spin dynamics in an effective microscopic magnetic field. The magnetic bipolar transistor is based on spin population differences, relying on ensemble spin (magnetization). Other types of spin transistors are discussed in this book in Chapters 4 and 6 and also reviewed in [1, 112].

2.2.4.1 *Datta–Das spin field effect transistor* The Datta–Das spin field effect transistor [113] uses a simple physical picture illustrated in Fig. 2.13. In a field effect transistor the source and the drain are connected by a conducting channel formed at a heterostructure interface. The width of the channel, and thus the conductance through it, is controlled by a gate separated from the channel by an insulating barrier. Imagine now the source and drain are ferromagnets (could be ferromagnetic metals or semiconductors), with parallel magnetizations. The source injects spin-polarized electrons into the channel. In a short channel the spin would be preserved until the electron enters the drain. What would happen if the electron's spin flipped? Then the electron would have a much larger probability to bounce off the drain, contributing to an increase of resistance. Instead of a spin flip, we can try to rotate the spin with a magnetic field. In addition, if this magnetic field could be controlled electronically, we would have full electronic control over conductance. The difference from the conventional field effect transistor is that (i) there is no need to change the channel width and thus no need to waste energy and time on charging (capacitance) effects, and (ii) the information about the electrical properties of the transistor can be reprogrammed by changing the magnetization orientations of the source and drain, giving the transistor the potential for nonvolatile random access memory.

The crucial point in the above picture is the effective field controlled by the gate. Datta and Das suggested using the Bychkov–Rashba field [74], which is a microscopic, electron momentum $\mathbf{k}$ dependent field giving rise to electron spin procession. This field is also called structure inversion asymmetry, since it appears only in low-dimensional systems whose macroscopic structure (confinement potentials, effective masses) lacks inversion symmetry. Semiconductor heterostructures used in field effect transistors are an example of a structure

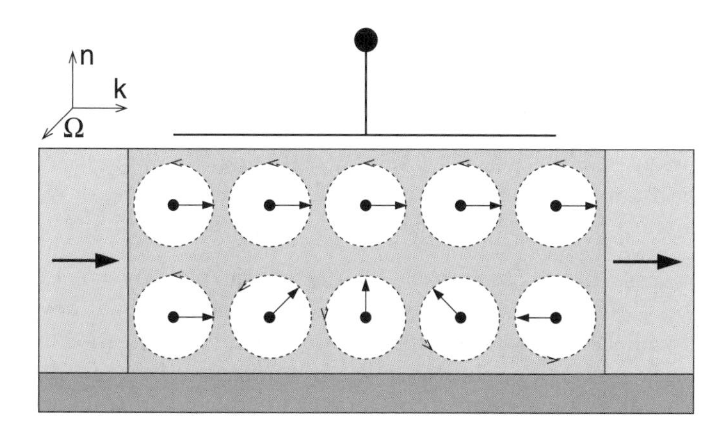

FIG. 2.13. Datta–Das spin field effect transistor. The source and drain are ferromagnetic metals with parallel magnetizations. The channel is formed at a heterostructure interface. The gate modifies the Bychkov–Rashba field Ω, which is perpendicular to both the growth direction $\mathbf{n}$ and electron momenta $\mathbf{k}$. The electrons either enter in the drain if their spin direction is unchanged (top) or bounce off if the spin has precessed (bottom), giving ON and OFF states, respectively. Adapted from [1].

inversion asymmetry. The Bychkov–Rashba field arises from spin-orbit coupling and is most pronounced in narrow gap semiconductors like InAs (where the coupling is enhanced). The corresponding Hamiltonian has a form usual for spin precession,

$$H(\mathbf{k}) = \frac{1}{2}\hbar\sigma \cdot \Omega(\mathbf{k}), \tag{2.72}$$

where σ is the vector of Pauli matrices and

$$\Omega(\mathbf{k}) = 2\alpha_{\mathrm{BR}}(\mathbf{k} \times \mathbf{n}). \tag{2.73}$$

Here α_{BR} is the Bychkov–Rashba parameter, the measured values of which are about 10^{-11} eV m, and $\mathbf{n}$ is the unit vector along the heterostructure growth. We emphasize that $\Omega(\mathbf{k})$ is not a real magnetic field (see Section 2.1.5.1). Indeed, the Bychkov–Rashba field does not lead to equilibrium spin polarization. Time reversal symmetry is preserved. What is exciting about the Bychkov–Rashba field is the fact that it can be tuned by the gate field, by changing the degree of asymmetry. There have been reports confirming this claim, although the matter is still not settled (see the discussion in [1], p. 355).

Is the Bychkov–Rashba field strong enough to cause at least a half precession of the spin? Take the channel width to be L, and electrons traveling ballistically parallel to the source-drain axis. The time of flight is $t = Lm/(\hbar k)$, where m is the electron effective mass. The precession frequency is $|\Omega| = 2\alpha_{\mathrm{BR}} k$ (see Eq. 2.73). Then the spin perpendicular to Ω precesses through angle $\phi = 2\alpha_{\mathrm{BR}} mL/\hbar$. This precession angle does not depend on the electron momentum. The probability to find such a spin in the initial state is then $\cos^2(\phi/2)$, the factor that also gives the current modulation. If we take $L \approx 0.1$ μm, $\hbar\alpha_{\mathrm{BR}} \approx 10^{-11}$ eV m, the effective electron mass 0.1 m_e (where m_e is the free electron mass), we get $\phi \approx 3$, enough to make a half precession. The value of α_{BR} sets the limit on L.

2.2.4.2 *Bipolar junction transistor* The magnetic bipolar transistor (MBT) builds on the bipolar junction transistor (BJT), a conventional device scheme introduced by Shockley [114] and widely used in signal amplification and processing, as well as in fast logic applications. We first introduce BJT and its formalism in order to recall some standard transistor terminology and to make a smooth transition to the magnetic case.

Conventional bipolar junction transistors comprise two *p*-*n* junctions in series, forming a three-terminal device. While such an arrangement may sound like a trivial extension of the *p*-*n* junction diode physics, the new structure has the remarkable novel functionality of amplifying small-current signals. The structure of a *npn* BJT is in Fig. 2.14. The emitter is doped with N_{de} donors, the base with N_{ab} acceptors, and collector with N_{dc} donors. The donor (acceptor) densities are also the electron (hole) majority densities in the respective regions. In equilibrium the minority densities are small. For example, the number of conduction electrons in the base is $n_{0b} = n_i^2/N_{ab}$, where n_i is the intrinsic carrier density in the semiconductor. External biases drive the current. In the most useful form of transistor operation, the forward active mode, in which the transistor is an amplifier, the emitter-base junction is forward biased with potential $V_{be} > 0$, while the collector-base junction is reverse biased with potential $V_{bc} < 0$. This means that the built-in potential in the emitter-base junction is reduced by V_{be}, allowing electron injection from the emitter to the base. The number of minority electrons in the base close to this junction increases exponentially to the non-equilibrium density

$$n_{be} = n_{0b} e^{qV_{be}/k_{\mathrm{B}}T}. \tag{2.74}$$

As in Section 2.2.1 we introduce the non-equilibrium (excess) density as

$$\delta n_{be} = n_{be} - n_{0b} = n_{0b}\left(e^{qV_{be}/k_{\mathrm{B}}T} - 1\right). \tag{2.75}$$

Similarly, the non-equilibrium electron density in the base at the base-collector junction is

$$\delta n_{bc} = n_{bc} - n_{0b} = n_{0b}\left(e^{qV_{bc}/k_{\mathrm{B}}T} - 1\right). \tag{2.76}$$

In the forward active mode δn_{bc} is small (and can be neglected) because $V_{bc} < 0$. It becomes important in other modes. The hole excess densities in the emitter and collector, close to the depletion region with the base, are

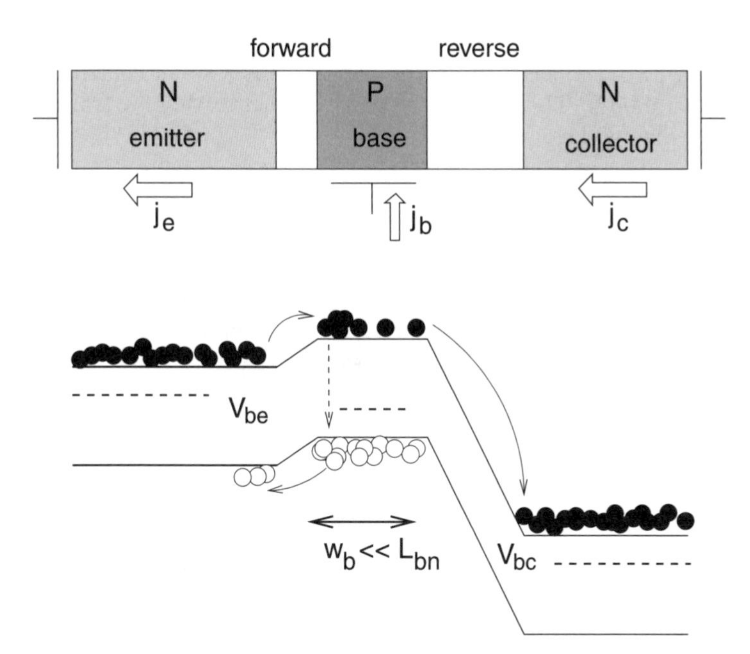

FIG. 2.14. Scheme of a conventional *npn* bipolar junction transistor in the forward active mode. The upper figure shows the overall structure, the lower figure shows the conduction and valence bands, populated with electrons (filled circles) and holes (empty), respectively. The dashed lines indicate the Fermi levels (chemical potentials). The emitter-base junction is forward biased with potential $V_{be} > 0$, while the collector-base junction is reverse biased with potential $V_{bc} < 0$. The magnitude of the corresponding applied potentials is given by the difference between the Fermi levels. The solid arrows indicate the carrier flow, while the dashed arrows illustrate recombination. For effective operation it is required that the base width w_b is smaller than the electron diffusion length in the base L_{nb}.

$$\delta p_e = p_{0e}\Big(e^{qV_{be}/k_B T} - 1\Big), \tag{2.77}$$

$$\delta p_c = p_{0c}\Big(e^{qV_{bc}/k_B T} - 1\Big). \tag{2.78}$$

Again, only δp_e needs to be considered in the forward active mode.

The emitter current j_e is formed by the electron injection current into the base, and the hole injection current into the emitter. As the injected electrons travel through the base, some of them recombine with the holes and leave the base through the valence band. Together with the flow of holes in the opposite direction, this constitutes the hole current j_b. Most electrons manage to reach the collector junction, where they are swept by the large electric field in the depletion region to the collector. Unless the electrons recombine inside the depletion region, they all reach the collector, forming the collector current j_c together with the holes injected to the collector from the base (this is a small contribution in the forward active mode). The most interesting characteristic of a transistor is the current gain, defined by

$$\beta = \frac{j_c}{j_b}. \tag{2.79}$$

Typically $\beta \approx 100$, meaning that the small-current signal brought in by varying j_b is amplified a hundred times in the collector circuit. In other words, taking away one electron (per unit time and area) from the base gives way to a hundred electrons reaching the collector. If there were no current drawn from the base, the electrons recombining there would oppose further injection from the emitter, stopping the current altogether. The sign convention for the current is specified in Fig. 2.14. The base current is

$$j_b = j_e - j_c. \tag{2.80}$$

We can calculate β by calculating the currents. A convenient way to write the currents in BJT is through the non-equilibrium densities of the minority carriers (see, for example, [115]):

$$j_e = j_{gb}^n \left[\frac{\delta n_{be}}{n_{0b}} - \frac{1}{\cosh(w_b/L_{nb})} \frac{\delta n_{bc}}{n_{0b}} \right] + j_{ge}^p \frac{\delta p_{eb}}{p_{0e}}, \tag{2.81}$$

$$j_c = j_{gb}^n \left[-\frac{\delta n_{bc}}{n_{0b}} + \frac{1}{\cosh(w_b/L_{nb})} \frac{\delta n_{be}}{n_{0b}} \right] - j_{gc}^p \frac{\delta p_{cb}}{p_{0c}}. \tag{2.82}$$

The base current is then calculated using Eq. (2.80). The generation currents j_g reflect the flow of thermally generated carriers in their majority regions close to the depletion region. Such carriers are then swept into the minority sides, irrespective of the applied bias. The electron generation current in the base is

$$j_{gb}^n = \frac{qD_{nb}}{L_{nb}} n_{0b} \coth\left(\frac{w_b}{L_{nb}}\right). \tag{2.83}$$

Here D_{nb} stands for the electron diffusion coefficient in the base whose width is w_b;[5] L_{nb} is the electron diffusion length in the base. The hole generation currents in the emitter, j_{ge}^p, and collector, j_{ge}^p, are

$$j_{ge}^p = \frac{qD_{pe}}{L_{pe}} p_{0e} \coth\left(\frac{w_e}{L_{pe}}\right), \tag{2.84}$$

$$j_{gc}^p = \frac{qD_{pc}}{L_{pc}} p_{0c} \coth\left(\frac{w_c}{L_{pc}}\right). \tag{2.85}$$

The notation is similar to the electron case.

[5] As in the case of diodes, the width of a region is an effective (rather than nominal) width of the neutral regions, excluding the depletion region whose size depends on the applied bias.

Equations (2.75)-(2.78) and (2.81)-(2.85) fully describe the electrical characteristics of ideal bipolar junction transistors. Let us calculate the gain β in the forward active mode. We will generalize this calculation in Section 2.2.4.3 for the magnetic case. The physics behind amplification becomes manifest by introducing three additional quantities: the transport factor α, the base transport factor α_t, and the emitter efficiency γ_e. They are related by

$$\alpha = \alpha_t \gamma_e = \frac{j_c}{j_e}, \tag{2.86}$$

where

$$\gamma_e = \frac{j_e^n}{j_e}, \tag{2.87}$$

$$\alpha_t = \frac{j_c}{j_e^n}. \tag{2.88}$$

The emitter efficiency γ_e measures the contribution of electrons to the emitter current. The higher it is, the more electrons (and less holes) are injected across the base-emitter junction. The base transport factor α_t shows how many of the injected electrons make it across the base to form the collector current. The current gain is

$$\beta = \frac{\alpha}{1-\alpha}. \tag{2.89}$$

Ideally, α is close to 1, so that β is large. For efficient current amplification both efficient emitter injection $\gamma_e \approx 1$ and base transport are needed.

In our *npn* BJT Eqs (2.81) and (2.82) give for the emitter efficiency[6]

$$\gamma_e = \frac{1}{1 + j_{ge}^p / j_{gb}^n}. \tag{2.90}$$

The base transport factor is

$$\alpha_t = \frac{1}{\cosh(w_b/L_{nb})}. \tag{2.91}$$

The emitter efficiency is usually increased by heavy emitter and small base doping, since $j_{ge}^p / j_{gb}^n \sim N_{ab}/N_{de}$. The greater the doping, the smaller the equilibrium number of minority carriers, and the smaller the corresponding generation current. The base transport factor can be increased by making the base narrower so that $w_b \ll L_{nb}$. In this limit the transistor amplification factor becomes

$$\beta = \frac{1}{w_b^2/2L_{nb}^2 + j_{ge}^p / j_{gb}^n}. \tag{2.92}$$

In Si transistors it is usually the emitter efficiency that determines amplification, since L_{nb} is rather large in Si, due to slow electron-hole recombination. In

[6]Note that δn_{bc} and δp_{bc} can be neglected in the forward active mode.

Table 2.2 Four (five) operating modes of conventional (magnetic) bipolar transistors. Forward (reverse) bias means positive (negative) voltage V. Symbols MA and GMA stand for magneto-amplification and giant magneto-amplification, while ON and OFF are modes of small and large resistance, respectively; SPSW stands for spin switch. The spin-voltaic mode applies only to MBT.

mode	V_{be}	V_{bc}	BJT	MBT
forward active	forward	reverse	amplification	MA, GMA
reverse active	reverse	forward	amplification	MA, GMA
saturation	forward	forward	ON	ON, GMA, SPSW
cut-off	reverse	reverse	OFF	OFF
spin-voltaic	null	null	N/A	SPSW

contrast, GaAs derived transistors have very small L_{nb} and the amplification is limited by the base transport factor. To reduce this factor, spatially modulated GaAs heterostructures are used to create electric drift in the base to boost the transport.

Table 2.2 summarizes different operating modes of both conventional BJT and magnetic transistors discussed in the next section. We have discussed the active forward mode where BJT amplify signals. The reverse active mode simply reverses the biases. In this mode a BJT can also amplify signals, but β is much smaller because the emitter efficiency is small. Usually transistors have small collector dopings to have large breakdown voltage in the reverse mode. The saturation mode is one with both junctions forward biased. The collector and base currents are similar in magnitude and amplification is inhibited. This mode, used in logic circuits, is denoted as ON, in contrast to the high-resistance cut-off (OFF) state in which both junctions are reverse biased and only small currents of the magnitudes of the generation currents flow. More discussion can be found in standard textbooks; see, for example, [116].

2.2.4.3 *Magnetic bipolar transistor* We propose the magnetic bipolar transistor (MBT) as a bipolar junction transistor that incorporates magnetic semiconductors as its active elements [121,122]. Simplified variations of our MBT not including the effects of non-equilibrium spin were later considered by Lebedeva and Kuivalainen [117], Flatté *et al.* [118], and Bandyopadhyay and Cahay [119]. Experimental realization of GaAs/(Ga,Mn)As-based MBT is currently in progress [120]. The magnetic semiconductors can be ferromagnetic or they can have giant g-factors and placed in a magnetic field. Either way there is a large, comparable to the thermal energy, spin splitting $2q\zeta_b$ of the carrier bands. Here we illustrate the properties of MBTs using electron spin polarization (leaving holes unpolarized). Only the base will have equilibrium spin polarization. The excit-

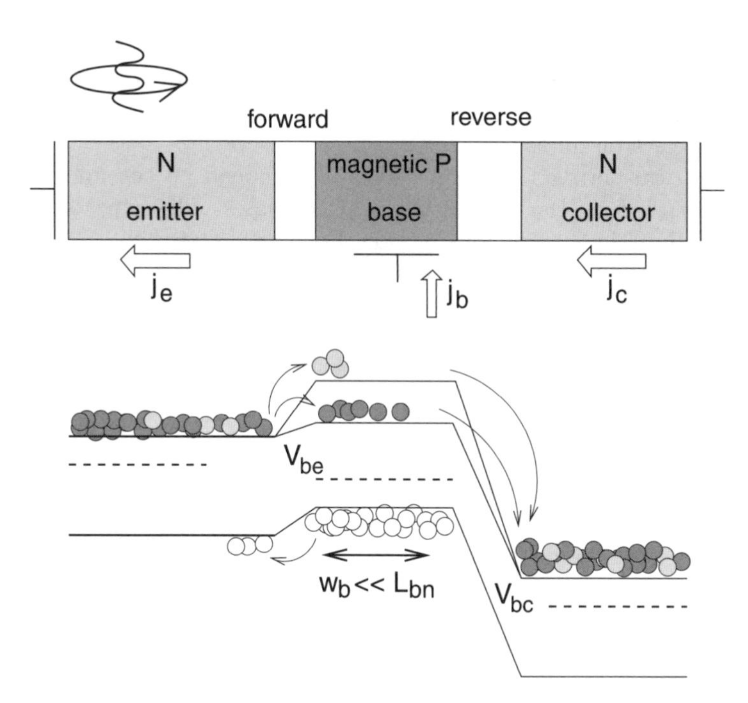

FIG. 2.15. Scheme of a magnetic *npn* bipolar transistor in the forward active mode. The notation is as in Fig. 2.14. Only the base has equilibrium electron spin polarization P_{0b}, illustrated by the spin-split conduction band. Spin up (down) electrons are pictured as dark (light) filled circles. Holes are unpolarized. The emitter has a source of spin polarization, here shown as circularly polarized light, giving rise to non-equilibrium spin polarization δP_e. The coupling between the equilibrium and non-equilibrium polarizations gives rise to many new functionalities as described in the text.

ing new features appear when we allow for a non-equilibrium spin to be added. This can be achieved by optical spin orientation, or electrical spin injection (see Sections 2.1.3 and 2.1.4). Here we assume that there is a non-equilibrium spin of polarization δP_e in the emitter, and equilibrium spin of polarization P_{0b} in the base. The magnetic field can modify P_{0b} by changing $q\zeta_b$, since (recall Eq. 2.51) $P_{0b} = \tanh(q\zeta_b/k_{\mathrm{B}}T)$. The scheme of a MBT is shown in Fig. 2.15.

The most important new feature of MBT is the spin dependent barrier for the electron injection from the emitter to the base. In Fig. 2.15 this barrier favors spin up electrons. More electrons are injected into the lower conduction level in the base than into the upper one. The equilibrium spin polarization in the base is preserved. We have learned in Section 2.2.4.2 that the emitter efficiency γ_e is a limiting factor in amplification. We can easily modify γ_e in our MBT. Indeed, just put more (or less) spin up electrons in the emitter, so that more (less) electrons are injected through the lower spin barrier. This tuning can be done by introducing non-equilibrium spin δP_e. In effect, the emitter efficiency can be controlled by spin-charge coupling. As for the base transport factor α_t, there is not much to be done by either spin or magnetic field. This factor is governed by electron diffusion.[7]

The electrical currents through MBTs also depend on the non-equilibrium minority carrier densities and on the applied biases. The expressions for the currents are those in the previous section, Eqs (2.81) and (2.82). The difference now is that the non-equilibrium electron densities depend on spin. The spin-charge coupling leads to the familiar $\delta P_e P_{0b}$ dependence for electron densities

$$\delta n_{be} = n_{0b}(\zeta_b)\left[e^{qV_{be}/k_BT}(1+\delta P_e P_{0b}) - 1\right], \tag{2.93}$$

$$\delta n_{bc} = n_{0b}(\zeta_b)\left(e^{qV_{bc}/k_BT} - 1\right). \tag{2.94}$$

The influence of the equilibrium spins is felt both in $n_{0b}(\zeta_b) = n_{0b}(0)\cosh(q\zeta_b/k_B T)$, which reflects the change of the equilibrium minority density in the magnetic region, and as well as in the spin-charge coupling factor. The non-equilibrium spin plays a role only in the latter. The expression for δn_{bc} remains Eq. (2.76). The excess hole densities are given by Eqs (2.77) and (2.78).

Substituting δn_{be} from Eq. (2.93) into the formula for j_e, Eq. (2.81), we obtain for the emitter efficiency

$$\gamma_e = \left[1 + \frac{j_{ge}^p}{j_{gb}^n(\zeta_b)}\frac{1}{1+\delta P_e P_{0b}}\right]^{-1}. \tag{2.95}$$

This generalizes Eq. (2.90) to the case of equilibrium spin polarization in the base and spin-charge coupling. We specify that j_{gb}^n depends on ζ_b through $n_{0b}(\zeta_b)$. The base factor α_t is given by Eq. (2.91). The gain β is then

$$\beta = \left[\frac{w_b^2}{2L_{nb}^2} + \frac{j_{ge}^p}{j_{gb}^n(\zeta_b)}\frac{1}{1+\delta P_e P_{0b}}\right]^{-1}, \tag{2.96}$$

where we assume the narrow base limit $w_b \ll L_{nb}$. The above formula generalizes Eq. (2.92). The amplification depends on both the equilibrium and non-equilibrium spin. The dependence on the equilibrium spin is through both $j_{gb}^n(\zeta_b)$, which is an even function of ζ_b and thus also P_{0b}, and through spin-charge coupling. We call this dependence *magneto-amplification* (MA), since it allows control over amplification by the magnetic field, which gives rise to equilibrium spin polarization. Magneto-amplification is present even without the non-equilibrium spin and can be used to detect a magnetic field or to measure the equilibrium spin polarization. In analogy with giant magnetoresistance, we call the effect of a relative change of β upon switching the sign of $\delta P_e P_{0b}$ *giant magneto-amplification*

[7] There is, however, the possibility that L_{nb} depends on B, leading to small magnetic effects, observed also with conventional transistors.

(GMA) [115]. The corresponding giant magneto-amplification factor GMA is defined as

$$GMA = \frac{\beta(\text{parallel}) - \beta(\text{antiparallel})}{\beta(\text{parallel})}, \tag{2.97}$$

where (anti)parallel refers to the relative orientation of the equilibrium and non-equilibrium spins P_{0b} and δP_e.

Spin and magnetic control of current amplification is optimized when the emitter efficiency dominates over base transport, as in Si-like transistors or specially tailored GaAs heterostructure transistors. In this case we can neglect the factor $w_b^2/2L_{nb}^2$ in Eq. (2.96) and write

$$\beta = \frac{j_{gb}^n(\zeta_b)}{j_{ge}^p}\left(1 + \delta P_e P_{0b}\right). \tag{2.98}$$

The relative change of β in a magnetic field or spin orientation change is now

$$GMA = 2\frac{|\delta P_e P_{0b}|}{1 + |\delta P_e P_{0b}|}, \tag{2.99}$$

which, for reasonable values of 50% spin polarizations, give about 40% giant magneto-amplification.

The above analysis of the forward active mode of MBT applies equally to the reverse active mode, with the same proviso that the amplification is usually much smaller due to the transistor design. The saturation mode of MBT differs from that of BJT. We have found that MBT can amplify signals also in this mode, solely due to spin-charge coupling which can significantly enhance j_c over j_b [123]. Suppose non-equilibrium spin is added to both the emitter (δP_e) and collector (δP_c). Then one can show that the current amplification can be controlled by the difference of the two spin polarizations [123]

$$\beta = \frac{P_{0b}(P_{0e} - P_{0c})}{w_b^2/L_{nb}^2 + j_{ge}^p/j_{gb}^n + j_{gc}^p/j_{gb}^n}. \tag{2.100}$$

The current gain is large because of the denominator which contains the ratio of the hole to electron equilibrium densities. It is remarkable that the gain can be made negative by switching from the emitter to collector spin polarization, or by changing the sign of the polarization. The ON and OFF logic states can also be tuned within this mode by spin charge coupling. MBT in the saturation mode can also act as a spin switch. In the cut-off mode MBT is in the OFF state. Spin effects are inhibited. We include also the spin-voltaic mode, where $V_{be} = V_{bc} = 0$. Conventional BJT is in equilibrium, with no currents flowing, but MBT can be active because of the presence of non-equilibrium spin. In this mode $\delta n_{be} = n_{0b}(\zeta_b)\delta P_e P_{0b}$, while the non-equilibrium hole densities vanish. All the activity is controlled by spin-charge coupling. The transistor can act as a spin switch, changing the direction of the currents by changing the direction of spin. Unfortunately, the currents are small, on the order of generation currents. Since only electrons flow, $\gamma_e = 1$ and amplification in this mode can be very large since it is solely due to α_t. These five modes are summarized in Table 2.2.

What is remarkable MA and GMA is that they do not depend on spin relaxation in the base.[8] The controlling factor is the carrier injection from the emitter into the base. Only spin relaxation in the depletion region between the emitter and base can mask the effect. Fortunately, the depletion region is rather small, especially in forward bias, and the built-in electric field causes fast spin drift. Magnetic bipolar transistors can also be used for electrical control of magnetism. In high injection limits (beyond the validity of our analytical theory), where the number of injected electrons from the emitter to the base is comparable to the base doping density, the presence of free carriers can induce ferromagnetism. Similar considerations would apply to the small regions of the depletion regions. When depleted more, for example by reverse biases, the regions can lose ferromagnetism as they are void of free carriers. Such electrical control of ferromagnetism could be used for magnetic storage.

2.2.4.4 *Ebers–Moll model of magnetic bipolar transistor* The Ebers–Moll model is an equivalent circuit to a BJT [116]. Let us introduce some new notation. Denote by j_{se} and j_{sc} the emitter and collector saturation currents:

$$j_{se} = j_{gb}^n + j_{ge}^p, \tag{2.101}$$

$$j_{sc} = j_{gb}^n + j_{gc}^p. \tag{2.102}$$

The emitter saturation current is the current that flows if $V_{be} < 0$, $V_{bc} = 0$, and only equilibrium spin present. Similarly for the collector saturation current. Denote next the forward and reverse currents (terminology from the forward active mode) as

$$j_f = j_{se}\left(e^{qV_{be}/k_BT} - 1\right), \tag{2.103}$$

$$j_r = j_{sc}\left(e^{qV_{bc}/k_BT} - 1\right). \tag{2.104}$$

Finally, we introduce spin-charge coupling forward and reverse currents

$$j_{mf} = j_{gb}^n \delta P_e P_{0b} e^{qV_{be}/k_BT}, \tag{2.105}$$

$$j_{mr} = j_{gb}^n \delta P_c P_{0b} e^{qV_{bc}/k_BT}. \tag{2.106}$$

Here the subscript "m" stands for magnetic to stress that the current appears only in magnetic junctions. These currents flow due to the presence of non-equilibrium spin polarization and are finite even at zero bias (recall our discussion of this spin-voltaic effect in Section 2.2.3). Here we include also the possibility

[8] Spin relaxation in the magnetic base is much faster than in the non-magnetic regions.

of non-equilibrium spin polarization δP_c in the collector. The Ebers–Moll model for MBT then reads [124]

$$j_e = j_f - \alpha_r j_r + j_{mf} - \alpha_t j_{mr} \tag{2.107}$$

$$j_c = \alpha_f j_f - j_r + \alpha_t j_{mf} - j_{mr}. \tag{2.108}$$

Here α_f has the meaning of the transport factor (see Eq. 2.86) in the forward active mode, while α_r is the transport factor in the reverse active mode in the absence of spin-charge coupling, as can be seen directly from Eqs (2.107) and (2.108). As in the conventional model, these two factors satisfy

$$\alpha_f j_{se} = \alpha_r j_{sc}, \tag{2.109}$$

which can be verified by requiring that

$$j_e(V_{be} = 0, V_{bc} = V) = -j_c(V_{be} = V, V_{bc} = 0), \tag{2.110}$$

for $\delta P_e = \delta P_c = 0$. In our ideal case it is straightforward to show that

$$\alpha_f j_{se} = \alpha_t j_{gb}^n, \tag{2.111}$$

where the base transport factor α_t is given by Eq. (2.91).

The equivalent circuit to Eqs (2.107) and (2.108) is shown in Fig. 2.16. The voltage sources are arranged for the forward active mode for illustration. The current flow is the same as in Fig. 2.14. The emitter circuit consists of four elements: (i) conventional diode with the directional current j_f that depends on V_{be}, (ii) conventional current source giving current $\alpha_r j_r$ that depends on V_{bc} and on the transport factor α_r measuring the amount of current injected into the emitter from the collector, (iii) spin diode with the forward current j_{mf}, and finally, (iv) spin current source $\alpha_t j_{mr}$. The first two elements already appear in conventional BJTs. The spin diode (iii) appears due to spin-charge coupling, and works similarly to a diode in the sense that its current is directional with $j_{mf} \sim \exp(qV_{be}/k_B T)$. The crucial difference from conventional diodes is that the direction of the current flow can be changed by changing the sign of $\delta P_e P_{0b}$. The symbol for the spin diode reflects this fact. The filled triangle shows the direction when $\delta P_e P_{0b}$ is positive. The new functionality of MBT lies in the ability to switch or modify the spin diode during its operation. There is also a spin current source (iv). This appears due to the electron current from spin-charge coupling, injected in the base from the collector, and diffusing towards the emitter through the base (this is why the transport factor α_t appears). The element is a current source because it does not depend on the voltage drop across it. It is, however, a controlled current source, similar to (ii), which can be controlled by V_{bc}. Because (iv) arises from spin-charge coupling, it can also be controlled by spin and magnetic field. A similar description applies to the collector circuit.

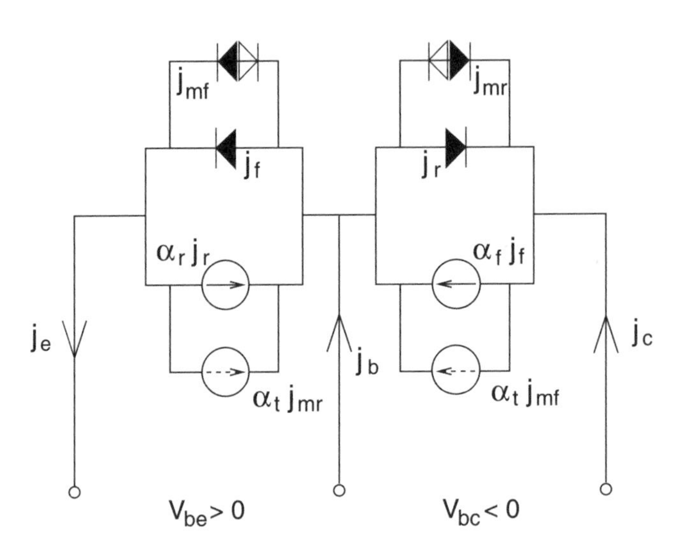

FIG. 2.16. The Ebers–Moll equivalent circuit of MBT in the forward active regime. Left is the emitter, right is the collector circuit. The emitter circuit has a diode for the forward current, and a current source which depends on the current in the collector circuit. In addition, there are two new elements. A spin diode whose direction can be flipped: its filled triangle points to the forward direction when the two polarizations are parallel. If they are antiparallel, the current direction changes. A new spin current source (dashed arrow), points in the direction of the current if the two polarizations are parallel. The direction of the current can also be switched. A similar notation applies for the right (collector) circuit.

2.2.5 *Outlook and future directions*

Most of the existing structures which rely on the effects of spin-polarized bipolar transport currently work at low temperature and future progress towards room-temperature operation will largely be driven by materials improvements. One can expect that reliable room-temperature operation will mainly be pursued in either all-semiconductor structures or in hybrid structures combining metallic ferromagnets and semiconductors. In the first approach the key issue will likely be an integration of ferromagnetic semiconductors with sufficiently high Curie temperature in relevant device structures. Experiments in (Ga,Mn)As [125] showing that selective doping can significantly increase the Curie temperature, as compared to the uniformly doped bulk samples, should also provide valuable guidance for optimizing ferromagnetism in other materials. Stringent tests of carrier-mediated ferromagnetism thus far have been performed only at relatively low temperatures and in a small number of materials (see also the discussion in Chapter 1). On the other hand, experimental reports of room temperature ferromagnetism, in an increasing number of semiconductors [126], remain controversial. It is often difficult to identify possible spurious effects. For example, early reports of 900 K ferromagnetism in La-doped $CaBa_6$ [127–129], were later revisited suggesting an extrinsic effect [130]. Carrier-mediated ferromagnetism in semiconductor junctions has been shown to strongly influence neighboring non-magnetic regions through spin injection and the properties discussed in Chapter 1. However, it remains to be examined what influence non-carrier-mediated ferromagnetic semiconductors are expected to have on neighboring nonmagnetic semiconductor regions. There also remain important theoretical challenges. Even in homogeneous bulk materials the dominant models of ferromagnetism are not capable of predicting trends in magnetic properties across different non-magnetic semiconductor hosts [14].

In the second approach, relying on metallic ferromagnets, there is a wide range of materials with desirably high Curie temperature. However, just knowing their behavior in the bulk is insufficient since the properties of spin-polarized transport will be predominantly determined by the nature of their interfaces with semiconductors. Indeed, as we mentioned in the introduction, the concept of spin polarization, relevant to potential device structures, is not a bulk property of a magnetic material [1]. A simple change in the choice of a neighboring non-magnetic semiconductor could lead to completely different interfacial properties (for example, GaAs could lead to a Schottky and InAs to an Ohmic contact with the same magnetic metal). Furthermore, it is known that the lattice mismatch between metals and semiconductors, as well as the presence of interfacial effects, can lead to strong effects on the efficiency of spin injection.

There are also several other possible routes to bipolar spin-polarized transport. In structures involving superconductors both electrons and holes could contribute to transport. A formal similarity with continuity equations for bipolar spin-polarized transport in *p*-*n* junctions can already be noticed in a pioneering proposal for spin injection in a superconductor by Aronov [131]. An interesting interplay between strongly correlated effects and bipolar spin-polarized transport can be realized in structures involving manganite perovskites. It has already been demonstrated that in these systems one can fabricate magnetic *p*-*n* junctions which show rectifying behavior [132] and electric modulation of ferromagnetism [133]. Furthermore, recent measurements of exponentially large magnetoresistance in manganite-titanite heterojunctions [134] show similar behavior to that predicted for semiconductor *p*-*n* junctions [84]. Perhaps, one of the next challenges could be to fabricate three-terminal structures which would lead to signal amplification.

Acknowledgements

We thank S. Das Sarma, S. C. Erwin, S. Garzon, M. Johnson, B. T. Jonker, S. Maekawa, I. I. Mazin, H. Munekata, S. S. P. Parkin, A. Petukhov, and E. I. Rashba for useful discussions. This work was supported by the US ONR, NSF, and DARPA. I. Ž. acknowledges financial support from the National Research Council.

References

[1] I. Žutić, J. Fabian, and S. Das Sarma, Rev. Mod. Phys. **76**, 323 (2004).
[2] B. T. Jonker, A. T. Hanbicki, D. T. Pierece, and M. D. Stiles, J. Magn. Magn. Mater. **277**, 24 (2004).
[3] I. I. Mazin, D. J. Singh, and C. Ambroasch-Draxl, Phys. Rev. B **59**, 411 (1999).
[4] P. M. Tedrow and R. Meservey, Phys. Rev. B **7**, 318 (1973).
[5] I. I. Mazin, Phys. Rev. Lett. **83**, 1427 (1999).
[6] S. Maekawa, S. Takahashi, and H. Imamura, in *Spin Dependent Transport in Magnetic Nanostructures*, edited by S. Maekawa and T. Shinjo (Taylor and Francis, New York, 2002), pp. 143–236.
[7] J. S. Moodera and G. Mathon, J. Magn. Magn. Mater. **200**, 248 (1999).
[8] T. Miyazaki, in *Spin Dependent Transport in Magnetic Nanostructures*, edited by S. Maekawa and T. Shinjo (Taylor and Francis, New York, 2002), pp. 113–142.
[9] B. Dieny, V. S. Speriosu, S. S. P. Parkin, B. A. Gurney, D. R. Wilhoit, and D. Maur, Phys. Rev. B **43**, 1297 (1991).
[10] M. Jullierè, Phys. Lett. **54 A**, 225 (1975).
[11] R. A. de Groot, F. M. Mueller, P. G. van Engen, and K. H. J. Buschow, Phys. Rev. Lett. **50**, 2024 (1983).
[12] W. E. Pickett and J. S. Moodera, Phys. Today **54 (5)**, 39 (2001).
[13] I. Galanakis and P. H. Dederichs, cond-mat/0408068.
[14] S. C. Erwin and I. Žutić, Nature Mat. **3**, 410 (2004).
[15] R. J. Soulen Jr., J. M. Byers, M. S. Osofsky, B. Nadgorny, T. Ambrose, S. F. Cheng, P. R. Broussard, C. T. Tanaka, J. Nowak, J. S. Moodera, *et al.*, *Science* **282**, 85 (1998).
[16] S. K. Upadhyay, A. Palanisami, R. N. Louie, and R. A. Buhrman, Phys. Rev. Lett. **81**, 3247 (1998).
[17] K. Kikuchi, H. Immamura, S. Takahashi, and S. Maekawa, Phys. Rev. B **65**, 020508 (2002).
[18] I. Žutić and O. T. Valls, Phys. Rev. B **60**, 6320 (1999).
[19] I. Žutić and O. T. Valls, Phys. Rev. B **61**, 1555 (2000).
[20] S. Kashiwaya, Y. Tanaka, N. Yoshida, and M. R. Beasley, Phys. Rev. B **60**, 3572 (1999).
[21] J.-X. Zhu, B. Friedman, and C. S. Ting, Phys. Rev. B **59**, 9558 (1999).
[22] P. A. Dowben and R. Skomski, J. Appl. Phys. **95**, 7453 (2004).
[23] S. S. P. Parkin, C. Kaiser, A. Panchula, P. Rice, M. Samant, and S.-H. Yang Nature Mater. **3**, 862 (2004).
[24] J. G. Braden, J. S. Parker, P. Xiong, S. H. Chun, and N. Samarth, Phys. Rev. Lett. **91**, 056602 (2003).
[25] I. Žutić and S. Das Sarma, Phys. Rev. B **60**, R16322 (1999).
[26] H. Saito, V. Zayets, S. Yamagata, and K. Ando, Phys. Rev. Lett. **90**, 207202 (2003).
[27] G. A. Medvedkin, T. Ishibashi, T. N. Hayata, Y. Hasegawa, and K. Sato, Jpn. J. Appl. Phys., Part 2 **39**, L949 (2000).

[28] S. Cho, S. Cho, G.-B. Cha, S. C. Hong, Y. Kim, Y.-J. Zhao, A. J. Freeman, J. B. Ketterson, B. J. Kim, and Y. C. Kim, Phys. Rev. Lett. **88**, 257203 (2002).
[29] S. Koshihara, A. Oiwa, M. Hirasawa, S. Katsumoto, Y. Iye, S. Urano, H. Takagi, and H. Munekata, Phys. Rev. Lett. **78**, 4617 (1997).
[30] A. Oiwa, Y. Mitsumori, R. Moriya, T. Supinski, and H. Munekata, Phys. Rev. Lett. **88**, 137202 (2002).
[31] H. Boukari, P. Kossacki, M. Bertolini, D. Ferrand, J. Cibert, S. Tatarenko, A. Wasiela, J. A. Gaj, and T. Dietl, Phys. Rev. Lett. **88**, 207204 (2002).
[32] H. Ohno, D. Chiba, F. Matsukura, T. O. E. Abe, T. Dietl, Y. Ohno, and K. Ohtani, Nature **408**, 944 (2000).
[33] Y. D. Park, A. T. Hanbicki, S. C. Erwin, C. S. Hellberg, J. M. Sullivan, J. E. Mattson, T. F. Ambrose, A. Wilson, G. Spanos, and B. T. Jonker, *Science* **295**, 651 (2002).
[34] A. Kastler, J. Phys. (Paris) **11**, 255 (1950).
[35] G. Lampel, Phys. Rev. Lett. **20**, 491 (1968).
[36] F. Meier and B. P. Zakharchenya (Eds.), *Optical Orientation* (North-Holand, New York, 1984).
[37] M. Oestreich, M. Brender, J. Hübner, D. H. W. W. Rühle, T. H. P. J. Klar, W. Heimbrodt, M. Lampalzer, K. Voltz, and W. Stolz, Semicond. Sci. Technol. **17**, 285 (2002).
[38] C. Kittel, *Quantum Theory of Solids* (Wiley, New York, 1963).
[39] R. R. Parsons, Phys. Rev. Lett. **23**, 1152 (1969).
[40] D. J. Hilton and C. L. Tang, Phys. Rev. Lett. **89**, 146601 (2002).
[41] R. Fiederling, M. Kleim, G. Reuscher, W. Ossau, G. Schmidt, A. Waag, and L. W. Molenkamp, Nature **402**, 787 (1999).
[42] B. T. Jonker, Y. D. Park, B. R. Bennett, H. D. Cheong, G. Kioseoglou, and A. Petrou, Phys. Rev. B **62**, 8180 (2000).
[43] D. K. Young, E. Johnston-Halperin, D. D. Awschalom, Y. Ohno, and H. Ohno, Appl. Phys. Lett. **80**, 1598 (2002).
[44] A. T. Hanbicki, O. M. J. van t'Erve, R. Magno, G. Kioseoglou, C. H. Li, B. T. Jonker, G. Itskos, R. Mallory, M. Yasar, and A. Petrou, Appl. Phys. Lett. **82**, 4092 (2003).
[45] X. Jiang, R. Wang, S. van Dijken, R. Shelby, R. Macfarlane, G. S. Solomon, J. Harris, and S. S. P. Parkin, Phys. Rev. Lett. **90**, 256603 (2003).
[46] S. M. Sze, *Physics of Semiconductor Devices* (John Wiley, New York, 1981).
[47] S. Takahashi and S. Maekawa, Phys. Rev. B **67**, 052409 (2003).
[48] M. Johnson and R. H. Silsbee, Phys. Rev. Lett. **55**, 1790 (1985).
[49] E. I. Rashba, Phys. Rev. B **62**, R16267 (2000).
[50] E. I. Rashba, Eur. Phys. J. B **29**, 513 (2002).
[51] B. T. Jonker, S. C. Erwin, A. Petrou, and A. G. Petukhov, MRS Bull. **28**, 740 (2003).
[52] M. A. M. Gijs and G. E. W. Bauer, Adv. Phys. **46**, 285 (1997).
[53] M. D. Stiles and A. Zangwill, Phys. Rev. B **66**, 014407 (2002).

[54] S. Hershfield and H. L. Zhao, Phys. Rev. B **56**, 3296 (1997).
[55] P. C. van Son, H. van Kempen, and P. Wyder, Phys. Rev. Lett. **58**, 2271 (1987).
[56] T. Valet and A. Fert, Phys. Rev. B **48**, 7099 (1993).
[57] G. Schmidt, D. Ferrand, L. W. Molenkamp, A. T. Filip, and B. J. van Wees, Phys. Rev. B **62**, R4790 (2000).
[58] K. M. Schep, J. B. A. N. van Hoof, P. J. Kelly, and G. E. W. Bauer, Phys. Rev. B **56**, 10805 (1997).
[59] M. D. Stiles and D. R. Penn, Phys. Rev. B **61**, 3200 (2000).
[60] A. G. Aronov and G. E. Pikus, Fiz. Tekh. Poluprovodn. **10**, 1177 (1976), [Sov. Phys. Semicond. **10**, 698 (1976)].
[61] M. Johnson and R. H. Silsbee, Phys. Rev. B **35**, 4959 (1987).
[62] M. Johnson and R. H. Silsbee, Phys. Rev. B **37**, 5312 (1988).
[63] L. Esaki, P. Stiles, and S. von Molnár, Phys. Rev. Lett. **19**, 852 (1967).
[64] J. S. Moodera, X. Hao, G. A. Gibson, and R. Meservey, Phys. Rev. B **42**, 8235 (1988).
[65] S. Garzon, I. Žutić, and R. A. Webb, Phys. Rev. Lett. **94**, 176601 (2005).
[66] D. Pines and C. P. Slichter, Phys. Rev. **100**, 1014 (1955).
[67] R. J. Elliott, Phys. Rev. **96**, 266 (1954).
[68] Y. Yafet, in *Solid State Physics, Vol. 14*, edited by F. Seitz and D. Turnbull (Academic, New York, 1963), p. 2.
[69] M. I. D'yakonov and V. I. Perel', Fiz. Tverd. Tela **13**, 3581 (1971), [Sov. Phys. Solid State **13**, 3023 (1971)].
[70] G. L. Bir, A. G. Aronov, and G. E. Pikus, Zh. Eksp. Teor. Fiz. **69**, 1382 (1975), [Sov. Phys. JETP **42**, 705 (1976)].
[71] M. I. D'yakonov and V. I. Perel', Zh. Eksp. Teor. Fiz. **38**, 362 (1973), [Sov. Phys. JETP **38**, 177 (1973)].
[72] J. Fabian and S. Das Sarma, J. Vac. Sci. Technol. B **17**, 1708 (1999).
[73] G. Dresselhaus, Phys. Rev. **100**, 580 (1955).
[74] Y. A. Bychkov and E. I. Rashba, Zh. Eksp. Teor. Fiz. Pisma Red. **39**, 66 (1984), [JETP Lett. **39**, 78 (1984)].
[75] Y. A. Bychkov and E. I. Rashba, J. Phys. C **17**, 6039 (1984).
[76] G. E. Pikus and A. N. Titkov, in *Optical Orientation, Modern Problems in Condensed Matter Science, Vol. 8*, edited by F. Meier and B. P. Zakharchenya (North-Holland, Amsterdam, 1984), p. 109.
[77] R. I. Dzhioev, K. V. Kavokin, V. L. Korenev, M. V. Lazarev, B. Y. Meltser, M. N. Stepanova, B. P. Zakharchenya, D. Gammon, and D. S. Katzer, Phys. Rev. B **66**, 245204 (2002).
[78] J. M. Kikkawa and D. D. Awschalom, Phys. Rev. Lett. **80**, 4313 (1998).
[79] K. V. Kavokin, Phys. Rev. B **64**, 075305 (2001).
[80] R. I. Dzhioev, V. L. Korenev, B. P. Zakharchenya, D. Gammon, A. S. Bracker, J. G. Tischler, and D. S. Katzer, Phys. Rev. B **66**, 153409 (2002).
[81] R. I. Dzhioev, K. V. Kavokin, V. Korenev, M. Lazarev, N. K. Poletaevx, B. P. Zakharchenya, E. A. Stinaff, D. Gammon, A. Bracker, and M. Ware, cond-mat/0407133.

[82] N. W. Ashcroft and N. D. Mermin, *Solid State Physics* (Saunders, Philadelphia, 1976).
[83] I. Žutić, J. Fabian, and S. Das Sarma, Phys. Rev. B **64**, 121201 (2001).
[84] I. Žutić, J. Fabian, and S. Das Sarma, Phys. Rev. Lett. **88**, 066603 (2002).
[85] J. Fabian, I. Žutić, and S. Das Sarma, Phys. Rev. B **66**, 165301 (2002).
[86] I. Žutić, J. Fabian, and S. C. Erwin, cond-mat/0412580.
[87] I. Žutić, J. Fabian, and S. Das Sarma, Appl. Phys. Lett. **79**, 1558 (2001).
[88] I. Žutić, J. Fabian, and S. Das Sarma, Appl. Phys. Lett. **82**, 221 (2003).
[89] W. Shockley, *Electrons and Holes in Semiconductors* (D. Van Nostrand, Princeton, 1950).
[90] D. T. Pierce and R. J. Celotta, in *Optical Orientation, Modern Problems in Condensed Matter Science, Vol. 8*, edited by F. Meier and B. P. Zakharchenya (North-Holland, Amsterdam, 1984), pp. 259–294.
[91] J. I. Pankove, *Optical Processes in Semiconductors* (Prentice-Hall, Inc. Englewood Cliffs, N.J., 1971).
[92] M. I. D'yakonov and V. I. Perel', **13**, 206 (1971), [JETP Lett. **13**, 144-146 (1971)].
[93] S. D. Ganichev and W. Prettl, J. Phys.: Condens. Matter. **15**, R935 (2003).
[94] W. Long, Q.-F. Sun, H. Guo, and J. Wang, Appl. Phys. Lett. **83**, 1397 (2003).
[95] A. Brataas, Y. Tserkovnyak, G. E. W. Bauer, and B. I. Halperin, Phys. Rev. B **66**, 060404 (2002).
[96] A. G. Mal'shukov, C. S. Tang, C. S. Chu, and K. A. Chao, Phys. Rev. B **68**, 233307 (2003).
[97] Y. V. Pershin and V. Privman, Phys. Rev. Lett. **90**, 256602 (2003).
[98] H. Munekata, private communications (2003).
[99] C. P. Wen, B. Hershenov, H. von Philipsborn, and H. L. Pinch, IEEE Trans. Magn. **4**, 702 (1968).
[100] E. Janik and G. Karczewski, Acta Physica Polonica A **73**, 439 (1988).
[101] Y. Ohno, I. Arata, F. Matsukura, K. Ohtani, S. Wang, and H. Ohno, Appl. Surf. Sci. **159-160**, 308 (2000).
[102] M. Kohda, Y. Ohno, K. Takamura, F. Matsukura, and H. Ohno, Jpn. J. Appl. Phys. **40**, L1274 (2001).
[103] E. Johnston-Halperin, D. Lofgreen, R. K. Kawakami, D. K. Young, L. Coldren, A. C. Gossard, and D. D. Awschalom, Phys. Rev. B **65**, 041306 (2002).
[104] I. Arata, Y. Ohno, F. Matsukura, and H. Ohno, Physica E **10**, 288 (2001).
[105] P. Van Dorpe, Z. Liu, W. V. Roy, V. F. Motsnyi, M. Sawicki, G. Borghs, and J. De Boeck Appl. Phys. Lett. **84**, 3495 (2004).
[106] F. Tsui, L. Ma, and L. He, Appl. Phys. Lett. **83**, 954 (2003).
[107] D. Schmeltzer, A. Saxena, A. Bishop, and D. L. Smith, Phys. Rev. B **68**, 195317 (2003).
[108] J. Stephens, J. Berezovsky, J. P. McGuire, L. J. Sham, A. C. Gossard, and D. D. Awschalom, Phys. Rev. Lett. **93**, 097602 (2004).
[109] A. M. Bratkovsky and V. V. Osipov, J. Appl. Phys. **96**, 4525 (2004).
[110] R. H. Silsbee, Bull. Magn. Reson. **2**, 284 (1980).

[111] I. Žutić and J. Fabian, Mater. Trans., JIM **44**, 2062 (2003).
[112] J. F. Gregg, R. P. Borges, E. Jouguelet, C. L. Dennis, I. Petej, S. M. Thompson, and K. Ounadjela, J. Magn. Magn. Mater. **265**, 274 (2003).
[113] S. Datta and B. Das, Appl. Phys. Lett. **56**, 665 (1990).
[114] W. Shockley, M. Sparks, and G. K. Teal, Phys. Rev. **83**, 151 (1951).
[115] J. Fabian and I. Žutić, Phys. Rev. B **69**, 115314 (2004).
[116] S. Dimitrijev, *Understanding Semiconductor Devices* (Oxford University Press, New York, 2000).
[117] N. Lebedeva and P. Kuivalainen, J. Appl. Phys. **93**, 9845 (2003).
[118] M. E. Flatté, Z. G. Yu, E. Johnston-Halperin, and D. D. Awschalom, Appl. Phys. Lett. **82**, 4740 (2003).
[119] S. Bandyopadhyay and M. Cahay, Appl. Phys. Lett. **86**, 133502 (2005).
[120] M. Field, private communication (2004).
[121] J. Fabian, I. Žutić, and S. D. Sarma, Appl. Phys. Lett. **84**, 85 (2004).
[122] J. Fabian, I. Žutić, and S. Das Sarma, cond-mat/0211639.
[123] J. Fabian and I. Žutić Acta Physica Polonica A **106**, 109 (2004).
[124] J. Fabian and I. Žutić Appl. Phys. Lett. **86**, 133506 (2005).
[125] A. M. Nazmul, S. Sugahara, and M. Tanaka, Phys. Rev. B **67**, 241308 (2003).
[126] S. J. Pearton, C. R. Abernathy, M. E. Overberg, G. T. Thaler, D. P. Norton, N. Theodoropoulou, A. F. Hebard, Y. D. Park, F. Ren, J. Kim, *et al.*, J. Appl. Phys. **93**, 1 (2003).
[127] D. P. Young, D. Hall, M. E. Torelli, Z. Fisk, J. L. Sarrao, J. D. Thompson, H.-R. Ott, S. B. Oseroff, R. G. Goodrich, and R. Zysler, Nature **397**, 412 (1999).
[128] H. R. Ott, J. L. Gavilano, B. Ambrosini, P. Vonlanthen, E. Felder, L. Degiorgi, D. P. Young, Z. Fisk, and R. Zysler, Physica B **281/282**, 423 (2000).
[129] H. J. Tromp, P. van Gelderen, P. J. Kelly, G. Brocks, and P. A. Bobbert, Phys. Rev. Lett. **87**, 016401 (2001).
[130] M. C. Bennett, J. van Lierop, E. M. Berkeley, J. F. Mansfield, C. Henderson, M. C. Aronson, D. P. Young, A. Bianchi, Z. Fisk, F. Balakirev, *et al.* Phys. Rev. B **69**, 132407 (2004).
[131] A. G. Aronov, Zh. Eksp. Teor. Fiz. **71**, 370 (1976), [Sov. Phys. JETP **44**, 193-196 (1976)].
[132] J. Zhang, H. Tanaka, and T. Kawai, Appl. Phys. Lett. **80**, 4378 (2002).
[133] H. Tanaka, J. Zhang, and T. Kawai, Phys. Rev. Lett. **88**, 027204 (2002).
[134] N. Nakagawa, M. Asai, Y. Mukunoki, T. Susaki, and H. Y. Hwang, Appl. Phys. Lett. **86**, 082504 (2005).

3 Probing and manipulating spin effects in quantum dots

S. Tarucha, M. Stopa, S. Sasaki, and K. Ono

3.1 Introduction and some history

Information technology and indeed all electronics have historically been based primarily on the charge-carrying properties of the electron more than its spin and magnetic properties. Such is also the case with quantum dots. Initially quantum dots were, essentially by definition, circumscribed regions of a discrete number of electrons (specifically conduction band electrons) whose sole good quantum number was the number of electrons N on the dot. The isolation of the electrons into one small, spatial region of a semiconductor crystal (see Fig. 3.1) was accomplished by means of sophisticated epitaxial techniques for growing very pure crystals with well-defined material interfaces and conduction band offsets, together with advanced microfabrication techniques for depositing metal gates on the surface of semiconductor wafers in order to modulate the potential landscape within the semiconductor. When, with these techniques, an isolated "box" of electrons is formed such that classically inaccessible (i.e. tunnel) barriers exist all around, the box has a most-stable electron number N_0, just as an atom, with a given nuclear charge Z, has a maximally stable electron number.

In the atomic case, of course, Z is always an integer and the stable electron number is equal to that nucleon number. One could imagine, however, Z taking fractional values whereupon the stable electron number would increase as Z passed through certain transition points close to half-integer values.[9] Further, when Z was exactly at one of those transition values, neighboring charge states of the atom, $N_0 = Z, Z+1$, would be degenerate. This is the principle behind the phenomena of "Coulomb blockade " and "Coulomb oscillations" in quantum dots.

For dots, the "nuclear" charge is actually a background positive charge consisting of ionized donor atoms and charges on nearby surface electrostatic gates as well as effective charges related to things like band offsets and piezoelectric effects. This effective background positive charge, often denoted ρ_0, (1) is not necessarily integer and (2) can be tuned by varying the voltages on the gates. When this effective background charge is not at a transition point, the addition or subtraction of an electron requires electrostatic energies which are the quan-

[9] Here we say "close to" half-integer because the spacing of the quantum levels and their energies also affect the charge degeneracy points.

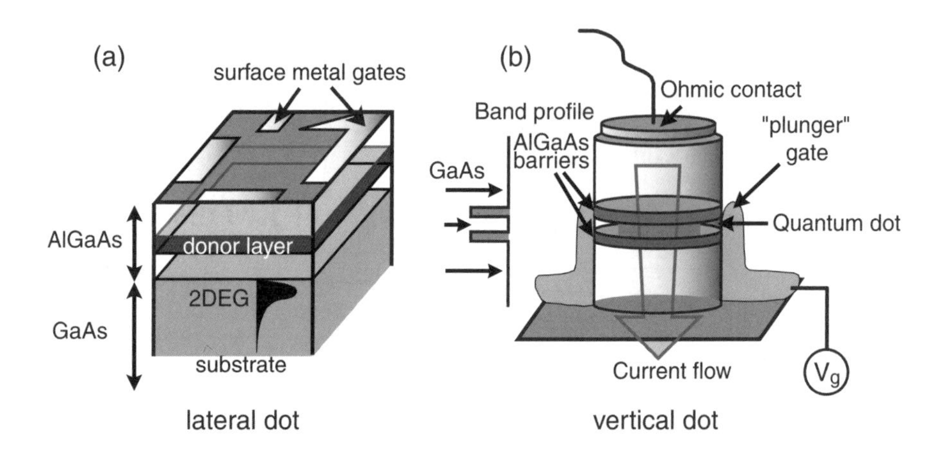

FIG. 3.1. Two standard configurations for semiconductor quantum dots. (a) Lateral dot: a two-dimensional electron gas (2DEG) heterostructure grown as a modulation doped GaAs wafer capped with 10 nm of AlGaAs is plated, via e-beam lithography, with surface metal gates. A negative bias applied to these gates depletes the electron gas underneath leaving a puddle of electrons connected, via quantum point contact tunnel barriers, to 2DEG leads. (b) Vertical dot: a double barrier heterostructure with n-doped, 3DEG leads is etched into a submicron size pillar and an ohmic contact is placed on top (via an air bridge or a narrow mesa). Current flows vertically between the top contact and a conducting substrate. A gate surrounding the pillar acts as a plunger to vary the electron number.

tum dot equivalents of the electron affinity and ionization potential, respectively, for atoms. In quantum dot physics, the addition energy is known as the charging energy E_C and is often parameterized in terms of some effective capacitance matrix for the dot [1]. For the effective background charge exactly at the transition point the ground state is degenerate with respect to N, modulo the chemical potential of the leads, i.e. $E(N_0+1) - E(N_0) = \mu_l$.

The principal experimental signatures of the underlying physics which we are interested here in describing fall into the category of *transport* measurements. In contrast to optical studies of quantum dots which typically operate by creating overall neutral electron pairs, transport studies necessarily place electrons onto dots and remove them, and the charge therefore fluctuates. Therefore, it is essential that, unlike atoms, quantum dots are typically surrounded by Fermi seas of electrons that constitute "leads" to the dot. At the degeneracy points of neighboring charge states, electrons are available to tunnel into and off the dot, changing the number N of electrons from $N = N_0$ to $N_0 + 1$ and back again. If a small bias exists across the dot ("source-drain bias"), a current will then preferentially flow, across the dot, in the direction of the bias. This is known as a "Coulomb oscillation" in contrast to the values of ρ_0 where a particular value of N is stable,[10] which are called Coulomb blockade regions. Note that the Coulomb blockade regions also require that the intrinsic quantum level *widths* due to tunneling to the leads, with typical element written as $\hbar\Gamma$, are smaller than the charging energy E_C (explained below). This is nothing more than the condition that the dot is closed.

The Coulomb blockade is central to the relevance of quantum dots for spintronics applications. It holds whenever the size of the quantum dot is small enough that the energy of adding or subtracting an electron, E_C, is greater than k_BT and $\hbar\Gamma$. For low-temperature experiments it is now easily possible for charging energies to be much larger than k_BT and since Γ is modulated by gates (at least in lateral dots, see below) it can also be made as small as desired.

As noted above, Coulomb oscillations probe the border between Coulomb blockade regions. In general transport studies examine the ways in which electrons transit from source to drain, through the "system," under a variety of conditions. We will introduce experiments relating to the variation of several experimental conditions, including a gate voltage V_g (standard Coulomb oscillations), a magnetic field B, the source-drain voltage, V_{sd}, (non-linear regime) and, in the case of double quantum dots, a second gate voltage that biases the second dot. These data can be visualized in various ways, some of which have become standard in the field, including (1) the usual conductance G $(= dI/dV_{sd})$ versus gate voltage V_g; (2) G as a gray scale versus V_g and B; (3) G as a gray scale versus V_g and V_{sd} (Coulomb diamonds) and, in the case of double dots, (4) G as a gray scale in the plane of the two gate voltages V_{g1} and V_{g2} (honeycomb stability diagram). In each case one is probing the chemical potential of the system (dot or dots) and whether it is between that of the source and the drain; or else whether higher order tunneling processes (cotunneling) are capable of connecting source and drain via intermediate virtual states in the system.

In many semiconductor quantum dots the individual (typical) quantum level spacing, Δ, is greater than the k_BT and $\hbar\Gamma$ but smaller than the charging energy: $k_BT, \hbar\Gamma < \Delta \ll E_C$. In this case the microscopic state is well-defined in quantum dots. The condition for electron transport must now include not only the electron numbers of the two states, but also the quantum states. In other words, when $E_p(N_0 + 1) - E_q(N_0) = \mu_l$, where p and q label all of the quantum numbers for the two states, a transition in the dot can occur that changes the electron number from N_0 to $N_0 + 1$. Nevertheless, Coulomb oscillations in particular are assumed to occur between the ground states (i.e. $p = 0$ and $q = 0$) of the dot for the two electron numbers. The reason for this is that while two excited states for the two charge numbers may be degenerate modulo the chemical potential of the leads, generally the relaxation time within the dot due to, for example, phonon coupling is assumed to be much smaller than the time between tunneling events between the leads and the dot. The result is that the initial state of a transition

[10] Here, by stable, we mean that the charging energy is greater than both the source-drain bias and the temperature.

is usually the ground state for a particular electron number. However, as we will see in the case of Coulomb diamonds below, transitions to excited states of the next electron number are possible and frequently observed in the non-linear regime. Also, there are exceptions to the above principle when a metastable state of the dot can occur and the signatures of excited to ground or excited to excited state transitions can be observed [2].

In this chapter we concentrate on the concept of spin electronics. The driving motivation behind "spintronics" is to control the spin state of a device or system and, further, to control its transport through the system. We generally think of the spin as a magnetic entity associated with some kind of internal current within the electron. Due to the tiny size of the single Bohr magneton, however, the interaction of this magnetic quantum with those of neighboring electrons is extremely weak. This property gives spin its long-lived nature, that is, an electron will preserve its spin quantum number in a disordered, inhomogeneous system much longer than it will preserve other quantum numbers (e.g. momentum and energy). This same isolation from its environment makes spin difficult to manipulate, however. Frequently, the best method for manipulating spin lies in the deep connection found in quantum electrodynamics between the spin of a particle and its statistics. Since the electron has spin 1/2, it is necessarily a Fermion and obeys Fermi statistics. This implies the existence of an exchange interaction between spins which is *electrostatic* (rather than magnetic) in origin. Many of the phenomena we discuss in this chapter, such as the transition between the singlet and triplet states for pairs of electrons, originate in this exchange effect.

Finally, while control of large numbers of electrons in bulk semiconductors via optical means have recently been demonstrated, we concentrate here on the most elementary systems which we can fabricate, i.e. the smallest quantum dots or sets of quantum dots, so as to explore the spins, to the extent possible, one at a time. While this approach may lead to concepts for future devices, its principal benefit and reason for existence is the understanding it provides of electron behavior at a fundamental level.

3.2 Charge and spin in single quantum dots

3.2.1 *Constant interaction model*

For spintronics applications, the principal problems are to distinguish and then manipulate the spin state of the dot. This was still difficult for intermediate size dots, however, since even for these somewhat smaller dots the confining potential contour defined by the gates is irregular and furthermore the dot devices invariably contained disorder due to the donor impurities employed to generate the electron gas. As a consequence much of the study of the quantum state of dots at this stage was undertaken from a statistical point of view. It was conjectured that the presumably chaotic nature of the underlying classical dynamics of quantum dots resulted in the spectral properties of these dots being equivalent to those of diffusive systems [3]. In this phase, the spectral properties of quantum dots were analyzed with "random matrix theory" (RMT), which eschews a computation of the specific properties of a given system in favor of statistical predictions about level and wave function *distributions*, based on the class of Hamiltonian (orthogonal, unitary or symplectic) to which the system belongs. Thus initial calculations concerned the distribution of spin states, and derivative properties, resulting from the interplay of fluctuating level spacing and fluctuating exchange interaction as some external parameter such as a gate voltage or magnetic field was varied; in effect the RMT, mesoscopic extension of the Stoner criterion [4, 5]. Experimentally the principal means of access to these and other properties of the quantum states of the dots was the analysis of the statistics of Coulomb peak spacings [6, 7].

While RMT and semiclassical methods applied to quantum dots provided a considerable amount of interesting physics, technologically these approaches were rendered obsolete by the development of "vertical" quantum dots; see Figure 3.1(b), whose confining potential profile was much better controlled than the earlier, 2DEG "lateral" dots (a) and which, furthermore, could be gated down to a few and even zero electrons [8, 9]. As shown in Fig. 3.1(b), vertical dots are defined between epitaxially grown barriers (InGaAs between layers of AlGaAs) which separate the dot from 3DEG n-GaAs leads. The confinement in the plane of the dot is achieved by etching the initial double barrier wafer into a pillar and, furthermore, surrounding it with metal gates. Since the heterostructure barriers are (nominally) perfect to within an atomic layer and since the lateral confinement is provided by the Schottky barriers and gates at distant surfaces, and is therefore very smooth, the potential profile of vertical quantum dots is unusually regular. Many experiments have shown, in fact, that the potential can be represented, to good accuracy, by a circularly parabolic potential and, in the growth direction, confined to a single vertical mode (the so-called "electric quantum limit"). In the case of significant, intentional anisotropy in the confining potential the lateral potential can be considered to be elliptically parabolic. The one-electron Schrödinger equation for a 2D parabolic potential can be solved analytically and the solution is known as the "Fock–Darwin" spectrum [10] (Fig. 3.3a). This will be employed and discussed below.

The chemical potential of a quantum dot, which is what is measured by Coulomb oscillations, is often expressed in the so-called "constant interaction" (CI) model [11]. This model employs effective capacitances to parameterize the electrostatic energy of the dot and assumes no coupling between the electrostatic structure and the quantum mechanical spectrum, hence it is fairly radical, particularly for small N. Nonetheless it is conceptually very useful. One writes:

$$E\left(N,\{n_p\}\right)=\sum_{p=1}^{\infty} n_p\epsilon_p+\frac{N^2e^2}{2C}+eN\alpha V_{\mathrm{g}}, \tag{3.1}$$

where n_p is the occupancy of single-particle level p with energy ϵ_p, C is the dot self-capacitance (total capacitance to ground) and $\alpha = C_{\mathrm{g}}/C$ is the so-called "lever arm" of the principal gate (often called the "plunger gate"). Note that

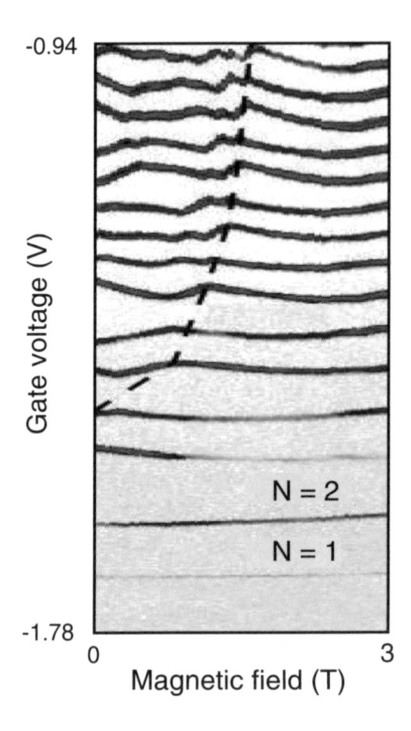

FIG. 3.2. Conductance ($G = dI/dV_{\mathrm{sd}}$) as a gray scale in the plane of magnetic field B and gate voltage V_{g}. Peaks of G are Coulomb oscillations which indicate when the chemical potential of the dot equals that of the leads (which have a very small source-drain bias). Spacing between oscillations includes a charging energy but a major portion of the evolution of the Coulomb oscillations comes from the single-particle level at the Fermi surface as approximated by the Fock–Darwin spectrum.

the quantum state is here given by the set of occupancies of a set of single-particle levels (which are assumed to be known). The chemical potential must be defined carefully, since addition and subtraction of an electron generally do not require the same energy. Further, one should, unless there is no question of ambiguity, specify the initial and final energies by their electron numbers and their quantum states as well (be they ground state to ground state or involve an excited state). In the zero temperature (T) limit, where $n_p = 1$ for $p = 1$ to N and zero otherwise, the chemical potential is

$$\mu(N, N+1) = \epsilon_N + 1 + \frac{(2N+1)e^2}{2C} + e\alpha V_{\mathrm{g}}. \tag{3.2}$$

Since the dot chemical potential is equal to that of the leads in the small V_{sd} limit at the Coulomb oscillation, this expression implies that the Coulomb oscillation traces the single-particle level energy, since the charging term (the second term)

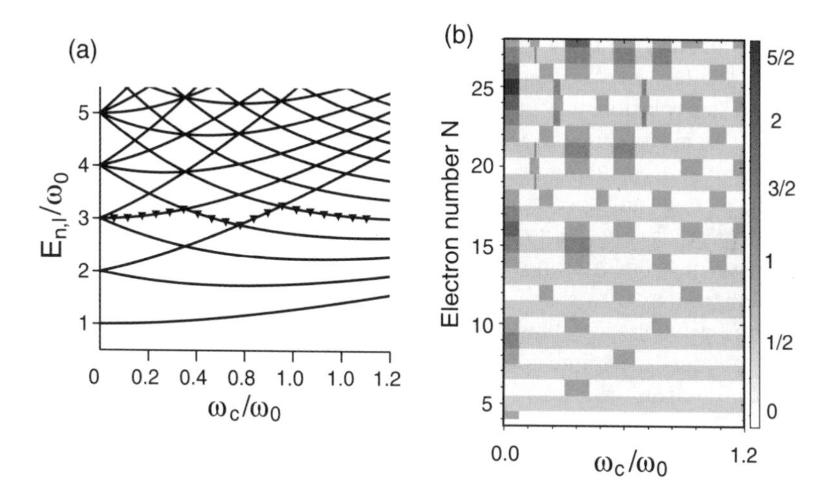

FIG. 3.3. (a) Fock–Darwin spectrum of a (2D) circularly parabolic potential in a transverse magnetic field. Each level is spin degenerate. Triangles indicate evolution of the Fermi level when $N = 10$. (b) Fock–Darwin + X (FD+X) plot of electron spin as a gray scale versus magnetic field and N, obtained from minimizing Eq. (3.5) with respect to $N_\uparrow$. For $B = 0$ spin maximizes for half-filled shells ($N = 4, 9, 16, 25$, etc.). Degeneracies in FD spectrum in (a) at non-zero B correspond to enhanced spin (beyond 0 or 1/2).

gives a constant increment with N. This is seen in Fig. 3.2, where a typical plot of conductance through a vertical dot as a function of V_g and B is shown [12,13]. As noted above, the single particle energy levels are given approximately by the Fock–Darwin spectrum, where each energy level, $E_{n,l}$, has a radial quantum number n and an angular momentum quantum number l:

$$E_{n,l} = -\frac{l}{2}\hbar\omega_c + \left(n + \frac{1}{2} + \frac{1}{2}l\right)\hbar\sqrt{4\omega_0^2 + \omega_c^2}, \tag{3.3}$$

where $\hbar\omega_0$ is the lateral confinement energy and $\hbar\omega_c = eB/m^*$ is the cyclotron frequency (m^* is the effective mass). The lowest few levels of the Fock–Darwin spectrum are plotted versus B in Fig. 3.3(a).

3.2.2 *Spin and exchange effect*

As we will discuss later, the spin state of electrons in quantum dots is very robust and the spin relaxation and decoherence times are surprisingly long. The reason for this is simply that the spin interacts very weakly with its environment. We will see that the main mechanisms that shorten spin lifetime in GaAs are spin-orbit interaction and hyperfine coupling of conduction electron spin to nuclear spins. Naturally, when a sufficiently large external magnetic field is present, the Zeeman effect splits the otherwise degenerate energy levels in a quantum dot.

Nevertheless, the magnetic field *produced* by the electron and therefore the magnetic dipole interaction between electrons is extremely small. The result of this is that by far the largest spin-dependent term in the interaction between electrons arises from the relationship between spin and statistics, namely the exchange interaction. As is well known, the exchange interaction depends on spin because the Pauli exclusion principle prevents electrons of parallel spin from approaching too closely to one another. This lowers the Coulomb interaction between electrons with parallel spins and can be considered to be the leading correction to mean field theory. For a homogeneous electron gas the gain in energy resulting from spin alignment competes with the loss in energy required to occupy higher energy plane-wave states. This leads to the "Stoner criterion" for spontaneous polarization where exchange energy must exceed level spacing. In a confined system such as an atom or a dot, the Pauli principle and the exchange energy also lead to a tendency for spins to align; however, this can generally only occur when there is a degeneracy, or near-degeneracy in the spectrum, since strong confinement makes the level spacing too large to be overcome by exchange. Thus, in atoms, an empirical rule known as Hund's rule, states that for a partially filled shell (of orbitally degenerate states) the spin will take on its maximum value consistent with the Pauli principle. As a concrete example, if a shell with eight degenerate levels is filled with five electrons, four will be spin up and one will be spin down, for a net spin $S = 3/2$. Precisely the same rule is expected to hold in quantum dots insofar as the confining potential is regular enough to produce sufficiently precise degeneracies (relative to the scale of the exchange energy). A look at Fig. 3.3(a) shows that the Fock–Darwin spectrum has many degeneracies occurring at particular values of ω_c/ω_0 including, notably, zero. The exchange interaction for two electrons in orbitals ψ_p and ψ_q is:

$$V_{ex} = \int d\mathbf{r}_1 \int d\mathbf{r}_2 \psi_p^*(\mathbf{r}_1)\psi_q^*(\mathbf{r}_2)V(\mathbf{r}_1,\mathbf{r}_2)\psi_q(\mathbf{r}_1)\psi_p(\mathbf{r}_2), \tag{3.4}$$

where $V(\mathbf{r}_1, \mathbf{r}_2)$ is the (generally screened) Coulomb interaction. A drastic approximation to the interacting electrons in a parabolic dot is to assume that the direct Coulomb interaction is entirely incorporated into the bare parabolicity of the dot and that the exchange term in Eq. (3.4) is independent of level indices. A total energy functional for the dot, employing the Fock–Darwin spectrum and a phenomenological exchange constant V_x, can then be written (at $T = 0$) as:

$$E\left(N_\uparrow, N_\downarrow\right) = \sum_{p=1}^{N} \epsilon_p - \frac{V_x}{2}\left(N_\uparrow(N_\uparrow + 1) + N_\downarrow(N_\downarrow + 1)\right), \tag{3.5}$$

where $N_{\uparrow(\downarrow)}$ is the total number of electrons in the dot with spin up (down) and $N = N_\uparrow + N_\downarrow$; ($p$ here is shorthand for the Fock–Darwin quantum numbers). Minimizing this expression with respect to $N_\uparrow$ produces a polarization (total spin) as a function of total energy and magnetic field which is illustrated in Fig. 3.3(b). While the symmetry of the potential is rarely perfect and verification of spin states with $S > 3/2$ is difficult, nevertheless this simple model, Fock–Darwin + Exchange (FD+X) provides a reasonable conceptual picture of where spin states with S different from zero or one-half are expected to be found. In particular, experimental evidence relating to the shape of the Coulomb oscillations has been found for various magnetic field-induced degeneracies with relatively small N, a case which we discuss next.

3.3 Controlling spin states in single quantum dots

3.3.1 *Singlet-triplet and doublet-doublet crossings*

Reproducible manipulation of interacting two-body states is essential to the realization of the quantum computing paradigm as well as being a means of studying electron-electron interaction [12–14]. The FD+X model suggests that spin states beyond $S = 0, 1/2$ should be common in quantum dots. How does the transition with gate voltage or magnetic field (or some other external parameter) between different spin states manifest itself in the transport data, specifically in the Coulomb oscillations? The signature of these states is best analyzed in the simplest case: the transition, for N even, between $S = 0$ and $S = 1$, which occurs when two spatial orbitals, *at the Fermi surface*, cross in energy as a function of magnetic field B (Fig. 3.4). Since N is even, when the levels are well separated one orbital is doubly filled and the other is empty. Near the degeneracy point, however (the anticrossing and Zeeman energies are negligible here) the direct Coulomb and exchange energies favor parallel spins, as in Hund's rules, and each orbital becomes occupied by a single electron.

Our purpose is to determine the change to the Coulomb oscillations traces that occur in response to this singlet-triplet cross-over and show how this behavior is verified experimentally. This requires an analysis that goes beyond the constant interaction model; so that now we do not describe the charging energy with capacitances but rather treat the various energies in terms of the direct and exchange Coulomb matrix elements that make up the many-body energies. To proceed, we treat all but the *two uppermost electrons* as an inert core; the electrochemical potentials are the differences between the one-electron ground state and the ground and excited states of two electrons

$$\mu_i = U_i(2, S) - U(1), \tag{3.6}$$

where S denotes the total spin. With E_a and E_b the bare (B-dependent) levels (Fig. 3.4), the possible values of $U_i(2, S)$ (i =1-4) are

$$U_1(2, 0) = 2E_a + C_{aa}, \tag{3.7a}$$

$$U_2(2, 0) = 2E_b + C_{bb}, \tag{3.7b}$$

$$U_3(2, 1) = E_a + E_b + C_{ab} - |K_{ab}|, \tag{3.7c}$$

$$U_4(2, 0) = E_a + E_b + C_{ab} + |K_{ab}|, \tag{3.7d}$$

where C_{ij} is the direct Coulomb interaction between electrons in levels i and j and K_{ab} is the exchange energy between a and b for electrons with parallel spins.

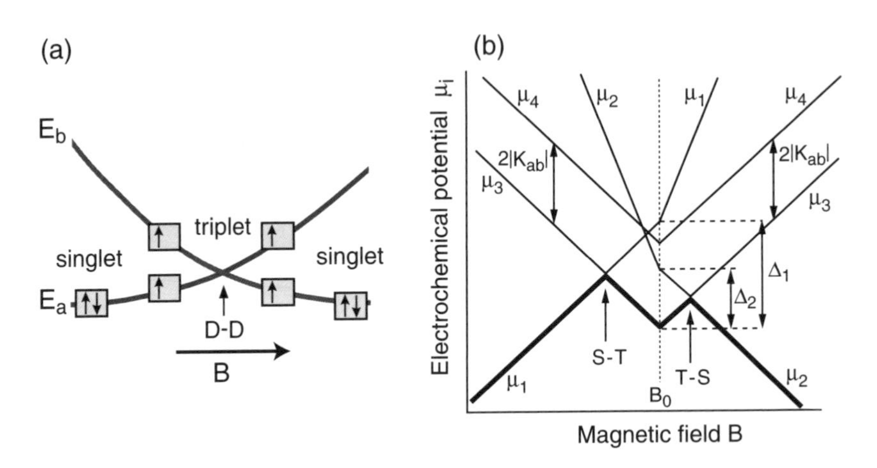

FIG. 3.4. (a) Schematic of level crossing between two dot orbitals with energies E_a and E_b induced by a changing magnetic field B. The two orbital states (single-orbital states) cross each other at $B = B_0$. This is the point of crossing between two doublet states (D-D crossing) for one-electron state. For two electrons, far from the crossing, both electrons occupy the lower orbital. Close to the crossing, the gain in direct Coulomb and exchange energy makes single occupancy of each orbital energetically favorable, with the spins aligned. (b) Electrochemical potential, $\mu_i(2) = U_i(2) - U(1)$, for two interacting electrons. The thick line depicts the ground state energy, whereas the thin lines show the excited states. Near the crossing point, the spin triplet state with $\mu_3(2)$ becomes the ground state.

A schematic illustration of the evolution of the chemical potentials as E_a and E_b vary with magnetic field B is shown in Fig. 3.4. The two orbital states with E_a and E_b cross with each other at $B = B_0$. This is the point of doublet-doublet (D-D) crossing. On both sides of B_0 the singlet-triplet (S-T) transition produces a downward cusp in the ground state chemical potential μ_1, which, as we have seen, is equivalent to the position of the Coulomb oscillation peak. An example of experimental data is shown in Fig. 3.5.

This figure shows Coulomb peaks for $N = 7$ to 16 evolving with B field measured for a vertical InGaAs quantum dot [13]. The Coulomb peaks generally shift in pairs with B, due to the lifting of spin degeneracy. So from the shift of the paired peaks on increasing B, quantum numbers (n, l) in the Fock–Darwin states are assigned to the respective pairs. The wiggles or anticrossings between pairs of peaks correspond to the crossings of the Fock–Darwin states (Fig. 3.5c). Clear signatures of a D-D crossing, and two S-T transitions, as illustrated in Fig. 3.4(a), and (b), respectively, are observed in each of the dotted ovals connecting pairs of peaks at non-zero B field (Fig. 3.5a). Two examples of the cusp like structure of the oscillations, found generally for even N as B induces level crossings, can

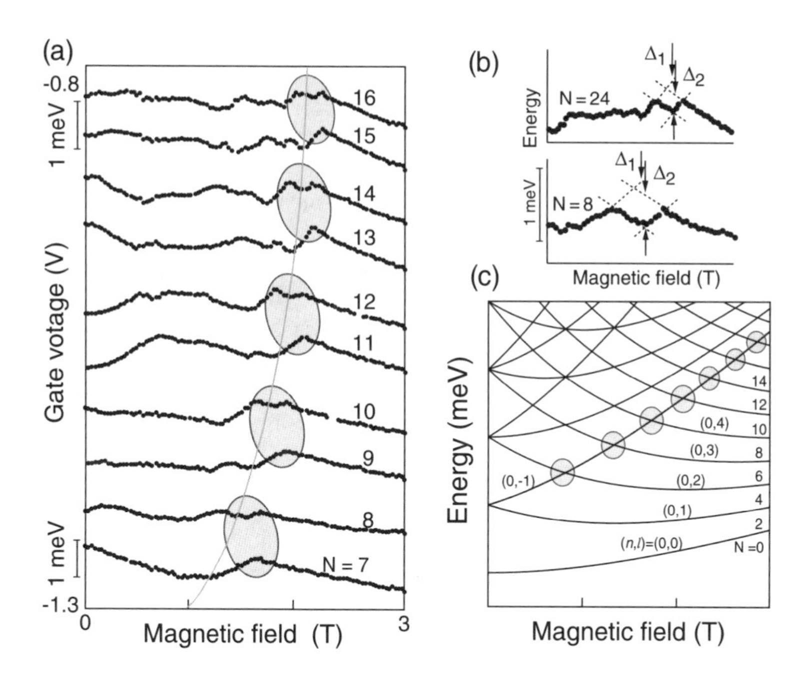

FIG. 3.5. (a) Evolution of the ground state energies from $N = 7$ to 16 as measured from the current peaks vs. magnetic field at $V_{\mathrm{sd}} = 120$ mV. The bars along the gate voltage axis show 1 meV energy scales calibrated at -1.26 and -0.85 V. The dotted curve indicates the last crossing between single-particle states. Dashed ovals correlate pairs of ground states for odd and even electron numbers. Spin transition in the ground states occurs in the ovals for $B \neq 0$ T. (b) Magnified plot of the $N = 8$ and 24 current peaks vs. magnetic field. The dashed lines illustrate how the interaction energy parameters Δ_1 and Δ_2 are determined. (c) Fock–Darwin single-particle states calculated for $\hbar\omega_0 = 2$ meV.

be seen in Fig. 3.5(b) at $N = 8$ and $N = 24$.

For further analysis, it is convenient to define two energies to characterize the cusp: $\Delta_1 \equiv \mu_1 - \mu_3 = C_{aa} - C_{ab} + |K_{ab}|$ and $\Delta_2 \equiv \mu_2 - \mu_3 = C_{bb} - C_{ab} + |K_{ab}|$. As shown in Fig. 3.5 (compare also Fig. 3.4 where the energies are defined graphically), these parameters can be determined from the extrapolation of the Coulomb oscillations in the vicinity of a S-T transition. We combine the experimental measure of these characteristic energies with a calculation of the spectrum based on the Fock–Darwin spectrum (Fig. 3.5c) and the direct and exchange Coulomb matrix elements to determine the relative strength of these effects as a function of the electron number N. The parameters Δ_1 and Δ_2 can also be calculated with the Fock–Darwin states. We find that the agreement between calculation and experiment of the N dependence of these parameters depends crucially on the assumption of screening in the Coulomb interaction in the dot [13]. An additional finding of note is that the exchange energy $|K_{ab}|$ falls off very rapidly with N, due to the shrinking overlap of different states.

3.3.2 *Non-linear regime for singlet-triplet crossing*

So far we have only considered the case where the source-drain bias is the smallest energy in the problem (or second smallest, after the temperature). In that case, the flow of current requires the near degeneracy of the two adjacent charge states (the ground states). In other words, the chemical potential to add an electron to the dot must be between the very closely spaced chemical potentials of the source and drain. If we allow for an expansion of the source-drain window then information on the excited states can be obtained from the transport measurement. In particular, in this so-called "non-linear" regime, if we have

$$\mu_s > \mu(N, N+1^*) > \mu(N, N+1) > \mu_d, \tag{3.8}$$

where $\mu(N, N+1^*)$ is the energy difference between the N electron ground state and some $N+1$ electron excited state, then we expect more current to flow than if only the ground-to-ground chemical potential is between the source and drain Fermi surfaces. Thus the feature to examine is the differential of the current with respect to the gate voltage.

Figure 3.6 shows such a $dI/dV_{\rm g}$ plot, taken for $\mu_s - \mu_d \equiv V_{\rm sd} = 2\,{\rm mV}$. This larger voltage opens a sufficiently wide window between the Fermi levels of the source and drain that both the ground state and the first few excited states can be detected. The spectrum for $N = 8$ compares well to Fig. 3.4(b) and we can clearly distinguish the parallelogram formed by the ground state and the first excited state. The downward cusp in the ground state for $N = 8$ is at slightly higher B than the upward cusp for the excited state. This asymmetry implies that $\Delta_1 > \Delta_2$, i.e., $C_{aa} > C_{bb}$.

The physics of level crossing for N even is more complicated than that for N odd. The reason is that, for N even, assuming the $N-2$ electrons below the Fermi surface are essentially an inert, spin-zero core, when the bare single-particle levels cross there are two electrons with four states to choose from. They tend to choose two spin parallel states to lower their exchange energy and also reduce the direct energy.[11] For N odd, however, the crossing of single-particle states at the Fermi surface (D-D crossing) results in four states again, but only a single "valence" electron that has to choose. The Coulomb oscillation in that case (again as a function of B) exhibits an upward cusp (see, for example, $N = 9$ in Fig. 3.5a), which is nothing more than the track of the lower single-particle orbital as the two orbitals cross. We will see later, however, that while the simple Coulomb oscillations exhibit an essentially trivial behavior for D-D crossing, the

[11] It should be noted that in addition to the exchange energy physics, the direct Coulomb repulsion of two electrons also depends on which spatial states they occupy and in particular the greatest repulsion will result from the two electrons occupying the same state (spin up and spin down).

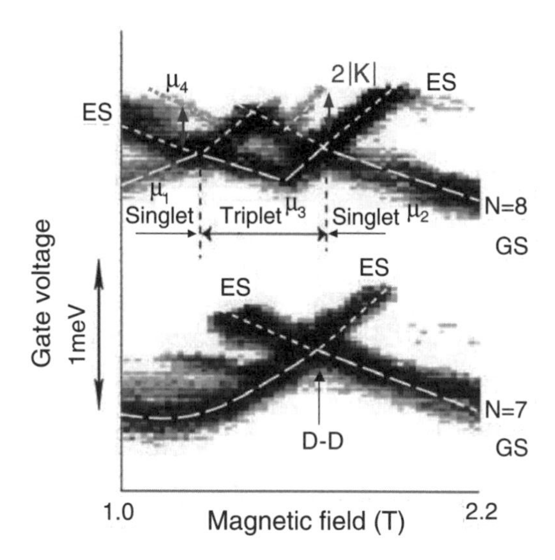

FIG. 3.6. dI/dV_g in the plane of V_g and B for $N = 7$ and 8 measured for $V_{sd} = 2$ mV. $dI/dV_g > 0$ for dark, $dI/dV_g < 0$ for white. The dashed lines indicate the evolution of the ground states with magnetic field, whereas the dotted lines show the excited states. The two dotted vertical lines indicate singlet-triplet (S-T) and triplet-singlet (T-S) transitions in the ground state for $N = 8$.

added degeneracy (from 2 to 4) at the crossing point does have a significant effect on the higher order tunneling processes that constitute the Kondo effect and its variations.

3.3.3 *Zeeman effect*

The coupling between an electron spin and an external magnetic field in a semiconductor quantum dot, and even in a bulk semiconductor, is more complicated than that of free electrons or electrons in atoms. For free electrons, the Pauli equation of relativistic quantum mechanics gives for the gyromagnetic ratio of the electron $\gamma_S = 2$ [15]. The experimentally observed value differs slightly from this due to radiative corrections. In atoms, the existence of spin-orbit coupling causes the energy splitting that results from a magnetic field to be proportional to the Landé g-factor:

$$g_J = \frac{J(J+1) + L(L+1) - S(S+1)}{2J(J+1)} + g_S \frac{J(J+1) - L(L+1) + S(S+1)}{2J(J+1)}, \tag{3.9}$$

where J, L and S are the total, orbital and spin angular momenta and the result is generally interpreted in the LS coupling scheme [16]. For s orbitals ($L = 0$) this expression reduces to $g_J = g_S$, however in crystals where the conduction band is formed from an s orbital (such as GaAs) spin-orbit coupling in the valence band can lead to a renormalization of g_c in the conduction band and the effect is not negligible. In particular, for the eight-band Kane model, which is used for most direct-gap semiconductors, the g-factor of the conduction band, denoted g_c, is energy-dependent and is given by

$$g_c(E) = g^* - \frac{2E_p}{3} \frac{\Delta}{(E_g + E)(E_g + \Delta + E)}, \tag{3.10}$$

where $g^* = g_S + \Delta g$ and Δg is a (generally small) contribution from remote bands; also, Δ is the spin-orbit splitting at the top of the valence band, E_g is the band gap, E_p is the so-called Kane parameter which is related to the coupling between the Bloch functions for the valence and conduction bands via the momentum operator and E is the energy measured relative to the conduction band edge [17]. In bulk GaAs, the g-factor for the conduction band is experimentally known to be $g_c = -0.44$, illustrating the considerable effect that the valence band plays.

While this result, Eq. (3.10), is established for bulk crystals, the significance to quantum dots arises from the energy dependence. In particular, the quantized energy levels of dots raise the electron's energy above the conduction band edge. Thus, quantum confinement alone is expected to affect the g-factor. In addition, the penetration of the electron wave function into material with a different intrinsic g-factor (such as AlGaAs) will cause an inevitable variation of the observed Zeeman splitting. Finally, the g-factor also contains terms related to the boundary conditions on the envelope functions at the interface of different crystalline materials. These latter effects are more pronounced in nanocrystals than in the lateral and vertical effective mass quantum dots which we study here. A more detailed discussion of these effects can be found in Ref. [17].

Direct experimental observation of Zeeman splitting of levels in quantum dots has only occurred relatively recently [18]. The principal significance of these observations to quantum information processing relates to the proposal to use the spin in a quantum dot as a scalable qubit [19]. In contrast to the previously described case of the S-T transition, the spin decay from $S_z = +1/2$ to $S_z = -1/2$ does not change the total angular momentum and therefore does not violate the selection rule $\Delta S = 0$. As we describe below, the relaxation time for the S-T transition has been measured to be of the order of hundreds of microseconds. In the case of Zeeman split levels a smaller, but still appreciable value of relaxation time T_1 has been found [18]. The experimental signature for the Zeeman splitting of a single level in an $N = 1$ quantum dot occurs in "Coulomb diamonds", which we now describe.

The non-linear bias regime described above provided data that was visualized as a plot of dI/dV_g in the plane of V_g and B. Coulomb diamonds, by contrast, designate a typically diamond-shaped feature that emerges when we plot dI/dV_{sd} (also called the differential conductance G) as a gray scale in the plane of V_{sd} and V_g. A typical Coulomb diamond plot for a vertical quantum dot is shown in Fig. 3.7. Along the $V_{sd} = 0$ axis the conductance is zero except at a set of

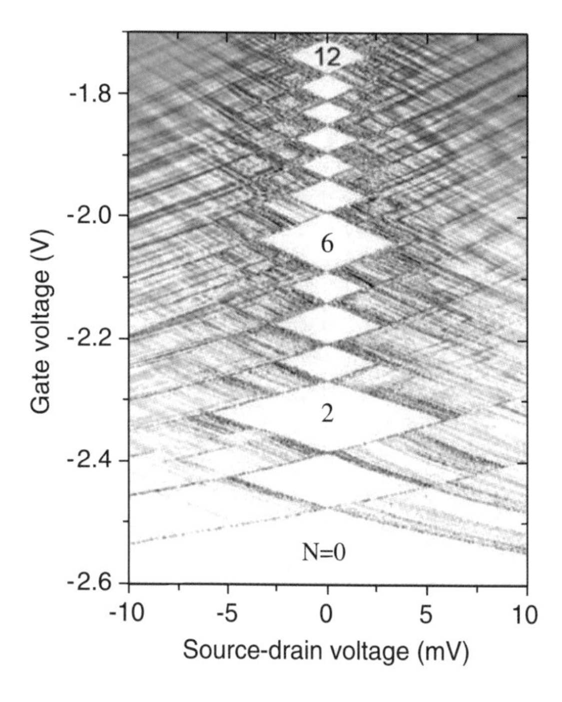

FIG. 3.7. Characteristic Coulomb diamond plot of conductance in the plane of the source-drain voltage $V_{\rm sd}$ and gate voltage $V_{\rm g}$. Along $V_{\rm sd} \approx 0$, G is non-zero only for isolated points of $V_{\rm g}$ corresponding to Coulomb oscillations. N changes in integer steps at these points. The threshold $V_{\rm sd}$ is a function of $V_{\rm g}$, greatest at the diamond centers. The diamond slopes reveal the capacitance ratio (see Eq. 3.11). Lines parallel to the diamond slopes reveal excited states.

isolated points which correspond to values of $V_{\rm g}$ where a degeneracy of two charge states (ground states) occurs. These are the standard linear regime Coulomb oscillations. For values of $V_{\rm g}$ away from the Coulomb oscillations, however, a current can be made to flow if a sufficient source-drain bias is provided, thereby overcoming the Coulomb blockade. It can be easily shown in the capacitance model for the charging energy (even completely ignoring the single-particle levels) that the onset of G as a function of $V_{\rm g}$ and $V_{\rm sd}$ is linear [1] with a slope given by

$$\frac{\partial V_{\rm g}}{\partial V_{\rm sd}} = \frac{C_{\rm dd} - C_{\rm ds}}{C_{\rm dg}}, \tag{3.11}$$

where $C_{\rm dd}$ is the dot self-capacitance, $C_{\rm ds}$ is the capacitance between the dot and a lead (here taken as the source) and $C_{\rm dg}$ is the dot-gate capacitance. The derivation of this expression begins from writing the full energy within the capacitance model [1] and demanding that $\mu(N, N+1) = \mu_s$. This means, in particular, that the capacitances refer to the energies of adding the $(N+1)$st electron and could, in principle, be dependent on the single-particle state into which this final electron is added. Specifically, the electron cloud for this final charge could be closer to or farther from the gate or lead than other electrons. The upshot is that the slopes of the Coulomb diamonds could be expected to show some variation according to the quantum state at the Fermi surface. In the following, however, we can take the diamond slopes as constant and given simply by the gross geometry of the dot.

As noted above, recent advances in lateral dot gate design[12] have led to the development of dots whose electron number can be tuned down to zero. The practical difficulty in achieving this result arises from the variation of the barrier thickness between the dot and the leads as a function of N in lateral dots. (In vertical dots, by contrast, the barriers between dot and leads are established during crystal growth.) Thus, while it was always possible to pump the electron number in a dot down to an arbitrarily small value, current would cease to flow at some point and no further evidence of the charge state could be observed.

Employing such a small N lateral dot, recent experiments by Hanson *et al.* [18] have revealed, in the $N = 1$ Coulomb diamond, the signature of an excited state whose energy, relative to the ground state, is proportional to B (Fig. 3.8). This excited state appears in the Coulomb diamonds as a line parallel to the diamond edge but outside the diamond (i.e. at greater absolute bias). The clear indication is that $dI/dV_{\rm sd}$ shows a jump when an additional, excited quantum state of the $N+1$ electron dot enters into the bias window. Such a parallel excited state line was clearly observed in Fig. 3.8 for $N = 1$. Of critical importance to the experiments, however, is the fact that the magnetic field was in plane. Since the dot is quantum mechanically two dimensional the orbital angular momentum is almost perfectly perpendicular to the plane of the dot. Therefore, the only coupling possible for B in the plane of the dot is to the spin of the electrons. Thus the evolution of the energy of the excited state as a function of B gives a direct measurement of the Landé g-factor for the dot. The results in Fig. 3.8 demonstrate that the g-factor has a clear B-dependence as

$$|g(B)| = (0.42 \pm 0.04) - (0.0077 \pm 0.0020)B\,[\mathrm{T}]. \tag{3.12}$$

The magnetic field dependence of g is not yet clearly understood. Equation (3.11) could provide such a dependence if the energy or the Kane parameter depended sufficiently on B. Other possible origins include nuclear polarization effects (resulting in a smaller than expected effective magnetic field) or else hyperfine coupling of the electron spin to nuclear spins. We refer the reader to the original paper for details [18]. Similar experiments of Coulomb diamonds have been performed on GaAs vertical quantum dots to derive the g-factor of $|g| = 0.23$ [20].

In addition to exhibiting in the non-linear Coulomb diamonds the Zeeman split level of the one-electron quantum dot and its dependence on B, the same

[12] Note that very recently control of lateral dot confinement has improved to the point where $N = 1$ or 0 are reproducibly accessible [9].

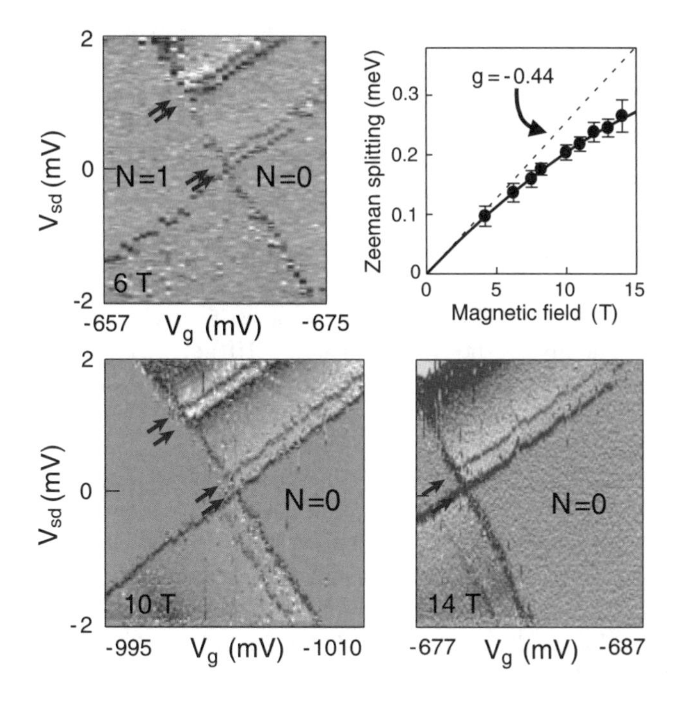

FIG. 3.8. dI/dV_{sd} as a function of V_{sd} and V_g near the $0 \sim 1$ transition in a lateral quantum dot, at in-plane B fields of $B = 6$ T (a), 10 T (b), and 14 T (c), respectively. The darker corresponds to larger dI/dV_{sd}. The zero-field spin degeneracy is lifted by the Zeeman energy as indicated by arrows. (d) Extracted Zeeman splitting as a function of B. At high B fields a clear deviation from the bulk GaAs g-factor of -0.44 (dashed line) is observed.

experiment explores, using a particular gate pulse technique, the lifetime of a spin in the excited state T_1. These and other issues related to spin lifetime and decoherence will be discussed further below.

3.4 Charge and spin in double quantum dots

3.4.1 *Hydrogen molecule model*

We have thus far considered only single quantum dots connected to leads. A natural extension to this system, and one which has great significance for potential quantum computing applications, is the quantum dot molecule, the simplest example of which is the double dot. Double dots have been fabricated in a variety of configurations including both vertical and lateral geometries [21], as well as a hybrid vertical/lateral structure [22]. There have also been studies on InAs self-organized double dots, grown on top of one another in the Stranski–Krastanov mode, which have been placed in a vertical dot pillar [23]. For double dots, the current from source to drain flows either in parallel or in series. The literature on double quantum dots is already fairly large. We are not interested here in a full review of all of the experiments and theory on double dots, but instead refer the reader to the review article by van der Wiel *et al.* [24] Here we are concerned with the relevance of double dots to spin physics.

While the spin in a double quantum dot is controlled principally by the exchange interaction, the spatial state is affected by the competition between tunneling between the dots and Coulomb repulsion. The physics of the spatial state is best illustrated by considering the hydrogen molecule [25]. In H_2 the two electrons interact with each other in a double well potential produced by the two protons. Two possibilities for the two-body wave function of the electrons can be envisioned. The first is the Heitler–London (HL) ansatz [26] which treats the ground state as strongly correlated so that two electrons do not co-exist on the same nucleus:

$$\Psi^S_{\mathrm{HL}} = \frac{1}{2}\left[\phi_1(\mathbf{r}_1)\phi_2(\mathbf{r}_2) + \phi_2(\mathbf{r}_1)\phi_1(\mathbf{r}_2)\right]\chi_S, \tag{3.13}$$

where $\phi_{1,2}(\mathbf{r})$ are centered on nuclei 1 and 2 and χ_S is the two-electron spin singlet state. In contrast to this wave function, one could imagine solving the single-particle Schrödinger equation with the external potential given by the two nuclei and then filling the ground spatial state with two electrons, one spin up and the other spin down. This single-electron ground state will be the symmetric state $[\phi_1(\mathbf{r}) + \phi_2(\mathbf{r})]/\sqrt{2}$ with probability of a half to occupy either nucleus, and the properly symmetrized two-body wave function is:

$$\Psi^S_{\mathrm{HF}} = \frac{1}{2^{3/2}}\left[\phi_1(\mathbf{r}_1)\phi_1(\mathbf{r}_2) + \phi_1(\mathbf{r}_1)\phi_2(\mathbf{r}_2) + \phi_2(\mathbf{r}_1)\phi_1(\mathbf{r}_2) + \phi_2(\mathbf{r}_1)\phi_2(\mathbf{r}_2)\right]\chi_S, \tag{3.14}$$

where "HF" denotes Hartree–Fock. The difference between the HL and HF wave functions consists of the first and last terms in Eq. (3.14): the "ionic" configurations where both electrons are near the same nucleus. The HL ansatz represents the correct asymptotic form when the nucleus are assumed to be far apart, whereupon each nucleus will contain only a single electron. The HF ansatz is reasonable in the opposite limit where the two nuclei merge together. Empirically, the hydrogen molecule is found to more faithfully represent the Heitler–London state. The significant fact of quantum dots, however, is that the effective distance of the "nuclear centers" (actually the interdot barrier height) can be modulated by metal gates and also by a magnetic field B. Increasing the magnetic field has the effect of localizing electrons on their respective dots and is therefore similar to increasing the barrier height.

3.4.2 *Stability diagram of charge states*

Experimentally realized double quantum dots have only reached the regime of one electron per dot in the vertical, series transport case, which we discuss further below. In this case it is difficult to significantly tune the interdot barrier, since it

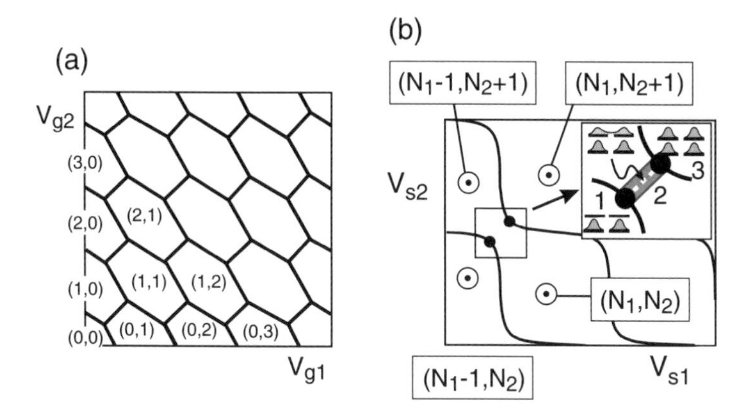

FIG. 3.9. (a) "Honeycomb" stability diagram for double dot in the quantized charge regime showing regions of stability for N_1 and N_2 (the electron numbers on the two dots) as a function of V_{g1} and V_{g2}, the gates biasing the dots. (b) Close-up of one vertex (closer still in the inset) is shown. The anticrossing region represents both a classical, Coulomb repulsion between the electrons on the two dots as well as a tunnel repulsion of the single-particle quantum levels in the two dots.

is established in the crystal growth. For multiple electrons per dot, a useful conceptual device for describing electron numbers N_1 and N_2 on dots 1 and 2 is the stability diagram, also called the honeycomb diagram. One assumes that each dot is biased by a gate and that that gate has the capacity to change the electron number on the dot (without appreciably affecting the other dot, although such "cross-capacitance" is in practice a real problem). If the dots are isolated from each other, the variation of each dot's gate, V_{g1} and V_{g2}, will result in the increase of N_1 and N_2 independently. If one plots the stable electron numbers as a function of V_{g1} and V_{g2}, then the stable islands will be simple rectangles. However, when the dots interact, either Coulombically or quantum mechanically, adding an electron to each dot simultaneously becomes energetically unfavorable. The result is (see Ref. [24] for details) that the areas of stability become hexagons and the pattern is that of a honeycomb (Fig. 3.9).

Note that the full honeycomb diagram does not emerge from transport data. Rather, the experimental signature depends on whether the dots are in parallel or in series. If they are in series, then in the linear source-drain bias regime, V_{g1} and V_{g2} must be set so that both dots are at a bi-stable point (i.e. a Coulomb oscillation) simultaneously. This only occurs at the vertices of the honeycomb. For the parallel configuration, on the other hand, current can flow through either dot when it alone is at a Coulomb oscillation, and so the data exhibit all the lines in the honeycomb where electron numbers on either dot are degenerate. This includes all lines except the diagonal, "anticrossing" lines, which represent charge transfer between the dots.

While each dot in a double dot can exhibit kinks in Coulomb oscillations corresponding to S-T crossings, we have not so far shown any behavior of spin, which is *intrinsic* to double dots. This requires an investigation of the exchange interaction *between* dots. Such physics is best illustrated in the artificial hydrogen molecule (two dots, two electrons), which, as noted, has not yet been realized in the lateral and/or parallel configuration. Nonetheless, calculations on such structures, which can be thought of as an artificial hydrogen molecule, have been done which demonstrate many of the relevant features of such structures.

3.4.3 *Exchange coupling in the scheme of quantum computing*

The main impetus for this study is the quantum computing proposal of Loss and DiVincenzo [19]. In this proposal, a single electron on a single dot is envisioned as the basic qubit. The spins on neighboring dots are made to interact in order to realize two-qubit operations. In particular, Loss and DiVincenzo showed that for a spin-spin interaction of the form $J\mathbf{S}_1 \cdot \mathbf{S}_2$, where J could be tuned by an interdot gate which modulates the interdot tunnel barrier, a spin-swap operation could be performed. Furthermore, when this two-dot operation was combined with rotations of single dot spins (via, for example, electron spin resonance) the so-called XOR (exclusive or) operation could be realized [27]. This XOR operation, combined with single qubit rotations, is sufficient to allow any quantum computing operation to be performed.

The form of the "exchange" interaction $J\mathbf{S}_1 \cdot \mathbf{S}_2$ is more easily understood by writing $\mathbf{S}_1 \cdot \mathbf{S}_2 = (\mathbf{S}_\mathrm{T}^2 - \mathbf{S}_1^2 - \mathbf{S}_2^2)/2 = (\mathbf{S}_\mathrm{T}^2 - 3/2)/2$, where $\mathbf{S}_\mathrm{T} = \mathbf{S}_1 + \mathbf{S}_2$. Thus, when the Hamiltonian is written with this as the only spin-dependent part, the J implicitly contains all terms that make the singlet energy (for $N = 2$) different from the triplet energy. While this term is called the "exchange" coupling, it actually includes terms other than the simple exchange integral (Eq. 3.4). This can be best illustrated within the "Hund–Milliken" formalism, which we now briefly describe.

We first assume that, as in real H_2, there is an external potential with two minima and a barrier in between. Various functional forms for this potential for the case of double dots have been used in calculations in the literature [27, 28]. Two single states localized about each dot are chosen as the basis, similar to ϕ_1 and ϕ_2 used for H_2 above. The Heitler–London state (Eq. 3.13) is used as one of three two-electron singlet basis states. Rewriting this with the other two basis states we have

$$\Psi^S_\mathrm{HL}(\mathbf{r}_1, \mathbf{r}_2) = \frac{1}{2}\left[\phi_1(\mathbf{r}_1)\phi_2(\mathbf{r}_2) + \phi_2(\mathbf{r}_1)\phi_1(\mathbf{r}_2)\right]\chi_S, \tag{3.15}$$

$$\Psi^S_\mathrm{L}(\mathbf{r}_1, \mathbf{r}_2) = \phi_1(\mathbf{r}_1)\phi_1(\mathbf{r}_2)\chi_S, \tag{3.16}$$

$$\Psi^S_\mathrm{R}(\mathbf{r}_1, \mathbf{r}_2) = \phi_2(\mathbf{r}_1)\phi_2(\mathbf{r}_2)\chi_S, \tag{3.17}$$

i.e. these latter two are the states with double filling of the "left" dot and double filling of the "right" dot. The triplet state, which must be antisymmetric in its spatial indices, cannot have double occupancy of a single orbital due to the Pauli

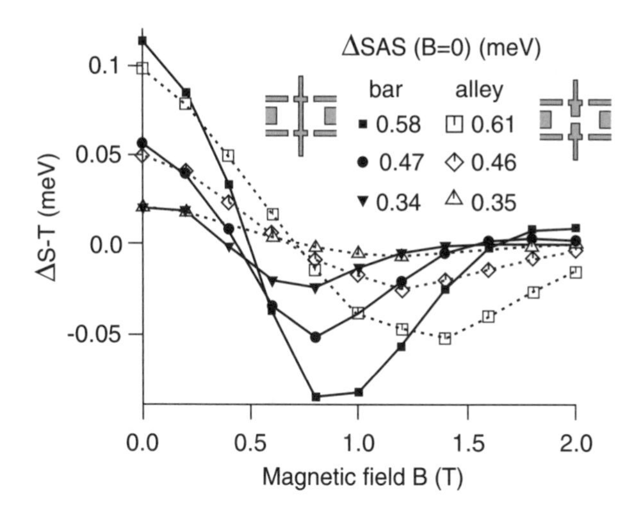

FIG. 3.10. The singlet-triplet splitting as a function of magnetic field for the $N = 2$ double dot. The two sets of curves, solid and dotted, are for two different gate patterns shown in insets ("bar", left, solid and "alley"). For the alley cross-over occurs at the same value of B for all values of the initial barrier height (at $B = 0$), whereas a dispersion is seen for the crossing of the bar. Numbers indicate symmetric-antisymmetric splitting at $B = 0$ in meV.

exclusion principle. Therefore, the only triplet state that can be formed with two orbitals is

$$\Psi_{\mathrm{T}}(\mathbf{r}_1, \mathbf{r}_2) = \frac{1}{2} \left[\phi_1(\mathbf{r}_1)\phi_2(\mathbf{r}_2) - \phi_2(\mathbf{r}_1)\phi_1(\mathbf{r}_2)\right] \chi_{\mathrm{T}}, \tag{3.18}$$

where χ_{T} represents the three spin triplet states. The total Hamiltonian of the system contains the kinetic energy term (with a vector potential if a magnetic field is included) and the external potential as the single-body effects. It then contains the Coulomb interaction, which consists of two-body integrals. The Coulomb integrals must usually be calculated numerically, but given these, the diagonalization of the singlet 3×3 matrix is simple and the triplet case contains only the single 1×1 term. Without writing down the entire solution (see Ref. [27] for details) approximate expressions for the singlet and triplet ground state energies are fairly intuitive:

$$E_{\mathrm{S}} = 2\epsilon + V_{\mathrm{inter}} + V_{\mathrm{ex}} - \frac{4t^2}{V_{\mathrm{intra}} - V_{\mathrm{inter}}}, \tag{3.19}$$

$$E_{\mathrm{T}} = 2\epsilon + V_{\mathrm{inter}} - V_{\mathrm{ex}}, \tag{3.20}$$

where the bare energies of the two dots (assumed equal) are ϵ, the tunnel coupling between the two dots is t, the direct intradot Coulomb matrix element is V_{intra} and the direct interdot Coulomb matrix element is V_{inter}. These results are easily interpreted. The singlet, which is always the ground state (if $B = 0$), is lowered by the delocalization afforded by the two double occupancy states, even though the charging energy cost may be high. By contrast, the triplet is lowered relative to the singlet by exchange. Considerable study has gone into the difference between the singlet and triplet energies, since this is the origin of J and it must therefore be tuned in order to perform two qubit operations. We conclude simply by noting that when a full calculation of J, including the realistic geometry of an actual lateral double dot and also including correlation via an exact diagonalization treatment, are performed, the evolution of J, as a function of B is non-trivial (Fig. 3.10) and, in particular, crosses to negative values as B increases [29]. Further, the nature of the crossing is affected by the exact shape of the interdot barrier. These and other issues of the interdot spin interaction will need to be further clarified before practical quantum computing operations can be implemented

3.5 Spin relaxation in quantum dots

3.5.1 *Transverse and longitudinal relaxation*

The concept of controling spin effects in semiconductor nanostructures is embodied in recent experiments for probing and manipulating electronic spins, such as the observation of long spin dephasing times [30], injection of spin-polarized current into semiconductor material [31], spin filtering using quantum dots [32], and spin rectification in double quantum dots [33]. Now it is established that spin is a robust, operable quantum number in semiconductor nanostructures. This feature is characterized by spin relaxation times – longitudinal relaxation time, T_1 and transverse relaxation time, T_2 –, which are much longer than corresponding times for the orbital relaxation. This has been well-studied both experimentally and theoretically for electrons in bulk and quantum wells [34]. These times can be even longer in quantum dots because there are no inhomogeneous broadenings as encountered by *mobile* electrons in 2D and 3D systems. Further, spectral discreteness reduces the number and efficiency of spin scattering mechanisms like phonon assisted spin-orbit interaction and hyperfine interaction.

Studies of the relaxation of the electronic spin degree of freedom in quantum dots have started only recently. Electronic spin coherence was measured for a collection (10^5) of CdSe nanocrystals using a pump-probe Faraday rotation technique [35]. The obtained transverse relaxation time or dephasing time, T_2^*, was a few nsec. This T_2^* is much shorter than that ($\sim$ 100 nsec) obtained for n-GaAs but not so short if we consider the large inhomogeniety resulting from the large size distribution and composition fluctuations. While the technique of electron spin resonance (ESR) allows direct measurement of spin coherence, the typical response to ESR is very weak. To this end there are no reports of successful ESR experiments for either collections of many quantum dots or single quantum dots.

On the other hand, a spin-flip energy relaxation for a single electronic spin in a quantum dot was first measured for a vertical quantum dot [36]. Here, we

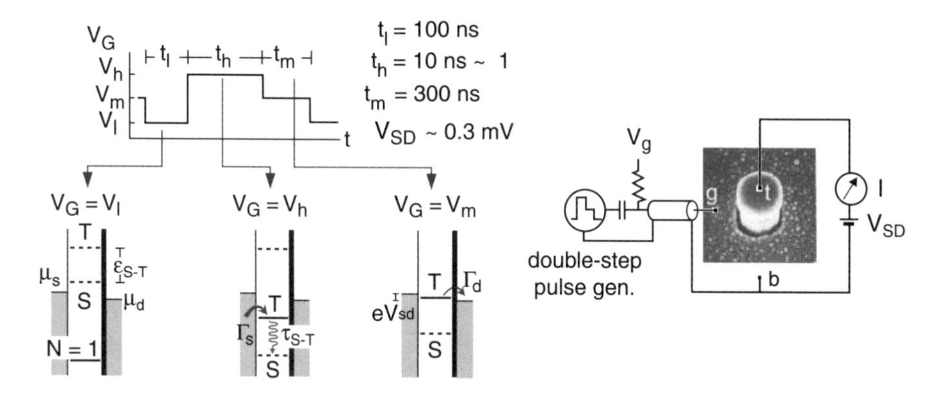

FIG. 3.11. (a) Schematic of double-step gating pulse waveform used for pump-probe measurement. (b), (c), and (d) Schematic energy diagram showing low (b), high (c) and intermediate (d) pulse situations.

review the experiment. For $N = 2$ the ground state of a vertical quantum dot is a singlet state ($1s^2$: $S = 0$) having two antiparallel spin electrons both in the lowest state $1s$. The first excited state is a spin triplet ($1s2p$: $S = 1$) having two parallel spin electrons in the $1s$ and $2p$ orbitals. The scheme for measuring relaxation from the first excited state to the ground state is illustrated in Fig. 3.11. A double-step pulse gating is employed to measure a spin-flip energy relaxation time much longer than 100 nsec. Suppose a dot is initially empty with the gate voltage at a large, negative value. Then the gate voltage is abruptly changed to the positive such that just two electrons are non-adiabatically injected from the source contact to form either a spin singlet state ($1s^2$ with $S = 0$, ground state) or a triplet state ($1s2p$ with $S = 1$, first excited state) in the $N = 2$ Coulomb blockade regime. Note, electrons are preferentially trapped by the dot through the source contact barrier if the tunnel barriers of the dot are asymmetric such that Γ_s ($\sim$ (3 nsec)$^{-1}$) $\gg \Gamma_d$ ($\sim$(100 nsec)$^{-1}$), where Γ_s (Γ_d) is the tunneling rate through the source (drain) contact barrier. If and only if the triplet state is populated, it starts to relax with a time constant, τ_{ST}, to the singlet state via a spin-flip and a momentum relaxation from $2p$ to $1s$. Then after a waiting time (t_h), the gate voltage is stepped down such that the triplet state only is located between the Fermi levels of the source and drain. If the triplet state is still populated after t_h, an electron in the triplet state escapes from the dot on the time scale of Γ_d^{-1} ($\sim$ 100 nsec) to generate a current. Note, single-electron transport through the triplet state between the contact leads continues in this stage before the singlet is populated, but this only gives a background current if $\tau_{ST} \gg \Gamma_d^{-1}$. Finally the gate voltage is restored to the initial level to make the dot empty again. In this gating cycle, at most one electron generates a net current with a probability of $\langle n_e \rangle = A \exp(-t_h/\tau_{ST})$ (< 1), where A is a constant related to the tunneling rates between the ground and excited states. By repeating this

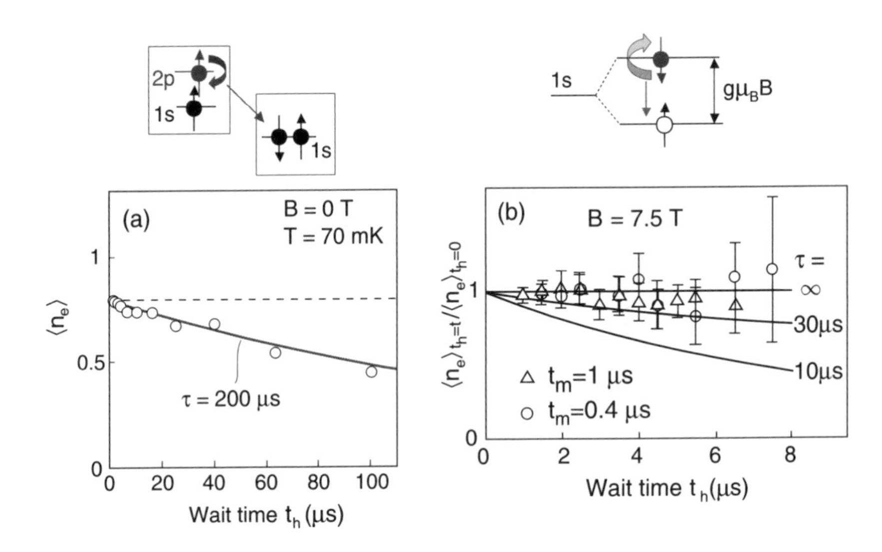

FIG. 3.12. Average number of tunneling electrons per double step pulse gating, $\langle n_{\mathrm{e}}\rangle$, measured for relaxation from a $1s2p$ triplet state to a $1s^2$ singlet state of the $N = 2$ vertical quantum dot at $B = 0$ T (a), and relaxation between the $S_Z = 1/2$ and $-1/2$ Zeeman sublevels of the $N = 1$ lateral quantum dot at $B = 7$ T (b). The relaxation times, τ_{ST} in (a) and τ_{Z} in (b), are obtained from the exponential decays (solid lines).

gating cycle a number of times ($\sim 10^7$), the current reaches a measurable level. Thus measured $\langle n_{\mathrm{e}}\rangle$ vs. t_{h} at 0.1 K for $B = 0$ T is shown in Fig. 3.12(a). The excitation energy of the $1s2p$ triplet state is 0.6 meV. The data points fall on the curve of $\exp(-t_{\mathrm{h}}/\tau_{\mathrm{ST}})$ with a time constant of $\tau_{\mathrm{ST}} = 0.2$ msec. This τ_{ST} is almost constant with B between 0 and 2 T, applied perpendicular to the dot plane, and much longer than the energy relaxation time, τ_{1s2p} without spin-flip measured for the same dot but for relaxation of a one-electron state from $2p$ to $1s$. The measured τ_{1s2p} ranges from 10 to 3 nsec for B from 1 to 4.5 T. In this B range the excitation energy, $\Delta\epsilon_{1s-2p}$, of the $2p$ state changes from 2.3 to 1.4 meV. Electron-acoustic phonon scattering is a major mechanism for τ_{1s2p}, and the experimental data of τ_{1s2p} compare well to calculations using Fermi's Golden rule, which includes scattering by piezoelectric and deformation potential effects. The increase of τ_{1s2p} with decreasing B or increasing $\Delta\epsilon_{1s-2p}$ is attributable to the reduced electron-phonon scattering in quantum dots, the so-called "phonon bottleneck effect". This effect is predicted to appear when the acoustic phonon wavelength λ_{ph} is shorter than the characteristic size of the system [37]. In this experiment $\lambda_{\mathrm{ph}} \sim 10$ nm for $\Delta\epsilon_{1s-2p} = a$ few meV is much shorter than the effective lateral size (50 to 100 nm) of the dot.

The measured τ_{ST} of 0.2 msec is influenced not only by intrinsic spin-flip relaxation in the quantum dot but also by inelastic cotunneling (see Section

3.7.1). Theory predicts that the cotunneling effect becomes strong as the occupied triplet state approaches the Fermi levels of the source and drain. In this experiment the measured $\tau_{\rm ST}$ becomes short as the energy difference between the Fremi level and the triplet state is reduced as a function of gate voltage, following inelastic cotunneling theory. Therefore, the measured $\tau_{\rm ST}$ is still a lower bound of the intrinsic spin-flip relaxation time T_1. Note the effect of thermalization can be ruled out, because the measured $\tau_{\rm ST}$ is constant with temperature for $T < 0.5$ K and starts to decrease for $T > 0.5$ K.

Using the same technique, the spin-flip relaxation time, $\tau_{\rm Z}$, was measured for Zeeman sublevels of a lateral quantum dot, as shown in Fig. 3.12 (b) [18]. The measured $\tau_{\rm ST}$ is 50 μsec at $B = 7.5$ T as a lower bound of T_1. In addition, the $\tau_{\rm Z}$ decreases with B^{-5} as B increases, reflecting a phonon-assisted spin-orbit effect [38]. $\tau_{\rm Z}$ for Zeeman sublevels has also been measured for InAs self-assembled dots, using a combined electrical and optical pump-probe technique [39]. The measured $\tau_{\rm Z}$ is 30 msec at an in-plane B field of 4 T, and much longer than that for a GaAs lateral dot. In addition, B^{-5} dependence of $\tau_{\rm Z}$ is observed, the same as for lateral GaAs dots, reflecting the effect of spin-orbit interaction. The long $\tau_{\rm Z}$ is assigned to the strong confinement of self-assembled dots and attendant large level spacing. The electron-phonon interaction can then be strongly reduced. Note in this experiment a large number (10^4) of quantum dots are optically excited, but the large distribution of dot size does not influence T_1.

3.5.2 *Effect of spin-orbit interaction*

The spin-flip relaxation time observed for quantum dots is five to six orders of magnitude longer than the orbital relaxation time. This indicates that the spin degree of freedom in quantum dots is well isolated from the environment, and also implies that the decoherence time is much longer for spin than for charge. To account for the experimentally observed long T_1, theories including spin-orbit coupling and hyperfine coupling have been presented for relaxation between a spin triplet and singlet states [40] and between Zeeman sublevels [41, 42].

Spin-orbit interaction, which mixes the spin and orbital degrees of freedom, is known to be a dominant spin-flip mechanism in a semiconductor 2DEG. This interaction arises from the broken inversion symmetry, either in the elementary crystal cell (bulk inversion asymmetry) or at the heterointerface (structural inversion asymmetry) [34]. However, as discussed in Ref. [41], the interaction is significantly weakened in quantum dots. The spin-orbit interaction term appears only if either the admixture of the higher states in the z-direction, or else higher orders in the expansion of the in-plane momentum, are considered. This further weakens the spin-orbit interaction, and finally results in extremely low spin-flip rates of electrons in the dots. For Zeeman sublevels, one-acoustic phonon scattering can mediate the spin-flip relaxation by mixing the Zeeman sublevels via the spin-orbit interaction. Perturbation theory [41, 42] predicts that the rate by such a spin relaxation process is given by

$$\Gamma_{\rm sf} \sim A(g\mu_{\rm B}B)^5/\hbar(\hbar\omega_0)^4, \tag{3.21}$$

where A is a dimensionless constant that represents the strength of the effective spin-piezophonon coupling in the heterostructure. Note the acoustic phonon has an energy equivalent to the Zeeman energy ($\Delta\epsilon_Z = g\mu_B B$). For a GaAs lateral quantum dot, Γ_{sf} of Eq. (3.21) is $\sim$ (1 msec)$^{-1}$ at $B = 1$ T, which is comparable to experiment [18, 38]. In addition, Γ_{sf} in Eq. (3.21) is proportional to B^5, and this is also consistent with experiment [38, 39]. Note for the Zeeman sublevels, the phonon wavelength is longer than the dot size in all directions, and no phonon bottleneck effect is expected. However, this is not the case for the spin-flip transition from triplet to singlet in the experiment of Fig. 3.12(a), because the transition energy is large and only gradually decreases with decreasing B.

Suppose an electronic spin precesses about a static DC B field. T_1 is influenced by fluctuations of B perpendicular to the precession axis, whereas T_2 is only influenced by fluctuations along the precession axis. Perturbation theory [42] predicts that spin-orbit interaction, in first order, can only cause B to fluctuate perpendicularly to the precession axis and therefore $T_2 = 2T_1$ in the dissipation limit. T_1 and T_2 can then be comparable for quantum dots, although this has not yet been experimentally confirmed. If so, such a long T_2 (> 0.1 msec) is well qualified for making qubits with quantum dots in the scheme of quantum computing.

3.6 Spin blockade in single-electron tunneling

3.6.1 *Suppression of single-electron tunneling*

Coulomb blockade discussed to date solely reflects charge discreteness. In this scheme, the number of electrons in a dot is varied one-by-one (single-electron tunneling), so that the N electron ground state can only be constructed by adding (subtracting) one electron to (from) the $N-1$ ($N+1$) electron ground state (see Section 3.2). However, in some cases the ground state cannot be constructed in this way because of the correlation effect, as we now describe. Particularly, in the presence of internal ferromagnetic or antiferromagnetic spin couplings, the total spin for the N electron ground state, $S(N)$, can differ from that for the $N-1$ electron ground state, $S(N-1)$ by more than 1/2: $|\Delta S(N)| = |S(N) - S(N-1)| > 1/2$. Thus the transition between these N and $N-1$ ground states is impossible via the tunneling of a single electron since the total spin can then only change by $\pm 1/2$. This leads to suppression of single-electron tunneling, which is called "spin blockade" [43, 44]. This blockade is usually not so strong as Coulomb blockade, because it can be lifted by spin relaxation in the quantum dot. A couple of experiments on spin blockade have been reported for lateral quantum dots [45]. In these experiments, however, it is difficult to isolate the spin effect because the electronic configuration is not well defined.

Pauli exclusion can also lead to a simple spin effect to block single-electron tunneling which arises directly from the Fermionic nature of electrons. The Pauli exclusion principle states that a single quantum state cannot be occupied by more than one electron. In other words two parallel spin electrons cannot occupy the same orbital state. This can lead to suppression of single-electron tunneling, for example, in a system such that electrons in the source contact are fully spin polarized, because parallel spin electrons cannot be consecutively filled into the same orbital state in the quantum dot even when $|\Delta S| = 1/2$. The 2DEG in the contact lead can be fully spin-polarized in the presence of a sufficiently high B field, reaching the $\nu = 1$ quantum Hall region. However, in such a strong B field the states in the quantum dot can also be confined to the spin-polarized lowest Landau state, and then no suppression of single-electron tunneling results. This makes it difficult to facilitate the experiment (ideally one would like polarized leads with a dot in zero magnetic field). Instead, we have recently used a weakly coupled *double* quantum dot to demonstrate the Pauli effect [33]. In the following subsections we show that in transport through a series of 0D sites, with symmetry under inversion appropriately broken, the Pauli effect, in combination with the Coulomb blockade, can be used to block current altogether in one direction while permitting it to flow in the opposite direction. Before we describe the Pauli effect, one other recently observed phenomenon that deserves mention is the suppression of current arising from the formation of the $\nu = 2$ edge channels in the contact leads [46]. In this experiment, the dot-lead tunneling probability is very much smaller for down-spin electrons in the inner edge channel than for up-spin electrons in the outer edge channel. Then single-electron tunneling for adding a down-spin electron to the dot is significantly reduced.

3.6.2 *Pauli effect in coupled dots*

Consider two sites, site 1 and site 2, weakly coupled with one electron permanently trapped on site 2 (see Fig. 3.13). Now consider serial transport of a second electron through the system between two contact leads. The number of electrons on site 1 (2), N_1 (N_2), varies between 0 and 1 (1 and 2). An electron can only be transported through the system when the transition between the two-electron states $(N_1, N_2) = (0, 2)$ and $(1, 1)$ is allowed. This condition is met when the necessary energy cost to add one more electron to the system is compensated for by the action of a nearby plunger gate voltage, or the voltage between the leads, which is a familiar single-electron tunneling phenomenon. Crucially, spin effects also markedly influence the electron transport. Because the tunnel coupling between the two sites is sufficiently weak, the $(N_1, N_2) = (1, 1)$ spin-singlet and spin-triplet states are practically degenerate. Additionally, for $(N_1, N_2) = (0, 2)$, only a spin singlet is permitted because of Pauli exclusion. Therefore, electron transport is only allowed for a channel made from the (1, 1) and (0, 2) singlet states. This always holds true for reverse bias when the Fermi level of the left lead, μ_l, nearest site 1, is lower than that of the right lead, μ_r, nearest site 2, because only an antiparallel spin electron can be injected onto site 2 from the right lead (Fig. 3.13). On the other hand, for forward bias, $\mu_l > \mu_r$, either the (1, 1) singlet or triplet can be populated with more or less the same probability by injection of an electron onto site 1 from the left lead. If the (1, 1) singlet is populated, this subsequently generates a single-electron tunneling current that can flow through the singlet state. Once the triplet is populated, however, sub-

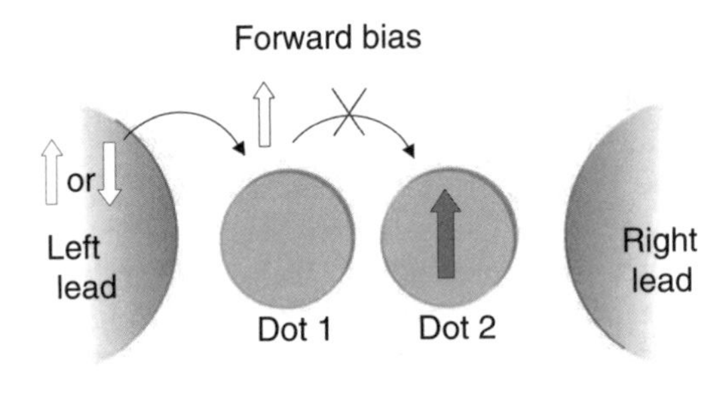

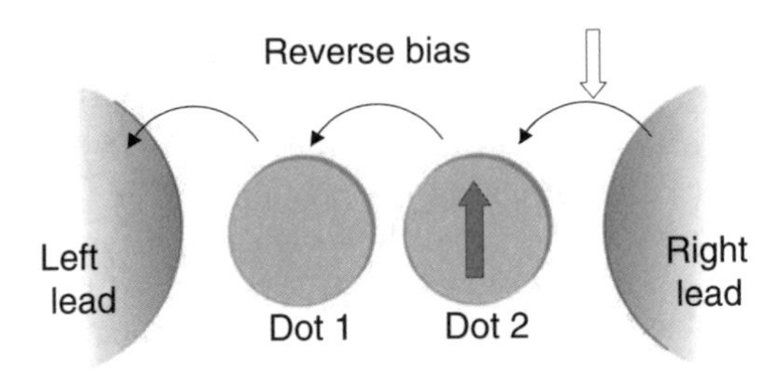

FIG. 3.13. Model for current rectification by the Pauli effect in a series of two quantum dots, dot 1 and 2. Each quantum dot holds just one spin degenerate orbital state, and an up-spin electron is permanently trapped in dot 2. Single-electron tunneling occurs via the two-electron state in the double dot: from the left (right) to the right (left) lead for forward (reverse) bias.

sequent electron transfer from site 1 to 2 is blocked by Pauli exclusion. Thus the (1,1) triplet will sooner or later be occupied on a time scale sufficiently longer than the electron tunneling time between the leads, and this should lead to clear current suppression, for example, in DC measurement.

Here experiments using a vertical double quantum dot device are explained. Two 2D harmonic $In_{.05}Ga_{.95}As$ quantum dots (labeled dot 1 and 2) are vertically weakly coupled (see Fig. 3.14a). The tunnel coupling energy ($= 0.3$ meV [48,49]) between two dots is much smaller than both the 2D harmonic potential energy ($\hbar\omega_0 \sim 4$ meV), as well as the charging energy for each dot ($U \sim 4$ meV). Note the total number of electrons in the double dot system, N $(= N_1 + N_2)$, where N_1 (N_2) is the number of electrons in dot 1(dot 2), can be varied one-by-one as a function of gate voltage, V_G, starting from $N = 0$. The transmission coefficients for all the tunnel barriers are sufficiently small that electrons are only sequentially transported between the source and drain.

The situation of Fig. 3.13 can be reproduced in this device (see Fig. 3.14c), if there is an appropriate potential offset, 2δ, between the two quantum dots at $V_{sd} = 0$ V. Then, just one electron is trapped in the $1s$ orbital state of dot 2, and the two-electron ground state is a spin singlet state with $(N_1, N_2) =$

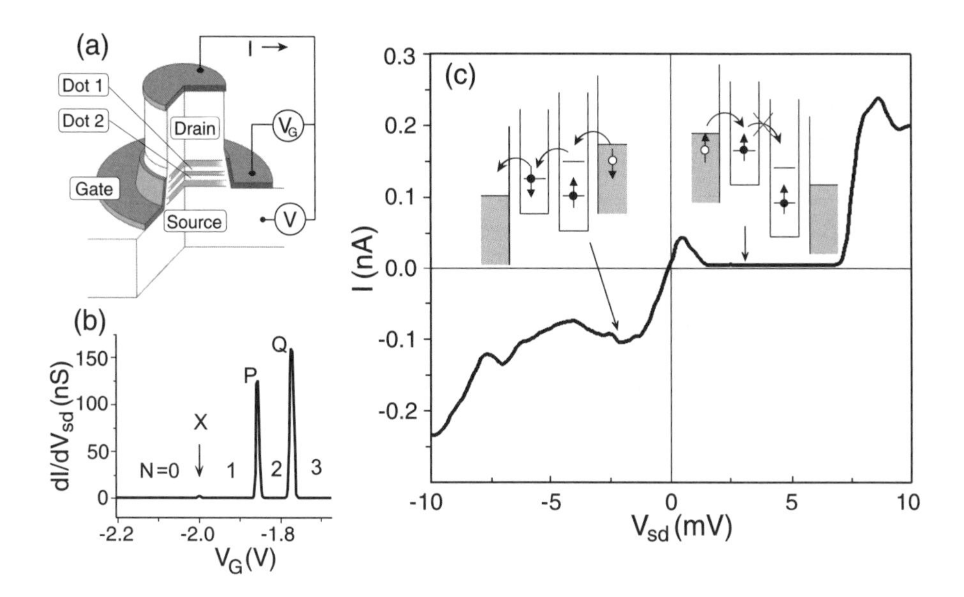

FIG. 3.14. (a) Schematic of a vertical device containing two quantum dots, which are made in a 0.6 micron diameter circular mesa of two $In_{0.05}Ga_{0.95}$ As wells, a $Al_{0.3}Ga_{0.7}As$ central barrier, two $Al_{0.3}Ga_{0.7}As$ outer barriers and two n-GaAs contact leads. Two two-dimensional harmonic quantum dots are connected, located between the heterostructure barriers. (b) Coulomb peaks (I vs. V_G) corresponding to adding $N = 1$ to 3 electrons to the double dot. (c) Spin blockade of single-electron tunneling current in a weakly coupled double dot system: I versus V_{sd} curve measured for V_G fixed at the second Coulomb peak in (b).

(0, 2). The first and second excited states are a spin triplet, and singlet state for (1, 1). These two states are closely spaced, split by the exchange coupling energy. Note well above these excited states a (0, 2) triplet excited state can be formed by putting two parallel spin electrons in the $1s$ and $2p$ states in dot 2. When viewing Fig. 3.14(c), we can see that electrons can be transported via the (0, 2) and (1, 1) singlet states for reverse bias, whereas for forward bias, the (1, 1) triplet can be populated, leading to the blockade of electron transport. This is valid unless the triplet configuration is not degraded. We now define the electrochemical potential of the (N_1, N_2) ground state to be $\mu(N_1, N_2)$, and the Fermi level or chemical potential of the source (drain) contact to be μ_r (μ_l).

In the experiment the situation in Fig. 3.13 can be realized by adjusting the source-drain voltage, V_{sd} and gate voltage, V_G and by using a device with an appropriate potential offset 2δ. Note about 10% of V_{sd} is dropped between the two dots. Potential offsets ranging from 0.5 to a few meV are usually present in all devices probably due to a small decrease in the effective electron density in the cylindrical mesa just above the two dots relative to that below [49]. For this experiment a double dot device having an appropriate value of 2δ, such that the condition $\mu(1,1) = \mu(0,2)$ is satisfied at $V_{sd} = 0$ V, is used.

Figure 3.14(b) shows the first three Coulomb peaks in the dI/dV_{sd} vs. V_G measured for $V_{sd} \sim 0$ (linear regime). The first peak (X) is very small but definitely present at $V_G \sim -2$ V, and this indicates the transport through the double-dot system for N fluctuating between 0 and 1. The second peak (P) and the third peak (Q) are much larger. This implies that tunneling is elastic between the source and drain leads for $N = 1 \leftrightarrow 2$ (second peak) and $2 \leftrightarrow 3$ (third peak). However, because of the potential offset 2δ, this is not the case for $N = 0 \leftrightarrow 1$ (first peak). Electrons tunneling through the lowest state in dot 2 must tunnel virtually through higher states in dot 1 in order to reach the drain [50]. We then study non-linear transport through the two-electron states by measuring the $I - V_{sd}$ for V_G fixed at the second Coulomb peak (Fig. 3.14c, main curve). The data reveal a large, nearly constant dI/dV_{sd} at $V_{sd} \sim 0$ for both bias polarities, and this is due to elastic tunneling in the linear response region. Tunneling here occurs via the (1, 1) and (0, 2) singlet states, which are aligned at $V_{sd} \sim 0$ by the potential offset. For non-linear transport ($|V_{sd}| \geq 1\,\text{mV}$), the current is clearly suppressed in forward bias because of spin blockade, whereas a large current flows in reverse bias because of inelastic tunneling via the singlet states. The spin blockade is lifted for $V_{sd} > 7\,\text{mV}$ when an electron can be ejected from the lowest state of dot 2 to the source. The spin blockade is further confirmed from measurements of non-linear transport for various V_G values [33].

From the level of the current ($I = 1 \sim 2$ pA) in the Pauli spin blockade region, we estimate the lifetime of the spin triplet to be $e/I \sim 100\,\text{ns}$. This is markedly longer than the time taken for just single-electron inelastic tunneling between the contact leads ($e/I \sim 16$ ns) for $V_{sd} < -1$ mV. As well as by the second-order tunneling processes, a small current can also be induced by spin-flips inside the dot. This, for example, can arise from spin-orbit interaction. Note that the current level in the Pauli spin blockade region is much higher than that in the region of Coulomb blockade ($< 100\,\text{fA}$), as described before in Section 3.6.1.

3.6.3 *Lifting of Pauli spin blockade by hyperfine coupling*

Spin-orbit interaction is discussed as a dominant source for spin relaxation in Section 3.5.2. Another important source for spin relaxation is hyperfine coupling to nuclei of the host material. For the present vertical quantum dots all of the constituent atoms of ^{71}Ga, ^{69}Ga, and ^{75}As have nuclear spin of $I = 3/2$, and can interact with electronic spins via hyperfine interaction. The interaction Hamiltonian includes a flip-flop term between an electronic spin and nuclear spin, conserving spin and energy between the systems of electrons and nuclei. This interaction is usually very weak in quantum dots due to the discreteness of the electron energy [51]. For example, the Zeeman energy is greater for electronic spin by more than two orders of magnitude compared to nuclear spin. However,

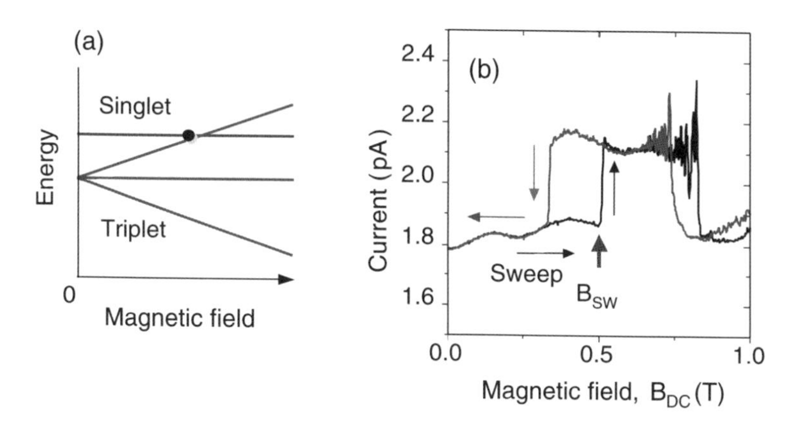

FIG. 3.15. (a) Zeeman effect on the $(N_1, N_2) = (1, 1)$ triplet state (first excited state) and singlet state (second excited state) in a weakly coupled double dot. (b) Leakage current at a $V_{\rm sd}$ in the middle of the Pauli spin blockade region (Fig. 3.14c) as a function of in-plane magnetic field ($B_{\rm DC}$) for the sweep up (black) and sweep down (gray). Detailed positions of the step and largest fluctuations depend on the $B_{\rm DC}$ sweep rate and values of $V_{\rm sd}$ and $V_{\rm G}$.

the triplet and singlet two-electron states (1,1) discussed in Section 3.6.2, can be tuned to near-degeneracy insofar as the triplet has three branches $S_z = 1, 0$ and -1. The energy separation between these states is calculated to be a few tens of μeV for the present device [52]. Application of B therefore leads to degeneracy of the singlet and one of the three branches ($S_z = -1$ state in this case) of the triplet state (Fig. 3.15a). Then transition between the singlet state and the triplet, $S_z = -1$ state costs an energy comparable to the nuclear Zeeman energy. We have examined this effect by measuring the leakage current changing with in-plane B field in the Pauli spin blockade region [53].

Figure 3.15(b) shows the result of such an experiment. The in-plane B field, $B_{\rm DC}$, is changed with a constant field sweep rate of 2 min/T. As the B field initially increases, the current I is nearly constant for $B_{\rm DC} < 0.5$ T, and rises with a sharp step at $B_{\rm DC} \sim$ T. Then it starts to fluctuate more strongly with increasing $B_{\rm DC}$ up to ~ 0.87 T, and suddenly decreases for $B_{\rm DC} > 0.9$ T. A similar characteristic, but shifted to lower field, is observed for sweeping down the B field. The B field shift is about 0.2 T, and becomes smaller for a slower $B_{\rm DC}$ sweep rate until it saturates at about 0.15 T for a sweep rate slower than 1 hour/T. Similar characteristics are observed at different ($V_{\rm sd}$, $V_{\rm G}$) within the Pauli spin blockade region. For any $B_{\rm DC}$ field in the current fluctuation regime (0.6 to 0.87 T), the current shows periodic oscillations as a function of time, lasting for longer than 15 hours. Both the period and amplitude of the current oscillations increase with $B_{\rm DC}$, and become maximal with a period as long as ~ 200 sec and amplitude of ~ 0.4 pA near 0.87 T. The time scales of minutes for the current oscillation period and also for the $B_{\rm DC}$ sweep over the range of

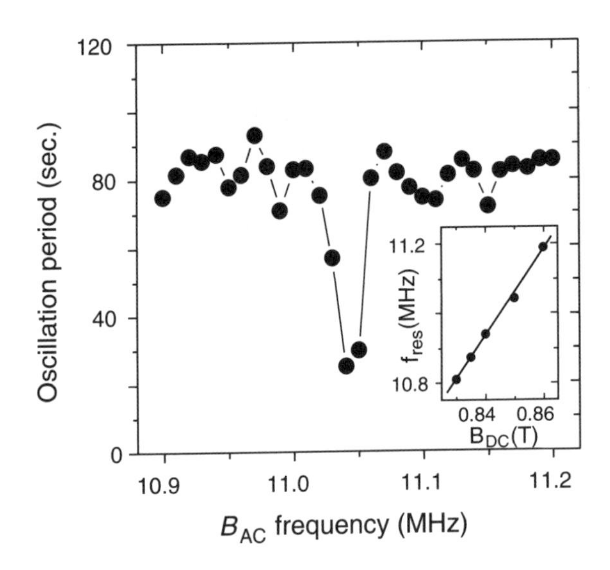

FIG. 3.16. Time period of the current oscillations measured for various frequencies of the AC magnetic field applied perpendicular to the plane of the dot. Inset: DC in-plane magnetic field (B_{DC}) dependence of the B_{AC} resonance frequency f_{res} observed in the oscillation period

the current hysteresis loop can be associated with the nuclear spins, which have an unusually long relaxation time T_1($>$ 10 min at low temperatures) [54].

As is well known in the context of NMR, nuclear spin effects can be examined with an AC magnetic field. We use a three-turn coil of 3 mm diameter located just above the device to apply a vertical AC magnetic field, B_{AC}, and measure the change in the oscillatory current for various frequencies of B_{AC} at $B_{DC} = 0.85$ T. A strong reduction in both the oscillation period and amplitude is observed when the B_{AC} frequency matches the ^{71}Ga and ^{69}Ga nuclear spin resonance (Fig. 3.16 for resonance of ^{71}Ga). The resonance frequency changes linearly with B_{DC} (inset).

It is interesting to note the oscillations are consistently described if we assume that larger nuclear polarization leads to larger period and amplitude of the current oscillations. The nuclear polarization grows with increasing magnetic field, gradually grows (decays) by turning on (off) the spin blockade, and resonantly decays under the NMR condition. Based on this knowledge we propose a model that accounts for only the dynamic polarization of nuclei in the spin blocked double dot at a certain B_{DC} field, where the $S_Z = -1$ triplet and the singlet states are degenerate as described before. Around this B_{DC} field, the hyperfine flip-flop scattering from $S_Z = -1$ triplet to the singlet states should be much favored compared to the scatterings from $S_Z = 0$ or $+1$. Thus a nuclear spin will be flopped from "down" to "up" rather than "up" to "down". Because of the long relaxation time of nuclear spins at low temperatures, the flopped nuclear spins are steadily accumulated over cycles where the triplet scatters to the singlet, leading eventually to dynamic polarization of nuclei. Assuming an electron g-factor of -0.44 the degeneracy of the $S_Z = -1$ triplet and singlet states occurs at $B_{\rm DC} = 0.4$–2 T. This agrees with the experimental $B_{\rm DC}$ field where a step and oscillations are observed.

The Overhauser effect due to the dynamic nuclear polarization is known to affect the electron spin and its transport characteristics [55]. However, this effect simply detunes the singlet-triplet cross-over and does not cause the oscillatory behavior. The results suggest the presence of more complicated feedback from the polarized nuclei to the electron spin. It has recently been predicted for double dot systems that the coupled electron-nuclear spin system exhibits instability near the singlet-triplet cross-over [56].

3.7 Cotunneling and the Kondo effect

3.7.1 *Cotunneling*

Thus far we have been considering the regime where the relaxation time of the dot to its ground state is much smaller than the time between tunneling events to and from the leads. We noted, however, that the possibility for long-lived metastable states existed which violates this condition. Away from those metastable states, however, the rapid relaxation of the dot implies that tunneling into and out of the dot, from source to drain, occurs sequentially, i.e. the dot is effectively coupled to a heat bath such that information relating to the tunneling across the first barrier is lost before the electron tunnels across the second barrier. This implies in particular that the peak conductance for a Coulomb oscillation is proportional to $\Gamma_{\ell p}\Gamma_{rq}/(\Gamma_{\ell p} + \Gamma_{rq})$ (i.e. the reciprocals of the rates add to the reciprocal of the total). Here ℓ and r refer to left and right leads and p and q are arbitrary quantum numbers. Now we wish to explicitly consider the regime where the tunnel junctions between dot and leads become increasingly transparent and the level widths of the various states, Γ_p, become increasingly large. The main physical consequence of this is that the dot begins to show transport characteristics beyond the sequential regime in what are referred to as "cotunneling" events [50]. In cotunneling, two tunneling events occur around a virtual intermediate state wherein the dot possesses one fewer or one greater electron than the stable number, that is: $N = N_0 \pm 1$. The energy of this state can be greater than the energy of the initial state and the duration of its occupancy is limited by the uncertainty principle.

Cotunneling events can be categorized as either "elastic" or "inelastic" depending on whether the two tunneling events occur between the leads and the same dot level (thereby leaving the dot in its initial state) or else the two tunneling events occur between the leads and two different levels (whereupon the final state of the dot is different from the initial state). As a concrete example, suppose the dot is initially in its $T = 0$ ground state with levels 1 through N_0 occupied and all other levels empty. Suppose further that the dot is in the

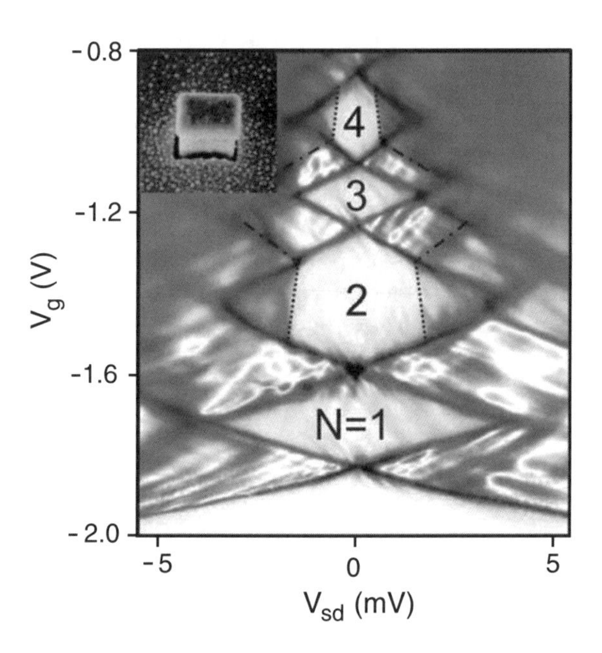

FIG. 3.17. Measured stability diagram of a quantum dot at 15 mK and zero magnetic field. dI/dV_{sd} is plotted in gray scale as a function of (V_{sd}, V_{g}). Dotted lines have been superimposed to highlight the onset of inelastic cotunneling. The dot-dashed lines indicate the onset of first-order tunneling via an excited state. Inset: scanning electron micrograph of the dot used in the study.

Coulomb blockade regime, i.e. it is not at the degeneracy point of two charge states. An inelastic cotunneling process could occur such that the (N_0+1)st level is filled by an electron from the source simultaneously (within the uncertainty principle) with the exiting to the drain of the electron that initially occupies the N_0th level. After this process the dot is in an excited state with levels 1 through $N_0 - 1$ and level $N_0 + 1$ filled and level N_0 empty. However, while the virtual intermediate, $N = N_0+1$ state can be "off the energy shell" via quantum uncertainty, the energy to excite the dot to the final excited $N = N_0$ state must come from somewhere. For sufficiently low temperature this energy must come from the source-drain chemical potential difference, i.e. the bias. The signature of this effect, seen in Fig. 3.17, is best seen in the Coulomb diamonds [57]. Here, it is assumed that an excited state of the dot exists (perhaps due to just a single particle-like excitation as we just described). At $V_{\mathrm{sd}} \approx 0$ the inelastic cotunneling process that fills this excited state does not have enough energy to occur. However, beyond a threshold value $|eV_{\mathrm{sd}}| = E^* - E$, where E^* and E are shorthand for the excited and ground state energies, the current can flow through this cotunneling process. The result is that the insides of the diamonds are shaded

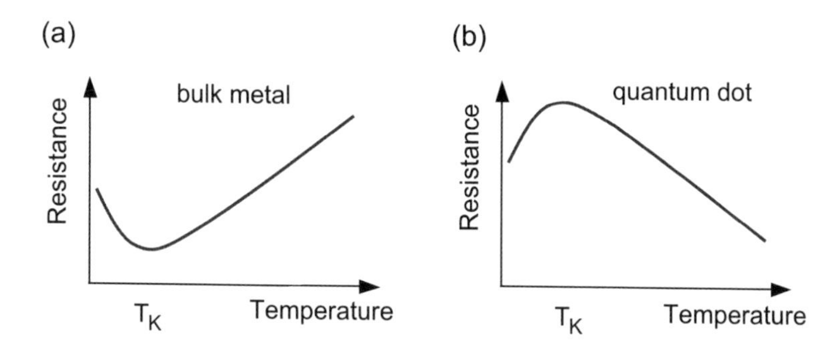

FIG. 3.18. Schematic temperature dependence of the resistance in the presence of the Kondo effect (a) in a bulk metal containing magnetic impurities and (b) in a quantum dot having a finite magnetic moment.

outside the threshold limits, as seen in the figure.

3.7.2 *The standard Kondo effect*

Because cotunneling involves higher order processes that relate specifically to the excited states of the dot, they are interesting in that they give an immediate experimental signature of properties beyond the ground state. However we might ask: what properties of cotunneling relate specifically to spin? The answer to this question, which forms the basis of a considerable amount of work on the physics of quantum dots, is that cotunneling is the lowest order element in the phenomenon of the Kondo effect.

The electrical resistance of a metal usually decreases monotonically as the temperature, T, decreases because phonon scattering is suppressed at low temperature. The resistance may eventually saturate due to crystal defects and impurities, or suddenly drop to zero if the metal undergoes a transition to a superconducting phase. However, the resistance of some metals containing a small amount of magnetic impurities was found to increase at some low temperature as depicted in Fig. 3.18(a). The mechanism of this anomalous resistance increase remained a mystery until Jun Kondo explained in 1964 that it arises from scattering by the magnetic impurities that are antiferromagnetically correlated with spins of the conduction electrons [58]. The temperature which marks the upturn of the resistance due to this spin interaction is called the Kondo temperature, T_{K}.

More recently, the Kondo effect was predicted to occur also in artificial nanostructures such as a quantum dot [59–61]. When the dot holds an odd number of electrons, the uppermost energy level has an unpaired electron while all the other underlying levels (ignored as an inert core) are occupied by electron pairs. Then, the dot acts as a single magnetic impurity with a spin 1/2, and its interaction with Fermi seas in the leads can be modeled as a Kondo system. The Hamiltonian for this system is the Anderson Hamiltonian:

$$H = \sum_{\sigma=\uparrow,\downarrow} \epsilon_{\rm d} n_{{\rm d},\sigma} + U n_{{\rm d},\uparrow} n_{{\rm d},\downarrow} + \sum_{\alpha,\mathbf{k},\sigma} \epsilon_{\alpha,\mathbf{k}} c^{\dagger}_{\alpha,\mathbf{k},\sigma} c_{\alpha,\mathbf{k},\sigma}$$
$$+ \sum_{\alpha,\mathbf{k},\sigma} \left(\Gamma_{\alpha,\mathbf{k}} c^{\dagger}_{{\rm d},\sigma} c_{\alpha,\mathbf{k}',\sigma} + {\rm h.c.} \right), \quad (3.22)$$

where the bare discrete level has energy ϵ_d, the number operator for discrete level d with spin σ $(=\uparrow, \downarrow)$ is $n_{{\rm d},\sigma}$ and that of the continuum level $\mathbf{k}$ in lead α with spin σ is $n_{\alpha,\mathbf{k},\sigma}$; the connection coefficients between the discrete level d and the leads is $\Gamma_{\alpha,\mathbf{k}}$ and the creation and annihilation operators for the lead states and the discrete states are $c^{\dagger}_{\alpha,\mathbf{k},\sigma}$, $c_{\alpha,\mathbf{k},\sigma}$ and $c^{\dagger}_{{\rm d},\sigma}$, $c_{{\rm d},\sigma}$, respectively. The Kondo effect proceeds from the Anderson Hamiltonian by developing a perturbation series in $\Gamma_{\alpha,\mathbf{k}}$. Since the initial state of the $N = 1$ dot (or impurity) is spin degenerate, and since the Hamiltonian contains no terms which explicitly couple spin up and spin down (such as spin-orbit interaction), the only way for the dot spin state to make a flip is via an intermediate virtual state with one greater or one less electron, i.e. a cotunneling process. The derivation of the Kondo effect begins by showing that such cotunneling processes produce an off-diagonal coupling in the $N = 1$ submatrix of the Hamiltonian. In so doing, the Hamiltonian is transformed into the so-called *s-d* Hamiltonian

$$H_{\rm sd} = \sum_{\mathbf{k},\mathbf{k}'} J_{\mathbf{k},\mathbf{k}'} \left[S^+ c^{\dagger}_{\mathbf{k},\downarrow} c_{\mathbf{k}',\uparrow} + S^- c^{\dagger}_{\mathbf{k},\uparrow} c_{\mathbf{k}',\downarrow}, + S_z \left(c^{\dagger}_{\mathbf{k},\uparrow} c_{\mathbf{k}',\uparrow} + c^{\dagger}_{\mathbf{k},\downarrow} c_{\mathbf{k}',\downarrow} \right) \right], \quad (3.23)$$

where we have suppressed the lead index α and where the internal dynamics of the dot have been reduced to operators representing its spin: spin raising S^+, spin lowering S^- and spin z-component S_z.

Without a prolonged derivation, the essential physics of the Kondo effect can be stated as follows. The cotunneling process produces a coupling between up and down spins which is absent when the Fermi seas don't exist. This reduces the Hamiltonian to one where the only relevant internal characteristic of the dot is its spin: up or down. What Kondo demonstrated was that a perturbation series in S^+ and S^- produces non-vanishing terms to all orders due to the non-commutativity of these operators. This in turn leads to a hybridization of the spin up and spin down states and the generation of a screening spin cloud in the leads.

Figure 3.19 shows an energy diagram of a quantum dot that has one spin-up electron at the uppermost level whose energy is $\epsilon_{\rm d}$. The dot is coupled with source- and drain-leads via tunnel barriers. The parameters $\Gamma_{\rm L}$ and $\Gamma_{\rm R}$ are the tunnel coupling strength (tunnel rate) for the left and right barriers, respectively. $\epsilon_{\rm d}$ is adjusted so that $\epsilon_{\rm d} < \mu < \epsilon_{\rm d} + U$, where μ is the Fermi energy and U is the charging energy. Then, the first-order tunneling through the dot is inhibited because of U (Coulomb blockade). However, when the temperature is lowered to $\sim T_{\rm K}$, the electron in the dot forms an antiferromagnetic coupling with a spin-down electron in the lead, and concurrent tunneling of these two electrons contributes to a net electron transfer between the source and drain leads as

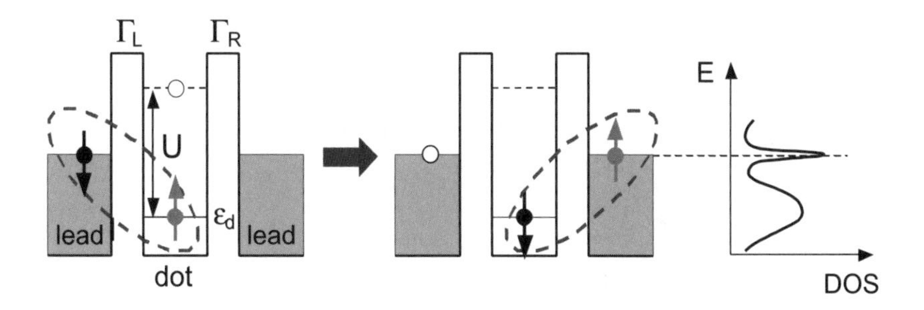

FIG. 3.19. Schematic energy diagram of a quantum dot coupled with the source and drain leads via tunnel barriers. The dot has spin $S = 1/2$ when an uppermost level has an unpaired electron. The Kondo effect enhances the conductance through higher order tunneling of antiferromagnetically correlated spin pairs at $T < T_K$.

depicted in Fig. 3.19. At the same time, a Kondo resonance peak appears in the local density of states (DOS) at the Fermi energy. Therefore, the resistance (conductance) starts to drop (increase) at $T \sim T_K$ as shown in Fig. 3.18(b). This behavior is qualitatively opposite to that in bulk metals because the current path is only through the "magnetic impurity" in the case of the quantum dot. T_K is given as

$$T_K = \frac{\sqrt{\Gamma U}}{2} \exp\left[-\pi(\mu - \epsilon_d)\left(U - \mu + \epsilon_d\right)/\Gamma U\right], \tag{3.24}$$

where $\Gamma = \Gamma_L + \Gamma_R$. The above second-order tunneling occurs coherently and tends to screen the initial localized magnetic moment in the dot through the dot-lead spin singlet formation.

The number of electrons, N, contained in the quantum dot can be changed one by one by adjusting the gate voltage as shown in Fig. 3.20(a). The conductance shows normal Coulomb blockade oscillations at $T \gg T_K$ (dashed line). The peaks appear when N changes by one. The conductance in the Coulomb blockade valleys of odd N increases as T is lowered below T_K due to the Kondo effect, while that of even N slightly decreases because of suppressed cotunneling (solid line). When the gate voltage is fixed at the odd N Kondo valleys (open triangles in Fig. 3.20a), a peak appears at zero source-drain bias, V_{sd}, in the dI/dV_{sd} vs. V_{sd} characteristic reflecting a peak in DOS as shown in Fig. 3.20(b). The conductance, G, of this zero-bias peak shows a temperature dependence

$$G = \frac{2e^2}{h}\frac{4\Gamma_L\Gamma_R}{\left(\Gamma_L + \Gamma_R\right)^2} f(T/T_K), \tag{3.25}$$

where $f(T/T_K)$ is a universal function that shows a linear $\log T$ dependence at $T \sim T_K$ and saturates to 1 at $T \ll T_K$ (the unitary limit), as schematically illustrated in Fig. 3.20 (c). An empirical form of f is given by

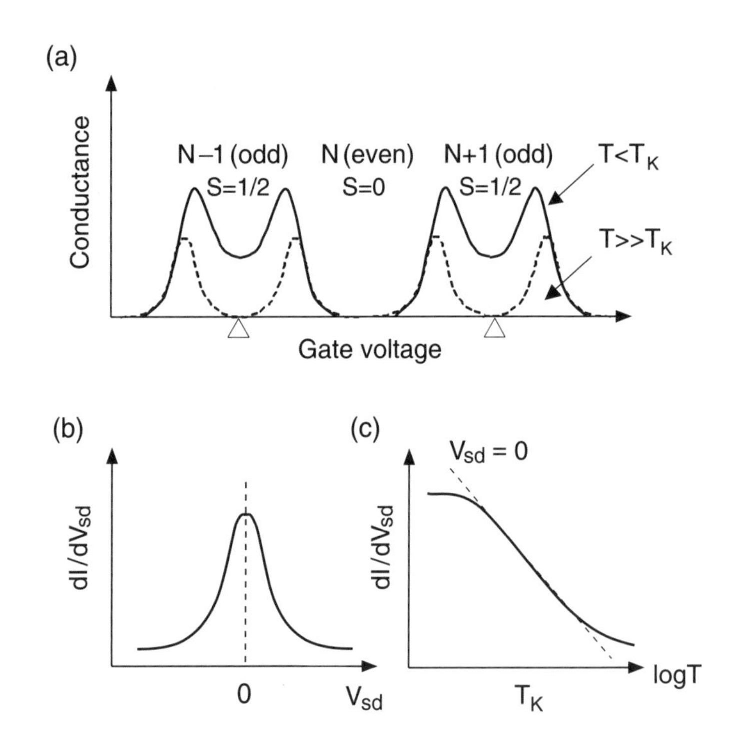

FIG. 3.20. (a) Coulomb blockade oscillations for $T \gg T_{\mathrm{K}}$ (dashed line) and for $T < T_{\mathrm{K}}$ (solid line). The conductance of only the odd N valleys increases due to the Kondo effect. (b) The differential conductance vs. source-drain bias characteristic when the gate voltage is fixed at the center of the odd N valleys. (c) Temperature dependence of the Kondo zero-bias peak showing a linear $\log T$ behavior at $T \sim T_{\mathrm{K}}$.

$$f(x) = \frac{1}{\left[x^2 \left(2^{1/s} - 1\right) + 1\right]^s}, \tag{3.26}$$

where s is a fitting parameter (close to 0.2 in the case of a spin 1/2 system) [62,63]. These are the experimental "fingerprints" of the Kondo effect in a quantum dot observed in the first reports in 1998 [64–66]. Since then, there has been renewed interest in the Kondo effect in artificial systems. One can tune relevant parameters by, for instance, gate voltage and external magnetic field, thus opening up unprecedented opportunities of studying rich aspects of Kondo physics.

One of the most remarkable findings in the quantum dot related Kondo physics is the observation of the unitary limit shown in Fig. 3.21 [67]. The inset shows the AFM image of the device structure: GaAs/AlGaAs 2DEG is constricted by combined dry-etching and surface Schottky gates to form a quantum dot of a size 100 nm embedded in each arm of a larger ring structure. In the following, the ring structure is irrelevant and we focus on only one of the dots. The

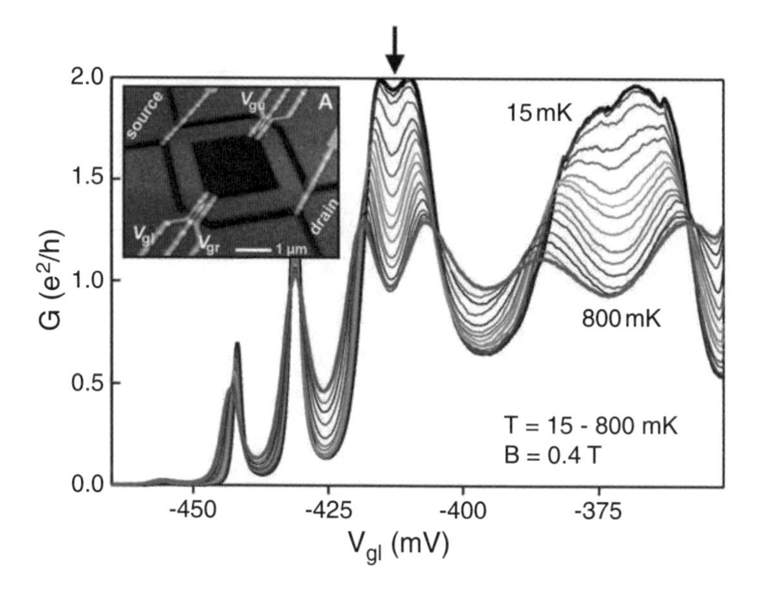

FIG. 3.21. Temperature dependence of the Coulomb oscillations in a lateral quantum dot between 15 mK and 800 mK. The unitary limit is reached at the odd N valley marked by an arrow. The inset shows an atomic force microscope image of the lateral dot device embedded in a ring.

dot has two long gates that define Γ_L and Γ_R, and another shorter gate (plunger gate) that mainly modifies the electrostatic potential of the dot. Figure 3.21 shows the temperature dependence of the Coulomb oscillation characteristics when one of the long gates is swept. Here, the long gate is used for modifying both Γ_L and the electrostatic potential. The conductance in the Coulomb blockade valley marked by an arrow increases as the temperature decreases, while that in the adjacent valleys shows the opposite temperature dependence. This indicates that the Kondo effect occurs in the marked valley because N is odd and spin S is 1/2 there. Moreover, the conductance saturates to a value close to $2e^2/h$ at the low-temperature limit, thus reaching the unitary limit. This corresponds to $\Gamma_L = \Gamma_R$ and $T/T_K \ll 1$ in Eq. (3.25).

3.7.3 *The S-T and D-D Kondo effect*

Since we have assumed no orbital degeneracy so far, no Kondo effect should occur for even N. However, an exceptional case was reported where the Kondo effect was observed for even N in addition to odd N [68]. This was followed by a more detailed report of the Kondo effect for even N [69] in a "vertical" quantum dot having well-defined N and S [12]. This novel Kondo effect was found to occur when a magnetic field induces a spin singlet ($S = 0$)-triplet ($S = 1$) degeneracy in the multi-orbital ground state [70–73].

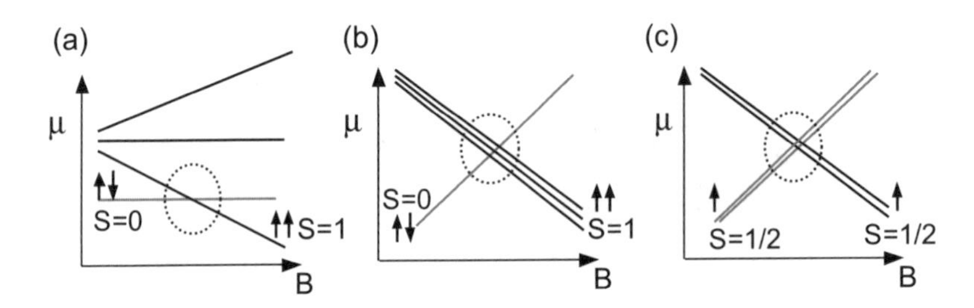

FIG. 3.22. Schematic energy diagrams involving two orbitals for (a) large Zeeman splitting with even N, (b) small Zeeman splitting with even N, and (c) small Zeeman splitting with odd N. Coulomb interaction between the two electrons is implicit in (a) and (b).

Figure 3.22 shows three cases of magnetic field induced degeneracy involving two orbital states. First, Fig. 3.22(a) is for even N and large Zeeman splitting compared to an orbital shift. In this case, one of the three Zeeman split sublevels of the triplet state becomes degenerate with the singlet state in a magnetic field (dotted circle). It was theoretically shown that the Kondo effect occurs for such a situation in spite of even N [71], and later experimentally realized in a carbon nanotube [74]. Figures 3.22(b) and (c) are relevant for GaAs-based quantum dots with a small Zeeman splitting. In Fig. 3.22(b) with even N, all three triplet sublevels can be regarded as degenerate. Then, there are four degenerate levels at the magnetic field induced singlet-triplet degeneracy (dotted circle). This enhanced degeneracy compared to the conventional $S = 1/2$ case (two-fold degeneracy) leads to enhanced T_{K} as experimentally [69] and theoretically [70] demonstrated. A similar four-fold degeneracy is realized for odd N when two doublet states associated with two different orbitals become degenerate as shown in Fig. 3.22(c). The enhanced Kondo effect for this case is observed [75] as described in more detail below.

The inset to Fig. 3.23 schematically shows the vertical quantum dot structure. It is in the form of a submicron-sized circular mesa fabricated from an AlGaAs/InGaAs/AlGaAs double barrier structure. Current flows vertically through the mesa, and N is changed with a gate electrode wrapped around the mesa. S and N can be unambiguously determined in this type of vertical quantum dot owing to the highly symmetric lateral confinement potential and a built-in dot-lead coupling Γ via AlGaAs barriers [12]. Γ for the present dot device is 400 μeV for the first electron entering the dot and gradually increases as N increases. We apply a magnetic field parallel to the current.

Figure 3.23 shows a gray-scale plot of the linear conductance, G, as a function of gate voltage, V_{g}, and magnetic field, B, at 60 mK. Bright stripes correspond to Coulomb oscillation peaks showing an evolution of the N-electron ground state with B. Large Coulomb blockade gaps at $N = 2$ and 6 at $B = 0$ reflect the shell filling of $1s$ and $2p$ orbital states, respectively, formed in a disk-shaped circular

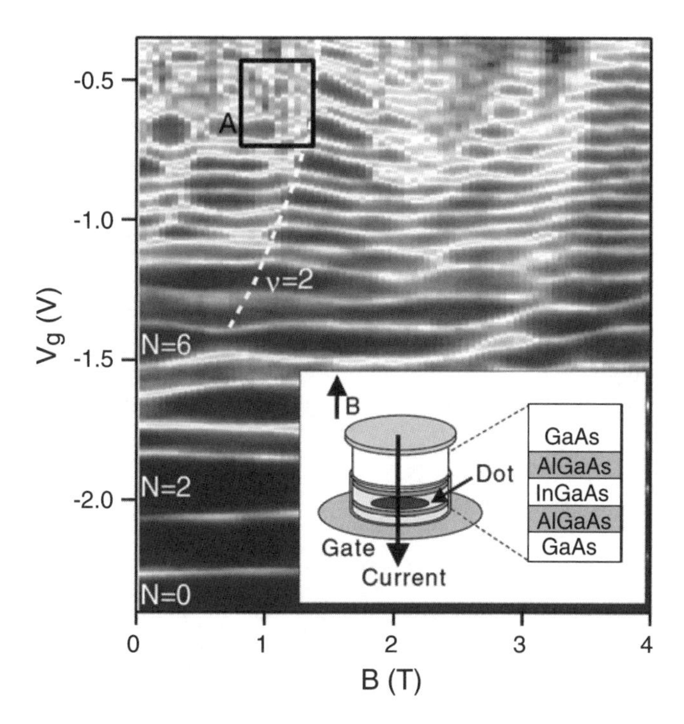

FIG. 3.23. Grey-scale plot of the linear conductance observed in a vertical quantum dot as a function of the gate voltage and magnetic field. Black corresponds to zero conductance, and white to $G = 50\,\mu\text{S}$. The inset shows a schematic diagram of the dot structure made from a AlGaAs/InGaAs/AlGaAs double barrier tunnel diode. The magnetic field is perpendicular to the dot plane.

artificial atom [12]. The overall evolution of stripes with B agrees with successive filling of the single-particle levels, or Fock–Darwin states, $E_{n,l}$, of Eq. (3.3) by spin-up and spin-down electrons. The pairwise motion of the stripes in Fig. 3.23 is due to the spin degeneracy of each $E_{n,l}$ state.

When the Kondo effect occurs at $T < T_{\text{K}}$, the conductance in the Coulomb blockade gap increases. In our vertical quantum dot, Γ gradually increases with increasing V_{g}. Therefore, T_{K} generally increases as V_{g} increases, and the enhancement of the conductance due to the Kondo effect is noticeable in the Coulomb blockade gaps for $V_{\text{g}} > -1.0$ V.

The solid lines in Fig. 3.24(b) schematically show a magnetic field dependence of the electrochemical potential, μ, for electron numbers from $N + 1$ (odd) to $N + 4$ (even) when electrons are added to the two crossing orbitals. Coulomb interaction favors a spin triplet state ($S = 1$) for $N + 2$ in the vicinity of the magnetic field where a single-particle level crossing occurs, and the ground state shows a downward cusp [13]. The S-T Kondo effect is expected on the dotted lines

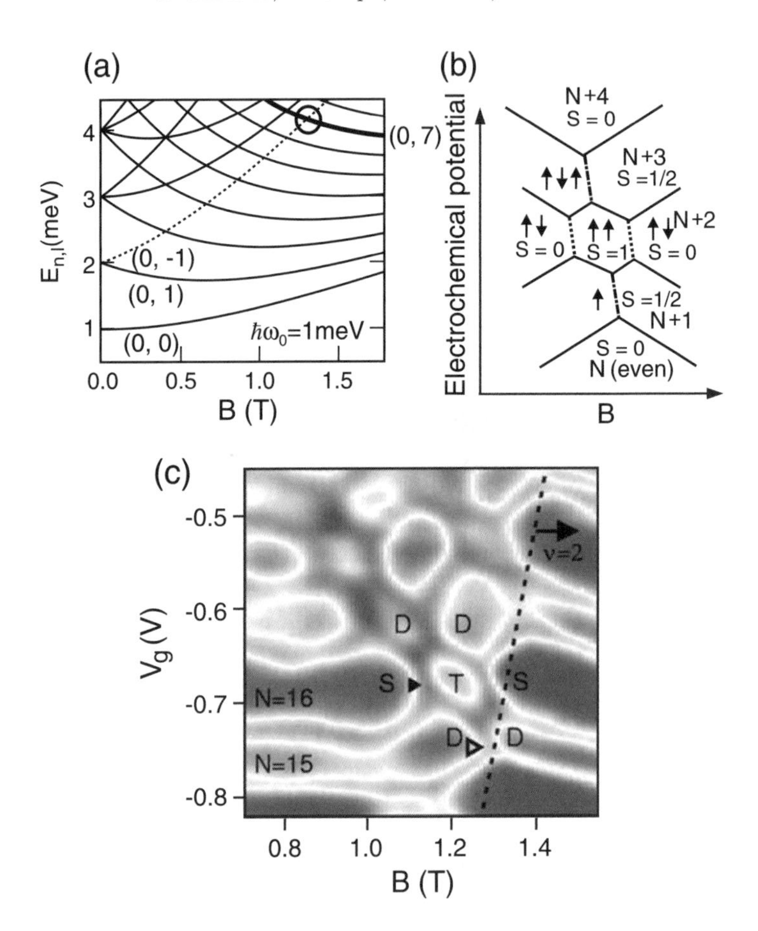

FIG. 3.24. (a) Fock–Darwin states calculated with $\hbar\omega_0 = 1$ meV. (b) Schematic magnetic field dependence of the electrochemical potential for electron numbers from $N + 1$ (odd) to $N + 4$ (even) occupying two crossing orbitals. (c) Detailed measurement conducted in region A in Fig. 3.23. S, T, and Δ denote states with $S = 0$ (singlet), $S = 1$ (triplet), and $S = 1/2$ (doublet).

with a considerably higher Kondo temperature, $T_{\mathrm{K}}^{\mathrm{S-T}}$, than the conventional $S = 1/2$ Kondo temperature, $T_{\mathrm{K}}^{\mathrm{D}}$, because of the larger degeneracy [69].

Figure 3.24(c) shows detailed measurements conducted in region A marked in Fig. 3.23. The last orbital crossings occur between $E_{n,l}$ states with $(n, l) = (0, -1)$ and $(0, l)(l > 1)$ on the dotted line, and all the electrons occupy the ground Landau level at higher B (filling factor $n = 2$). A spin triplet state is observed at $N = 16$ and $B = 1.2$ T, where states $(n, l) = (0, -1)$ and $(0, 7)$ are occupied by electrons having parallel spins (see the circle in Fig. 3.24a). In the $N = 16$ Coulomb blockade gap, the conductance is enhanced by the S-T Kondo effect at $B = 1.1$ and $B = 1.3$ T, corresponding to the dotted lines in Fig. 3.24

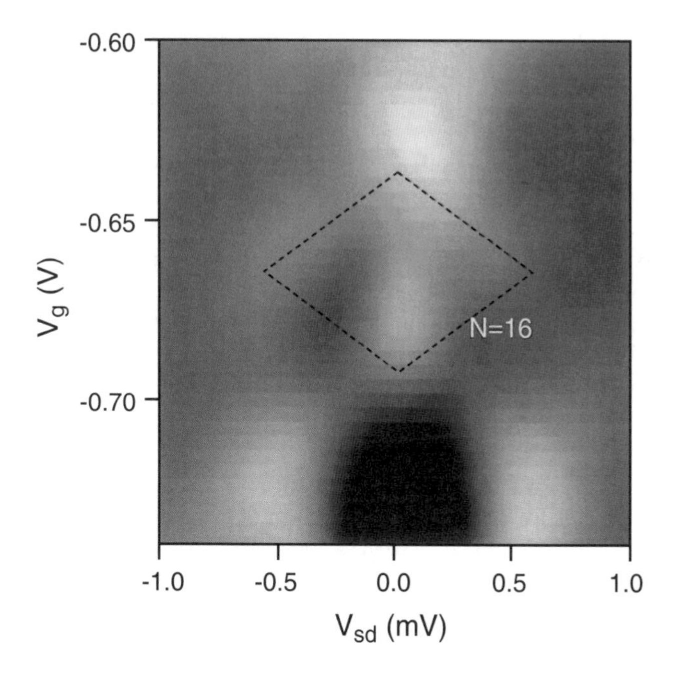

FIG. 3.25. Grey-scale plot of the conductance in the V_g–V_{sd} plane (Coulomb diamonds) at $B = 1.12$ T. Dashed lines mark the boundary of the $N = 16$ Coulomb blockade region.

(b). As for $N = 15$ and 17, the conventional $S = 1/2$ Kondo effect is expected. However, the conductance enhancement in the Coulomb blockade gaps is not clearly observed except in the regions corresponding to the dash-dotted lines in Fig. 3.24(b), where two $S = 1/2$ states with different total angular momentum, M, are degenerate. When such an orbital degeneracy is present for odd N, a total of four states, i.e., $M = M_1, M_2$ ($M_1 \neq M_2$), $S_Z = \pm 1/2$, are involved in forming the Kondo singlet state (if one can neglect the Zeeman splitting), and enhancement of T_K similar to the S-T Kondo effect is expected. We refer to this new multilevel mechanism for odd N as the "doublet-doublet" (D-D) Kondo effect. Because T_K^D is much lower than the D-D Kondo temperature, T_K^{D-D}, only a slight conductance enhancement is observed in the odd N Coulomb blockade regions when there is no orbital degeneracy. Then, a honeycomb pattern is formed in a B-N diagram, as is clearly captured in Fig. 3.24(c), due to the S-T and D-D Kondo effects that occur consecutively for different orbital crossings, provided $T_K^D < T < T_K^{S-T}, T_K^{D-D}$.

Figure 3.25 shows a gray-scale plot of G as a function of V_g and source-drain bias, V_{sd} (Coulomb diamonds). B is tuned to 1.12 T where the S-T Kondo effect occurs at $N = 16$. The edges of the $N = 16$ Coulomb blockade region are shown by the dashed lines. A clear conductance peak, or ridge, is observed along

$V_{\rm sd} = 0$ within the $N = 16$ diamond, further evidence of the Kondo effect. The peak height decreases with increasing temperature as expected for the Kondo effect. By fitting this temperature dependence to Eq. (3.25), we have deduced $T_{\rm K}^{\rm S-T} = 700$ mK. A similar curve fitting conducted for the D-D Kondo effect at $N = 15$ yielded $T_{\rm K}^{\rm D-D} = 490$ mK. The higher $T_{\rm K}$ for the S-T Kondo effect may be due to the larger Γ, and it is difficult to experimentally determine which is larger, $T_{\rm K}^{\rm S-T}$ or $T_{\rm K}^{\rm D-D}$, for the same Γ. The unitary limit conductance of $2e^2/h$ [67] is not reached in our device at the low-temperature limit probably because of the asymmetry in the two tunnel barriers; it is impossible to tune Γ for the two barriers separately as is commonly done in a lateral quantum dot. The expected Zeeman splitting of ~ 30 μeV at $B = 1.12$ T is smaller than $T_{\rm K}$ estimated above. Therefore, no Zeeman splitting of the Kondo peak is observed, and we treat all four S-T and D-D states as quasidegenerate.

The strong S-T and D-D Kondo effect is expected to disappear when the degeneracy is lifted by changing B. Figures 3.26(a) and (b) show a gray-scale plot of $dI/dV_{\rm sd}$ in the B-$V_{\rm sd}$ plane for the D-D Kondo effect ($N = 15$) and for the S-T Kondo effect ($N = 16$), respectively, with $V_{\rm g}$ fixed in the center of the respective Coulomb blockade gap. Conductance peaks at $V_{\rm sd} = 0$ are observed near the degeneracy field, B_0. The two zero-bias S-T Kondo peaks in Fig. 3.26(b) at $B_0 = 1.12$ T and 1.25 T correspond to the two conductance maxima in the $N = 16$ Coulomb blockade gap (see Fig. 3.24c). Because the S-T or D-D degeneracy is lifted as $|\Delta B| = |B - B_0|$ increases, the Kondo effect is broken and the zero-bias peak is suppressed. At large $|\Delta B|$, a peak or step is observed at $eV_{\rm sd} = \pm\Delta$, where the brightness suddenly changes. Here, Δ is the B-dependent energy difference between the singlet and triplet states, or between the two doublet states. This peak/step is due to cotunneling associated with the two states separated by Δ [69], and is therefore observed within the Coulomb blockade gap (~ 0.6 meV).

Figure 3.26(c) shows $V_{\rm sd}$ values of the conductance peak/step as a function of ΔB. Peak/step positions for the S-T ($B_0 = 1.12$ T) and the D-D ($B_0 = 1.25$ T) Kondo effect almost coincide, indicating that they involve the same orbital states, namely, $(n, l) = (0,-1)$ and $(0, 7)$. Figure 3.26(d) shows the relative conductance, ΔG, measured from the degeneracy ($\Delta B = 0$) at $V_{\rm sd} = 0$, as a function of ΔB. In the S-T Kondo effect, the conductance drops more quickly on the singlet side ($\Delta B < 0$) compared to the triplet side ($\Delta B > 0$) because there is no Kondo effect for $S = 0$. The conductance drops even more slowly and symmetrically in the D-D Kondo effect.

Figure 3.26(e) shows the Δ dependence of $T_{\rm K}$ (Δ) for both the S-T and D-D Kondo effect obtained by scaling calculation [70, 75]. $T_{\rm K}$ (D) and Δ are normalized by $T_{\rm K}(0)$. $T_{\rm K}$ drops according to the power law $T_{\rm K}(\Delta) \propto 1/\Delta^{\gamma}$ with $\gamma = 2 + \sqrt{5}$ for the triplet side [70, 71] in the S-T Kondo effect. On the other hand, $\gamma = 1$ is obtained by considering an SU(4) model in the case of the D-D Kondo effect [75]. Thus, $T_{\rm K}$ drops more slowly than in the S-T Kondo effect. These are in qualitative agreement with the above experimental results.

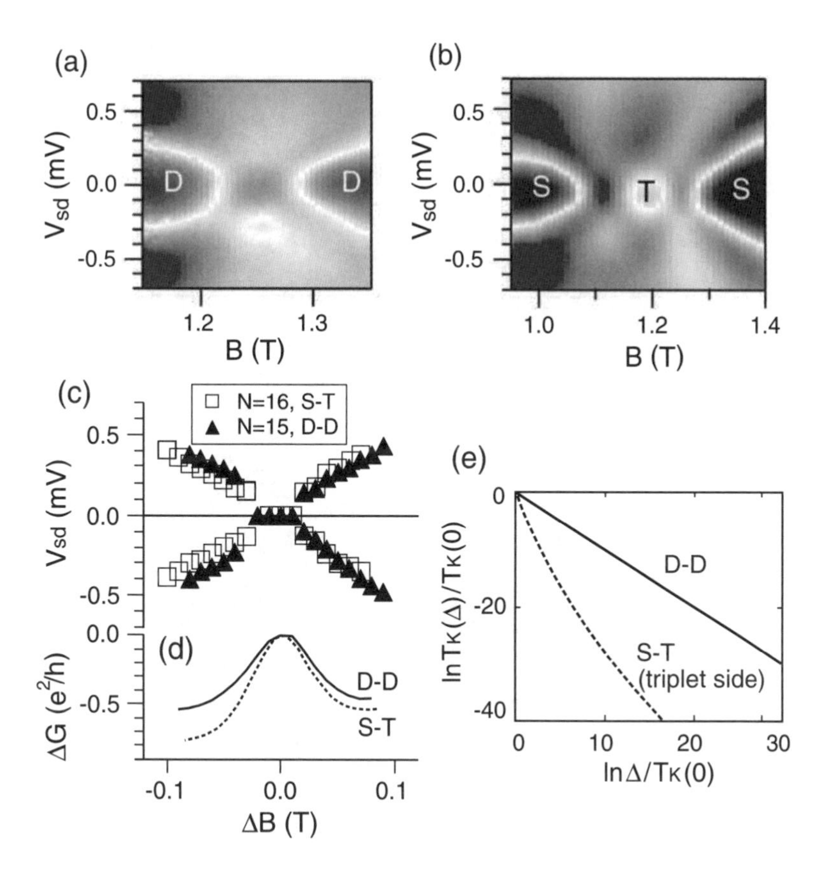

FIG. 3.26. Grey-scale plot of $dI/dV_{\rm sd}$ in the B-$V_{\rm sd}$ plane for (a) the $N = 15$ D-D Kondo effect and (b) the $N = 16$ S-T Kondo effect with $V_{\rm g}$ fixed in the center of the respective Coulomb blockade gap. Black corresponds to $G = 10\,{\rm mS}$ and white to $G = 35\,\mu{\rm S}$. (c) $V_{\rm sd}$ values of the conductance peak or step obtained from (a) and (b) as a function of the magnetic field difference $\Delta B = B - B_0$, where B_0 is the degeneracy magnetic field. (d) Relative conductance measured from the degeneracy ($\Delta B = 0$) at $V_{\rm sd} = 0$, for the D-D (solid line) and the S-T (dotted line) Kondo effect. (e) Calculated results of $T_{\rm K}/T_{\rm K}(0)$ as a function of $\Delta/T_{\rm K}K(0)$, on a log-log scale.

The honeycomb pattern in Fig. 3.24(c) may be reminiscent of the "chessboard pattern" discussed in a "lateral" quantum dot [76–79]. Figure 3.27(b) shows a gray scale plot of the linear conductance, G, as a function of B and V_{gl} observed in a lateral quantum dot shown in Fig. 3.27(a) [79]. The dark regions correspond to large or small G. For the most negative values of V_{gl}, the coupling of the dot to the leads is weak. This results in relatively sharp Coulomb peaks and low valley conductance. However, if V_{gl} is increased, the valley conductance reaches

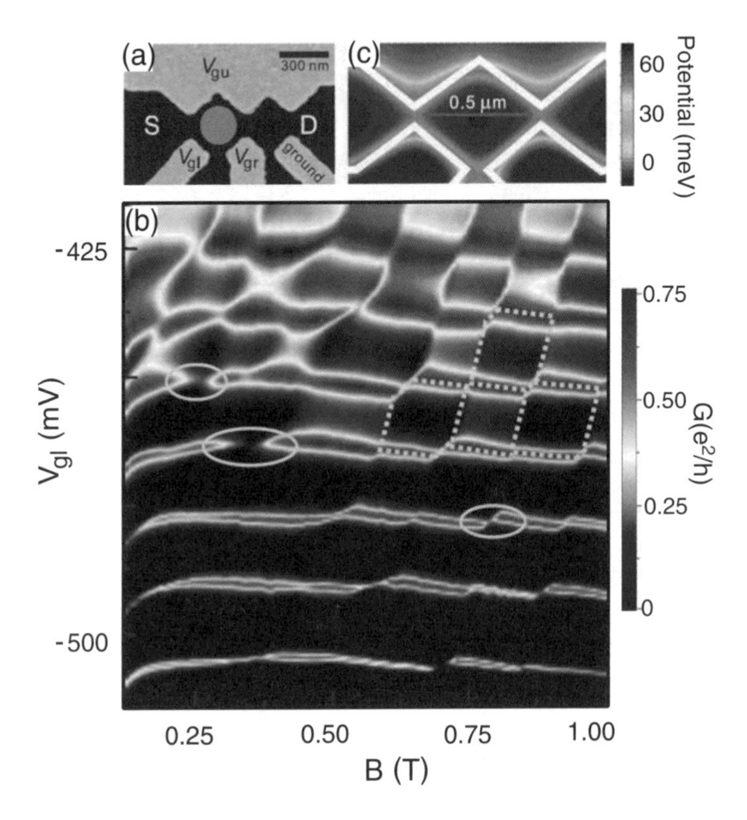

FIG. 3.27. (a) Scanning electron micrograph of the lateral dot device. Metal gates are light gray; the dot is indicated by a circle. (b) Gray scale plot of the experimental linear conductance G through the dot as function of B and V_{gl} at 10 mK. The dotted hexagons highlight the shape of a few chessboard fields; ellipses indicate some regions where Coulomb peak suppression occurs. (c) Calculated self-consistent potential landscape of device; white lines denote contours of the metal gates.

considerable values ($\sim e^2/h$) in certain regions due to the Kondo effect. Most strikingly, the regions of low and high valley conductance alternate both along the V_{gl} and the B axis in a regular fashion, resulting in the aforementioned chessboard pattern.

The transition from low to high valley conductance is associated with an abrupt jump of the V_{gl} position of the Coulomb peaks. This jump occurs every time rearrangement of the electrons within the dot occurs between the inner orbital associated with the second Landau level, and the outer orbital associated with the first Landau level. In a lateral quantum dot, a "single channel" in the leads preferentially couples to the outer orbital. Then, the Kondo effect occurs when the number of electrons in the "outer" orbital, rather than the "total" number of electrons, is odd. In a vertical quantum dot, on the other hand, one can assume that the orbital quantum numbers are conserved in tunnel processes between the dot and leads due to their same rotational symmetry. Hence we expect "two channels" of conduction electrons in the leads when two orbitals are relevant in the quantum dot; each channel couples to only one of the two orbitals. Therefore, the chessboard pattern is not observed there.

3.8 Conclusions

In this chapter we have described how to probe and manipulate spin effects in quantum dots. The electronic configuration in quantum dots is tunable with various parameters such as number of electrons, magnetic field, and symmetry in the confinement, and also with coupling strength for systems of coupled double quantum dots and coupled quantum dot-contact leads. This is particularly true for spin configurations as we have typically discussed for Zeeman splitting, crossings of spin states and exchange coupling. These spin-related phenomena are sufficiently robust that they manifest themselves in the electronic properties.

We have also shown that control over the spin configurations enables a new approach to spin-related phenomena such as Pauli spin blockade and the novel Kondo effect in single-electron transport, and in addition, can provide a novel concept of spintronics, i.e., spin-based quantum computing. Long decoherence time and good tunability of spin in quantum dots are embodied both experimentally and theoretically. These are quite advantageous for implementing qubits and quantum gates for quantum computing. Probing decoherence for single electron spin in individual quantum dots and correlation between two spins in two quantum dots are currently two major subjects. The next step is to facilitate local, temporal access to single spins for making one or more qubits, and exchange coupling for making quantum gates. For this purpose, new techniques such as a local, pulsed ESR and a local, pulsed gating of the tunnel coupling between two dots will be necessary. In addition, readout of a single spin orientation and control of entanglement between two spins are also key technologies.

Acknowledgements

The authors thank T. Fujisawa, T. Hatano, D. G. Austing, Y. Tokura, W. van der Wiel, L. P. Kouwenhoven, S. De Franceschi, R. Hanson, and J. M. Elzerman for valuable contributions to the work described here. Part of this work is financially supported from a Grant-in-Aid for Scientific Research A (No. 40302799) from the Japan Society for the Promotion of Science, from Focused Research and Development Project for the Realization of the World's Most Advanced IT Nation, IT Program, MEXT and from the DARPA grant number DAAD19-01-1-0659 of the QuIST program.

References

[1] M. Stopa, Y. Aoyagi, and T. Sugano, Phys. Rev. B **51**, 5494-5497 (1995).

[2] K. Yamada M. Stopa, T. Hatano, T. Ota, T. Yamaguchi, and S. Tarucha, Superlattices Microstruct. **34**, 185 (2003).

[3] O. Bohigas, M. J. Giannoni, and C. Schmit, Phys. Rev. Lett. **52**, 1 (1984).

[4] M. Stopa, Semiconductor Science and Technology, **13**, A55-A58 (1998).

[5] Y. Alhassid and T. Rupp, Phys. Rev. Lett. **91**, 056801 (2003)

[6] U. Sivan, Y. Berkovits, Y. Aloni, O. Prus, A. Auerbach, and G. BenYoseph, Phys. Rev. Lett. **77**, 1123 (1996).

[7] S. R. Patel, S. M. Cronenwett, D. R. Stewart, A. G. Huibers, C. M. Marcus, C. I. Duruoz and J. S. Harris, Jr., K. Campman, and A. C. Gossard, Phys. Rev. Lett. **80**, 4522 (1998).

[8] D. G. Austing, T. Honda, Y. Tokura, and S. Tarucha, Jpn. J. Appl. Phys. **34**, 1320 (1995); D. G. Austing, T. Honda, and S. Tarucha, Semicond. Sci. Technol. **11**, 388 (1996).

[9] M. Ciorga1, A. S. Sachrajda1, P. Hawrylak1, C. Gould, P. Zawadzki1, S. Jullian, Y. Feng1, and Z. Wasilewski, Phys. Rev. B **61**, R16315 (2000).

[10] V. Fock, Z. Phys. **47**, 446 (1928); C. G. Darwin, Proc. Cambridge Philos. Soc. **27**, 86 (1930).

[11] P. L. McEuen, E. B. Foxman, Jari Kinaret, U. Meirav, M. A. Kastner, Ned S. Wingreen, and S. J. Wind, Phys. Rev. B **45**, 11419 (1992).

[12] S. Tarucha, D. G. Austing, T. Honda, R. J. van der Hage, and L. P. Kouwenhoven, Phys. Rev. Lett. **77**, 3613 (1996).

[13] S. Tarucha, D. G. Austing, Y. Tokura, W. G. van der Wiel, and L. P. Kowenhoven, Phys. Rev. Lett. **84**, 2485 (2000).

[14] D. R. Stewart, D. Sprinzak, C. M. Marcus, C. I. Duruoz, and J. S. Harris, Science **278**, 1784 (1997).

[15] See, for example, Claude Itzykson, and Jean-Bernard Zuber, *Quantum Field Theory*, (McGraw-Hill, New York, 1980).

[16] See, for example, G. K. Woodgate, *Elementary Atomic Struture* (McGraw-Hill, London 1970).

[17] A. V. Rodina, Al. L. Efros, and A. Yu. Alekseev, Phys. Rev. B **67**, 155312 (2003).

[18] R. Hanson, B. Witkamp, L. M. K. Vandersypen, L. H. Willems van Beveren, J. M. Elzerman, and L. P. Kouwenhoven, Phys. Rev. Lett. **91**, 196802 (2003).

[19] D. Loss and D. P. DiVincenzo, Phys. Rev. A **57**, 120 (1998).

[20] T. Kodera, W. G. van der Wiel, T. Maruyama, Y. Hirayama, and S. Tarucha, to appear in World Scientific Publishing (2005).

[21] T. Fujisawa and S. Tarucha, Appl. Phys. Lett. **68**, 526 (1996); T. Fujisawa and S. Tarucha, Superlattices Microstruct. **21**, 247 (1977); T. Fujisawa and S. Tarucha, Jpn. J. Appl. Phys. Part 1, **36**, 4000 (1977); C. Livermore, C. H. Crouch, R. M. Westervelt, K. L. Campman, and A. C. Gossard, Science, **274**, 1332 (1996).

[22] T. Hatano, M. Stopa, T. Yamaguchi, T. Ota, K. Yamada, and S. Tarucha, Phys. Rev. Lett. **93**, 066806 (2004).
[23] T. Ota, K. Ono, M. Stopa, T. Hatano, S. Tarucha, H. Z. Song, T. Nakata, T. Miyazawa, T. Oshima and N. Yokoyama, Phys. Rev. Lett. **93**, 066801 (2004).
[24] W. G. van der Wiel, S. De Franceschi, J. M. Elzerman T. Fujisawa, S. Tarucha and L. P. Kouwenhoven, Rev. Mod. Phys. **75**, 1 (2003).
[25] P. Fulde, *Electron Correlations in Molecules and Solids*, (Springer, Berlin, Heidelberg 1995).
[26] W. Heitler and F. London, Z. Phys. **44**, 455 (1927).
[27] G. Burkard, D. Loss, and D. P. DiVincenzo, Phy. Rev. Lett. **59**, 2070 (1999).
[28] X. Hu, and S. Das Sarma, Phy. Rev. A **61**, 062301 (2000).
[29] M. Stopa, S. Tarucha and T. Hatano, preprint.
[30] J. M. Kikkawa and D. D. Awschalom, Phys. Rev. Lett. **80**, 4313 (1998).
[31] Y. Ono, D. K. Young, D. Beschoten, F. Matsukura, H. Ohno, and D. D. Awschalom, Nature **402**, 790 (1999).
[32] J. A. Folk, R. M. Potok, C. M. Marcus, and V. Umansky, Science **299**, 679 (2003).
[33] K. Ono, D. G. Austing, Y. Tokura, and S. Tarucha, Science **297**, 1313 (2002).
[34] M. E. Flatte, J. M. Byers, and W. H. Lau, pp. 107-145 in *Semicondutor Spintronics and Quantum Computation*, edited by D. D. Awschalom, D. Loss and N. Samarth (Springer, 2002).
[35] J. A. Gupta and D. D. Awschalom, Phys. Rev. B **63**, 085303 (2001).
[36] T. Fujisawa, D. G. Austing, Y. Tokura, Y. Hirayama, and S. Tarucha, Nature **419**, 278 (2002).
[37] U. Bockelmann, Ph. Roussignol, A. Filoramo, W. Heller, G. Abstreiter, K. Brunner, G. Bohm, and G. Weimann, Phys. Rev. Lett. **76**, 3622 (1996).
[38] J. M. Elzerman, R. Hanson, L. H. Willems van Beveren, B. Witkamp, L. M. K. Vandersypen, and L. P. Kouwenhoven, Nature **430**, 431 (2004).
[39] M. Kroutvar, Y. Ducommun, D. Heiss, M. Bichler, D. Schuh, G. Abstreiter, and J. J Finley, Nature **432**, **81** (2004).
[40] A. V. Khaetskii, and Yu. V. Nazarov, Phys. Rev. B **61**, 12639 (2000).
[41] A. V. Khaetskii, and Yu. V. Nazarov, Phys. Rev. B **64**, 125316 (2001).
[42] V. N. Golovach, A. Khaetskii, and D. Loss, Phys. Rev. Lett. **93**, 016601 (2004).
[43] For single dot, D. Weimann, H. Hausler, and B. Kramer, Phys. Rev. Lett. **74**, 984 (1995); Y. Tanaka and H. Akera, Phys. Rev. B **53**, 3091 (1996).
[44] For double dot, H. Imamura, H. Aoki, and P.A. Maksym, Phys. Rev. B **57**, R4259 (1998); Y. Tokura, D. G. Austing, and S. Tarucha, J. Phys.: Condens. Matter **11**, 6023 (1999).
[45] L. P. Rokhinson, L. J. Guo, S. Y. Chou, and D. C. Tsui, Phys. Rev. B 63, 035321 (2001); A. H. Huettel, H. Qin, A. W. Holleithner, R. H. Blick, K. Neumaier, D. Weimann, K. Eberl, and J. P. Kotthaus, Europhys. Lett. **62**, 712 (2003).
[46] M. Ciorga, A. S. Sachrajda, P. Hawrylak, C. Gould, P. Zawadzki, S. Jullian, Y. Feng, and Z. Wasilewski, Phys. Rev. B **61**, R16315 (2000).

[47] D. G. Austing, T. Honda, K. Muraki, Y. Tokura, and S. Tarucha, Physica B **249-251**, 206 (1998).
[48] S. Amaha, D. G. Austing, Y. Tokura, K. Muraki, K. Ono, and S. Tarucha, Solid State Commun. **119**, 183 (2001).
[49] M. Pi, A. Emperador, M. Barranco, F. Garcias, K. Muraki, S. Tarucha, and D. G. Austing, Phys. Rev. Lett. **87**, 066801 (2001).
[50] D. V. Averin, Yu. V. Nazarov, in: *Single Charge Tunneling, in Coulomb Blockade Phenomena in Nanostructures*, edited by H. Grabert and M. H. Devoret, 217-247 (Plenum Press and NATO Scientific Affairs Division, 1992).
[51] S. I. Erlingsson Yu. V. Nazarov, and V. I. Fal'ko, Phys. Rev. B **64**, 195306 (2001); A. V. Khaetskii, D. Loss, and L. Glazman, Phys. Rev. Lett. **88**, 186802 (2002).
[52] Y. Tokura, unpublished. The energy separation is calculated to be $10-40\,\mu$eV depending on V_{sd}.
[53] K. Ono and S. Tarucha, Phys. Rev. Lett. **92**, 256803 (2004).
[54] There are many experiments on nuclear spin relaxation in a 2DEG; M. Dobers *et al.*, Phys. Rev. Lett. **61**, 1650 (1988); A. Berg *et al.*, Phys. Rev. Lett. **64**, 2653 (1990); K. R. Wald *et al.*, Phys. Rev. Lett. **73**, 1011 (1994); D. D. C. Dixon *et al.*, Phys. Rev. B **56**, 4743 (1997); S. Kronmuller *et al.*, Phys. Rev. Lett. **81**, 2526 (1998); S. Kronmuller *et al.*, *ibid.* **82**, 4070 (1999); K. Hashimoto *et al.*, Phys. Rev. Lett. **88**, 176601 (2002); J. H. Smet *et al.*, Nature (London) **415**, 281 (2002); T. Machida *et al.*, Appl. Phys. Lett. **80**, 4178 (2002); S. Teraoka *et al.*, Physica E **21**, 928 (2004).
[55] *Optical Orientation*, edited by F. Meier and B. P. Zakharchenya (North-Holland, Amsterdam, 1984).
[56] T. Inoshita, K. Ono, and S. Tarucha, J. Phys. Soc. Jpn. **72**, Suppl. A 183 (2003).
[57] S. De Franceschi, S. Sasaki, J. M. Elzerman, W. G. van der Wiel, S. Tarucha, and L. P. Kouwenhoven, Phys. Rev. Lett. **86**, 878 (2001).
[58] J. Kondo, Prog. Theor. Phys. **32** (1964) 37.
[59] L. I. Glazman and M. E. Raikh, JETP Lett. **47**, 452 (1988).
[60] T. K. Ng and P. A. Lee, Phys. Rev. Lett. **61**, 1768 (1988).
[61] A. Kawabata, J. Phys. Soc. Jpn. **60**, 3222 (1991).
[62] T. A. Costi and A. C. Hewson, Phil. Mag. B **65**, 1165 (1992).
[63] D. Goldhaber-Gordon, J. Gores, M. A. Kastner, H. Shtrikman, D. Mahalu, and U. Meirav, Phys. Rev. Lett. **81**, 5225 (1998).
[64] D. Goldhaber-Gordon, H. Shtrikman, D. Mahalu, D. Abusch-Magder, U. Meirav, and M. A. Kastner, Nature **391**, 156 (1998).
[65] S. M. Cronenwett, T. H. Oosterkamp, and L. P. Kouwenhoven, Science **281**, 540 (1998).
[66] J. Schmidt, J. Weis, K. Ebel, and K. von Klitzing, Physica B **256-258**, 182 (1998).
[67] W. van der Wiel, S. De Franseschi, T. Fujisawa, J. M. Elzerman, S. Tarucha, and L. P. Kouwenhoven, Science **289**, 2105 (2000).

[68] J. Schmid, J. Weiss, K. Ebel, and K. von Klitzing, Phys. Rev. Lett. **84**, 5824 (2000).
[69] S. Sasaki, S. De Franceschi, J. M. Elzerman, W. G. van der Wiel, M. Eto, S. Tarucha, and L. P. Kouwenhoven, Nature **405**, 764 (2000).
[70] M. Eto and Yu. V. Nazarov, Phys. Rev. Lett. **85**, 1306 (2000); Phys. Rev. B 64, 85322 (2001); Phys. Rev. B **66**, 153319 (2002).
[71] M. Pustilnik, Y. Avishai, and K. Kikoin, Phys. Rev. Lett. **84**, 1756 (2000); Phys. Rev. Lett. **85**, 2993 (2000); Phys. Rev. B **64**, 045328 (2001).
[72] D. Giuliano and A. Tagliacozzo, Phys. Rev. Lett. **84**, 4677 (2000); Phys. Rev. B **63**, 125318 (2001).
[73] W. Izumida, O. Sakai, and S. Tarucha, Phys. Rev. Lett. **87**, 216803 (2001).
[74] J. Nygard, D. H. Cobden and P. E Lindelof, Nature **408**, 342 (2000).
[75] S. Sasaki, S. Amaha, N. Asakawa, M. Eto, and S. Tarucha, Phys. Rev. Lett. **93**, 17205 (2004).
[76] M. Keller, U. Wilhelm, J. Schmid, J. Weis, K. von Klitzing, and K. Ebel, Phys. Rev. B **64**, 033302 (2001).
[77] C. Fuhner, U. F. Keyser, R. J. Haug, D. Reuter, and A. D. Wieck, Phys. Rev. B **66**, 161305(R) (2002).
[78] U. F. Keyser, C. Fuhner, S. Borck, R. J. Haug, M. Bichler, G. Abstreiter, and W. Wegscheider, Phys. Rev. Lett. **90**, 196601 (2003).
[79] M. Stopa, W. G. van der Wiel, S. De Franceschi, S. Tarucha, and L. P. Kouwenhoven, Phys. Rev. Lett. **91**, 046601 (2003).

4 Spin-dependent transport in single-electron devices

Jan Martinek and Józef Barnaś

Electron tunneling in ferromagnetic junctions [1] is of current interest due to possible applications in magnetic storage technology and in spin-electronics devices [2–6]. The key effect for these applications is the tunnel magnetoresistance (TMR) in simple planar junction, i.e., a decrease (increase) in the junction resistance when its magnetic configuration changes from antiparallel to parallel. Tunneling in complex junctions, particularly in mesoscopic ones, where charging effects become important, is still purely explored. Specific kinds of such systems are double-barrier junctions with a small central electrode, known also as single-electron transistors (SETs). Tunneling in such devices has been extensively studied in the past decade, but only in the non-magnetic limit [7–10].

Recent experiments on magnetic nanostructured materials revealed new phenomena associated with the interplay of ferromagnetism and discrete charging effects. A typical example is a ferromagnetic single-electron transistor, i.e., a small grain or quantum dot coupled by tunnel junctions to ferromagnetic electrodes. The interplay of charge and spin degrees of freedom in such a system was studied only very recently both experimentally and theoretically. First ferromagnetic single-electron transistors were fabricated by Ono *et al.* [11] and later by Brückl *et al.* [12]. Recently ferromagnetic granular film systems, where the Coulomb interaction also plays an important role, were investigated by several groups [13–16]. Transport through a ferromagnetic single-electron transistor with a non-magnetic metallic island, both normal and superconducting was also measured [17–19]. From the technical point of view it is easier to attach a metallic island to ferromagnetic leads than a semiconducting quantum dot. One strategy is to use ferromagnetic semiconductor materials as electrodes [20] for a semiconducting dot. Another possibility is related to ultrasmall metallic nanoparticles attached to metallic ferromagnets, where due to the small size the quantum electronic structure becomes important [21]. Also magnetic impurities in the middle of the tunnel barrier of a ferromagnetic tunnel junction [22] can be considered as quantum dots with very strong Coulomb interaction. Another group of single-electron devices are molecular ferromagnetic transistors [23] and especially ferromagnetically contacted carbon nanotubes [24]. In this chapter we will review basic transport characteristics of a ferromagnetic single-electron transistor in the sequential tunneling, cotunneling and strong coupling (Kondo) regimes. In particular, we will discuss such properties of the device like electric and spin currents, tunnel magnetoresistance, spin and charge accumulation and fluctuations, shot noise, exchange interaction, RKKY interaction, spin torque and precession, as well as the Kondo effect.

In the first part of this chapter we introduce the basic principles of single-electron transport (Section 4.1) and present an effective model Hamiltonian (Section 4.2) describing the metallic and/or ferromagnetic island (or quantum dot) in the presence of strong Coulomb interaction and attached to ferromagnetic leads. In the next section different transport regimes are introduced (Section 4.3). This gives us the framework for further discussion of the system properties. In the weak coupling regime such transport properties like spin accumulation, spin precession in the non-collinear geometry, magnetoresistance, shot noise and others are thoroughly described (Section 4.4). The next two sections deal with the cotunneling (Section 4.5) and Kondo effect in the strong coupling regime (Section 4.6). In the former case a scheme of direct detection of the chemical potential spin-splitting is analyzed, whereas in the latter case the properties of the Kondo effect in the presence of ferromagnetism are discussed in terms of various techniques beyond perturbation theory. In the last part (Section 4.7) we review recent results on the RKKY interaction between semiconducting quantum dots.

4.1 Single-electron transport

During the last few decades continuous progress in micro- and nanofabrication techniques have allowed workers to fabricate small metallic or semiconducting tunnel junctions and small islands with capacitances in the range of $C \approx 10^{-15}$ F or less [7–10]. In such systems, the classical electrostatic energy associated with addition or removal of a single electron charge e from the island, given by

$$E_{\mathrm{C}} = e^2/2C \, , \tag{4.1}$$

can be as high as of the order of 10^{-4}–10^{-5} eV, which corresponds to the temperature range $T \approx 1$–10 K. This implies that electron transport in the low temperature regime is strongly dominated by the charging effects. At low temperatures and small bias voltage V, $(k_{\mathrm{B}}T, |eV| \ll E_{\mathrm{C}})$, a complete suppression of transport through the island, the so-called *Coulomb blockade*, occurs far away from the degeneracy points. The linear response conductance then shows *Coulomb oscillation* as a function of gate voltage V_{g}, i.e., a series of peaks with a uniform spacing. The low-temperature transport between the degeneracy points is possible only with a large bias voltage V, which generally also leads to some non-equilibrium effects.

The electrostatic energy required to add N excess electrons (corresponding to the total charge $Q = eN$) to the island, while keeping constant the voltages V_{L} and V_{R} in the left and right electrodes and the gate voltage V_{g}, is given by

$$E_{\mathrm{ch}}(N, n_x) = E_{\mathrm{C}}(N - n_x)^2 \, . \tag{4.2}$$

In the above equation the term $-E_{\mathrm{C}} n_x^2$ has been dropped because it is independent of N. Here, the total island capacitance is the sum of the capacitances of

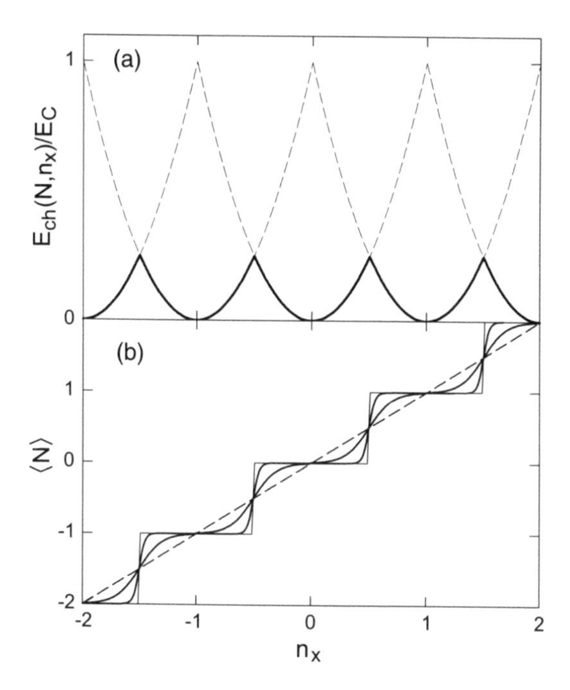

FIG. 4.1. (a) The charging energy $E_{\mathrm{ch}}(N, n_x)$ of a single-electron transistor as a function of n_x (controlled also by a gate voltage V_{g}) for different numbers N of excess electrons on the island. The solid line depicts the ground state energy. (b) The average number of excess electrons on the island as a function of n_x for different temperatures $k_{\mathrm{B}}T/E_{\mathrm{C}} = 0.4$ (dashed line), 0.2, 0.05, and 0.

the left and right junctions and of the gate, $C = C_{\mathrm{L}} + C_{\mathrm{R}} + C_{\mathrm{g}}$. The external charge $en_x \equiv C_{\mathrm{L}}V_{\mathrm{L}} + C_{\mathrm{R}}V_{\mathrm{R}} + C_{\mathrm{g}}V_{\mathrm{g}}$ accounts for the effect of applied voltages, and can be continuously tuned be means of V_{g}. The charging energy $E_{\mathrm{ch}}(N, n_x)$ for different numbers of excess electrons N is plotted in Fig. 4.1(a). With tuning the gate voltage V_{g}, the electron number on the island changes in discrete steps from N to $N + 1$ at the degeneracy points (near half-integer values of n_x). At finite temperatures and for weak coupling to the external electrodes, the average occupancy of the island is given by the Boltzmann distribution

$$\langle N(n_x) \rangle = \frac{1}{Z_{\mathrm{ch}}} \sum_N N e^{-E_{\mathrm{ch}}(N,n_x)/k_{\mathrm{B}}T} , \tag{4.3}$$

where Z_{ch} denotes the partition function. The results for different temperatures are presented in Fig. 4.1(b).

For a very small island, whose spatial dimensions are comparable with the Fermi wavelength $\lambda_{\mathrm{F}} = 2\pi/k_{\mathrm{F}}$, the size quantization effects become important, too. Electrons residing on the dot then occupy quantized energy levels ε_j with some average level spacing $\Delta\varepsilon$. The total energy is then given by

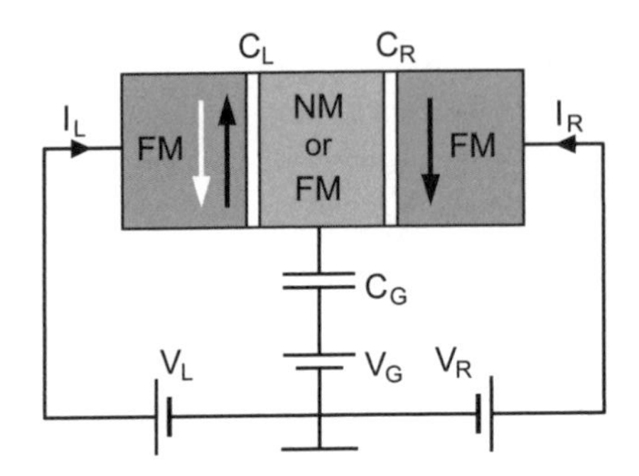

FIG. 4.2. Geometry of the ferromagnetic single-electron transistor. The central electrode (island) can be non-magnetic or magnetic. The orientation of magnetization of each element can be independently controlled. For a small island discrete energy levels become resolved and the island becomes a quantum dot.

$$E(N, n_x) = E_{\rm ch}(N, n_x) + \sum_{j=1}^{N} \varepsilon_j \, . \tag{4.4}$$

Usually the following two limits are considered:

(i) a metallic island with a continuous density of states $\rho(\omega)$ for $\Delta\varepsilon \ll k_{\rm B}T$, and

(ii) the quantum dot (QD) limit with a discrete spectrum and large level spacing, $\Delta\varepsilon \gtrsim k_{\rm B}T$.

In this chapter we will discuss transport properties in both situations assuming that the island is attached to ferromagnetic leads.

4.2 Model Hamiltonian

The ferromagnetic single-electron transistor consists of a small ferromagnetic or non-magnetic metallic grain or quantum dot (island) connected via tunnel barriers to two ferromagnetic leads, as shown schematically in Fig. 4.2. It can be described by the Hamiltonian

$$H = \sum_{\rm r=L,R} H_{\rm r} + H_{\rm I} + H_{\rm ch} + H_{\rm T} \equiv H_0 + H_{\rm T} \, . \tag{4.5}$$

Here H_0 describes the system decoupled from the leads, and $H_{\rm T}$ takes into account tunneling processes between the island and leads.

The ferromagnetic leads are described by

$$H_{\rm r} = \sum_{k\sigma} \varepsilon_{rk\sigma} c^{\dagger}_{rk\sigma} c_{rk\sigma} \, , \tag{4.6}$$

where $c_{rk\sigma}$ are the Fermi operators for electrons with a wavevector k and spin σ in the lead r ($r = L, R$). Within the Stoner model of itinerant ferromagnetism [25] we assume a spin-split electron band in the leads, with a strong spin asymmetry in the density of states, $\rho_{r+}(\omega) \neq \rho_{r-}(\omega)$, for majority (+) and minority (−) electrons. For the density of states, which is assumed to be independent of energy, $\rho_{r\sigma}(\omega) = \rho_{r\sigma}$, it is convenient to introduce an effective spin polarization factor of the leads, defined usually as $P_r \equiv (\rho_{r+} - \rho_{r-})/(\rho_{r+} + \rho_{r-})$. For non-magnetic leads the spin polarization vanishes, $P_{\rm L} = P_{\rm R} = P = 0$, whereas for half-metallic ones $P = 1$ (only one type of spins is present at the Fermi level).

In this chapter we restrict considerations mainly to collinear lead magnetizations, and assume that by means of a weak magnetic field it is possible to change magnetic the configuration from the parallel to antiparallel alignment. Non-Collinear magnetic configurations will be considered in Section 4.4.2.

4.2.1 *Metallic or ferromagnetic island*

For a metallic island the density of states are continuous, and the number of electrons on the island is large. The Hamiltonian then reads

$$H_{\rm I} + H_{\rm ch} = \sum_{q\sigma} \epsilon_{q\sigma} d^\dagger_{q\sigma} d_{q\sigma} + E_{\rm C}(N - n_x)^2 \,, \tag{4.7}$$

where $d_{q\sigma}$ are the Fermi operators for electrons on the island, described by the wavenumber q and spin σ. The last part of the Hamiltonian,

$$H_{\rm T} = \sum_{r={\rm L,R}} \sum_{kq\sigma} T_{rkq\sigma} c^\dagger_{{\rm r}k\nu\sigma} d_{q\nu\sigma} {\rm e}^{-i\hat{\varphi}} + {\rm h.c.} \,, \tag{4.8}$$

describes tunneling processes. The phase operator $\hat{\varphi}$ is the conjugate to the charge eN on the island, and the operator $e^{\pm i\hat{\varphi}}$ describes changes of the island charge by $\pm e$. When writing Eq. (4.8) it has been assumed that the electron spin is conserved during tunneling. One usually assumes that the tunnel matrix elements $T_{rkq\sigma}$ depend only on the junction index r, $T_{rkq\sigma} = T_r$, and are related to the spin-dependent tunneling resistance of the barriers via the relation

$$1/R_{r\sigma} = (2\pi e^2/\hbar) N_c \left|T_r\right|^2 \rho_{r\sigma} \rho_{{\rm I}\sigma} \,, \tag{4.9}$$

where $\rho_{r\sigma}$ is the spin-dependent density of electron states at the Fermi level in the electrode r, $\rho_{{\rm I}\sigma}$ the density of states per spin in the island, and N_c the number of transverse channels.

4.2.2 *Quantum dot – Anderson model*

In the limit of $\Delta\varepsilon \gtrsim k_{\rm B}T, |eV|$ effectively only a single energy level $\epsilon_{{\rm d}\sigma}$ participates in the transport and the level can be occupied by zero, one, or two electrons. The physical properties of such a single-level quantum dot coupled to ferromagnetic electrodes can be effectively described by the Anderson model:

$$H_{\rm I} + H_{\rm ch} = \sum_{\sigma} \epsilon_{{\rm d}\sigma} d^\dagger_\sigma d_\sigma + U d^\dagger_\uparrow d_\uparrow d^\dagger_\downarrow d_\downarrow \,, \tag{4.10}$$

where the charging energy U is associated with a double occupation. In the presence of an external magnetic field, the energy level experiences a Zeeman splitting, $\epsilon_{\mathrm{d}\uparrow} - \epsilon_{\mathrm{d}\downarrow} = g\mu_{\mathrm{B}}B$.

Tunneling between the leads and dot is described by the standard tunneling Hamiltonian

$$H_{\mathrm{T}} = \sum_{rk\sigma} \left(T_{rk\sigma} c^{\dagger}_{rk\sigma} d_{\sigma} + T^{*}_{rk\sigma} d^{\dagger}_{\sigma} c_{rk\sigma} \right) . \tag{4.11}$$

The tunneling processes lead to a broadening of the dot level, defined as

$$\Gamma_{r\sigma}(\omega) = \pi \sum_{k} |T_{rk\sigma}|^2 \delta(\omega - \epsilon_{k\sigma}) = \pi \rho_{r\sigma}(\omega) |T_r|^2 , \tag{4.12}$$

where in the last step we assumed $T_{rk\sigma} = T_r$. The level width is then spin-dependent due to the spin asymmetry of the density of states in the leads.

4.3 Transport regimes

The transport properties of a ferromagnetic single-electron transistor or a quantum dot strongly depend on the coupling strength to the leads and on how far the system is from resonance (degeneracy point). It is convenient to distinguish different transport regimes, for which one can expect significantly different behaviors.

Sequential tunneling. In the limit of weak dot–lead coupling, $\Gamma \ll k_{\mathrm{B}}T$, for quantum dots and high tunnel barriers for metallic islands, referred to as the sequential-tunneling regime, transport is dominated by processes of the first order in the coupling (unless we are not far from resonance). In this limit one can determine the transition rates by the Fermi Golden-Rule. The rate of tunneling of an electron from one of the states k in the left lead into one of the available states q in the island, which changes the electron number from N to $N+1$, is given by

$$\begin{aligned} \alpha^{\pm}_{r\sigma}(\Delta_N) = & \int_{-\infty}^{\infty} d\varepsilon_k d\varepsilon_q |T_{rkq}|^2 \rho_{r\sigma}(\varepsilon_{rk\sigma}) \rho_{\mathrm{I}\sigma}(\varepsilon_{q\sigma}) \\ & \times f^{\pm}_{r}(\varepsilon_{rk\sigma}) f^{\mp}_{I}(\varepsilon_{q\sigma}) \delta(\Delta_N - \mu_{r\sigma} + \varepsilon_{q\sigma} - \varepsilon_{rk\sigma}) , \end{aligned} \tag{4.13}$$

where $f^{+}(x)$ denotes the Fermi function and $f^{-}(x) = 1 - f^{+}(x)$. In the tunneling process considered the charging energy changes by $\Delta_N = E_{\mathrm{ch}}(N+1, n_x) - E_{\mathrm{ch}}(N, n_x)$. The integrals over the electron states in Eq. (4.13) can be performed analytically. The resulting single-electron tunneling rate is [7]

$$\alpha^{\pm}_{r\sigma}(\omega) = \pm \alpha^{0}_{r\sigma} \frac{\omega - \Delta\mu_r}{\exp[\pm\beta(\omega - \Delta\mu_r)] - 1} , \tag{4.14}$$

where the dimensionless conductance of the junction r for spin σ is given by $\alpha^{0}_{r\sigma} = h/(4\pi^2 e^2 R_{r\sigma})$, ω is the energy of the tunneling electron, $\beta \equiv 1/k_{\mathrm{B}}T$, and $\Delta\mu_r = \mu_r - \mu_{\mathrm{I}}$.

In a similar way one can calculate the tunnel rates for a quantum dot

$$2\pi\hbar\,\gamma_{r\sigma}^{\pm}(\omega) = \Gamma_{r\sigma}(\omega) f_r^{\pm}(\omega)\,, \tag{4.15}$$

where $f_r^+(\omega) \equiv [1 + \exp((\omega - \mu_r)/k_BT)]^{-1}$ and $f_r^-(\omega) \equiv 1 - f_r^+(\omega)$ with the electrochemical potential $\mu_r = -eV_r$

Cotunneling. Sequential tunneling is exponentially suppressed in the Coulomb blockade regime and transport is dominated by cotunneling [26], i.e., second-order processes. But the second-order corrections also become important on resonance for intermediate coupling strengths, $\Gamma \sim k_BT$. If a transport voltage V is applied, higher order tunneling processes transferring an electron charge through the system are energetically allowed, and the states with an excess charge in the island exist only virtually. Second-order perturbation theory gives the cotunneling rate [26]

$$\alpha_{i\to f} = \frac{2\pi}{\hbar}\left|\sum_q \frac{\langle i|H_{\rm T}|q\rangle\langle q|H_{\rm T}|f\rangle}{\varepsilon_q - \varepsilon_i}\right|^2 \delta(\varepsilon_i - \varepsilon_f)\,. \tag{4.16}$$

This tunneling rate is only algebraically suppressed rather than exponentially as for the sequential tunneling regime. Because of that, even at low temperatures and in the strong Coulomb blockade regime the higher-order rates do not vanish. For a metallic island the cotunneling process leaves an electron-hole excitation in the island, making this process inelastic. Since only a single level participates in transport through quantum dots, cotunneling is a coherent elastic process unless the tunneling is accompanied with a spin-flip in the dot.

Strong coupling. For strong coupling of the metallic island to electrodes, the tunneling of electrons leads to logarithmic corrections and perturbation theory fails at the degeneracy points of two consecutive charge states. In the case of quantum dots additionally the Kondo effect appears at low temperatures (below the Kondo temperature $T \lesssim T_{\rm K}$) leading to an enhanced conductance in the linear-response regime [27]. In Section 4.6 we will discuss how the Kondo effect is modified by the presence of ferromagnetic leads.

4.4 Weak coupling – sequential tunneling

4.4.1 *Quantum dot*

In the sequential tunneling limit, $k_BT > \Gamma_{r\sigma}$, the dynamics of the system can be modeled by the time evolution of the occupation probabilities $\hat{p} \equiv [p_{\rm d}, p_\uparrow, p_\downarrow, p_0]$ described by the master equation

$$\frac{d\hat{p}}{dt} = \hat{M}\hat{p}\,. \tag{4.17}$$

Here $p_{\rm d}$, $p_\uparrow$, $p_\downarrow$, and p_0 denote the probabilities that there are respectively two electrons on the dot, a single electron with spin $\uparrow$ or $\downarrow$, and the dot is empty.

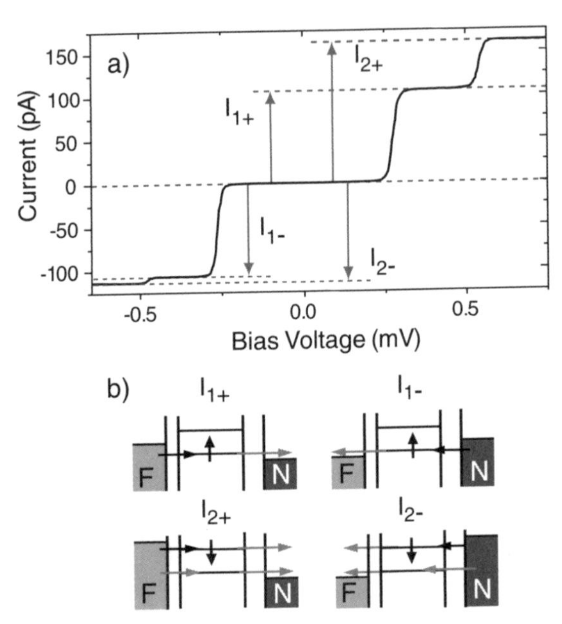

FIG. 4.3. (a) Current vs. voltage curve for an aluminum quantum dot coupled to cobalt and aluminum electrodes for $B = 1$ T, showing the range of V where tunneling occurs via one pair of Zeeman-split energy levels. (b) Energy-level diagrams at each current step. Black horizontal arrows show the threshold tunneling transition. Gray arrows depict other transitions which contribute to the current. From Deshmukh and Ralph [21].

The total probability obeys the normalization condition $\sum_\chi p_\chi = 1$. The matrix $\hat{M}$ is given by [28, 29]

$$\hat{M} = \begin{bmatrix} \sum_\sigma \gamma_{\sigma U}^- & \gamma_{\downarrow U}^+ & \gamma_{\uparrow U}^+ & 0 \\ \gamma_{\downarrow U}^- & -\gamma_\uparrow^- - \gamma_{\downarrow U}^+ & 0 & \gamma_\uparrow^- \\ \gamma_{\uparrow U}^+ & 0 & -\gamma_\downarrow^- - \gamma_{\uparrow U}^+ & \gamma_\downarrow^+ \\ 0 & \gamma_\uparrow^- & \gamma_\downarrow^- & \sum_\sigma \gamma_\sigma^+ \end{bmatrix} . \tag{4.18}$$

The stationary probabilities obtained from $\hat{M}\hat{p} = 0$ can be used to calculate the average value of the tunneling current through the r barrier as

$$I_r = \sum_\sigma I_{r\sigma} = -e \sum_\sigma \left[\gamma_{r\sigma}^+ p_0 - \gamma_{r\sigma}^+ p_\sigma + \gamma_{r\sigma U}^+ p_\sigma - \gamma_{r\sigma U}^- p_{\rm d} \right] , \tag{4.19}$$

where $\gamma_{r\sigma}^\pm \equiv \gamma_{r\sigma}^\pm(\epsilon_{\rm d})$ and $\gamma_{r\sigma U}^\pm \equiv \gamma_{r\sigma}^\pm(\epsilon_{\rm d} + U)$ are given by Eq. (4.15).

Using Eqs (4.17), (4.18, and (4.19), one can model the transport measurements of individual energy levels in an aluminum quantum dot attached to one ferromagnetic (cobalt) and one non-magnetic (aluminum) electrodes obtained by Deshmukh and Ralph [21]. The $I - V$ characteristics corresponding to the

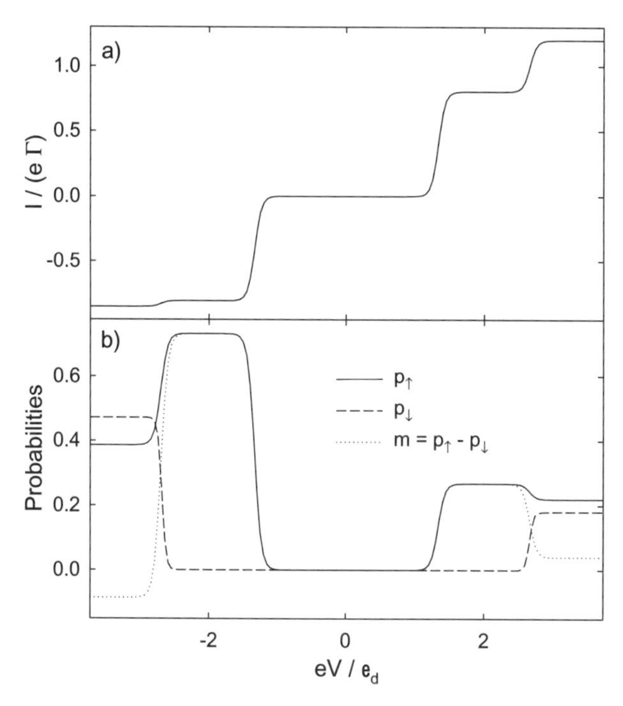

FIG. 4.4. Current (a), the occupation probabilities $p_\uparrow$ and $p_\downarrow$, and the spin accumulation $m \equiv p_\uparrow - p_\downarrow$ (b) vs. bias voltage for a quantum dot coupled to a ferromagnetic and normal electrode calculated for the same parameters as in the experimental situation shown in Fig. 4.3. Here we use $P_{\text{Co}} = 0.1$, $\Gamma_{\text{Al}} = 1.5\,\Gamma_{\text{Co}}$ and $k_{\text{B}}T/\epsilon_{\text{d}} = 0.025$.

Zeeman splitting $B = 1\ T$ are presented in Fig. 4.3(a), whereas in Fig. 4.4 the calculated current, occupancy probabilities, and spin accumulation $m \equiv p_\uparrow - p_\downarrow$ are shown for the same parameters as in Fig. 4.3. The first step in current for either sign of bias V corresponds to electron transport through only a spin-$\uparrow$ state. The current is then fully spin-polarized and the above model predicts that the two currents I_{1+} and $|I_{1-}|$ (see Fig. 4.3a) should have the same magnitude, $I_{1+} = |I_{1-}|$. The difference in the occupancy $p_\uparrow$ for these two currents (Fig. 4.4b) is related to a charge accumulation due to different coupling between the dot and the Al (Γ_{Al}) and Co ($\Gamma_{\text{Co}} = \Gamma_\uparrow + \Gamma_\downarrow$) electrodes ($\Gamma_{\text{Al}} \approx 1.5\Gamma_{\text{Co}}$). When $|V|$ increases to permit tunneling through either the spin-$\uparrow$ or spin-$\downarrow$, the currents for positive and negative V are different $I_{2+} \neq I_{2-}$. Using the values of I_{1+}, I_{2+}, and I_{2-}, it is possible to determine Γ_{Al}, $\Gamma_\uparrow$, and $\Gamma_\downarrow$ as well as the resulting spin polarization $P_{\text{Co}} \approx 0.08 - 0.12$ [21]. When both spin states participate in transport, the presence of a ferromagnetic electrode induces a non-zero spin accumulation $m \neq 0$, with different sign for positive and negative bias V (Fig. 4.4b). For the case of a quantum dot coupled to both normal leads the spin accumulation is

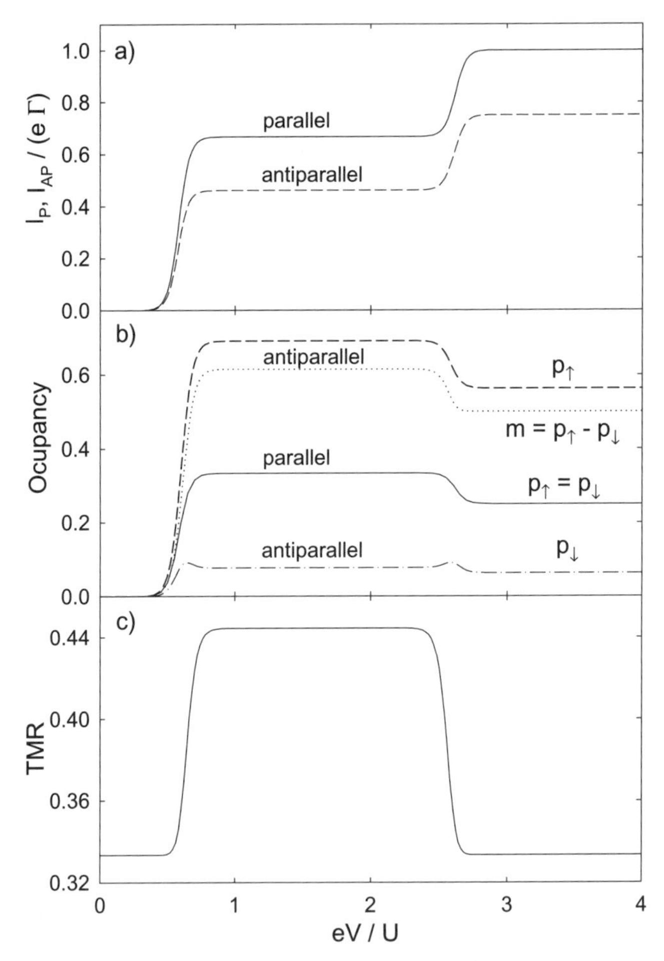

FIG. 4.5. Current (a), occupation probabilities $p_\uparrow$, $p_\downarrow$, spin accumulation $m \equiv p_\uparrow - p_\downarrow$ (b), and TMR (c) vs. bias voltage for a quantum dot symmetrically coupled to two ferromagnetic electrodes, calculated for $k_\mathrm{B}T/U = 0.02$, $\epsilon_\mathrm{d}/U = 0.3$, and $P = 0.5$.

absent, $m = 0$.

Let us consider a symmetrical junction, i.e., the case when both barriers are ferromagnetic and identical $P_\mathrm{L} = P_\mathrm{R}$ and $\Gamma_\mathrm{L} = \Gamma_\mathrm{R}$ [29]. Since now $I(-V) = I(V)$, the discussion will be restricted to positive bias only, $V > 0$. Figure 4.5(b) shows typical variations of the occupation numbers $p_\uparrow$ and $p_\downarrow$ with the bias voltage V, calculated for both parallel and antiparallel configurations. In both cases, the first step in $p_\uparrow$ and $p_\downarrow$ occurs at the bias, when the discrete level ϵ_d crosses the Fermi level of the left (source) electrode. On the other hand, the step at a higher voltage corresponds to the case when $\epsilon_\mathrm{d} + U$ crosses this Fermi level. It is also interesting to note that in the parallel configuration $p_\uparrow = p_\downarrow$, whereas in the antiparallel configuration $p_\uparrow \neq p_\downarrow$. The situation is similar to that in the case of symmetrical junctions with large central electrodes (islands), where the difference in spin asymmetry for tunneling rates across the left and right barriers (which takes place in the antiparallel configuration only) gives rise to a spin splitting of the island Fermi level, and consequently leads to spin accumulation [29]. No such effect occurs when the spin asymmetry for both barriers is the same, $P_\mathrm{L} = P_\mathrm{R}$, which for symmetrical junctions occurs only in the parallel configuration. Figure 4.5(a) shows the tunneling current calculated for both magnetic configurations. There are two steps in the current, which correspond to the steps in the occupation numbers. At each step a new channel for tunneling becomes open. Figure 4.5(c) shows the corresponding tunnel magnetoresistance (TMR), defined quantitatively as

$$\mathrm{TMR} \equiv \frac{I_\mathrm{P} - I_\mathrm{AP}}{I_\mathrm{AP}}\,, \tag{4.20}$$

where I_AP and I_P denote the current in the antiparallel and parallel configurations, respectively. The magnetoresistance is significantly enhanced in the bias range bounded by the voltages corresponding to the two steps in the current, i.e., when the dot is occupied by a single electron.

4.4.2 *Non-Collinear geometry*

In this section we will discuss transport properties of a quantum dot attached to ferromagnetic leads with non-collinear directions of magnetizations. There are different quantization axes $\hat{\mathbf{n}}_\mathrm{L}$ and $\hat{\mathbf{n}}_\mathrm{R}$ in both electrodes. It turned out to be convenient to quantize the dot spin $\sigma =\uparrow, \downarrow$ along the z-direction of the coordinate system in which the basis vectors $\hat{\mathbf{e}}_x$, $\hat{\mathbf{e}}_y$, and $\hat{\mathbf{e}}_z$ are along $\hat{\mathbf{n}}_\mathrm{L} + \hat{\mathbf{n}}_\mathrm{R}$, $\hat{\mathbf{n}}_\mathrm{L} - \hat{\mathbf{n}}_\mathrm{R}$, and $\hat{\mathbf{n}}_\mathrm{R} \times \hat{\mathbf{n}}_\mathrm{L}$, respectively (Fig. 4.6). The tunneling Hamiltonian for the left tunneling barrier is then given by [30–32]

$$H_{\mathrm{T,L}} = \frac{T_\mathrm{L}}{\sqrt{2}} \sum_k \left(a^\dagger_{\mathrm{L}k+}, a^\dagger_{\mathrm{L}k-} \right) \begin{pmatrix} e^{i\phi/4} & e^{-i\phi/4} \\ -e^{i\phi/4} & e^{-i\phi/4} \end{pmatrix} \begin{pmatrix} c_\uparrow \\ c_\downarrow \end{pmatrix} + \mathrm{h.c.} \tag{4.21}$$

The tunneling Hamiltonian for the right junction, $H_{\mathrm{T,R}}$, is of the same form, but with L $\to$ R and $\phi \to -\phi$.

The state of the dot is described by the reduced density matrix

$$\rho_\mathrm{dot} = \begin{bmatrix} p_0^0 & 0 & 0 & 0 \\ 0 & p_\uparrow^\uparrow & p_\downarrow^\uparrow & 0 \\ 0 & p_\uparrow^\downarrow & p_\downarrow^\downarrow & 0 \\ 0 & 0 & 0 & p_\mathrm{d}^\mathrm{d} \end{bmatrix}, \tag{4.22}$$

where the diagonal elements $p_\chi \equiv p_\chi^\chi$ are the probabilities of finding the dot in the corresponding state χ. The non-vanishing complex off-diagonal elements,

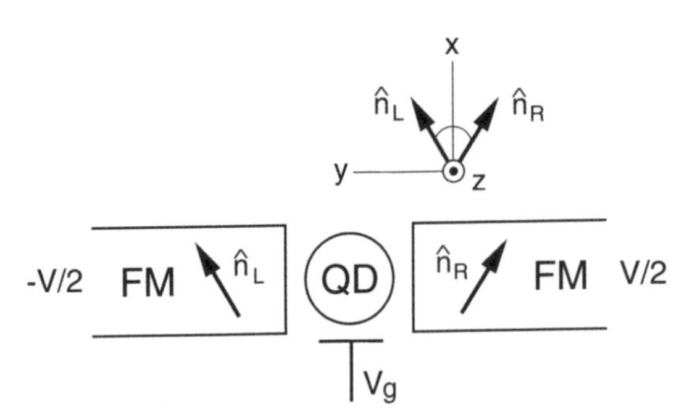

FIG. 4.6. A quantum dot is connected to two ferromagnetic leads (FM) with magnetizations $\hat{\mathbf{n}}_{\mathrm{L}}$ and $\hat{\mathbf{n}}_{\mathrm{R}}$, respectively. The choice of the coordinate system: the z-axis is perpendicular to both $\hat{\mathbf{n}}_{\mathrm{L}}$ and $\hat{\mathbf{n}}_{\mathrm{R}}$. Here, the tunneling Hamiltonian in the form of Eq. (4.21) is used. The quantum-dot spin is quantized along the z-axis.

$p_{\downarrow}^{\uparrow} = (p_{\uparrow}^{\downarrow})^*$, reflect the fact that spin accumulation in the dot is not restricted to the z-direction but can also have a finite x- and y-component. A finite spin can only emerge for a single occupancy. The average spin $\hbar\mathbf{S}$ with $\mathbf{S} = [S_x, S_y, S_z]$ is related to the density matrix elements via $S_x = \frac{1}{2}(p_{\downarrow}^{\uparrow} + p_{\uparrow}^{\downarrow})$, $S_y = \frac{i}{2}(p_{\downarrow}^{\uparrow} - p_{\uparrow}^{\downarrow})$, and $S_z = \frac{1}{2}(p_{\uparrow}^{\uparrow} - p_{\downarrow}^{\downarrow})$.

The reduced density matrix is given by the generalized master equation in Liouville space,

$$(\epsilon_{\chi_1} - \epsilon_{\chi_2})\, p_{\chi_2}^{\chi_1} + \sum_{\chi_1', \chi_2'} p_{\chi_2'}^{\chi_1'} \Sigma_{\chi_2', \chi_2}^{\chi_1', \chi_1} = 0\,, \tag{4.23}$$

where χ_1 and χ_2 label the quantum dot states, and ϵ_{χ_1} and ϵ_{χ_2} are the corresponding energies. The terms $\Sigma_{\chi_2' \chi_2}^{\chi_1' \chi_1}$ act as generalized transition rates in Liouville space. They are defined as irreducible self-energy parts of the dot propagator on a Keldysh contour, and can be expanded in powers of the dot-lead coupling strength Γ. In the weak dot-lead coupling, only the terms linear in Γ can be retained.

The transport current through the dot can be expressed in terms of Keldysh Green's functions [33]

$$I_r = \frac{-e\, i}{2h} \int d\omega \; \mathrm{Tr}\left[\hat{\Gamma}_r f_r^+(\omega) \mathbf{G}^>(\omega) + \hat{\Gamma}_r f_r^-(\omega) \mathbf{G}^<(\omega)\right]. \tag{4.24}$$

The functions $\mathbf{G}^<(\omega) = \begin{pmatrix} G_{\uparrow\uparrow}^<(\omega) & G_{\uparrow\downarrow}^<(\omega) \\ G_{\downarrow\uparrow}^<(\omega) & G_{\downarrow\downarrow}^<(\omega) \end{pmatrix}$ and $\mathbf{G}^>(\omega)$ are the Fourier transforms of the usual lesser and greater Green's functions, respectively. The coupling matrices are defined as $\hat{\Gamma}_{\mathrm{L}} = \frac{\Gamma_{\mathrm{L}}}{2} \begin{pmatrix} 1 & P_{\mathrm{L}}\, e^{-i\phi/2} \\ P_{\mathrm{L}}\, e^{+i\phi/2} & 1 \end{pmatrix}$, and for $\hat{\Gamma}_{\mathrm{R}}$ with the substitutions $\mathrm{L} \to \mathrm{R}$ and $\phi \to -\phi$.

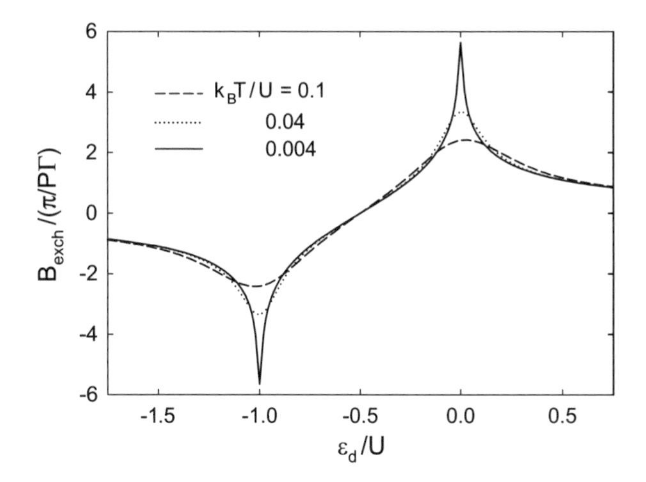

FIG. 4.7. Normalized exchange field B_{exch} as a function of the level position ϵ_{d} for three different values of temperature.

It is possible to obtain self-energies $\Sigma^{\chi_1' \, \chi_1}_{\chi_2' \, \chi_2}$ and Green's functions using, e.g., a real-time diagrammatic technique [30–32] and then calculate transport properties of the system. At a non-zero bias voltage electric current flows through the system and a certain average spin accumulates on the dot. To describe the dot state in the linear-response regime, $eV \ll k_{\text{B}}T$, it is sufficient to expand the steady-state solution of Eqs (4.23) and (4.24) up to terms linear in V. The spin accumulation is then

$$\left.\frac{\partial|\mathbf{S}|}{\partial(eV)}\right|_{V=0} = \frac{P(p_\uparrow + p_\downarrow)}{4k_{\text{B}}T} \cos\alpha(\phi) \sin\frac{\phi}{2} \,. \tag{4.25}$$

In Ref. [30, 31], it has been shown that due to the exchange interaction between the quantum dot spin and ferromagnetic leads, the dot spin experiences a torque, which results in spin precession. This exchange interaction can be described by the effective exchange field

$$B_{\text{exch}} = -\frac{P\Gamma}{\pi}\text{Re}\left[\Psi\left(\frac{1}{2} + i\frac{\beta(\epsilon_{\text{d}})}{2\pi}\right) - \Psi\left(\frac{1}{2} + i\frac{\beta(\epsilon_{\text{d}}+U)}{2\pi}\right)\right], \tag{4.26}$$

where $\Psi(x)$ denotes the digamma function. The exchange field B_{exch} is shown in Fig. 4.7 as a function of the level position for several values of temperature. Out of resonance ($\epsilon_{\text{d}}/U \neq 0,\ 1$), it depends only weakly on temperature. In the absence of an exchange field, the accumulated spin is oriented along y. The exchange field B_{exch} leads to a precession of the spin in the yz-plane about the x-axis by the angle α with the y-axis, where

$$\tan\alpha(\phi) = -\frac{B_{\text{exch}}}{\Gamma[1 - f(\epsilon_{\text{d}}) + f(\epsilon_{\text{d}}+U)]} \cos\frac{\phi}{2} \,. \tag{4.27}$$

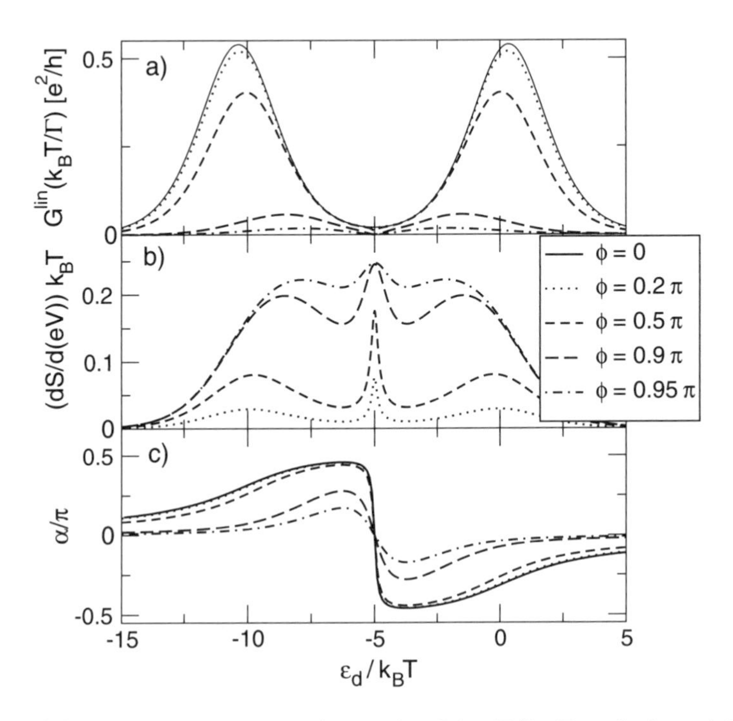

FIG. 4.8. (a) Linear conductance (normalized by $\Gamma/k_{\rm B}T$ and plotted in units of e^2/h) as a function of level position $\epsilon_{\rm d}$ for five different angles ϕ. (b) The derivative of accumulated spin S with respect to bias voltage V normalized by $k_{\rm B}T$. (c) The angle α between the quantum-dot spin and the y-axis. The charging energy is $U/k_{\rm B}T = 10$ and $P = 1$.

The factor $(\Gamma[1 - f(\epsilon_{\rm d}) + f(\epsilon_{\rm d} + U)])^{-1} \equiv \tau$ in Eq. (4.27) can be identified as the lifetime of the dot spin, limited by electron tunneling out of the dot or by tunneling in of a second electron with opposite spin. Since both the lifetime and exchange field are of first order in Γ, the angle α acquires a finite value.

The spin accumulation $dS/d(eV)$ is plotted in Fig. 4.8(b). It is high for a single occupation of the dot, in the valley between the two conductance peaks. Figure 4.8(c) presents the evolution of the rotation angle α as a function of the level energy $\epsilon_{\rm d}$. At the special point, $\epsilon_{\rm d} = -U/2$, the exchange interaction vanishes due to particle-hole symmetry and changes sign. As a consequence, α shows a sharp transition from positive to negative values. For a large angle α both the magnitude of the accumulated spin and the relative angle to the magnetization of the drain electrode are reduced. Both effects enhance transport through the dot. Figure 4.8(a) shows the linear conductance for several values of the angle ϕ. With increasing angle ϕ, transport is suppressed due to the spin-valve effect.

The angular dependence of the linear conductance, given by

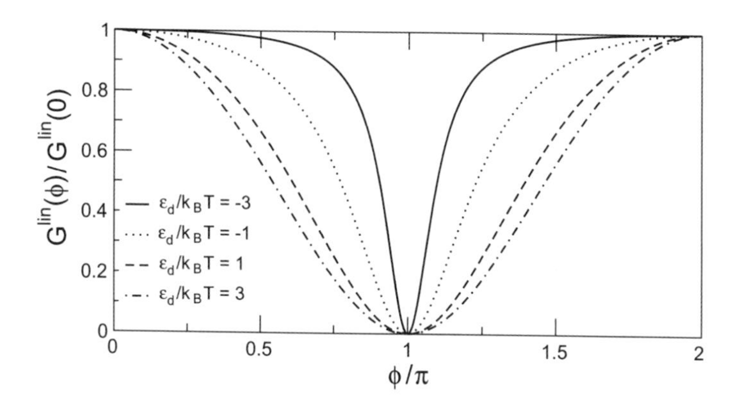

FIG. 4.9. Normalized linear conductance as a function of ϕ for $U/k_{\mathrm{B}}T = 10$, $P = 1$, and four different values of the level position.

$$G^{\mathrm{lin}}(\phi) = G^{\mathrm{lin}}(0)\left(1 - P^2\cos^2\alpha(\phi)\sin^2\frac{\phi}{2}\right), \tag{4.28}$$

is presented in Fig. 4.9. For values of the level position ϵ_{d} at which the rotation angle α is small, $\epsilon/k_{\mathrm{B}}T = 3$ and 1, the ϕ-dependence of the conductance follows the cosine law, as predicted by Slonczewski [34] and observed experimentally [35] for a single magnetic tunnel junction. For $\epsilon_{\mathrm{d}}/k_{\mathrm{B}}T = -1$ and -3, however, the spin-valve effect is strongly reduced, and conductance is enhanced, except in the regime close to antiparallel magnetization, $\phi = \pi$.

The detailed analysis of the lowest order transport through a quantum dot attached to ferromagnetic leads with a non-collinear geometry is presented in Ref. [30–32]. There is also a discussion of the non-linear response regime, where a spin blockade and negative differential conductance due to spin precession were predicted.

In Ref. [36] Braun *et al.* demonstrate, using the formalism described in this section, the possibility of the Hanle effect in a dot weakly coupled to ferromagnetic leads. In this geometry the magnetizations of the electrodes are collinear but the direction of the external magnetic field is perpendicular. Due to precession of a spin located in the dot around the magnetic field the spin accumulation is reduced and the transport enhanced.

4.4.3 *Ferromagnetic island*

For ferromagnetic single-electron transistors with ferromagnetic islands the spin-flip relaxation time is usually short because of typically strong spin-orbit coupling in such systems. The device can be then described by a straightforward extension of the formalism used for non-magnetic single-electron transistors, simply by taking into account the fact that the resistances of both tunnel barriers are spin dependent $R_{r\uparrow} \neq R_{r\downarrow}$ and also depend on the magnetic configuration of the junction. Using tunneling rates defined by Eq. (4.14) one can write a master

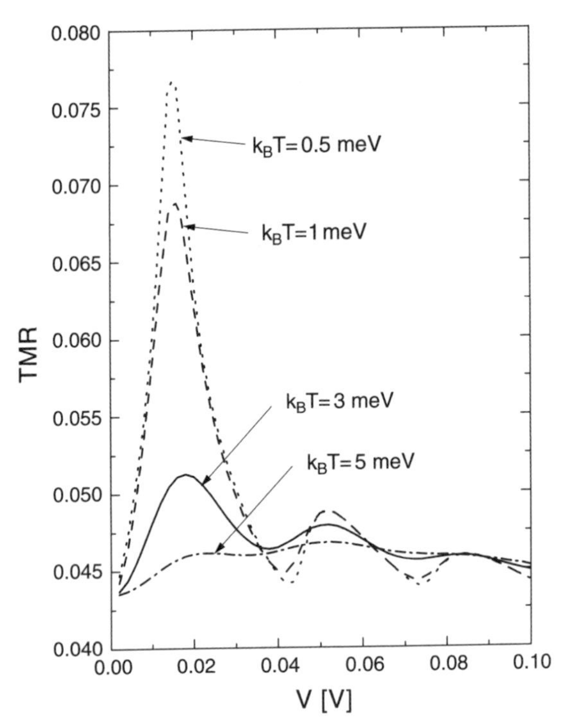

FIG. 4.10. TMR $\equiv (R_{\rm AP} - R_{\rm P})/R_{\rm P}$, shown for several values of the thermal energy. The parameters assumed for the parallel configuration are $R^{\rm P}_{{\rm L}\uparrow} = (5/3)R^{\rm P}_{{\rm L}\downarrow} = 250\,{\rm M}\Omega$, $R^{\rm P}_{{\rm R}\uparrow} = (7/3)R^{\rm P}_{{\rm R}\downarrow} = 70\,{\rm M}\Omega$, $C_{\rm L}/C_{\rm R} = 2$, $E_{\rm C} = 10$ meV, and for the antiparallel alignment $R^{\rm AP}_{r\uparrow} = R^{\rm AP}_{r\downarrow} = (R^{\rm P}_{r\uparrow}R^{\rm P}_{r\downarrow})^{1/2}$.

equation for probabilities $p(N)$ to find the island in the state with N excess electrons

$$\frac{d}{dt}p(N) = -\sum_{r\sigma}[\alpha^+_{r\sigma}(\Delta_N) + \alpha^-_{r\sigma}(\Delta_N)]p(N) + \sum_{r\sigma}\alpha^+_{r\sigma}(\Delta_{N-1})p(N-1) + \alpha^-_{r\sigma}(\Delta_{N+1})p(N+1)\,, \quad (4.29)$$

with the normalization condition $\sum_N p(N) = 1$. The rates and probabilities determine the current, which for the r-th junction is given by

$$I_r = \sum_\sigma I_{r\sigma} = -\frac{2\pi e}{h}\sum_{N\sigma}[\alpha^+_{r\sigma}(\Delta_N) - \alpha^-_{r\sigma}(\Delta_N)]p(N)\,, \quad (4.30)$$

where $I = I_{\rm L} = -I_{\rm R}$ due to charge conservation.

Using this approach, an enhancement of the TMR effect and also oscillations in the TMR with increasing bias voltage (Fig. 4.10) have been predicted [37, 38] and were recently observed experimentally [15, 16].

4.4.4 *Metallic island*

Ferromagnetic single-electron transistors with non-magnetic islands attract attention, too, since in such systems one can expect spin accumulation on the normal island when the spin-flip relaxation time is sufficiently long. To account for this effect one can also apply the approach described above, but now one should allow for a spin-dependent splitting of the chemical potential of the island due to spin accumulation, $\mu_{I\uparrow} \neq \mu_{I\downarrow}$. The resulting single-electron tunneling rate is

$$\alpha^{\pm}_{r\sigma}(\omega) = \pm\alpha^{0}_{r\sigma}\frac{\omega - \Delta\mu_{r\sigma}}{\exp[\pm\beta(\omega - \Delta\mu_{r\sigma})] - 1}\,, \tag{4.31}$$

where $\Delta\mu_{r\sigma} = \mu_r - \mu_{I\sigma}$ is now spin-dependent. In such an approach, the spin accumulation can be determined only self-consistently using the condition of spin current conservation, $I_{L\sigma} = I_{R\sigma}$ [39–41].

Recently, a more general approach has been developed [42], which takes into account spin degrees of freedom explicitly and allows us to calculate the system properties without additional assumptions. In such an approach the electron transport in the stationary state is governed by the solution of the generalized master equation [42, 43]

$$\begin{aligned} 0 = \; & -\{\Gamma(N_\uparrow, N_\downarrow) + \Omega_{\uparrow,\downarrow}(N_\uparrow, N_\downarrow) + \Omega_{\downarrow,\uparrow}(N_\uparrow, N_\downarrow)\}\, p(N_\uparrow, N_\downarrow) \\ & + \Gamma^{+}_{\uparrow}(N_\uparrow - 1, N_\downarrow) p(N_\uparrow - 1, N_\downarrow) + \Gamma^{+}_{\downarrow}(N_\uparrow, N_\downarrow - 1) p(N_\uparrow, N_\downarrow - 1) \\ & + \Gamma^{-}_{\uparrow}(N_\uparrow + 1, N_\downarrow) p(N_\uparrow + 1, N_\downarrow) + \Gamma^{-}_{\downarrow}(N_\uparrow, N_\downarrow + 1) p(N_\uparrow, N_\downarrow + 1) \\ & + \Omega_{\uparrow,\downarrow}(N_\uparrow - 1, N_\downarrow + 1) p(N_\uparrow - 1, N_\downarrow + 1) \\ & + \Omega_{\downarrow,\uparrow}(N_\uparrow + 1, N_\downarrow - 1) p(N_\uparrow + 1, N_\downarrow - 1)\,. \end{aligned} \tag{4.32}$$

Here $p(N_\uparrow, N_\downarrow)$ denotes the probability to find $N_\uparrow$ and $N_\downarrow$ excess electrons on the island ($N = N_\uparrow + N_\downarrow$ is the total number of excess electrons). The first term in Eq. (4.32) describes how the probability of a given configuration decays due to electron tunneling to or from the island, whereas other terms describe the rate at which this probability increases. The Ω terms account for spin-flip relaxation processes. The coefficients entering Eq. (4.32) are defined as $\Gamma^{\pm}_{\sigma}(N_\uparrow, N_\downarrow) = \sum_{r=\mathrm{L,R}} \Gamma^{\pm}_{r\sigma}(N_\uparrow, N_\downarrow)$ and $\Gamma(N_\uparrow, N_\downarrow) = \sum_{q=\pm}\sum_{\sigma}\Gamma^{q}_{\sigma}(N_\uparrow, N_\downarrow)$, where $\Gamma^{\pm}_{r\sigma}(N_\uparrow, N_\downarrow)$ are the tunneling rates for electrons with spin σ, which tunnel to (+) the grain from the lead $r = \mathrm{L, R}$ or back (−). These coefficients are given by

$$\begin{aligned} \Gamma^{\pm}_{r\sigma}(N_\uparrow, N_\downarrow) &= \sum_{i} \gamma^{r}_{i\sigma} F^{\mp}_{\sigma}(E_{i\sigma}|N_\uparrow, N_\downarrow) f^{\pm}(E_{i\sigma} + E^{\pm}_{r}(N) - E_{\mathrm{F}}), \\ \Omega_{\sigma\bar{\sigma}}(N_\uparrow, N_\downarrow) &= \sum_{i}\sum_{j} \omega_{i\sigma, j\bar{\sigma}} F^{+}_{\sigma}(E_{i\sigma}|N_\uparrow, N_\downarrow) F^{-}_{\bar{\sigma}}(E_{j\bar{\sigma}}|N_\uparrow, N_\downarrow). \end{aligned} \tag{4.33}$$

Here, $f^{+}(E)$ ($f^{-} = 1 - f^{+}$) is the Fermi function, whereas $F^{+}_{\sigma}(E_{i\sigma}|N_\uparrow, N_\downarrow)$ ($F^{-}_{\sigma} = 1 - F^{+}_{\sigma}$) describes the probability that the energy level $E_{i\sigma}$ is occupied by an electron with spin σ for the particular configuration $(N_\uparrow, N_\downarrow)$. The parameter $\gamma^r_{i\sigma}$ is the tunneling rate of electrons between the lead r and the energy level $E_{i\sigma}$ of the island, and $\omega_{i\sigma,j\bar{\sigma}}$ is the transition probability from the state $i\sigma$ to $j\bar{\sigma}$ due to the spin-flip processes. The energies $E^{\pm}_{\rm L}(N)$ and $E^{\pm}_{\rm R}(N)$ are given by $E^{\pm}_{\rm L}(N) = (C_{\rm R}/C)eV + U^{\pm}(N)$ and $E^{\pm}_{\rm R}(N) = -(C_{\rm L}/C)eV + U^{\pm}(N)$, where $U^{\pm}(N) = E_{\rm C}[2(N - N_x) \pm 1]$ and $N_x = C_{\rm g}V_{\rm g}/e$.

From the solution $p(N_\uparrow, N_\downarrow)$ of the master equation (Eq. 4.32) one can obtain the current flowing through the island,

$$I_r = -e\sum_{\sigma}\sum_{N_\uparrow,N_\downarrow} p(N_\uparrow, N_\downarrow)\left[\Gamma^{+}_{r\sigma}(N_\uparrow, N_\downarrow) - \Gamma^{-}_{r\sigma}(N_\uparrow, N_\downarrow)\right]. \tag{4.34}$$

For further evaluation one may assume that the discrete energy levels $E_{i\sigma}$ are equally separated with the level spacing $\Delta\varepsilon$. The tunneling rates $\gamma^r_{i\sigma}$ are then given by the formula $\gamma^r_{i\sigma} = \Delta\varepsilon/e^2 R_{r\sigma}$ [44,45]. For spin-flip processes one usually assumes $\omega_{i\sigma,j\bar{\sigma}} = 1/\tau_{sf}\,\delta_{i,j}$.

To emphasize the role of spin accumulation we assume that the intrinsic spin relaxation time on the island is long enough to neglect all intrinsic spin-flip processes. In Fig. 4.11(a) we show the *I-V* characteristics for the parallel and antiparallel alignments. In both cases the electric current is blocked below a threshold voltage ($\approx$ 13 mV). Above this voltage a typical "Coulomb staircase" appears with additional small steps due to the discrete levels. The effects due to discrete charging and discrete electronic structure are more clearly seen in the differential conductance shown in part (b). The difference between the *I-V* characteristics for the parallel and antiparallel configurations is due to a different spin accumulation in both geometries. In Fig. 4.11(c) we present the average value of the difference between the numbers of spin-up and spin-down excess electrons on the island, $\langle M\rangle \equiv \langle N_\uparrow - N_\downarrow\rangle$, i.e., the spin accumulation. As shown in part (c), there is no significant spin accumulation in the parallel configuration. The number $M \equiv N_\uparrow - N_\downarrow$ of spins accumulated on the island fluctuates around its average value $\langle M\rangle$, as shown in Fig. 4.11(d), where the standard deviation $(\langle M^2\rangle - \langle M\rangle^2)^{1/2} \equiv [\langle(N_\uparrow - N_\downarrow)^2\rangle - \langle N_\uparrow - N_\downarrow\rangle^2]^{1/2}$ is plotted against the voltage V. It is worth noting that although there is almost no spin accumulation in the parallel configuration, the corresponding fluctuations are relatively large.

The difference between the *I-V* curves in the parallel and antiparallel configurations leads to the tunnel magnetoresistance. The TMR ratio is shown in Fig. 4.11(e), where the broad peaks correspond to the Coulomb steps, while the fine structure originates from the discrete structure of the density of states.

It is possible to extend the presented formalism for the case of a ferromagnetic island [46] and calculate non-equilibrium spin accumulation on the ferromagnetic grain as well. Using this approach Inoue and Brataas [47] analyzed the possibility of magnetization reversal induced by spin accumulation rather than by means of spin current torque. They found that magnetization reversal is possible when the free energy change due to non-equilibrium spin accumulation is comparable to the anisotropy energy.

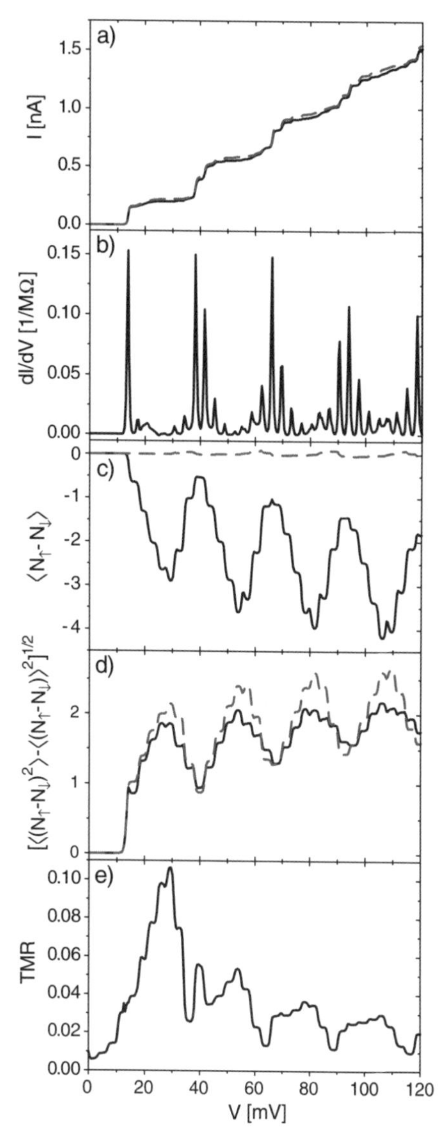

FIG. 4.11. Voltage dependence of the tunnel current I (a), the differential conductance $G \equiv dI/dV$ (b), spin accumulation $\langle N_\uparrow - N_\downarrow \rangle$ (c), standard deviation $[\langle (N_\uparrow - N_\downarrow)^2 \rangle - \langle N_\uparrow - N_\downarrow \rangle^2]^{1/2}$ (d), and tunnel magnetoresistance TMR (e), calculated at $T = 2.3$ K. The solid and dashed curves in (a), (c) and (d) correspond to the antiparallel and parallel configurations, respectively. The other parameters are: $\Delta E = 3\,\mathrm{meV}$, $C_\mathrm{L}/C_\mathrm{L} = 5$, $E_\mathrm{C} = 10\,\mathrm{meV}$, $R_{\mathrm{L}\uparrow} = 2R_{\mathrm{L}\downarrow} = 200\,\mathrm{M\Omega}$, $R_{\mathrm{R}\uparrow} = 2R_{\mathrm{R}\downarrow} = 4\,\mathrm{M\Omega}$ and $R_{\mathrm{R}\uparrow} = 2\,\mathrm{M\Omega}$ for the parallel alignment ($2R_{\mathrm{R}\uparrow} = R_{\mathrm{R}\downarrow} = 4\,\mathrm{M\Omega}$ for the antiparallel alignment).

4.4.5 *Shot noise*

From the application point of view, an important characteristics of the system is the current noise. The noise can also be an additional source of information on the system properties [48]. The shot noise in ferromagnetic single-electron transistors and quantum dots attached to ferromagnetic leads in the weak coupling regime was studied theoretically in a recent paper [49], where the generation-recombination approach [50] for multi-electron channels was extended by generalization of the method developed for spinless electrons in single-electron transistors [51–53]. The time correlation function of the quantities X and Y is expressed as [50]

$$\langle X(t)Y(0)\rangle = \sum_{N'_\uparrow,N'_\downarrow;N_\uparrow,N_\downarrow} X_{N'_\uparrow,N'_\downarrow} p(N'_\uparrow,N'_\downarrow;t|N_\uparrow,N_\downarrow;0) Y_{N_\uparrow,N_\downarrow} p^0(N_\uparrow,N_\downarrow). \quad (4.35)$$

Here, $p(N'_\uparrow,N'_\downarrow;t|N_\uparrow,N_\downarrow;0)$ is the conditional probability of finding the system in the final state with $N'_\uparrow$ and $N'_\downarrow$ excess electrons at time t, if there was $N_\uparrow$ and $N_\downarrow$ excess electrons in the initial time $t=0$. The probability p^0 is determined from Eq. (4.32) written in the matrix form $d\hat{p}/dt = \hat{M}\hat{p}$ and in the stationary condition, $\hat{M}\hat{p}^0 = 0$, where $\hat{M}$ is the matrix which enters the master equation.

According to this procedure the Fourier transform of the charge-charge S_{NN} (upper sign) and spin-spin S_{MM} (lower sign) correlation functions are given by [49]

$$S_{NN(MM)}(\omega) = 4 \sum_{N'_\uparrow,N'_\downarrow;N_\uparrow,N_\downarrow} (N'_\uparrow \pm N'_\downarrow)(N_\uparrow \pm N_\uparrow) \times \mathrm{Re}\left[\frac{1}{i\omega\hat{1} - \hat{M}}\right]_{N'_\uparrow,N'_\downarrow;N_\uparrow,N_\downarrow} p^0(N_\uparrow,N_\downarrow)\,, \quad (4.36)$$

while the current-current correlation function is given by

$$S_{II}(\omega) = S_{II}^{\mathrm{Sh}} + S_{II}^{\mathrm{c}}(\omega)\,, \quad (4.37)$$

where the Schottky value (the frequency independent part) is

$$S_{II}^{\mathrm{Sh}} = \frac{2e^2}{C^2} \sum_{N_\uparrow,N_\downarrow,r} \left(\frac{C_{\mathrm{L}}C_{\mathrm{R}}}{C_r}\right)^2 \Gamma_r(N_\uparrow,N_\downarrow) p^0(N_\uparrow,N_\downarrow) \quad (4.38)$$

and the second term in Eq.(4.37) is given by

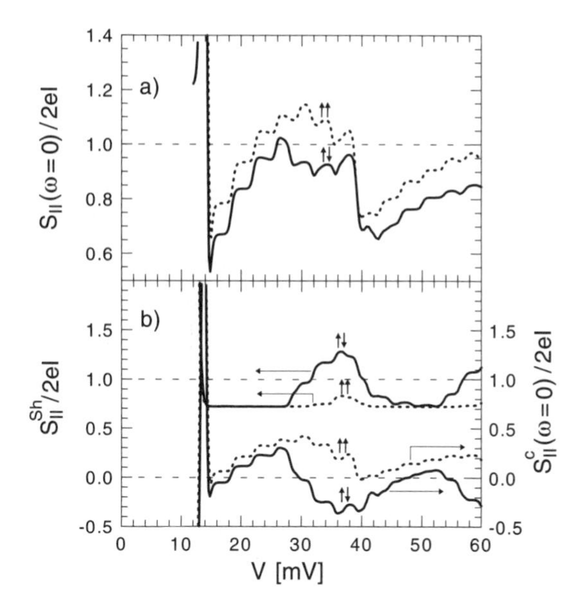

FIG. 4.12. Voltage dependence of the current shot noise at $\omega = 0$ (a) for $4R_{\mathrm{L}\uparrow} = R_{\mathrm{L}\downarrow} = 8\,\mathrm{M}\Omega$, $R_{\mathrm{R}\uparrow} = 4R_{\mathrm{R}\downarrow} = 240\,\mathrm{M}\Omega$ for the antiparallel alignment ($R_{\mathrm{L}\uparrow} = 4R_{\mathrm{L}\downarrow} = 8\,\mathrm{M}\Omega$, for the parallel configuration); other parameters as in Fig. 4.11. In part (b) $S_{II}(\omega = 0)$ is split into two components: S_{II}^{Sh} (upper curves) and $S_{II}^{\mathrm{c}}(\omega = 0)$ (lower curves).

$$
\begin{aligned}
S_{II}^{\mathrm{c}}(\omega) = {} & \frac{4e^2}{C^2} \sum_{N'_\uparrow, N'_\downarrow; N_\uparrow, N_\downarrow} \sum_r \left(\frac{C_{\mathrm{L}}C_{\mathrm{R}}}{C_r}\right) \left[\Gamma_r^+(N'_\uparrow, N'_\downarrow)\right. \\
& \left. - \Gamma_r^-(N'_\uparrow, N'_\downarrow)\right] \mathrm{Re}\left[\frac{1}{i\omega\hat{1} - \hat{M}}\right]_{N'_\uparrow, N'_\downarrow; N_\uparrow, N_\downarrow} \\
& \times \sum_{q=\pm 1} q \left\{ \left[C_{\mathrm{R}}\Gamma_{\mathrm{L}\uparrow}^{\pm}(N_\uparrow - q, N_\downarrow) \; - C_{\mathrm{L}}\Gamma_{\mathrm{R}\uparrow}^{\pm}(N_\uparrow - q, N_\downarrow)\right] p^0(N_\uparrow - q, N_\downarrow) \right. \\
& \left. + \left[C_{\mathrm{R}}\Gamma_{\mathrm{L}\downarrow}^{\pm}(N_\uparrow, N_\downarrow - q) \; - C_{\mathrm{L}}\Gamma_{\mathrm{R}\downarrow}^{\pm}(N_\uparrow, N_\downarrow - q)\right] p^0(N_\uparrow, N_\downarrow - q) \right\} \quad (4.39)
\end{aligned}
$$

Here, the Green's function $\hat{G}(\omega) = [i\omega\hat{1} - \hat{M}]^{-1}$ is defined in the two-dimensional space of states $(N_\uparrow, N_\downarrow)$, in contrast to spinless single-electron transistors [51–53], where the corresponding Green's function is defined in the one-dimensional space.

Figure 4.12(a) shows the bias dependence of the zero-frequency current noise $S_{II}(\omega = 0)$. The current noise is always smaller in the antiparallel configuration than in the parallel one. This is because in the presence of spin accumulation (which is significant only in the antiparallel alignment) the amplitude of fluctuations is smaller. In Fig. 4.12(b) $S_{II}(\omega = 0)$ is split into two parts; the frequency-independent component S_{II}^{Sh} and the contribution $S_{II}^{\mathrm{c}}(\omega = 0)$ arising from the

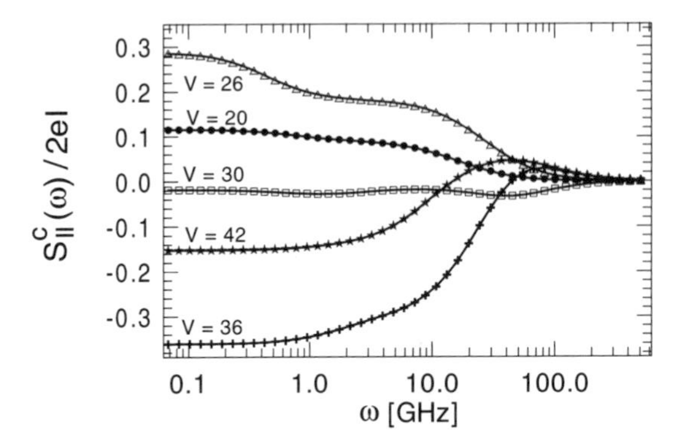

FIG. 4.13. Frequency dependence of the current shot noise $S^{c}_{II}(\omega)$ in the system defined in Fig. 4.12 for the antiparallel configuration.

frequency dependent part of the current noise (see Eq. 4.37). The component $S^{\rm Sh}_{II}$ is almost constant $\approx 2eI(C_1^2 + C_2^2)/C^2$ at the plateaux of the I-V curve and increases with opening of new channels. Dynamical correlations between the currents are described by $S^{c}_{II}(\omega)$. Its value in the limit $\omega \to 0$ can be positive between the I-V steps and negative when new channels become open. This is evident for the antiparallel alignment at $V \approx 26$ mV, when opening a tunneling channel for electrons with $\sigma =\downarrow$ leads to negative dynamical correlations. This effect is almost compensated by an increase in $S^{\rm Sh}_{II}$, and therefore one gets only a small reduction of the current noise $S_{II}(\omega = 0)$.

In the power spectrum of the current $S^{c}_{II}(\omega)$ (Fig. 4.13) [49], one can distinguish two distinct relaxation times, one in the high and another one in the low frequency ranges. In a wide voltage range the corresponding relaxation times are very close to the effective relaxation times for the charge and spin noise.

The asymmetry between the tunneling channels for electrons with opposite spins leads to activation of the spin component in the current noise. We extracted from S^{c}_{II} the components $S^{c}_{II\,\rm charge}$ and $S^{c}_{II\,\rm spin}$ corresponding to the charge and the spin noise, respectively. The results are presented in Fig. 4.14 as a function of the spin polarization P. It can be seen that the charge component is almost constant whereas the spin component increases with P and for $P \to 1$ can be much larger than the charge component. This analysis shows that both charge and spin fluctuations are relevant for the shot noise of the current in ferromagnetic single-electron transistors.

It is important to point out that both in the results presented in Fig. 4.12 for a multilevel grain and in a single-level quantum dot contacted to ferromagnetic leads [28] there is the possibility of a super-Poissonian Fano factor, $S_{II}(0)/2eI > 1$, due to lifting of the spin degeneracy. Recently an interacting three-terminal quantum dot with ferromagnetic leads was considered [54]. The dot operated as a beam splitter: one contact was a source and the other two acted as drains. The authors found a dynamical spin blockade (spin-dependent bunching of tunneling

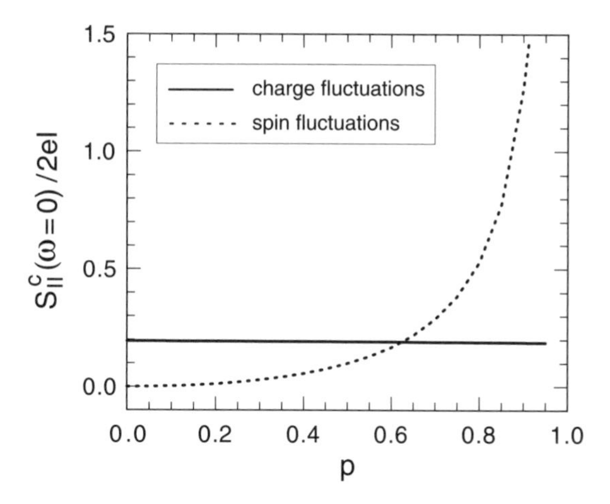

FIG. 4.14. Two components of $S^{\rm c}_{II}(\omega = 0)$ corresponding to the charge and spin noise plotted as a function of $P \equiv (R_{r\uparrow} - R_{r\downarrow})/(R_{r\uparrow} + R_{r\downarrow})$. The other parameters assumed here are: $R_{\rm R\downarrow} = 60\,{\rm M}\Omega$, $R_{\rm R\uparrow} = (1+P)/(1-P)R_{\rm R\downarrow}$, $R_{\rm L\downarrow} = 2\,{\rm M}\Omega$, $R_{\rm L\uparrow} = (1+P)/(1-P)R_{\rm L\downarrow}$, and other parameters as in Fig. 4.11.

events) and positive zero-frequency cross-correlations of the current in the drain electrodes.

4.5 Cotunneling

4.5.1 *Ferromagnetic island*

In single-electron transistors consisting of ferromagnetic metals [11, 12] and hybrid tunnel junctions containing ferromagnetic granules of nanometer size in the tunnel barrier [13–16] an enhancement of tunnel magnetoresistance has been observed in the Coulomb blockade regime. A similar effect was also observed in highly resistive magnetic granular systems. This effect is related to higher-order tunneling processes present in a small ferromagnetic double-tunnel junction. Since cotunneling is a higher-order process, where two electrons tunnel in a correlated way through both barriers, such tunneling events probe the spin asymmetry of both junctions and are very sensitive to the relative orientation of magnetization between electrodes and an island. Takahashi *et al.* [5, 55] have found that in this system the TMR ratio is given by ${\rm TMR} = 2P^2/(1-P^2)$ in the sequential-tunneling regime (for $k_{\rm B}T \gtrsim E_{\rm C}$) and ${\rm TMR} = 4P^2/(1-P^2)^2$ in the cotunneling regime (for $k_{\rm B}T \ll E_{\rm C}$). Therefore, the TMR is enhanced by the factor $2/(1-P^2)$ in the Coulomb blockade regime. A similar mechanism, also related to higher than second order tunneling processes, leads to a large enhancement of magnetoresistance. Wang and Brataas [56] have analyzed transport through a ferromagnetic single-electron transistor in the strong-coupling limit beyond the low-order sequential tunneling and cotunneling regimes using Monte Carlo simulations. They found further enhancement of the tunnel magnetoresistance ratio at low temperatures.

4.5.2 *Metallic island*

In the preceding sections we discussed transport characteristics and spin accumulation in the sequential-tunneling regime. In the Coulomb blockade regime, where sequential tunneling is suppressed, there are still tunneling processes of higher order (cotunneling) [55] which contribute to the electric current and also to spin accumulation [57,58]. Close to resonance both sequential and cotunneling currents may be comparable.

To calculate the tunneling current and associated spin accumulation (spin splitting of the chemical potential) the real-time diagrammatic formalism[59, 60] has been used. This technique is applicable for arbitrary temperature and arbitrary transport voltage.

The dominant second-order (cotunneling) contribution to the electric current can be divided into three parts, $I^{(2)} = \sum_{i=1}^{3} I_i^{(2)}$, with $I_i^{(2)} = \sum_\sigma I_{i\mathrm{L}\sigma}^{(2)} = -\sum_\sigma I_{i\mathrm{R}\sigma}^{(2)}$. These terms are given by the following equations:

$$I_{1r\sigma}^{(2)} = \frac{4\pi^2 e}{h}\sum_N p_N^{(0)} \int d\omega \left[\alpha^-(\omega)\,\alpha_{r\sigma}^+(\omega) - \alpha^+(\omega)\,\alpha_{r\sigma}^-(\omega)\right] \mathrm{Re}\, R_N(\omega)^2, \tag{4.40}$$

$$I_{2r\sigma}^{(2)} = -\frac{1}{2}\frac{4\pi^2 e}{h}\sum_N \left(p_N^{(0)} + p_{N+1}^{(0)}\right) \frac{\alpha^-(\Delta_N)\,\alpha_{r\sigma}^+(\Delta_N) - \alpha^+(\Delta_N)\,\alpha_{r\sigma}^-(\Delta_N)}{\alpha(\Delta_N)}$$
$$\times \int d\omega\, \alpha(\omega)\mathrm{Re}\left[R_N(\omega)^2 + R_{N+1}(\omega)^2\right], \tag{4.41}$$

$$I_{3r\sigma}^{(2)} = -\frac{1}{2}\frac{4\pi^2 e}{h}\sum_N \left(p_N^{(0)} + p_{N+1}^{(0)}\right) \frac{\partial}{\partial \Delta_N} \left[\frac{\alpha^-(\Delta_N)\alpha_{r\sigma}^+(\Delta_N) - \alpha^+(\Delta_N)\alpha_{r\sigma}^-(\Delta_N)}{\alpha(\Delta_N)}\right]$$
$$\times \int d\omega\, \alpha(\omega)\mathrm{Re}\left[R_N(\omega) - R_{N+1}(\omega)\right], \tag{4.42}$$

where $R_N(\omega) = 1/(\omega - \Delta_N + i0^+) - 1/(\omega - \Delta_{N-1} + i0^+)$. Here $\Delta_N = E_{\mathrm{ch}}(N+1) - E_{\mathrm{ch}}(N) = E_{\mathrm{C}}\left[1 + 2(N - n_x)\right]$, and $\alpha_{r\sigma}^{\pm}(\epsilon)$ are the forward and backward propagators $\alpha_{\mathrm{r}\sigma}^{\pm}(\epsilon)$ on the Keldysh contour in Fourier space given by Eq. (4.14). Apart from this we introduced $\alpha^{\pm}(\omega) = \sum_{r\sigma} \alpha_{r,\sigma}^{\pm}(\omega)$ and $\alpha(\omega) = \alpha^+(\omega) + \alpha^-(\omega)$. The probabilities $p_N^{(0)}$ obey the equation $p_N^{(0)}\alpha^+(\Delta_N) - p_{N+1}^{(0)}\alpha^-(\Delta_N) = 0$ (equivalent to Eq. 4.29) with $\sum_N p_N^{(0)} = 1$.

The terms $I_2^{(2)}$ and $I_3^{(2)}$ describe renormalization of the tunneling conductance and energy gap, respectively, and become important at resonance. The spin accumulation on the island (or equivalently spin splitting of the electrochemical potential) is determined from the condition of spin current conservation

$$\sum_r \left(I_{r\sigma}^{(1)} + I_{r\sigma}^{(2)}\right) - e\mu_\sigma D_{\mathrm{I}}/\tau_{\mathrm{sf}} = 0\,. \tag{4.43}$$

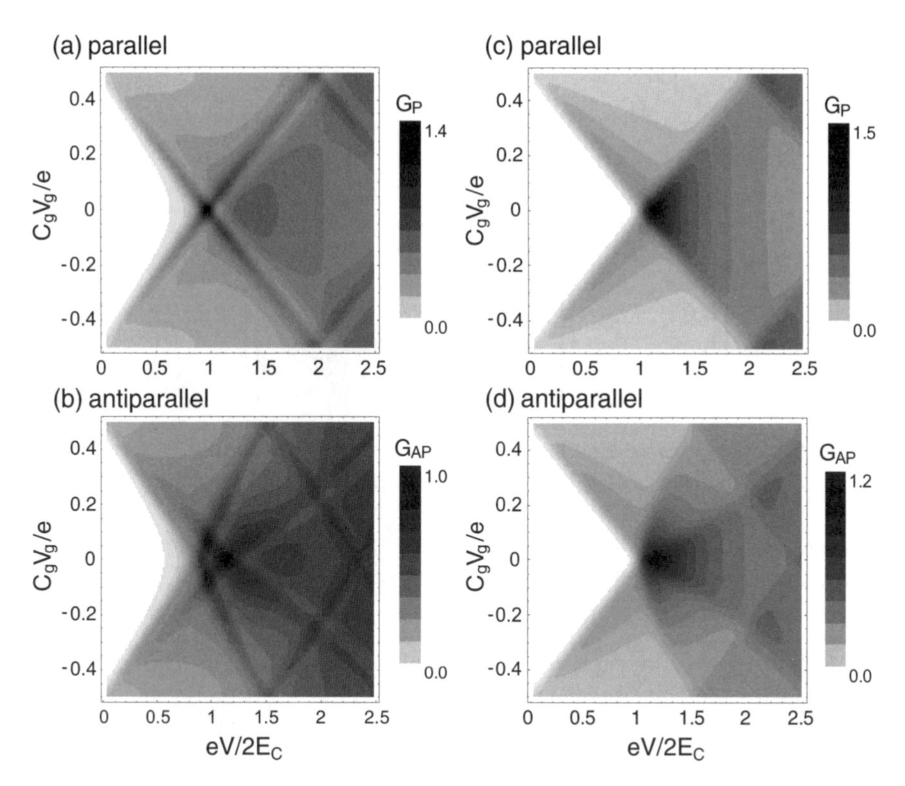

FIG. 4.15. The differential conductance [multiply by $(R_R + R_L)$] versus gate V_g and transport V voltages in a gray-scale representation in the (a) parallel and (b) antiparallel configurations calculated for Co electrodes in the cotunneling and (c) and (d) for the sequential-tunneling regime.

Here $I_{r\sigma}^{(1)}$ is obtained from Eq. (4.30). The spin-flip processes in the island are taken into account, and characterized by the relaxation time τ_{sf}.

In Fig. 4.15 we show the differential conductance for Co electrodes versus gate V_g and transport V voltages in a gray-scale representation for both parallel and antiparallel magnetic configurations in the sequential tunneling and cotunneling limits, calculated for $T/E_C = 0.03$ and for no spin-flip processes. We find well-resolved splitting of the conductance peak in the antiparallel alignment, which is a result of the spin splitting of the corresponding electrochemical potential of the island (see Fig. 4.16).

The above described splitting in the transport characteristics is directly related to the spin splitting of the electrochemical potential, and therefore can be used to detect spin accumulation. Generally, there are several experimental techniques by which the spin accumulation can be detected indirectly [61, 62]. However, the question whether spin splitting of the electrochemical potential can be observed directly, for instance by spectroscopic methods analogous to tunnel-

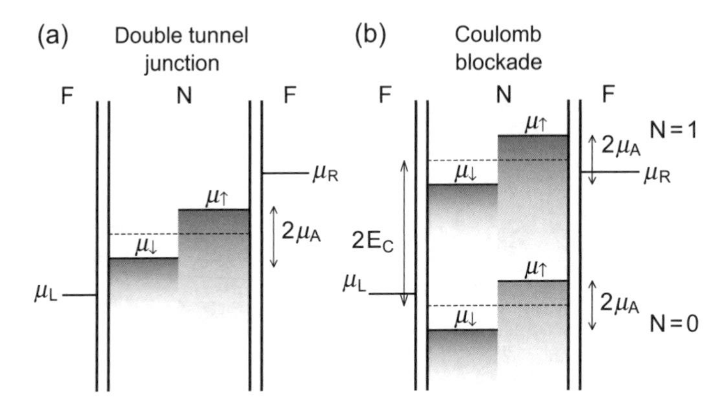

FIG. 4.16. Energy diagrams for a symmetric magnetic double tunnel junction with a normal metallic island for the antiparallel configuration (a) without and (b) with the Coulomb blockade. Here μ_L, μ_R are the electrochemical potentials for the left and right electrodes and $\mu_\uparrow$, $\mu_\downarrow$ are the electrochemical potentials for spin-up and spin-down electrons on the island. $N = 1\,, 0$ denotes the island charge.

ing spectroscopy for the superconducting gap, is still open. One can expect that the peculiarities in the transport characteristics of ferromagnetic single-electron transistors with a normal metallic island offer new possibilities.

4.5.3 *Quantum dot*

A detailed analysis of this transport regime for ferromagnetically contacted quantum dots can be found in Ref. [63], which includes, among others, the prediction and explanation of a peculiar zero-bias behavior for the antiparallel alignment. In Ref. [63], the cotunneling transport through a single-level and singly occupied quantum dot attached to ferromagnetic leads were study in the Coulomb blockade regime, far from resonance, where transport is dominated by cotunneling processes. When source and drain electrodes are magnetized antiparallel to each other, there is a pronounced zero-bias anomaly that is not related with Kondo correlations. It is rather a consequence of the interplay of non-equilibrium spin accumulation and spin relaxation due to spin-flip cotunneling. A finite spin accumulation on the quantum dot partially suppresses transport. Single-barrier spin-flip cotunneling provides a channel of spin relaxation and, hence, reduces the spin accumulation. In the absence of a magnetic field it plays a role only in linear response, $|eV| \ll k_B T$, but is negligible in the opposite limit. This leads to a zero-bias anomaly in the differential conductance. For parallel alignment, the exchange field gives rise to a finite and gate-voltage dependent spin polarization of the dot, $n_\uparrow \neq n_\downarrow$, even at zero bias; however in the Coulomb blockade regime it does not affect the transport.

4.6 Strong coupling – Kondo effect

The Kondo effect [27] in electron transport through a quantum dot predicted in Ref. [64] is experimentally well established [65, 66]. A dot with an odd number of electrons possesses a local spin which at low temperatures, $k_BT \leq k_BT_K \ll \Gamma$, and in the presence of strong coupling to the electrodes, behaves effectively like a magnetic Kondo impurity and leads to the Kondo effect. Screening of the dot spin due to the exchange coupling with the lead electrons leads then to the formation of a singlet state, which manifests itself as a Kondo resonance in the impurity density of states. This Kondo-correlated state is accompanied with an increased transmission through the dot, and gives rise to a sharp zero-bias anomaly in the conductance-voltage characteristics. The successful observation of the Kondo effect in molecular quantum dots like carbon nanotubes [67, 68] and single molecules [69] attached to metallic electrodes opened the possibility to study the influence of many-body correlations in the leads (superconductivity [70] or ferromagnetism [23, 71]) on the Kondo effect.

In this section we discuss how the presence of ferromagnetic leads influences the Kondo effect. In the extreme case of half-metallic leads, minority-spin electrons are completely absent, i.e., the screening of the dot spin is not possible, and no Kondo-correlated state can form. What happens, however, for the case of partially spin polarized leads?

In heavy-fermion systems [72] there is strong competition between the Kondo physics and magnetic RKKY interactions. The former and latter are characterized by the energy scales T_K and T_{RKKY}, respectively. Depending on the relation between the relevant energy scales T_{RKKY} and T_K, either the local spin is quenched and no magnetic order occurs (for $T_K > T_{RKKY}$), or the local molecular field removes spin degeneracy and suppresses the Kondo effect (for $T_K < T_{RKKY}$). A similar competition appears for a quantum dot attached to ferromagnetic leads; however in special situations ferromagnetism and the Kondo effect can coexist.

The possibility of the Kondo effect in a quantum dot attached to ferromagnetic electrodes was discussed in a number of publications [73–79], and it was shown that the Kondo resonance is split and suppressed in the presence of ferromagnetic leads [77–79]. However, it was also demonstrated that this splitting can be compensated by an appropriately tuned external magnetic field to restore the Kondo effect [77, 78], as will be discussed in detail below.

In the following we consider the two cases of parallel and antiparallel alignment of the lead magnetic moments. For the antiparallel configuration and zero magnetic field and bias voltage, the model is equivalent (by canonical transformation [64]) to a quantum dot coupled to a single lead with density of states $\rho_{L\uparrow} + \rho_{R\uparrow} = \rho_{L\downarrow} + \rho_{R\downarrow}$. In this case, the usual Kondo resonance forms, which is the same as for non-magnetic electrodes [27]. The situation changes for the parallel configuration, where there is an overall asymmetry for up and down spins, say $\rho_{L\uparrow} + \rho_{R\uparrow} > \rho_{L\downarrow} + \rho_{R\downarrow}$. To understand how this asymmetry affects the Kondo physics we discuss first the results obtained by the poor man's scaling technique

[80,81].

4.6.1 *Perturbative-scaling approach*

It is possible to analyze basic properties of a quantum dot attached to ferromagnetic leads in the Kondo regime using the perturbative-scaling approach. For details of the following calculations we refer to Ref. [77]. We consider first the result obtained with the poor man's scaling technique [80], performed in two stages [81]. In the first stage, when high-energy degrees of freedom are integrated out, charge fluctuations are dominant. In the second stage, the resulting model is mapped to a Kondo Hamiltonian, and the degrees of freedom involving spin fluctuations are integrated out.

First one reduces the energy scale of the effective bandwidth D from D_0, which is the smaller value of the bar bandwidth and the onsite repulsion U [81]. Charge fluctuations lead to a renormalization of the level position $\epsilon_{d\sigma}$ according to the scaling equations

$$\frac{d\epsilon_{d\sigma}}{d\ln(D_0/D)} = \frac{\Gamma_{\bar{\sigma}}}{2\pi}, \tag{4.44}$$

where $\bar{\sigma}$ is opposite to σ. Since the renormalization is spin dependent, a spin splitting of the level is generated. In the presence of a magnetic field, this spin splitting simply adds to the initial Zeeman splitting $\Delta\epsilon_d$. One then obtains the solution

$$\Delta\widetilde{\epsilon}_d = \widetilde{\epsilon}_{d\uparrow} - \widetilde{\epsilon}_{d\downarrow} = -(1/\pi)P\Gamma\ln(D_0/D) + \Delta\epsilon_d\,. \tag{4.45}$$

The scaling of Eq. (4.44) is terminated [81] at $\widetilde{D} \sim -\widetilde{\epsilon}_d$. When plugging in $D_0 = U$ and $D = \epsilon_d$, one finds that the generated level splitting exactly reflects the zero-temperature limit of the exchange field B_{exch}, Eq. (4.26).

To reach the strong-coupling limit, one can tune the external magnetic field B such that the total effective Zeeman splitting vanishes, $\Delta\widetilde{\epsilon}_d = 0$. In the second stage [80], spin fluctuations are integrated out. To accomplish this, one can perform the Schrieffer–Wolff transformation [27] to map the Anderson model (with renormalized parameters $\widetilde{D}$ and $\widetilde{\epsilon}_d$) to the effective Kondo Hamiltonian

$$\begin{aligned} H_{\mathrm{Kondo}} = {} & J_+ S^+ \sum_{rr'kq} a^\dagger_{rk\downarrow} a_{r'q\uparrow} + J_- S^- \sum_{rr'kq} a^\dagger_{rq\uparrow} a_{r'k\downarrow} \\ & + S^z \left(J_{z\uparrow} \sum_{rr'qq'} a^\dagger_{rq\uparrow} a_{r'q'\uparrow} - J_{z\downarrow} \sum_{rr'kk'} a^\dagger_{rk\downarrow} a_{r'k'\downarrow} \right), \end{aligned} \tag{4.46}$$

plus terms independent of either dot spin or lead electron operators, with $J_+ = J_- = J_{z\uparrow} = J_{z\downarrow} = |T|^2/|\widetilde{\epsilon}_d| \equiv J_0$ in the large-U limit. Although initially identical, the three coupling constants $J_+ = J_- \equiv J_\pm$, $J_{z\uparrow}$, and $J_{z\downarrow}$ are renormalized differently during the second stage of scaling. The scaling equations are

$$\frac{d(\rho_\pm J_\pm)}{d\ln(\widetilde{D}/D)} = \rho_\pm J_\pm(\rho_\uparrow J_{z\uparrow} + \rho_\downarrow J_{z\downarrow}) \ , \tag{4.47}$$

$$\frac{d(\rho_\sigma J_{z\sigma})}{d\ln(\widetilde{D}/D)} = 2(\rho_\pm J_\pm)^2 \ , \tag{4.48}$$

with $\rho_\pm = \sqrt{\rho_\uparrow \rho_\downarrow}$, $\rho_\sigma \equiv \sum_r \rho_{r\sigma}$. To solve these equations we observe that $(\rho_\pm J_\pm)^2 - (\rho_\uparrow J_{z\uparrow})(\rho_\downarrow J_{z\downarrow}) = 0$ and $\rho_\uparrow J_{z\uparrow} - \rho_\downarrow J_{z\downarrow} = J_0 p(\rho_\uparrow + \rho_\downarrow)$ is constant as well. There is only one independent scaling equation. All coupling constants reach the stable strong-coupling fixed point $J_\pm = J_{z\uparrow} = J_{z\downarrow} = \infty$ at the Kondo energy scale, $D \sim k_\mathrm{B} T_K$. For the parallel configuration, the Kondo temperature in leading order,

$$T_\mathrm{K}(P) \approx \widetilde{D} \exp\left[-\frac{1}{(\rho_\uparrow + \rho_\downarrow)J_0} \frac{\mathrm{arctanh}(P)}{P}\right] \ , \tag{4.49}$$

depends on the polarization P in the leads. It is maximal for non-magnetic leads, $P = 0$, and vanishes for $P \to 1$.

We point out here that the occurrence of the Kondo effect requires spin fluctuations in the dot as well as zero-energy spin-flip excitations in the leads. Indeed, a Stoner ferromagnet without full spin polarization $-1 < P < 1$ provides zero-energy Stoner excitations [25], even in the presence of an external magnetic field.

4.6.2 *Numerical renormalization group*

We will now discuss the results obtained by the numerical renormalization group (NRG) technique [27] – one of the most accurate methods available to study strongly correlated systems in the Kondo regime. Recently, it was adapted to the case of a quantum dot coupled to ferromagnetic leads [78, 79].

The simple way of modeling the ferromagnetic leads in the standard NRG procedure is to take the density of states in the leads to be constant and spin-independent, $\rho_{r\sigma}(\omega) \equiv \rho$, the bandwidths to be equal $D_\uparrow = D_\downarrow$, and lump all spin-dependence into the spin-dependent hybridization function, $\Gamma_{r\sigma}(\omega)$, which can be taken as ω-independent, $\Gamma_{r\sigma}(\omega) \equiv \Gamma_{r\sigma}$. The NRG method, with recent improvements related to high-energy features and finite magnetic field [82–84], is a well-established method to study Kondo impurity (quantum dot) physics. It allows one to calculate the level occupation $n_{\mathrm{d}\sigma} \equiv \langle d_\sigma^\dagger d_\sigma \rangle$ (a static property), the quantum dot spin spectral function, $\mathrm{Im}\ \chi_s^z(\omega) = \mathcal{F}\{i\Theta(t)\langle[S_z(t), S_z(0)]\rangle\}$, where $\mathcal{F}$ denotes the Fourier transform, and the spin-resolved single-particle spectral density $A_\sigma(\omega, T, B, P) = -\frac{1}{\pi}\mathrm{Im}\mathcal{G}^{\mathrm{ret}}_{\mathrm{d},\sigma}(\omega)$ for arbitrary temperature T, magnetic field B and polarization P. Here $\mathcal{G}^{\mathrm{ret}}_{\mathrm{d},\sigma}(\omega)$ denotes a retarded Green function. Using the spectral function one can find the spin-resolved linear conductance

$$G_\sigma = \frac{e^2}{\hbar} \frac{\Gamma_{\mathrm{L}\sigma}\Gamma_{\mathrm{R}\sigma}}{(\Gamma_{\mathrm{L}\sigma} + \Gamma_{\mathrm{R}\sigma})} \int_{-\infty}^{\infty} d\omega A_\sigma(\omega) \left(-\frac{\partial f(\omega)}{\partial \omega}\right) \tag{4.50}$$

with $f(\omega)$ denoting the Fermi function.

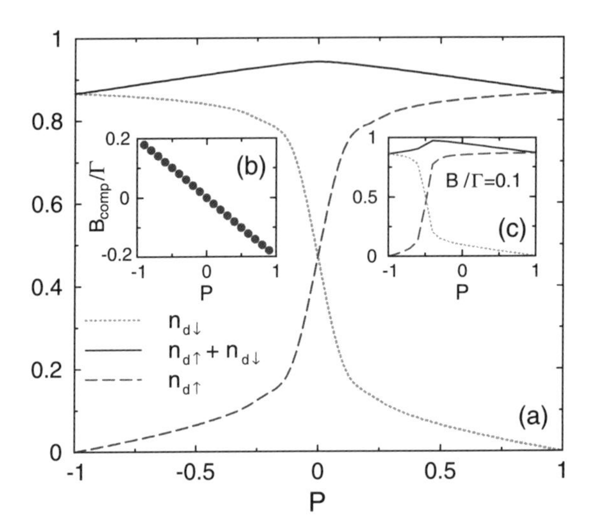

FIG. 4.17. Spin-dependent occupation of the dot level at (a) $B = 0$ and (b) $B = -0.1\Gamma$, as a function of spin polarization P. (a) For $B = 0$, the condition $n_{\mathrm{d}\uparrow} = n_{\mathrm{d}\downarrow}$ only holds at $P = 0$. (c) For finite P it can be satisfied if a finite, fine-tuned magnetic field, $B_{\mathrm{comp}}(P)$, is applied, whose dependence on P is shown in (b). Here $U = 0.12D$, $\epsilon_{\mathrm{d}} = -U/3$, $\Gamma = U/6$, and $T = 0$.

4.6.2.1 *Generation and restoration of spin splitting.* In the following we focus on the properties of the system at $T = 0$ in the local moment – Kondo regime, where the total occupancy of the level obeys the condition $n_{\mathrm{d}} = \sum_\sigma n_{\mathrm{d}\sigma} \approx 1$. The occurrence of charge fluctuations broadens and shifts the position of the dot levels (for both up and down spin orientations), and hence changes their occupation. For $P \neq 0$, the charge fluctuations, and hence level shifts and level occupations, become *spin-dependent*, causing the dot level to split [77] and the dot magnetization $n_{\mathrm{d}\uparrow} - n_{\mathrm{d}\downarrow}$ to be finite (Fig. 4.17). As a result, the Kondo resonance is spin-split as well and weakened (Fig. 4.18), similarly to the effect of an applied magnetic field [83,84]. This means that the Kondo correlations are reduced or even suppressed in the presence of ferromagnetic leads. However, for any fixed P, it is possible to compensate the splitting of the Kondo resonance (Figs 4.18c and 4.19c) by fine-tuning the magnetic field to an appropriate value, $B_{\mathrm{comp}}(P)$, defined as the field which maximizes the height of the Kondo resonance. This field is found to depend linearly on P (Fig. 4.17c) as predicted by Eq. (4.45) [77]. Remarkably, it was also found that at B_{comp} the local occupancies satisfy $n_{\mathrm{d}\uparrow} = n_{\mathrm{d}\downarrow}$ (Fig. 4.17b). The fact that this occurs simultaneously with the disappearance of the Kondo resonance splitting suggests that the local spin is fully screened at B_{comp}.

Below the magnetic field is fixed at $B = B_{\mathrm{comp}}(P)$. The spin spectral function $\mathrm{Im}\{\chi_s^z(\omega)\}$ at $T = 0$ is shown in Fig. 4.19. Its behavior is characteristic of the formation of a local Kondo singlet: as a function of decreasing frequency, the spin

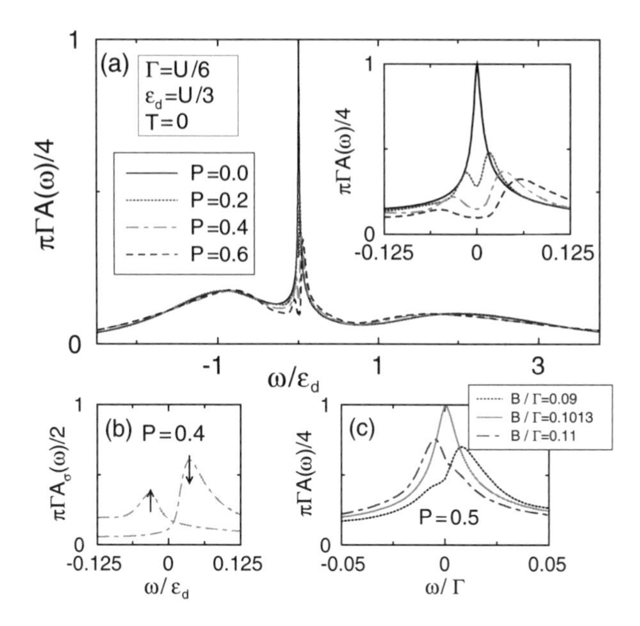

FIG. 4.18. (a) Quantum dot spectral function $A(\omega) = \sum_\sigma A_\sigma(\omega)$ for several values of spin polarization P; inset: expanded scale of the spectral function around ϵ_F. (b) The spin-resolved spectral function for fixed P. (c) Compensation of the spin splitting by fine-tuning an external magnetic field. Other parameters are as in Fig. 4.17.

spectral function shows a maximum at a frequency $\omega_{\max}$ which can be associated with the Kondo temperature [i.e. $k_B T_K \equiv \hbar\omega_{\max}$ at $B = B_{\text{comp}}(P)$], and then decreases linearly with ω, indicating the formation of the Fermi liquid state [27]. By determining $T_K(P)$ (from $\omega_{\max}$) for different P-values, one finds that T_K decreases with increasing P (Fig. 4.19b). For metals like Ni, Co, and Fe, where $P = 0.24, 0.35$, and 0.40, respectively [5], the decrease of T_K is rather weak, so the Kondo effect should still be experimentally accessible. Both $\text{Im}\{\chi_s^z(\omega)\}$ and the spectral function $A_\sigma(\omega)$ collapse rather well onto a universal curve if plotted in appropriate units (Figs 4.19a and 4.19c). This indicates that an applied magnetic field B_{comp} restores the universal behavior characteristic for the isotropic Kondo effect, in spite of the presence of spin-dependent coupling to the leads.

4.6.2.2 *Friedel sum rule.* Fig. 4.20(a) shows that the amplitude of the Kondo resonance is strongly spin dependent in the presence of ferromagnetic leads, which is unusual and unique; however the total conductance G_σ is not spin dependent (Fig. 4.20b). Further insights can be gained from the Friedel sum rule; an exact $T = 0$ relation [85] that holds for arbitrary values of P and B. The interacting retarded Green's function can be expressed as [27]

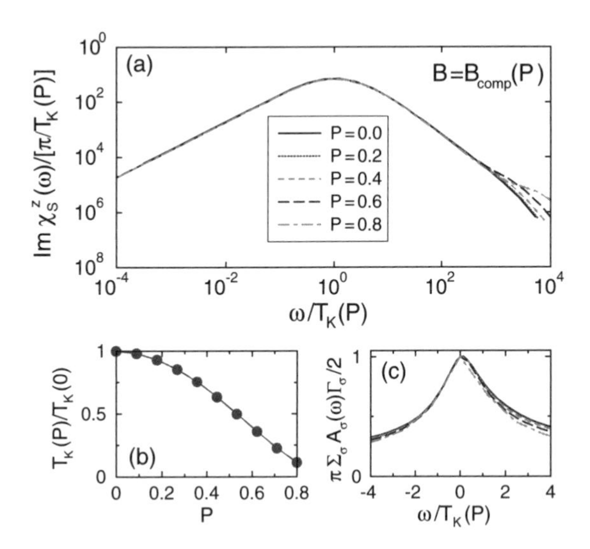

FIG. 4.19. (a) The spin spectral function $\mathrm{Im}\,[\chi_s^z(\omega)]$ as a function of energy. (b) Dependence of T_{K} on spin polarization P. The solid line shows the prediction from Eq. (4.45). (c) Quantum dot spectral function for several values of P. Other parameters are as in Fig. 4.17, $B = B_{\mathrm{comp}}(P)$.

$$\mathcal{G}_{d,\sigma}^{\mathrm{ret}}(\omega) = \frac{1}{\omega - \epsilon_{\mathrm{d}\sigma} + i\Gamma_\sigma - \Sigma_\sigma(\omega)}\,, \tag{4.51}$$

with spin-dependent $\epsilon_{\mathrm{d}\sigma}$ and Γ_σ; the former due to Zeeman splitting ($\epsilon_{\mathrm{d}\sigma} = \epsilon_{\mathrm{d}} - 1/2\ \sigma g \mu_B B$) and the latter due to the ferromagnetic leads. Here $\Sigma_\sigma(\omega)$ denotes the spin-dependent self-energy. Now, the Friedel sum rule [85] implies that at $T = 0$, the occupancy $n_{\mathrm{d}\sigma}$ and spectral functions can be written as

$$n_{\mathrm{d}\sigma} = \varphi_\sigma/\pi = \frac{1}{2} - \frac{1}{\pi}\tan^{-1}\left(\frac{\epsilon_{\mathrm{d}\sigma} - \epsilon_{\mathrm{F}} + \mathrm{Re}\ \Sigma_\sigma(\epsilon_{\mathrm{F}})}{\Gamma_\sigma}\right), \tag{4.52}$$

$$A_\sigma(\epsilon_{\mathrm{F}}) = \frac{\sin^2(\pi n_{\mathrm{d}\sigma})}{\pi\Gamma_\sigma}\,, \tag{4.53}$$

where $\Sigma_\sigma^{\mathrm{R}}(\omega) \equiv \mathrm{Re}\ \Sigma_\sigma(\omega)$, and $\varphi_\sigma(\omega)$ denotes the phase of the function $\mathcal{G}_{d,\sigma}^{\mathrm{ret}}(\omega)$. Since $\mathrm{Re}\ \Sigma_\uparrow(\epsilon_{\mathrm{F}}) \neq \mathrm{Re}\ \Sigma_\downarrow(\epsilon_{\mathrm{F}})$, an equal spin occupation, $n_{\mathrm{d}\uparrow} = n_{\mathrm{d}\downarrow}$, is possible only for $(\epsilon_{\mathrm{d}\uparrow} - \epsilon_{\mathrm{F}} + \mathrm{Re}\ \Sigma_\uparrow(\epsilon_{\mathrm{F}}))/\Gamma_\uparrow = (\epsilon_{\mathrm{d}\downarrow} - \epsilon_{\mathrm{F}} + \mathrm{Re}\ \Sigma_\downarrow(\epsilon_{\mathrm{F}}))/\Gamma_\downarrow$, which can be obtained only for an appropriate external magnetic field $B = B_{\mathrm{comp}}$. For the latter, in the local moment regime ($n \approx 1$) we have $n_\uparrow = n_\downarrow \approx 0.5$, so that $\varphi_\uparrow = \varphi_\downarrow \approx \pi/2$, which implies that the peaks of $A_\uparrow$ and $A_\downarrow$ are aligned. Thus, the Friedel sum rule clarifies why the magnetic field B_{comp} at which the splitting of the Kondo resonance disappears, coincides with that for which $n_{\mathrm{d}\uparrow} = n_{\mathrm{d}\downarrow}$. For $B = B_{\mathrm{comp}}$, the spin-dependent amplitude $A_\sigma(\epsilon_{\mathrm{F}})$ of Eq. (4.53) and the conductance $G_\sigma \sim \Gamma_\sigma A_\sigma(\epsilon_{\mathrm{F}})$ agree well with the above-mentioned NRG results (Figs 4.20a, 4.20b).

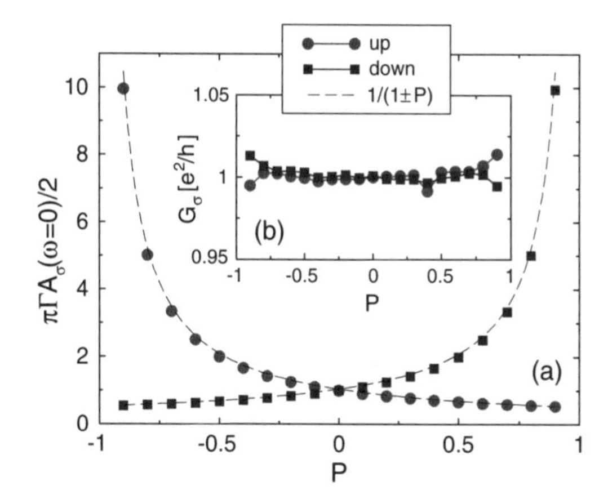

FIG. 4.20. (a) Spin resolved quantum dot spectral function amplitude $A_\sigma(\omega = 0)$ at the Fermi level, and (b) the quantum dot conductance G_σ, as functions of the spin polarization P, for $B = B_{\text{comp}}(P)$ and symmetric couplings ($\Gamma_{L\sigma} = \Gamma_{R\sigma}$), with U, ϵ_d, Γ, and T as in Fig. 4.17, implying $n_{\text{d}\sigma} \approx 0.5$. The dashed line in (a) is $1/(1 \pm P)$ (Eq. 4.53 with $n_{\text{d}\sigma} = 0.5$). As expected $G_\sigma = e^2/h$, with a numerical error less than 1 %.

4.6.3 *Gate-controlled spin-splitting in quantum dots*

In previous studies of quantum dots attached to ferromagnetic leads [73–79] an idealized, flat, spin-independent density of states (DOS) with spin-dependent tunneling amplitudes was considered. However, since the spin-splitting arises from renormalization effects, i.e., is a many-body effect, it depends on the *full* DOS structure of the involved material, and not only on its value at the Fermi surface. In realistic ferromagnetic systems, the DOS shape is strongly asymmetric due to the Stoner splitting and the different hybridization between the electronic bands [86].

In this section we discuss the gate voltage dependence of the spin-splitting of a dot level, resulting in a splitting and suppression of the Kondo resonance, and its dependence on the DOS structure. The numerical renormalization group technique was recently extended [87] to handle bands of arbitrary shape. For one class of DOS shapes one finds almost no V_g-dependence of the spin-splitting, while for another class the induced spin-splitting, which can be interpreted as the effect of a local exchange field, can be controlled by V_g. The spin-splitting can be fully compensated and its direction can even be reversed within this class. The physical mechanism that leads to this behavior is related to the compensation of the renormalization of the spin-dependent dot levels induced by the electron-like and hole-like quantum charge fluctuations.

In order to discuss the gate voltage dependence of the dot level spin-splitting, it is important to consider a more realistic, both energy and spin dependent band structure, $\rho_{r\uparrow}(\omega) \neq \rho_{r\downarrow}(\omega)$, violating particle-hole (p-h) symmetry $\rho_{r\sigma}(\omega) \neq$

$\rho_{r\sigma}(-\omega)$, which leads to an energy-dependent hybridization function $\Gamma_{r\sigma}(\omega)$ (Eq. 4.12). The NRG method [27, 82] was recently extended to handle arbitrary DOS shapes and asymmetry [87]. The standard logarithmic discretization of the conduction band was performed for *each* spin component separately, with the bandwidths, $D_\uparrow = D_\downarrow = D_0$. Within each interval $[-\omega_n, -\omega_{n+1}]$ and $[\omega_{n+1}, \omega_n]$ (with $\omega_n = D_0\Lambda^{-n}$) of the logarithmically discretized conduction band the operators of the continuous conduction band are expressed in terms of a Fourier series. Even though there is a non-constant conduction electron DOS, it is still possible to transform the Hamiltonian such that the impurity couples *only* to the zeroth-order component of the Fourier expansion of each interval [88]. Dropping the non-constant Fourier components of each interval [27, 82] then results in a discretized version of the Anderson model with the continuous spectrum in each interval replaced by a single fermionic degree of freedom (independently for both spin directions). Since we allow for an arbitrary DOS for *each* spin component σ of the conduction band this mapping needs to be performed for each σ separately. This leads to the Hamiltonian:

$$H = \sum_\sigma \epsilon_{d\sigma}\hat{n}_{d\sigma} + U\hat{n}_\uparrow\hat{n}_\downarrow + \sqrt{\xi_{0\sigma}/\pi}\sum_\sigma \left[d_\sigma^\dagger f_{0\sigma} + f_{0\sigma}^\dagger d_\sigma\right] + \sum_{\sigma n=0}^{\infty}\left[\varepsilon_{n\sigma} f_{n\sigma}^\dagger f_{n\sigma} + t_{n\sigma}\left(f_{n\sigma}^\dagger f_{n+1\sigma} + f_{n+1\sigma}^\dagger f_{n\sigma}\right)\right], \tag{4.54}$$

where $f_{n\sigma}$ are fermionic operators at the nth site of the Wilson chain, $\xi_{0\sigma} = 1/2\int_{-D_0}^{+D_0}\Gamma_\sigma(\omega)d\omega$, $t_{n\sigma}$ denotes the hopping matrix elements, and $\epsilon_{d\sigma} \equiv \epsilon_d - BS_z$. The absence of particle-hole symmetry leads to the appearance of non-zero on-site energies, $\varepsilon_{n\sigma}$, along the chain. In this *general* case no closed expression for the matrix elements $t_{n\sigma}$ and $\varepsilon_{n\sigma}$, both depending on the particular structure of the DOS via $\Gamma_\sigma(\omega)$, is known, therefore they have to be determined recursively [89].

4.6.3.1 *Spectral function and conductance.* Here, we again focus our attention on $T = 0$ properties. There are three typical classes of the V_g-dependence of the Kondo resonance splitting. In Ref. [87] calculations were performed for a parabolic band shape, namely $\rho_\sigma(\omega) = \frac{1}{2}\frac{3\sqrt{2}}{8}D^{-3/2}(1+\sigma Q)\sqrt{\omega + D + \sigma\Delta}$, where $\omega \in [-D-\sigma\Delta, D-\sigma\Delta]$, $D_0 = D+\Delta$, with Stoner splitting Δ [25], and some additional spin asymmetry Q, which modifies the amplitude of the DOS (see Fig. 4.21 insets). In Fig. 4.21 the weighted spectral function $\tilde{A}(\omega) \equiv \pi e^2/h\sum_\sigma \Gamma_\sigma(0)A_\sigma(\omega)$ is presented, normalized in a such way that for $\omega = 0$ it corresponds to the linear conductance $G = \tilde{A}(0)$, as a function of energy ω and ϵ_d. We focus on a narrow energy window around the Fermi surface where the Kondo resonance appears; charge resonances are visible when ϵ_d or $U + \epsilon_d$ approach the Fermi surface, namely at energies $\epsilon_d/U \gtrsim -0.1$ or $\lesssim -0.9$. Although the NRG method is designed to calculate equilibrium transport, one can still roughly deduce, from the spin-splitting of the Kondo resonance of the equilibrium spectral function $\tilde{A}(\omega)$,

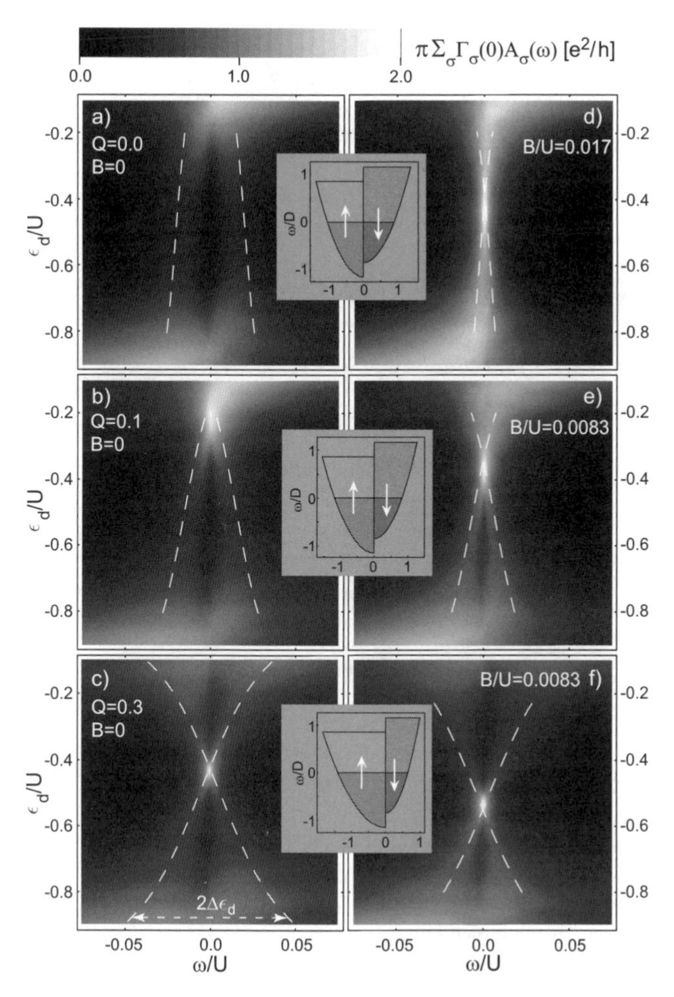

FIG. 4.21. V_g-dependence of the spin-splitting. Normalized spectral function $\pi \sum_\sigma \Gamma_\sigma(0) A_\sigma(\omega)$ as a function of energy ω and gate voltage ϵ_0, for the three different DOS shapes (depicted in insets) characterized by a different Q, which modifies both the spin and particle-hole asymmetry for different values of magnetic field as indicated. The white dashed lines are obtained using Eq. (4.55). Here $U = 0.12D_0$, $\pi V_0^2 = UD/6$, $\Delta = 0.15D$ and $T = 0$. Inset: the scheme of the parabolic DOS shape for spin $\uparrow$ and $\downarrow$.

the splitting of the zero-bias anomaly ΔV in the non-equilibrium conductance $G(V)$, since $e\Delta V \sim 2\Delta\epsilon_\mathrm{d}$ [77] ($\Delta\epsilon_\mathrm{d} \equiv \tilde{\epsilon}_{\mathrm{d}\uparrow} - \tilde{\epsilon}_{\mathrm{d}\downarrow}$ is the splitting of the renormalized levels). The spectra $\tilde{A}(\omega)$ are presented in Fig. 4.21 for three DOS shapes depicted in the insets, leading to the three typical behaviors. (i) For (a) there is hardly any ϵ_d-dependence of the spin-splitting; (ii) in (b) a strong ϵ_d-dependence

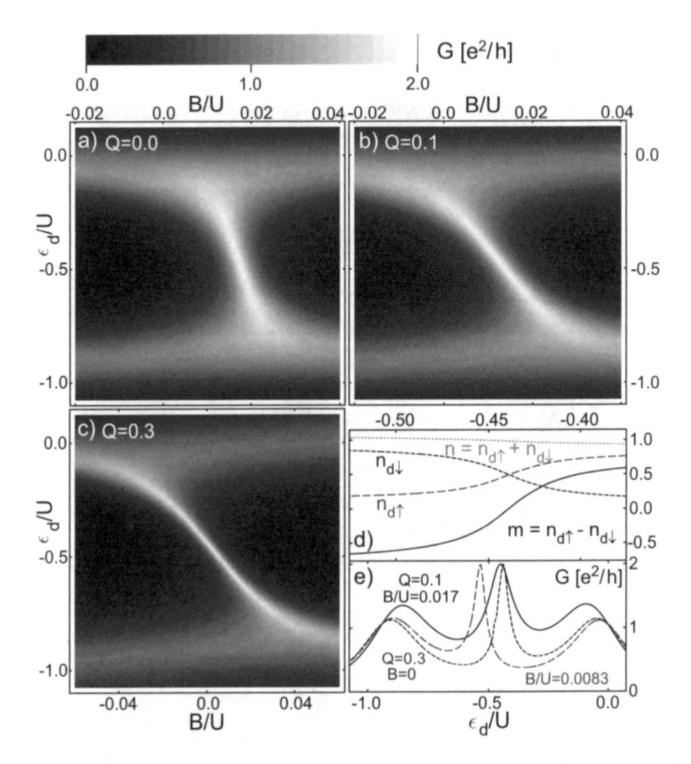

FIG. 4.22. The quantum dot linear conductance G as a function of gate voltage $\epsilon_{\rm d}$ and external magnetic field B for the DOS shapes (a), (b), and (c) as for Fig. 4.21(a), (b), and (c), respectively. (d) The $\epsilon_{\rm d}$-dependence of the spin-dependent n_σ, total occupancy of the dot $n_{\rm d} \equiv n_{\rm d\uparrow} + n_{\rm d\downarrow}$ and magnetization $m \equiv n_{\rm d\uparrow} - n_{\rm d\downarrow}$ for the situation from Fig. 4.21(c). (e) The conductance G for the situations from Fig. 4.21(c – dashed), (d – solid), and (f – long dashed). Parameters U, Γ, and T as in Fig. 4.21.

without compensation of the spin-splitting (i.e. no crossing), and (iii) for (c) a strong $\epsilon_{\rm d}$-dependence with a compensation (i.e. a crossing) and a change of the direction of the dot magnetization. The compensation (crossing) corresponds to the very peculiar situation where the Kondo effect (strong coupling fixed point) can be recovered in the presence of ferromagnetic leads without any external magnetic field.

4.6.3.2 *Effect of a magnetic field.* In Fig. 4.21(d,e,f) it is shown how a magnetic field B modifies the results of Fig. 4.21(a,b,c). The spin splitting can be compensated at a particular magnetic field $B_{\rm comp}$, whose magnitude depends on the $\epsilon_{\rm d}$-value, $B_{\rm comp}(\epsilon_{\rm d})$. Since $B_{\rm comp}$ can be viewed as a measure of the zero-field splitting, $\Delta\epsilon_{\rm d}(B=0,\epsilon_{\rm d}) \simeq -B_{\rm comp}(\epsilon_{\rm d})$, the $\epsilon_{\rm d}$-dependence of $\Delta\epsilon_{\rm d}$ can be measured by studying that of $B_{\rm comp}$, for which one needs to measure the linear conductance $G(\epsilon_d, B)$ as a function of both B and ϵ_d. In Fig. 4.22(a–c) $G(\epsilon_d, B)$ is plotted for the three bands of Fig. 4.21. The two horizontal ridges (resonances) in Fig. 4.22(a–c) correspond to quantum charge fluctuations (broadened dot level) of width $\sim \Gamma$. The lines with finite slope in Fig. 4.22(a–c) reflect the restored Kondo resonance and hence map out the ϵ_d-dependence of $B_{\text{comp}}(\epsilon_d) = -\Delta\epsilon_d(\epsilon_d)$ when the magnetic field compensates the spin-splitting. Interestingly the spin-splitting and the corresponding B_{comp} tend to diverge ($|\Delta\epsilon_d| \to \infty$) when approaching the charging resonance, as is best visible in Fig. 4.22(c).

In Fig. 4.22(d) it was shown how the occupation n_σ and the magnetic moment (spin) of the dot $m \equiv n_{d\uparrow} - n_{d\downarrow} \equiv 2\langle S_z \rangle$ change as a function of ϵ_d for the situation of Fig. 4.21(c). One finds that even though $B = 0$, it is possible to control the level spin-splitting of the dot, i.e. its spin, and thereby change the average spin direction of the dot from the parallel to antiparallel alignment w.r.t. the lead's magnetization. This can open the possibility of controlling the dot's spin state by means of a gate voltage without further need of an external magnetic field, which is difficult to apply locally in practical devices.

4.6.3.3 *Perturbative analysis.* One can understand the behavior presented in Fig. 4.21(a–c) by using Haldane's scaling method [81] discussed in Section 4.6.1, where charge fluctuations are integrated out. This leads to a spin-dependent renormalization of the dot's level position $\tilde{\epsilon}_{d\sigma}$ and a level broadening Γ_σ. In contrast to Section 4.6.1 we consider here the case of finite Coulomb interactions $U < \infty$, which means that also the doubly occupied state $|2\rangle$ is of importance. The spin-splitting is then given by $\Delta\epsilon_d \equiv \delta\epsilon_{d\uparrow} - \delta\epsilon_{d\downarrow} + B$, where

$$\delta\epsilon_{d\sigma} \simeq -\frac{1}{\pi}\int d\omega \left\{ \frac{\Gamma_\sigma(\omega)[1 - f(\omega)]}{\omega - \epsilon_{d\sigma}} + \frac{\Gamma_{-\sigma}(\omega) f(\omega)}{\epsilon_{d-\sigma} + U - \omega} \right\}. \tag{4.55}$$

The first term in the curly brackets corresponds to electron-like processes, namely charge fluctuations between a single occupied state $|\sigma\rangle$ and the empty $|0\rangle$ one, and the second term to hole-like processes, namely charge fluctuations between the states $|\sigma\rangle$ and $|2\rangle$. The amplitude of the charge fluctuations is proportional to Γ, which for $\Gamma \gg T$ determines the width of the dot's levels. Equation (4.55) shows that $\Delta\epsilon_d$ depends on the shape of $\Gamma_\sigma(\omega)$ for all ω, not only on its value at the Fermi surface. The dashed lines in Fig. 4.21(a–c) show $\pm\Delta\epsilon_d$ as a function of ϵ_d (from Eq. 4.55) for the same set of parameters as in the NRG calculation, and are in good agreement with the position of the (split) Kondo resonances observed in the latter. Equation (4.55) shows that the dramatic changes observed in Fig. 4.21 upon changing Q are due to the modification of the p-h and spin asymmetry.

Equation (4.55) predicts that even for systems with spin-asymmetric bands $\Gamma_\uparrow(\omega) \neq \Gamma_\downarrow(\omega)$, the integral can give $\Delta\epsilon = 0$, which corresponds to a situation where the renormalization of ϵ_σ due to electron-like processes are compensated by hole-like processes. For a flat band $\Gamma_\sigma(\omega) = \Gamma_\sigma$, Eq. (4.55) can be integrated analytically. For $D_0 \gg U, |\varepsilon_0|$ one finds: $\Delta\epsilon_d \simeq (P\,\Gamma/\pi)\ \mathrm{Re}[\,\phi(\epsilon_d) - \phi(U + \epsilon_d)\,]$,

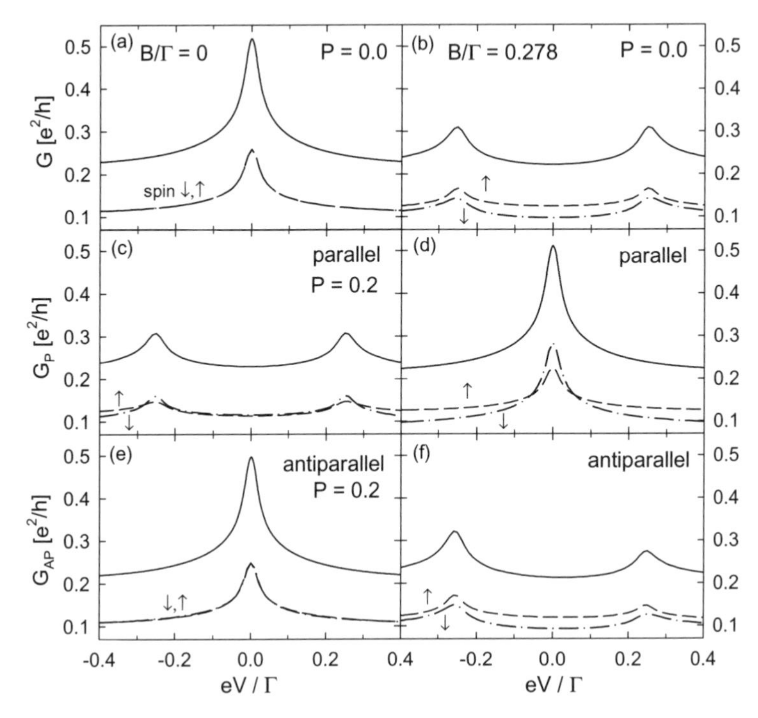

FIG. 4.23. Total differential conductance (solid lines) as well as the contributions from the spin-up (dashed) and the spin-down (dotted-dashed) channel vs. applied bias voltage V at zero magnetic field $B = 0$ (a,c,e) and at finite magnetic field (b,d,f) for normal (a,b) and ferromagnetic leads with parallel (c,d) and antiparallel (e,f) alignment of the lead magnetizations. The degree of spin polarization of the leads is $P = 0.2$ and the other parameters are $k_B T/\Gamma = 0.005$ and $\epsilon_d/\Gamma = -2$.

where $P \equiv (\Gamma_\uparrow - \Gamma_\downarrow)/\Gamma$, $\phi(x) \equiv \Psi(\frac{1}{2} + i\frac{x}{2\pi T})$, and $\Psi(x)$ denotes the digamma function. For $T = 0$, the spin-splitting is given by

$$\Delta\epsilon_d \simeq (P\,\Gamma\,/\pi)\ln(|\epsilon_d|/|U + \epsilon_d|)\,, \tag{4.56}$$

showing a logarithmic divergence for $\epsilon_d \to 0$ or $U + \epsilon_d \to 0$.

4.6.4 *Non-equilibrium transport properties*

In measurements of the Kondo effect usually the non-linear conductance is investigated. To get a qualitative prediction of non-linear transport properties, the equation-of-motion (EOM) method was employed with the usual decoupling scheme [90], but generalized by a self-consistent determination of the level energy to account for the exchange field. We skip all technical details here (they are given in Ref. [77]), and go directly to the discussion of the results.

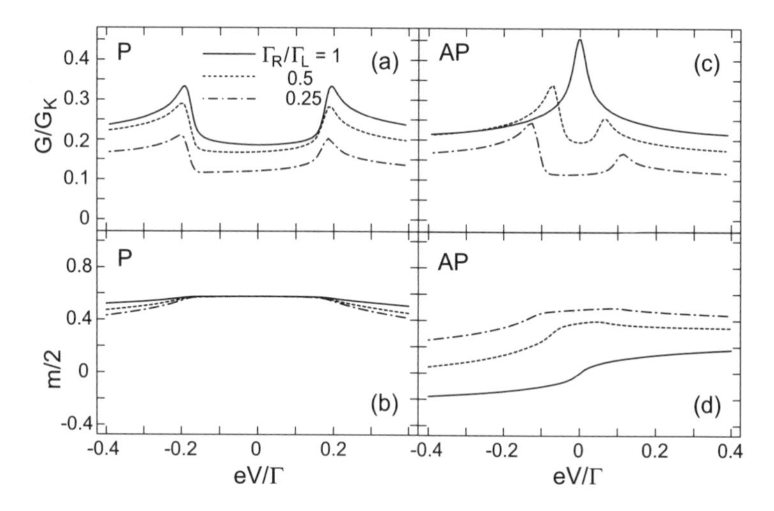

FIG. 4.24. The differential conductance (a) and the local magnetization (b) for the parallel alignment ($P = 0.2$). Solid lines, dashed lines and dot-dashed lines are results for $\Gamma_R/\Gamma_L = 1, 0.5$ and 0.25. Panels (c) and (d) are the corresponding plot of panels (a) and (b) with the antiparallel alignment. The other parameters are as in Fig. 4.23.

In Fig. 4.23 the differential conductance is shown as a function of the transport voltage. For non-magnetic leads, there is a pronounced zero-bias maximum (Fig. 4.23a) which splits in the presence of a magnetic field (Fig. 4.23b). For magnetic leads and parallel alignment, there is a splitting of the peak in the absence of a magnetic field (Fig. 4.23c), which can be tuned away by an appropriate external magnetic field (Fig. 4.23d). In the antiparallel configuration, the opposite happens, no splitting at $B = 0$ (Fig. 4.23e) but finite splitting at $B > 0$ (Fig. 4.23f) with an additional asymmetry in the peak amplitudes as a function of the bias voltage.

In the EOM approach the effect of spin-dependent quantum charge fluctuations is accounted for in a self-consistent but intuitive manner. The real-time diagrammatic technique [91,92] enables one to construct a systematic approach, where the effect of ferromagnetic electrodes can be analyzed without any additional assumptions. Recently the resonant tunneling approximation (RTA) was extended by Utsumi *et al.* [93] in order to account the presence of ferromagnetism. This technique gives more reasonable results and allows for further systematic insight into the physics.

This technique was used to analyze the effect of asymmetric coupling, $\Gamma_L \neq \Gamma_R$ [93]. Figure 4.24(a) and (b) shows the non-linear differential conductance and the local magnetization for the parallel alignment. In each panel the curves for various ratios of the left and right coupling strength Γ_R/Γ_L ($\Gamma = \Gamma_R + \Gamma_L$) are plotted. The splitting is not sensitive for this asymmetry because, as Eq. (4.45) indicates, $\Delta\epsilon_d$ depends only on the total coupling strength, which is independent of the ratio of coupling strength for the parallel alignment. The local magnetization of dot is not affected by the asymmetric coupling so much either. For the antiparallel alignment, there is a splitting of the zero bias anomaly for asymmetric coupling (Fig. 4.24c) accompanying by the spin accumulation (Fig. 4.24d). The splitting is due to the fact that for the asymmetric coupling exchange interactions from both leads do not cancel each other. The strong asymmetry in two split peaks' amplitudes is related with the fact that the intensities of split Kondo resonance out of equilibrium are different for the antiparallel alignment.

4.6.5 *Relation to experiment*

Recently Pasupaty *et al.* [23] studied the transport through a single C_{60} molecule attached to ferromagnetic nickel electrodes in the Kondo regime (see also Chapter 5). It was shown that the Kondo correlations appear even in the presence of ferromagnetic leads. The zero-bias anomaly in the non-equilibrium conductance was split for the parallel alignment of the leads magnetization in agreement with theoretical calculations presented in Fig. 4.24. For the antiparallel alignment there was no splitting of the zero-bias anomaly in one case, similarly to Fig. 4.24(e), and some residual splitting in the other case, which can be interpreted as an effect of asymmetric coupling $\Gamma_{\mathrm{L}} \neq \Gamma_{\mathrm{R}}$, similarly to Fig. 4.24(c). The measurement of the non-equilibrium conductance for the parallel alignment for several temperatures demonstrate that the splitting of the Kondo resonance does not depend on temperature, in agreement with Eq. (4.45).

In a recent paper, Nygard *et al.* [71] measured transport through a carbon nanotube attached to normal leads, which acts as a quantum dot exhibiting the zero bias anomaly at low temperatures. This anomaly is split probably due to interaction between the carbon nanotube and magnetic catalyst particles. The behavior presented in Fig. 4.21(a,b) was recently observed experimentally [71], where indeed a variation of the gate voltage results in two split conduction lines $G(V, V_{\mathrm{g}})$ which are parallel for one case and converging for the other case.

The lines (related to the Kondo resonance) with a finite slope in $G(\epsilon_{\mathrm{d}}, B)$ presented in Fig. 4.22(a,b,c) were observed for a singlet-triplet transition Kondo effect in a two-level quantum dot (Fig. 2d, Ref. [94]). The corresponding transition leads to a characteristic maximum in the valley between two charging resonances (Fig. 3c, Ref. [94]), similarly to Fig. 4.22(e). In that system the effective spin asymmetry is realized by the asymmetry in the coupling of two dot levels [95].

4.7 RKKY interaction between quantum dots

Up to now we discussed only spin polarized transport through a single quantum dot. New phenomena and transport characteristics can be observed in systems including more quantum dots, like for instance in double quantum dot systems. An example of such new phenomena is the RKKY interaction between two quantum dots, recently studied theoretically and also observed experimentally.

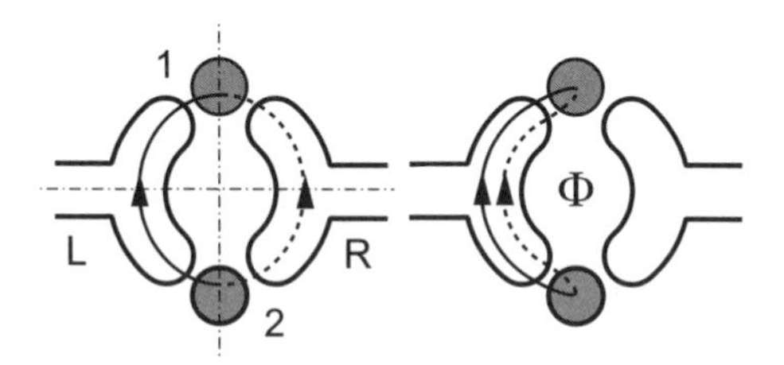

FIG. 4.25. Aharonov–Bohm ring embedded with one quantum dot (denoted by 1 or 2) in each arm. The magnetic flux Φ dependent (left panel) and independent (right panel) particle-hole excitation. The directed solid and dashed lines show the particle propagation and the hole propagation.

It is well known that two magnetic moments embedded in a metal induce spin polarization of the conduction electron gas, which leads to indirect coupling between the moments. This coupling is known as the RKKY interaction, first discussed in the 1950s [96]. Utsumi *et al.* [97] theoretically analyzed the possibility of the RKKY interaction in semiconducting nanostructures. The considerations were performed for an Aharonov–Bohm (AB) ring embedded with a quantum dot (in the Coulomb blockade regime with odd numbers of electrons) in each arm (Fig. 4.25) [98, 99]. Because the Fermi wavelength $\lambda_\mathrm{F} \equiv 2\pi/k_\mathrm{F}$, where k_F is the Fermi wave number, of a two-dimensional (2D) electron gas at a GaAs/AlGaAs interface is typically of the order 100 nm and the RKKY interaction decays weakly as $1/(k_\mathrm{F} l)$ for a 1D lead, where l is the distance between two dots, it could be possible to fabricate a sample with the long-range RKKY interaction. Actually, very recently, the RKKY interaction has been demonstrated for two dots coupled indirectly thorough an open big dot [100], which has stimulated intensive theoretical discussions [101–103]. The RKKY interaction in the AB ring depends on the flux and dominates AB oscillations in a characteristic manner, which could be a clear demonstration of the RKKY interaction.

4.7.1 *Flux-dependent RKKY interaction*

The model consists of two local spins coupled antiferromagnetically with electrons in two leads. An electron wave function acquires the "orbital phase" factor, $\mathrm{e}^{ik_\mathrm{F} l}$, during the propagation between two dots. When the magnetic field is applied, an additional phase factor, $\mathrm{e}^{\pm i\phi/2}$ ($\phi = 2\pi\,\Phi/\Phi_0$, where $\Phi_0 = hc/e$ is the flux quantum), is also counted for an electron tunneling through a dot in the clockwise/anticlockwise direction. The phase factor modifies the interference of a wave function through two paths (AB effect). The AB flux is written with the vector potential $\boldsymbol{A}$ as $\Phi = \oint \boldsymbol{A} \cdot d\boldsymbol{l}$, where the line integral is performed along the ring in the clockwise direction.

The Hamiltonian of the RKKY interaction was obtained by second-order perturbation theory in terms of J [96], $H_\mathrm{RKKY} = (J_\mathrm{RKKY}(\phi)/2)\ \mathbf{S}_1 \cdot \mathbf{S}_2$, where the RKKY coupling constant J_RKKY is given by

$$J_{\mathrm{RKKY}}(\phi) = (J^2/2)\,\chi\,(2 + 2\cos\phi)\,. \tag{4.57}$$

The phase-dependent term, $2\cos\phi$, is related to two configurations of particle-hole excitations, which enclose the flux (the left panel of Fig. 4.25), picking up a phase factor $e^{i\phi}$ or $e^{-i\phi}$. The two other configurations (the right panel) are independent of the flux and give the term 2. The physical interpretation may be as follows. At integer values of flux, the electron wave functions constructively interfere and thus the maximum coupling is induced. At half-integer values of flux, because of the destructive interference, the interaction is switched off. Thus it is possible to control the amplitude of the RKKY interaction by means of the flux. The susceptibility function χ counts details of electron states in leads. By considering the energy dependence of the orbital phase within the linearized dispersion around k_{F}, one obtains the RKKY oscillations for long distance, $l \gg \lambda_{\mathrm{F}}$, as

$$J_{\mathrm{RKKY}}(\phi) \simeq -\pi(J\rho)^2 D(1+\cos\phi)\frac{\cos(2k_{\mathrm{F}}l)}{(2k_{\mathrm{F}}l)}\,, \tag{4.58}$$

where ρ and D denote the density of states and the band-width of lead electrons, respectively.

4.7.1.1 *Linear conductance.* Due to the RKKY interaction, the two dot spins are entangled and probabilities of the singlet P_{S} and triplet P_{T} states depend on the flux through the flux-dependent RKKY interaction, $J_{\mathrm{RKKY}}(\phi)$. Thus the spin state, consequently also the RKKY interaction, can leave footprints in the flux dependence of the linear conductance. In Ref. [97] the conductance was calculated perturbatively in terms of J by using the real-time diagrammatic technique [91, 92].

First, we will discuss three cases classified by the value of the RKKY coupling constant J_{RKKY} and temperature:

(i) the uncorrelated local spins $[|J_{\mathrm{RKKY}}(\phi)| \ll T]$,
(ii) the ferromagnetic (F) coupling $[-J_{\mathrm{RKKY}}(\phi) \gg T]$, and
(iii) the antiferromagnetic (AF) coupling $[J_{\mathrm{RKKY}}(\phi) \gg T]$ cases.

For the case (i), the local-spin state is equally distributed among one singlet state and three triplet states. The conductance is the sum of the component showing AB oscillations, which is related with the cotunneling process preserving the spin, and the background of AB oscillations related with spin-flip processes enhanced by Kondo correlations (by $\ln T$):

$$G/G_{\mathrm{K}} \simeq (\pi J\rho)^2\left[1+\cos\phi\cos^2(k_{\mathrm{F}}l)+3\left(1+4J\rho\ln\frac{D}{k_{\mathrm{B}}T}\right)\right]\,, \tag{4.59}$$

where $G_{\mathrm{K}} \equiv h/e^2$ is the conductance quantum. For the case (ii), two local spins form a triplet state, and the conductance is that of the spin-1 Kondo model:

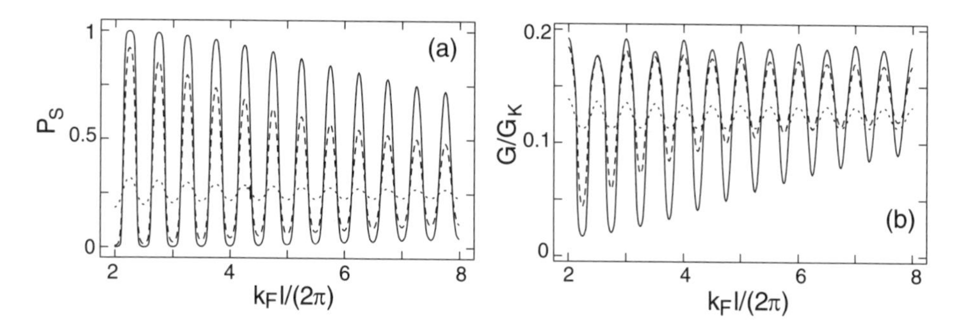

FIG. 4.26. (a) The probability of singlet state P_S and (b) the conductance as a function of the distance l between dots for $\phi=0$. Parameters: $T/D = 5\times10^{-5}$ (solid line), 10^{-4} (dashed line) and 10^{-3} (dotted line). $J\rho = 0.04$.

$$G/G_K \simeq (\pi J\rho)^2 \left\{ 8J\rho \cos^2(k_F l) \cos^2 \frac{\phi}{2} \ln \frac{2T^*}{\pi T} + \left(3 + 4J\rho \ln \frac{D}{k_B T}\right) \left[1+\cos\phi \cos^2(k_F l)\right] \right\} \tag{4.60}$$

for $k_F l \gg 1$. It is striking that contrary to case (i), Kondo correlations enhance the oscillatory component as shown in the second term. This is because the two spins are no longer independent phase-breaking scatterers, but become correlated. The first term appears because the spin-1 moment stretches over l and temperature T^* is related with the time of an electron traveling between two dots. For case (iii), two local moments form a singlet state. As the singlet state is decoupled from lead electrons, i.e., electrons flowing through dots cannot excite the system from the singlet state to a triplet state, only the cotunneling processes preserving spin contribute to the conductance:

$$G/G_K \simeq (\pi J\rho)^2 \left[1+\cos\phi \cos^2(k_F l)\right] . \tag{4.61}$$

In the following, we discuss the conductance for the whole rage of the flux ϕ and l. Figure 4.26(a) shows the l dependence of the probability of the singlet state. At low temperature $T \leq |J_{RKKY}(0)|$ (solid line), P_S oscillates with the same period as the RKKY interaction, i.e., when $k_F l/\pi$ is close to half-integer (integer), P_S is suppressed (enhanced). As the temperature increases (the dashed line for $T \sim |J_{RKKY}(0)|$ and the dotted line for $T \gg |J_{RKKY}(0)|$), the amplitude of oscillations is suppressed and the system approaches a uniform distribution between the singlet and triplet states. One can see also oscillations of the conductance (Fig. 4.26b) with the period of $k_F l/\pi = 1$. In the same way as in Fig. 4.26(a), the amplitude of oscillations is suppressed for $T \gg |J_{RKKY}(0)|$, which indicates that for $T < |J_{RKKY}(0)|$, the conductance oscillations are governed by the RKKY interaction.

Now we discuss the conductance in the full range of the flux ϕ for various temperatures (Fig. 4.27). The panels (a) and (b) are for the case of F coupling

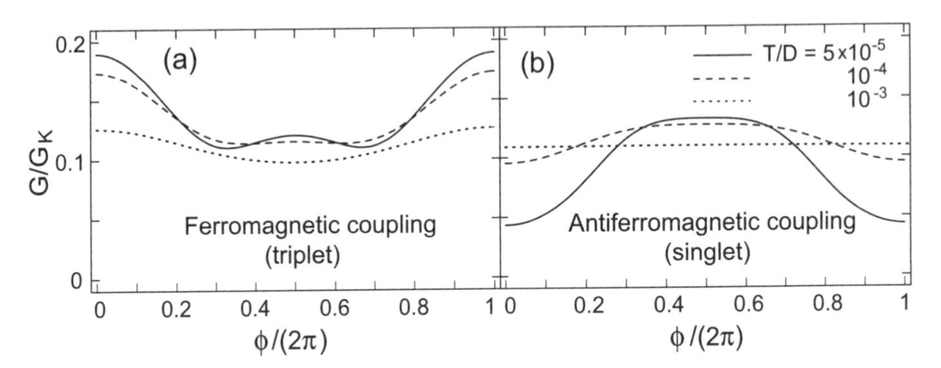

FIG. 4.27. Flux-dependent conductance for (a) $k_F l/(2\pi) = 50$ (the ferromagnetic coupling, spins form the triplet state) and (b) 50.25 (the antiferromagnetic coupling, spins form the singlet state). Parameters: $T/D = 5 \times 10^{-5}$ (solid line), 10^{-4} (dashed line) and 10^{-3} (dotted line). $J\rho = 0.04$.

($k_F l/\pi$ is an integer for Eq. 4.58) and the case of AF coupling ($k_F l/\pi$ is a half-integer for Eq. 4.58), respectively. For the parameters used in Fig. 4.27, the Kondo temperature is $T_K/D \approx 10^{-6}$. In the vicinity of zero or one quantum flux, the maximum ferromagnetic RKKY interaction is induced (Eq. 4.57). For the F coupling a triplet state is formed at low temperatures and the conductance is enhanced (case ii). For the AF coupling the singlet is formed at low temperatures and the conductance is suppressed (case iii). At half quantum flux, the RKKY interaction is switched off (Eq. 4.57). Surprisingly, one can observe the maximum in the conductance for both panels, where usually one expects the suppression of the conductance because of destructive interference. According to the discussion in (i), this maximum corresponds to incoherent transport thought the two independent spin-1/2 local moments enhanced by Kondo correlations. By combining the above behaviors, for the AF coupling case, the phase of AB oscillations is shifted by π, and for the F coupling case, the amplitude of AB oscillations is enhanced by Kondo correlations and an additional maximum appears at half-integer values of the flux. Such characteristic behavior is clear evidence of the RKKY interaction in the system.

4.7.2 *RKKY interaction – experimental results*

In this section we will discuss some recent experimental observations of the indirect RKKY interaction between semiconducting quantum dots, obtained by the group of C. M. Marcus. Following the discussion from Craig *et al.* [100], the device constructed from the two-dimensional electron gas consists of two smaller peripheral quantum dots connected to a larger open central dot, as shown in Fig. 4.28(a). Setting the bottom point contact to one fully conducting mode ($2e^2/h$) configured the central dot to act as a confined but open conducting region coupling the two peripheral dots.

The left and right dots were tuned to contain either an odd number or an even

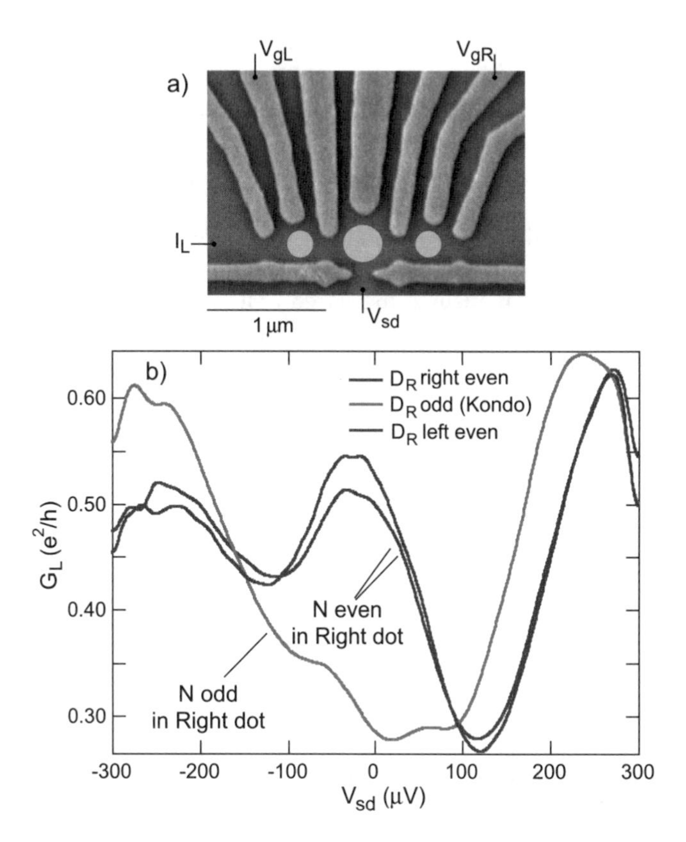

FIG. 4.28. (a) Scanning electron micrograph of a device, with schematic circles indicating locations of dots upon gate depletion. Gate voltages V_{gL} and V_{gR} change the energies and occupancies of the left and right dots, respectively. I_{L} indicates the current measured in the left (L) electrode. (b) Differential conductance G_{L} of the left dot for an odd number of electrons, N. When the right dot contains an even number of electrons, a zero-bias peak in G_{L} is seen, indicating a Kondo state. When the right dot contains an odd number of electrons, the Kondo state in the left dot is suppressed. From Craig *et al.* [100].

number of electrons by changing the voltage applied to the gates, V_{gL} and V_{gR}, respectively. Each dot individually showed characteristics of the Kondo effect – a zero-bias peak in the differential conductance in odd-occupancy valleys when the other peripheral dot has been pinched entirely off from the central region. When either dot was in an even-occupancy valley, turning on its coupling to the central dot (set initially at zero) did not qualitatively affect the signatures of the Kondo effect in the other dot. A more interesting situation arises when both dots contain odd numbers of electrons and the dots start to interact. Figure 4.28(b) shows the relevant comparison. When the right dot contains an even number of electrons, the odd Coulomb blockade valley in the left dot exhibits Kondo signatures, including a pronounced zero-bias peak in $G_{\rm L}$; however, when the right dot contains an odd number of electrons, the Kondo signatures in the left dot, including the zero-bias peak, are absent. The suppression of Kondo signatures in the odd-odd case can be interpreted as indicating that the RKKY interaction between the dots dominates the Kondo effect, forming either an overall spin-zero state (which has no Kondo effect) or a spin-one state with a much weaker Kondo effect at the temperatures measured. The zero-bias Kondo peak in the left dot first splits before being suppressed entirely as the right dot is coupled to the central dot. The splitting of the zero-bias peak is a signature of quantum coherence between Kondo states on the peripheral dots [101]. The magnitude of the splitting is related to the new energy scale of the RKKY interaction, $J_{\rm RKKY}$.

Acknowledgements

This work is based on joint publications with L. Borda, M. Braun, P. Bruno, R. Bulla, B. Bułka, J. von Delft, A. Fert, H. Imamura, J. König, S. Maekawa, G. Michałek, A. Pasupathy, D. Ralph, W. Rudziński, G. Schön, M. Sindel, Y. Utsumi, and I. Weymann, all of whom we thank for fruitful collaboration.

We thank G. Bauer, A. Brataas, T. Costi, L. Glazman, W. Hofstetter, B. Jones, C. Marcus, Yu. V. Nazarov, J. Nygård, A. Rosch, S. Takahashi, and M. Vojta for discussions. This work was supported by the Deutsche Forschungsgemeinschaft under the Center for Functional Nanostructures, and by the European Community under the "Spintronics" RT Network of the EC RTN2-2001-00440, Project PBZ/KBN/044/P03/2001, and the Centre of Excellence for Magnetic and Molecular Materials for Future Electronics within the EC Contract G5MA-CT-2002-04049.

References

[1] M. Jullière, Phys. Lett. A **54**, 225 (1975).
[2] *Spin Electronics*, edited by M. Ziese and M. J. Thornton (Springer, Berlin, 2001).
[3] S. A. Wolf, D. D. Awschalom, R. A. Buhrman, J. M. Daughton, S. von Molnar, M.L. Roukes, A. Y. Chtchelkanova, and D. M. Treger, Science **294**, 1488 (2001).
[4] *Semiconductor Spintronics and Quantum Computation*, edited by D. D. Awschalom, D. Loss, and N. Samarth (Springer, Berlin, 2002).
[5] *Spin Dependent Transport in Magnetic Nanostructures*, edited by S. Maekawa and T. Shinjo (Taylor & Francis, 2002).
[6] I. Žutić, J. Fabian, and S. Das Sarma, Rev. Mod. Phys. **76**, 323 (2004).
[7] D. V. Averin and K. K. Likharev, in *Mesoscopic Phenomenon in Solids*, Eds. B. L. Altshuler, P. A. Lee, and R. A. Webb (Amsterdam: North-Holland 1991).

[8] *Single Charge Tunneling: Coulomb Blockade Phenomena in Nanostructures*, edited by. H. Grabert and M. H. Devoret, NATO ASI Series B: Physics 294 (Plenum Press, New York, 1992).
[9] *Mesoscopic Electron Transport*, edited by L. L. Sohn, L. P. Kouwenhoven, and G. Schön (Kluwer, Dordrecht, 1997).
[10] J. von Delft and D. C. Ralph, Phys. Rep. **345**, 61 (2001).
[11] K. Ono, H. Shimada, S. Kobayashi and Y. Outuka, J. Phys. Soc. Jpn. **65**, 3449 (1996); K. Ono, H. Shimada and Y. Outuka, J. Phys. Soc. Jpn. **66**, 1261 (1997).
[12] H. Brückl, G. Reiss, H. Vizelberg, M. Bertram, I. Mönch and J. Schumann, Phys. Rev. B **58**, 8893 (1998).
[13] L. F. Schelp, A. Fert, F. Fettar, P. Holody, S. F. Lee, J.L. Maurice, F. Petroff and A. Vaures, Phys. Rev. B **56**, 5747 (1997); F. Fettar, S.-F. Lee, F. Petroff, A. Vaures, P. Holody, L. F. Schelp, and A. Fert, Phys. Rev. B **65**, 174415 (2002).
[14] S. Mitani, S. Takahashi, K. Takanashi, K. Yakushiji, S. Maekawa, and H. Fujimori, Phys. Rev. Lett. **81**, 2799 (1998); H. Imamura, J. Chiba, S. Mitani, K. Takanashi, S. Takahashi, S. Maekawa, and H. Fujimori, Phys. Rev. B **61**, 46 (2000); K. Yakushiji, S. Mitani, K. Takanashi, S. Takahashi, S. Maekawa, H. Imamura, and H. Fujimori, Appl. Phys. Lett. **78**, 515 (2001).
[15] K. Yakushiji, S. Mitani, K. Takanashi, and H. Fujimori, J. Appl. Phys. **91**, 7038 (2002); F. Ernult, K. Yamane, S. Mitani, K. Yakushiji, K. Takanashi, Y. K. Takahashi, and K. Hono, Appl. Phys. Lett. **84**, 3106 (2004).
[16] K. Yakushiji, F. Ernult, H. Imamura, K. Yamane, S. Mitani, K. Takanashi, S. Takahashi, S. Maekawa, and H. Fujimori, Nature Materials **4**, 57 (2005).
[17] C. D. Chen, W. Kuo, D. S. Chung, J. H. Shyu, and C. S. Wu, Phys. Rev. Lett. **88**, 047004 (2002); C. D. Chen, Y. D. Yao, S. F. Lee, and J. H. Shyu, J. Appl. Phys. **91**, 7469 (2002); J H. Shyu, Y. D. Yao, C. D. Chen, and S. F. Lee, J. Appl. Phys. **93**, 8421 (2003).
[18] J. Johansson, M. Urech, D. Haviland, and V. Korenivski, Phys. Rev. Lett. 91, 149701 (2003).
[19] D. Wang and J. G. Lu, J. Appl. Phys. **97**, 10A708 (2005).
[20] Y. Chye, M. E. White, E. Johnston-Halperin, B. D. Gerardot, D. D. Awschalom, and P. M. Petroff, Phys. Rev. B **66**, 201301 (2002).
[21] M. M. Deshmukh and D. C. Ralph, Phys. Rev. Lett. **89**, 266803 (2002).
[22] R. Jansen and J. S. Moodera, Appl. Phys. Lett. **75**, 400 (1999); S. Tanoue and A. Yamasaki, J. Appl. Phys. **88**, 4764 (2000).
[23] A. N. Pasupathy, R. C. Bialczak, J. Martinek, J. E. Grose, L. A. K. Donev, P. L. McEuen, and D. C. Ralph, Science **306**, 86 (2004).
[24] K. Tsukagoshi, B. W. Alphenaar, and H. Ago, Nature **401**, 572 (1999); B. Zhao, I. Mönch, H. Vinzelberg, T. Mühl, and C. M. Schneider, Appl. Phys. Lett. **80**, 3144 (2002); J. Appl. Phys. **91**, 7026 (2002); A. Jensen, J. Nygård and J. Borggreen in *Proceedings of the International Symposium on Mesoscopic Superconductivity and Spintronics*, edited by H. Takayanagi and J. Nitta

(World Scientific 2003), pp. 33-37.
[25] K. Yosida, *Theory of Magnetism* (Springer, 1996).
[26] D. V. Averin and A. A. Odintsov, Phys. Lett. A **140**, 251 (1989); D. V. Averin and Yu. V. Nazarov, Phys. Rev. Lett. **65**, 2446 (1990).
[27] A. C. Hewson, *The Kondo Problem to Heavy Fermions* (Cambridge Univ. Press, 1993).
[28] B. R. Bułka, Phys. Rev. B **62**, 1186 (2000).
[29] W. Rudziński and J. Barnaś, Phys. Rev. B **64**, 085318 (2001).
[30] J. König and J. Martinek, Phys. Rev. Lett. **90**, 166602 (2003).
[31] M. Braun, J. König, and J. Martinek, Phys. Rev. B **70**, 195345 (2004).
[32] J. König, J. Martinek, J. Barnas, and G. Schön, cond-mat/0404509.
[33] Y. Meir and N. S. Wingreen, Phys. Rev. Lett. **68**, 2512 (1992).
[34] J. C. Slonczewski, Phys. Rev. B **39**, 6995 (1989).
[35] J. S. Moodera and L. R. Kinder, J. Appl. Phys. **79**, 4724 (1996); H. Jaffrès, D. Lacour, F. Nguyen Van Dau, J. Briatico, F. Petroff, and A. Vaurès, Phys. Rev. B **64**, 064427 (2001).
[36] M. Braun, J. König, and J. Martinek, Phys. Rev. B **70**, 195345 (2004)
[37] J. Barnaś and A. Fert, Phys. Rev. Lett. **80**, 1058 (1998).
[38] K. Majumdar and S. Hershfield, Phys. Rev. B **57**, 11521 (1998).
[39] J. Barnaś and A. Fert, Europhys. Lett. **44**, 85 (1998); J. Magn. Magn. Mater. **192**, L 391 (1999).
[40] A. Brataas, Yu. V. Nazarov, J. Inoue and G. E. W. Bauer, Phys. Rev. B **59**, 93 (1999); European Phys. J. B **9**, 421 (1999).
[41] A. N. Korotkov and V. I. Safarov, Phys. Rev. B **59**, 89 (1999).
[42] J. Martinek, J. Barnaś, G. Michałek, B. R. Bułka and A. Fert, J. Magn. Magn. Mater. **207**, L 1 (1999); J. Barnaś, J. Martinek, G. Michałek, B. R. Bułka, and A. Fert, Phys. Rev. B **62**, 12363 (2000).
[43] W. Kuo and C. D. Chen, Phys. Rev. B **65**, 104427 (2002).
[44] D. V. Averin and K. K. Likharev, J. Low Temp. Phys. **62**, 345 (1986); D. V. Averin and A. N. Korotkov, Zh. Eksp. Teor. Fiz. **97**, 1661 (1990); D. V. Averin, A. N. Korotkov and K. K. Likharev, Phys. Rev. B **44**, 6199 (1991).
[45] C. W. J. Beenakker, Phys. Rev. B **44**, 1646 (1991).
[46] A. Brataas, M. Hirano J. Inoue, Yu. V. Nazarov, and G. E. W. Bauer, Jpn. J. Appl. Phys., Part 1 **40**, 2329 (2001).
[47] J. Inoue and A. Brataas, Phys. Rev. B **70**, 115315 (2004).
[48] Ya. M. Blanter and M. Büttiker, Phys. Rep. **336**, 1 (2000).
[49] B. R. Bułka, J. Martinek, G. Michałek and J. Barnaś, Phys. Rev. B **60**, 12246 (1999).
[50] K. M. van Vliet and J. R. Fassett, in *Fluctuation Phenomena in Solids*, edited by R. E. Burgess (Academic Press, 1965), p. 267.
[51] A. N. Korotkov, Phys. Rev. B **49**, 10381 (1994).
[52] S. Hershfield, J. D. Davies, P. Hyldgaard, C. J. Stanton and J. W. Wilkins, Phys. Rev. B **47**, 1967 (1993).
[53] U. Hanke, Y. M. Galperin, K. A. Chao and N. Zou, Phys. Rev. B **48**, 17209

(1993); U. Hanke, Y. M. Galperin and K. A. Chao, Phys. Rev. B **50**, 1595 (1994); A. Imamoglu and Y. Yamamoto, Phys. Rev. Lett. **70**, 3327 (1993).
[54] A. Cottet, W. Belzig, and C. Bruder, Phys. Rev. Lett. **92**, 206801 (2004); Phys. Rev. B **70**, 115315 (2004).
[55] S. Takahashi and S. Maekawa, Phys. Rev. Lett. **80**, 1758 (1998).
[56] X. H. Wang and A. Brataas, Phys. Rev. Lett. **83**, 5138 (1999); A. Brataas and X. H. Wang, Phys. Rev. B **64**, 104434 (2001).
[57] A. G. Aronov, Pis'ma Zh. Eksp. Teor. Fiz. **24**, 37 (1976) [JETP Lett. **24**, 32 (1976)].
[58] H. Imamura, S. Takahashi, and S. Maekawa, Phys. Rev. B **59**, 6017 (1999).
[59] J. Martinek, J. Barnaś, S. Maekawa, H. Schoeller, and G. Schön, Phys. Rev. B **66**, 014402 (2002).
[60] J. König, H. Schoeller, and G. Schön, Phys. Rev. Lett.**78**, 4482 (1997); Phys. Rev. B **58**, 7882 (1998).
[61] M. Johnson and R. H. Silsbee, Phys. Rev. Lett. **55**, 1790 (1985).
[62] F. J. Jedema, A. T. Filip, and J. van Wees, Nature **410**, 345 (2001).
[63] I. Weymann, J. Martinek, J. König, J. Barnaś, and G. Schön, Phys. Rev. B **72**, 113301 (2005)
[64] L. I. Glazman and M. E. Raikh, JETP Lett. **47**, 452 (1988); T. K. Ng and P. A. Lee, Phys. Rev. Lett. **61**, 1768 (1988).
[65] D. Goldhaber-Gordon, H. Shtrikman, D. Mahalu, D. Abusch-Magder, U. Meirav, and M. A. Kastner, Nature **391**, 156 (1998).
[66] S. M. Cronenwett, T. H. Oosterkamp, and L. P. Kouwenhoven, Science **281**, 540 (1998); F. Simmel, R. H. Blick, J. P. Kotthaus, W. Wegscheider, and M. Bichler, Phys. Rev. Lett. **83**, 804 (1999); J. Schmid, J. Weis, K. Eberl, and K. v. Klitzing, Phys. Rev. Lett. **84**, 5824 (2000); W. G. van der Wiel, S. De Franceschi, T. Fujisawa, J. M. Elzerman, S. Tarucha, and L. P. Kouwenhoven, Science **289**, 2105 (2000).
[67] J. Nygård, D. H. Cobden, and P. E. Lindelof, Nature **408**, 342 (2000).
[68] M. R. Buitelaar, A. Bachtold, T. Nussbaumer, M. Iqbal, and C. Schenberger, Phys. Rev. Lett. **88**, 156801 (2002).
[69] J. Park, A. N. Pasupathy, J. I. Goldsmith, C. Chang, Y. Yaish, J. R. Petta, M. Rinkoski, J. P. Sethna, H. D. Abru, P. L. McEuen, and D. C. Ralph, Nature **417**, 722 (2002); W. Liang, M. P. Shores, M. Bockrath, J. R. Long, and H. Park, Nature **417**, 725 (2002).
[70] M. R. Buitelaar, T. Nussbaumer, and C. Schönenberger, Phys. Rev. Lett. **89**, 256801 (2002).
[71] J. Nygård, W. F. Koehl, N. Mason, L. DiCarlo, and C. M. Marcus, cond-mat/0410467.
[72] A. H. Castro Neto and B. A. Jones, Phys. Rev. B **62**, 14975 (2000) and references there in.
[73] N. Sergueev, Q. F. Sun, H. Guo, B. G. Wang, and J. Wang, Phys. Rev. B **65**, 165303 (2002).

[74] P. Zhang, Q. K. Xue, Y. Wang, and X. C. Xie, Phys. Rev. Lett. **89**, 286803 (2002).
[75] B. R. Bułka and S. Lipinski, Phys. Rev. B **67**, 024404 (2003).
[76] R. Lopez and D. Sanchez, Phys. Rev. Lett. **90**, 116602 (2003).
[77] J. Martinek, Y. Utsumi, H. Imamura, J. Barnaś, S. Maekawa, J. König, and G. Schön, Phys. Rev. Lett. **91**, 127203 (2003).
[78] J. Martinek, M. Sindel, L. Borda, J. Barnaś, J. König, G. Schön, and J. von Delft, Phys. Rev. Lett. **91**, 247202 (2003).
[79] M. S. Choi, D. Sanchez, and R. Lopez, Phys. Rev. Lett. **92**, 056601 (2004).
[80] P. W. Anderson, J. Phys. C **3**, 2439 (1970).
[81] F. D. M. Haldane, Phys. Rev. Lett. **40**, 416 (1978).
[82] K. G. Wilson, Rev. Mod. Phys. **47** 773 (1975); T. A. Costi, A. C. Hewson and V. Zlatic, J. Phys.: Condens. Matter **6**, 2519 (1994).
[83] W. Hofstetter, Phys. Rev. Lett. **85**, 1508 (2000).
[84] T. A. Costi, Phys. Rev. Lett. **85**, 1504 (2000); Phys. Rev. B **64**, 241310 (2001).
[85] D. C. Langreth, Phys. Rev. **150**, 516 (1966).
[86] *Handbook of the Band Structure of Elemental Solids*, edited by D. A. Papaconstantopoulos (Plenum Press, 1986).
[87] J. Martinek, M. Sindel, L. Borda, J. Barnaś, R. Bulla J. König, G. Schön, S. Maekawa, and J. von Delft, Phys. Rev. B **72**, 121302 (2005)
[88] R. Bulla, T. Pruschke, and A. C. Hewson, J. Phys.: Condens. Matter **9**, 10463 (1997).
[89] R. Bulla, H.-J. Lee, N.-H. Tong, M. Vojta, Phys. Rev. B **71**, 045122 (2005)
[90] Y. Meir, N. S. Wingreen and P. A. Lee, Phys. Rev. Lett. **70**, 2601 (1993); N. S. Wingreen and Y. Meir, Phys. Rev. B **49**, 11040 (1994).
[91] H. Schoeller and G. Schön, Phys. Rev. B **50**, 18436 (1994).
[92] J. König, J. Schmid, H. Schoeller and G. Schön, Phys. Rev. B **54**, 16820 (1996).
[93] Y. Utsumi, J. Martinek, G. Schön, H. Imamura, and S. Maekawa, Phys. Rev. B **71**, 245116 (2005).
[94] S. Sasaki, S. De Franceschi, J. M. Elzerman, W. G. van der Wiel, M. Eto, S. Tarucha, and L. P. Kouwenhoven, Nature **405**, 764 (2000).
[95] D. Boese, W. Hofstetter, and H. Schoeller, Phys. Rev. B **64**, 125309 (2001).
[96] C. Kittel, Vol. 22 of *Solid State Physics*, (Academic Press, 1968).
[97] Y. Utsumi, J. Martinek, P. Bruno, H. Imamura, Phys. Rev. B **69**, 155320 (2004).
[98] A. Yacoby, M. Heiblum, D. Mahalu, and H. Shtrikman, Phys. Rev. Lett. **74**, 4047 (1995).
[99] K. Kobayashi, H. Aikawa, S. Katsumoto, and Y. Iye, Phys. Rev. Lett. **88**, 25806 (2002).
[100] N. J. Craig, J. M. Taylor, E. A. Lester, C. M. Marcus, M. P. Hanson, and A. C. Gossard, Science **304**, 565 (2004).
[101] M. G. Vavilov, and L. I. Glazman, Phys. Rev. Lett. **94**, 086805 (2005)
[102] P. Simon, R. López, and Y. Oreg, Phys. Rev. Lett. **94**, 086602 (2005).
[103] H. Tamura, K. Shiraishi, and H. Takayanagi, Jpn. J. Appl. Phys. **43**, L691 (2004).

5 Spin-transfer torques and nanomagnets

Daniel C. Ralph and Robert A. Buhrman

This chapter is divided into two parts. The first will deal with the phenomenon of spin-transfer torques. When a spin-polarized electrical current interacts with a ferromagnet, a portion of the spin angular momentum carried by the electrons can be transferred to the magnet. By elementary mechanics, the time rate of change of angular momentum is equal to torque, so this means that the spin-polarized electrons may apply a torque directly to the magnet. In small magnetic devices, smaller than about 250 nm in diameter, experiments have demonstrated that this spin-transfer torque can be much stronger per unit current than the torque on the magnet due to the magnetic field that is generated by the current. The spin-transfer torque is therefore of interest as an alternative mechanism, potentially more efficient than using magnetic fields, for manipulating the orientation of magnets in memory devices and nanoscale magnetic oscillators. This chapter will describe intuitive ways of understanding the microscopic mechanism and the dynamical consequences of spin-transfer torques, and will review experimental progress in the field to date for spin valves and magnetic tunnel junctions. The topic will then be explored further in Chapter 7, with a more formal discussion of the theory of spin-transfer torques and how these torques may be used to shift domain walls in magnetic wires.

The second part of this chapter will deal with the separate topic of how the transport of spin and charge in magnetic devices changes when the structures are made even smaller, extending from magnetic particles a micron in diameter, to a few nanometers, and down to a single molecule. As the size of the magnet shrinks, effects such as Coulomb blockade and energy-level quantization become dominant. It will become necessary to move beyond simple independent-electron models which work successfully in larger devices, to consider the true correlated many-electron quantum states at the root of ferromagnetism. We will argue that several interesting mysteries remain to be explored.

5.1 Spin-transfer torques

Research into the possibility that spin-polarized currents might apply torques to magnets by means of the exchange interaction (rather than by a magnetic field) dates at least as far back as the mid-1980s, with the work of Luc Berger and collaborators on current-induced domain wall motion [1–3]. In 1989, John Slonczewski calculated that a current traversing a magnetic tunnel junction should apply torques on the magnetic electrodes if their moments are non-collinear

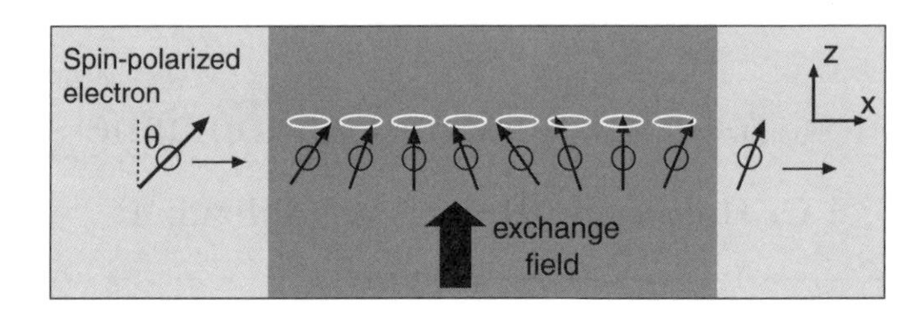

FIG. 5.1. Semiclassical picture of a spin-polarized electron interacting with a ferromagnetic thin film. The electron spin precesses as it travels through the magnet, due to the exchange field of the magnet. The changing angular momentum of the electron must be compensated by a corresponding change in angular momentum for the ferromagnet.

[4]. However, the current densities that would be required to produce torques large enough to reorient the magnetic moments were much larger than could be achieved through tunnel junctions at that time. Therefore it was an important step when in 1996 Slonczewski and Berger independently considered spin-valve devices consisting of a ferromagnet/normal-metal/ferromagnet multilayer, and predicted that spin-polarized currents could generate significant torques at achievable current densities [5,6]. Prior to this time, magnetic reconfigurations as a function of current had been observed in point-contact devices that could not be explained by magnetic fields, but the mechanism producing the reconfigurations was unclear [7]. Subsequently, with the theoretical framework of Slonczewski and Berger in place for guidance, experiments and theory have both moved forward rapidly [8–14], to the stage where it is now reasonable to consider what applications the spin transfer effect may have.

5.1.1 *Intuitive picture of spin-transfer torques*

We will begin with a discussion of the mechanism underlying the spin-transfer effect. We will first consider a simplified picture in which one spin-polarized electron travels through a non-magnetic metal toward a thin ferromagnetic layer (Fig. 5.1) [5,14,15]. Assume that the electron has already been polarized by some process upstream. Let the moment of the magnetic layer be oriented in the $\hat{z}$ direction, and assume that the incoming electron is traveling in the $\hat{x}$ direction and has its spin polarized at an angle θ with respect to $\hat{z}$ in the $\hat{x}\hat{z}$ plane. What happens as the electron encounters the magnetic layer?

First, the interface between the non-magnetic metal and the magnet can act like a filter for the spin states. For a materials combination like copper and cobalt, spin-up (majority) electrons have a larger transmission amplitude through the interface than spin-down electrons. The relative amplitudes of the spin-up and spin-down components in the transmitted part of the spin wavefunction will therefore be changed relative to the incident state, and for the Cu/Co case the transmitted spin can be thought of as being tilted at a smaller angle θ than the initial incident tilt. If the spin-up and spin-down transmission amplitudes have different phases, then the spin vector could also have a component in the $\hat{y}$ direction, corresponding to a precession about the $\hat{z}$ axis.

Next, as the electron then continues to travel through the magnet, it will be subject to a strong exchange interaction with the magnet's moment. Because we are assuming that the spin of the electron is not aligned with the magnetization of the magnet, the electron spin will precess about the direction of the exchange field. To conserve angular momentum, the moment of the magnet must also precess about the electron spin. (Strictly speaking, if there is no external field applied, both vectors will precess about their sum. However, because the moment of the magnetic layer is much larger than for a single electron, to a very good approximation it is correct to view the electron's spin vector as precessing about the $\hat{z}$ direction as it moves along the $\hat{x}$ direction.) This precession can be extremely rapid – for exchange fields in a material like Co, the electron will travel only a few lattice spacings per precession period. Therefore, even for a film as thin as 3 or 4 nm, the electron in this picture may precess several times before exiting out the rear side of the magnetic layer. In the end, depending on exactly how many fractions of a turn the electron has precessed, it could exit the magnet with either positive or negative spin components in the $\hat{x}$ and $\hat{y}$ directions. However, the $\hat{z}$ component of the spin should not change during the precessional part of the motion. In a rigorous quantum-mechanical treatment, it would be necessary to sum over all of the multiple reflections that the electron might experience inside the magnetic layer to determine its final quantum-mechanical state, but for our purposes this simple semiclassical picture of transmission and precession will be sufficient to build an intuitive picture.

So far, we have considered the physics of just a single electron, but in a real device many electrons would be propagating through the magnetic layer simultaneously, traveling with different energies and different incident angles in real space. Even if every electron started with exactly the same initial angle for its spin polarization, each would undergo a different degree of precession within the magnet because each would take a different amount of time traversing the magnetic layer. As long as the thickness of the magnetic layer is sufficiently greater than the precession length, approximately the same fractions of electrons will exit the magnet with positive and negative components in the $\hat{x}$ direction, and similarly with positive and negative components in the $\hat{y}$ direction. Therefore, if one sums the total spin angular momentum of all the electrons that exit the magnetic layer, the only non-negligible component will point in the $\hat{z}$ direction, parallel to the moment of the magnetic layer. The perpendicular components should average to zero.

This simple result is sufficient to understand the nature of the torque felt by the magnetic layer. We have assumed that the incoming electrons all have spin angular momenta oriented at the angle θ relative to $\hat{z}$ in the $\hat{x}\hat{z}$ plane. We have argued that the sum total of the electrons exiting the rear of the magnetic film will have some non-zero average component of angular momentum in the

$\hat{z}$ direction, but not in the $\hat{x}$ or $\hat{y}$ directions. Angular momentum is conserved. This means that the component of incoming spin angular momentum transverse to the magnet's moment (the component in the $\hat{x}$ direction) that is lost from the electrical current must be absorbed by the magnetic film. If the magnet responds as a single domain, rather than having the angular momentum transferred to short-wavelength spin waves, then one can expect the entire moment to begin to rotate. The direction of the torque is such as to rotate the moment of the magnetic layer toward the direction of the incident spin polarization. If the incident angle θ is large, the angular momentum transferred per electron can be a significant fraction of $\hbar$.

Now, as we have already mentioned, this discussion is not rigorous. A fully correct calculation should consider the possibility of multiple reflections within the magnetic layer and should also take into account that some angular momentum will be carried away by electrons reflected from the magnetic layer, as well as electrons transmitted through. However, in the end, more rigorous calculations do find the same result – to a very good approximation the magnetic layer will absorb the transverse component of the incoming spin angular momentum and feel a torque acting to rotate the magnet's moment toward the direction of the incoming spin polarization [14,16]. The leading correction arises when one takes into account that the reflection amplitudes for spin-up and spin-down electrons may have different phases. This can result in a component of torque in the $\pm\hat{y}$ direction in Fig. 5.1. However, band-structure calculations indicate that this correction is generally less than 10% the size of the main torque in the $\hat{x}$ direction. Experimentally, measurements of the dynamical effects that are caused by spin transfer confirm that the component of torque in the $\hat{x}$ direction is dominant by far [17–20], although recent work indicates that it may be possible to measure the $\pm\hat{y}$ component, as well [21].

5.1.2 *The case of two magnetic layers*

The spin-transfer devices that have been studied most thoroughly in experiments contain two magnetic layers, separated by a normal metal spacer or a tunnel junction. An analysis of this type of device can demonstrate one of the most important signatures of the spin-transfer effect, an asymmetry with respect to the direction of current. Samples can be constructed so that a current passing perpendicular to the layers in one direction provides a torque to turn the two magnetic moments parallel to each other, while current of the opposite sign can turn the moments toward an antiparallel configuration.

Consider the device geometry shown in Fig. 5.2, a N/F/N/F/N multilayer pillar, where N is a non-magnetic metal such as Cu, and F is a magnetic metal like Co. One of the magnetic layers is depicted as thinner than the other. We will call the thinner layer the "free layer", anticipating that it will be more readily reoriented by spin-transfer torques. The other "fixed" magnetic layer can be made less susceptible to the torques by making it thicker, by leaving it connected to an extended film rather than as part of the etched pillar structure,

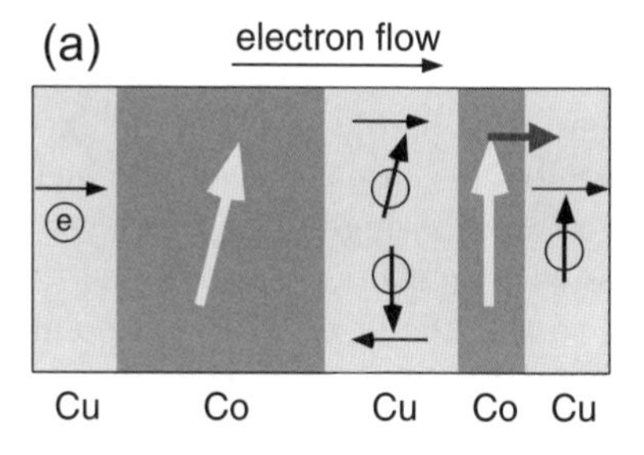

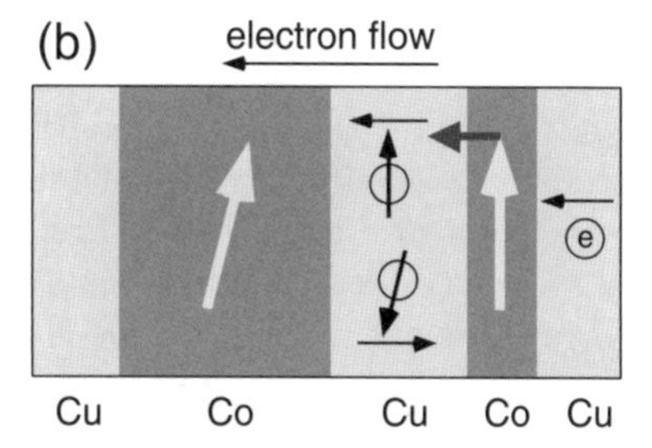

FIG. 5.2. Spin-transfer torque in a sample with two magnetic layers. (a) When electrons flow from the thicker "fixed" magnetic layer to the thinner "free" layer (negative current), spin transfer acts to turn the free layer moment in the direction of the fixed layer moment, thereby stabilizing the parallel configuration. (b) When the current is reversed (positive current), the spin-transfer torque acts to turn the free layer moment away from the fixed layer moment. For large enough currents in this direction, the parallel configuration can be destabilized.

or by making it from a material that gives it a larger total magnetic moment than the free layer. The fixed layer will serve as the polarizer. Electron spins can be filtered as they are transmitted through or reflected from this layer, so that this layer can provide the spin-polarized electrons that will act on the free layer. We will assume initially that the magnetic layers are positive polarizers, preferentially transmitting majority-spin electrons. This is the case for Cu/Co multilayers.

We will wish to analyze the torques on the layers caused by current flow. It happens that the torque is zero if the moments in the two layers are aligned exactly, so we will assume that there is initially a misalignment angle θ, perhaps due merely to thermal fluctuations. We can then conduct a linear stability analysis, asking whether the torque from spin transfer will push the moments together to stabilize the parallel alignment, or whether it will push them further apart, possibly destabilizing this parallel orientation.

We analyze first the case when the net electron flow is from the thicker layer toward the thinner layer. This case will correspond to negative currents in the experimental plots shown later in this chapter. If unpolarized electrons are incident on the fixed layer from the left in Fig. 5.2(a), they will undergo spin-filtering, and emerge into the central normal-metal spacer with an average spin moment directed parallel to the moment of the fixed layer. This filtering is not 100% efficient, but for materials systems like Co and Cu the resulting polarization can be 35–40%. When these polarized electrons next interact with the free layer, we have exactly the situation analyzed in the previous section. The free layer will absorb the component of the incoming spin current transverse to its magnetic moment, and will feel a torque tending to turn its moment toward the orientation of the incoming spins. This means that the torque on the free layer will turn the moment in the direction of the fixed layer's moment, thus stabilizing the parallel orientation of the two moments.

Next consider the case of positive current (Fig. 5.2b), with the net flow of electrons from the free layer to the fixed layer. In this case the electrons incident from the right in Fig. 5.2(b) will first undergo spin filtering by the free layer, and will flow into the central normal-metal spacer with an average spin parallel to the free-layer moment. No net torque is applied to the free layer by the unpolarized electrons at this step. The resulting spin-polarized electrons will then flow to the fixed layer, transferring angular momentum and applying a torque to that layer. However, we are assuming that the fixed layer moment is held rigidly in place, so that it cannot respond to this torque. Very importantly, another consequence of the interaction between the spin-polarized current and the fixed layer is that a fraction of electrons will be reflected back toward the free layer from the normal-metal/fixed-layer boundary. These reflected electrons will have an average spin polarization *antiparallel* to the fixed-layer moment, because we are assuming that the parallel spins are more readily transmitted. The reflected spin current therefore will have a polarization opposite to the case analyzed above for negative currents. When the free layer absorbs the transverse component of the incident spin current now, the torque it feels will have the sign to turn the free-layer moment *away from* the fixed-layer moment. This means that a large-enough positive current might destabilize the parallel orientation of the two moments. This is the case that can lead to the types of magnetic dynamics that we will analyze in the next few sections.

One can use the same types of diagrams and arguments to analyze other symmetry cases for the spin-transfer effect [15]. If one analyzes the case of initially antiparallel moments in the fixed and free layers, a positive current will stabilize the antiparallel alignment while a negative current will destabilize it. If the free layer acts as a negative polarizer, preferentially transmitting minority-spin electrons rather than majority spins, then there is actually no change in the sign of the torque on the free layer. No matter what the nature of its spin filtering, the free layer absorbs the transverse angular momentum coming from the spin current and it feels the same torque. However, if the fixed layer is a negative polarizer, the sign of the torque on the free layer will be reversed. These effects have been confirmed experimentally by using FeCr and NiCr layers as negative polarizers [22].

For more quantitative calculations that include effects such as electron scattering from disorder, or for calculations of torques in devices that contain more complicated geometries than simply two magnetic layers, a very useful "circuit theory" can be applied [23–25].

5.1.3 *Simple picture of spin-transfer-driven magnetic dynamics*

In the previous section, we discussed the spin-transfer torque on the free layer in a spin-valve device. Depending on the sign of the current, this torque acts primarily to turn the free-layer moment either toward or away from the direction of the fixed layer moment, within the plane defined by the two moment vectors. If the sign of the current is such as to turn the free layer moment away from the fixed-layer moment, this may destabilize the state in which moments are aligned parallel, and drive the free layer into some other type of state. In this section, we wish to discuss the types of dynamics that may result after this instability.

In order to understand the dynamics of the free layer, it is necessary to take into account not only the spin-transfer torque, but also all of the other torques that act on the free layer, including the applied magnetic field, magnetic anisotropies, and damping, and possibly thermal fluctuations. All of these torques can be modeled together using the Landau–Lifshitz–Gilbert (LLG) equation of motion [5, 26–29]. For purposes of illustration, we will consider, pictorially, a particularly simple and symmetric case. We assume that any applied magnetic field is in the $\hat{z}$ direction, that the free layer is subject to a purely uniaxial magnetic anisotropy with easy axis also along $\hat{z}$, and that the magnetic moment of the fixed layer points in the same direction. (A more complicated anisotropy function will be needed to describe the experimental samples discussed below.) We also assume that the free layer can always be described as a single-domain magnet, locked together to act as one large spin. We will initially ignore effects of non-zero temperature, except that we will assume that the fixed and free magnetic layers are not exactly aligned at the start.

At some instant, suppose that the free layer moment points at an angle θ with respect to $\hat{z}$. The torques acting on the free layer at that instant are depicted in Fig. 5.3. The effect of the combined torque from the magnetic field and magnetic anisotropy is to make the free-layer moment precess about the $\hat{z}$ direction. The damping torque, which accounts for energy loss from the free layer to its environment, points toward the $\hat{z}$ axis. Therefore in the absence of any spin-transfer term, the free layer will spiral toward the $\hat{z}$ axis, eventually ending up in the lowest-energy configuration pointing along $\hat{z}$. In the presence of the spin-transfer term, there is an additional torque that can point either in the same direction as the damping torque, or opposite to it, depending on the sign of the current. For negative currents, by our convention, the spin-transfer torque simply reinforces the damping torque, making the free-layer relax toward the $\hat{z}$ direction faster than it would ordinarily. For small positive currents, the spin-transfer torque weakens the effective damping, so that the free-layer relaxes toward $\hat{z}$ more slowly. A dynamical instability arises at $T = 0$ only when the spin-transfer torque is larger in magnitude than the damping torque, so that instead of relaxing toward $\hat{z}$ the free layer begins to spiral away from this direction. At this threshold, the torque from spin transfer can add energy to the free layer, so that it can move to higher-energy orientations away from $\theta = 0$.

Beyond the point of the instability, the dynamics can take several forms, depending on the magnitude of the applied field and the detailed angular dependence of the spin-transfer and damping torques. One possibility is that the free layer will spiral to ever-increasing values of θ all the way to $\theta = \pi$, where the free layer can remain in a stable static state antiparallel to the fixed layer. Ex-

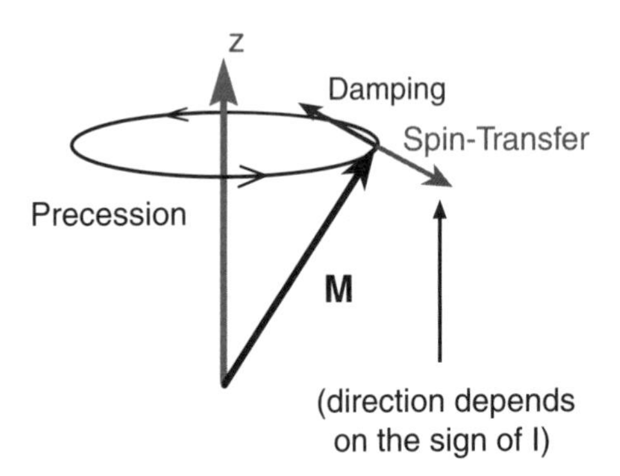

FIG. 5.3. Depiction of the torques acting on a single-domain nanomagnet. The torque due to an applied magnetic field causes precession. The damping torque causes the moment to spiral toward the applied-field direction. Depending on the sign of the current, the torque due to spin transfer can reinforce the damping torque, or act in the opposite direction from the damping.

perimentally, this type of simple magnetic spin-transfer-driven magnetic reversal is observed at low magnetic fields and is under investigation for applications in magnetic memory devices [11, 12]. The free layer can be controllably switched back to the parallel configuration by applying a sufficiently large negative current.

For other parameter values, it is also possible that the free layer may not spiral all the way to $\theta = \pi$, but might reach some state of dynamical equilibrium at an intermediate angle, where the energy gained from spin transfer and the energy lost to damping are balanced over each cycle. In this case, the spin-transfer torque allows a DC applied current to generate steady-state magnetic precession oscillations at GHz or tens of GHz frequencies. Such dynamical modes have been observed in a variety of different device geometries [19, 30–32]. In the simple symmetrical case depicted in Fig. 5.3, the only allowed class of dynamical mode corresponds to precession of the free-layer moment about the $\hat{z}$ axis at constant angle. However, in real devices containing more complicated biaxial magnetic anisotropies, several different types of dynamical modes are possible, depending on the strength and direction of the applied magnetic field and the magnitude of the applied current [19, 32].

The question of whether a dynamically precessing state can be stable or whether a magnet will simply switch between static states depends very sensitively on the exact angular dependence of the spin-transfer torque and the damping. Finding ways to calculate [13, 33] and measure these angular variations is turning out to be one of the primary challenges in this field.

Our simple picture has so far largely ignored the effects of temperature. At non-zero temperatures, the free-layer moment will also be subject to additional torques that produce thermal fluctuations in the orientation of the free-layer moment. These fluctuations can be particularly important at low magnetic fields, in the regime where spin-transfer torques cause hysteretic switching. Because the spin-transfer torque for positive currents can be thought of as producing a reduced effective magnetic damping, spin transfer can enhance the effect of thermal fluctuations by reducing the rate at which they relax. Consequently, even at current levels significantly smaller than needed to produce magnetic switching at $T = 0$, in the presence of spin-transfer torque the thermal fluctuations can be enhanced enough to produce thermally activated magnetic switching [17,20,34]. The effects of thermal fluctuations will be explored in more detail later in this chapter.

The simple picture has also assumed that the free layer always moves like one large spin locked together in a single domain, neglecting the possible importance of shorter-wavelength modes. The extent to which this is accurate is still a matter of debate [34–37]. The available experimental evidence suggests that the single-domain assumption gives an accurate representation of the dynamical phase diagram in many samples, at least for currents comparable to the threshold level needed to produce magnetic excitations. However, at larger currents there is some evidence for additional dynamics beyond the single-domain approximation. This is another issue to which we will return later in this chapter.

5.1.4 *Experimental results*

5.1.4.1 *Device geometries* In order to measure the effects of spin-transfer torques, it is necessary to fabricate devices in which current flow is restricted to a small diameter. This is true for several reasons. First, the amount of current needed to excite magnetic excitations using spin transfer scales with the total magnetic moment of the free layer. By making the free layer a few nm thick and a few hundreds of nm in diameter, the necessary current scale can be reduced to a mA or less. Second, it is necessary to utilize small devices in order that the effects of spin transfer dominate over the effects of the magnetic field that is generated by the current. The magnetic field produced by the current flowing through a magnetic multilayer sample tends to generate a different type of reorientation than the spin-transfer effect (e.g., producing vortex states rather than simple reversal) and its effects can also be distinguished from spin transfer by a different symmetry upon reversing the direction of current flow. For a circular wire, the current necessary to generate a given magnetic field (e.g., the coercive field of a magnetic sample) scales linearly with the wire diameter. The current necessary to generate spin-transfer-induced reorientation scales linearly with the volume of the free layer, or for a thin-film layer as (diameter)^2. Therefore, for a sample with small enough diameter the spin transfer effect will dominate, because the current necessary to generate spin-transfer effects will be less than the current required to produce a magnetic field large enough to reorient the magnet. For cobalt layers a few nm thick within a pillar geometry, the cross-over diameter has been measured to be approximately 250 nm [38]. Finally, spin-transfer effects are also easiest to interpret in devices small enough that the free-layer moment moves as

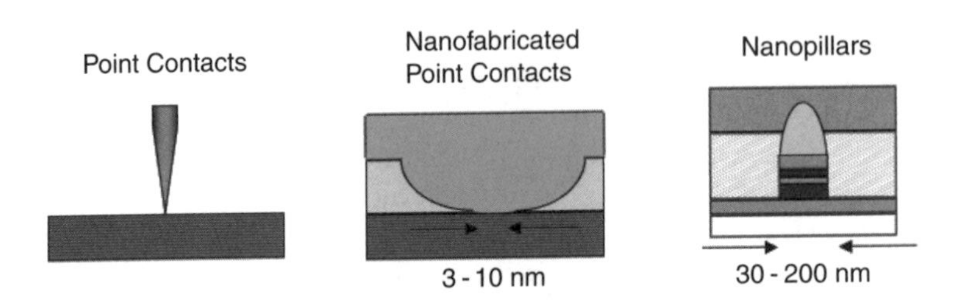

FIG. 5.4. Sample geometries used for studying spin-transfer torques.

a single domain, which also typically requires devices close to the 100-nm scale.

Three primary types of magnetic-multilayer devices have been used successfully to demonstrate spin-transfer effects (Fig. 5.4). One is the mechanical point contact device, in which a magnetic multilayer is contacted by a sharp metal tip to form a contact region on the scale of tens of nm [9, 30]. Lithographically fabricated point contact devices have also been used to produce a similar device geometry [11, 31]. In both varieties of point contacts, excitations of the magnetic layers can be generated within a nanoscale region in the vicinity of the contact. However, because the excited region is connected to a macroscopic magnetic film in the point contacts, the required current densities are rather large, typically greater than 10^9 A/cm^2. The third variety of device is the multilayer nanopillar, in which the free magnetic layer is patterned to a small cross-section so that it is not exchange-coupled to a larger film. Nanopillar devices have been made by lithography and ion milling [12, 35], by stencil techniques [39, 40], and by electrodeposition within a cylindrical nanopore [41]. The nanopillar geometry has the advantages that (a) the current densities required to generate spin-transfer excitations are reduced significantly compared to the point-contact samples, to values less than 10^7 A/cm^2 [34], (b) the lack of exchange coupling to an extended film simplifies modeling and thereby enables quantitative tests of theoretical predictions, and (c) the geometry allows improved control over the magnetic anisotropy properties of the free and fixed layer magnets. We will therefore focus primarily on data from nanopillars in our discussion.

5.1.4.2 *Experimental measurements of spin-transfer-driven magnetic dynamics*

The relative orientations of the fixed- and free-layer magnetic moments in the nanopillar devices can be monitored using the sample resistance. Due to the giant magnetoresistance effect (GMR) [42], magnetic multilayer samples generally have the highest resistance for the antiparallel (AP) state and the lowest resistance for the parallel (P) state (assuming that both layers act as spin filters of the same sign). Typically, the nanopillar devices are fabricated with an approximately elliptical cross-section, so that only the P and AP states are stable in zero applied magnetic field. Most of the data we will show will be from devices with elliptical cross-sections approximately 60 nm $\times$ 130 nm. Signals from dynamical magnetic states can be measured directly by contacting the sample with

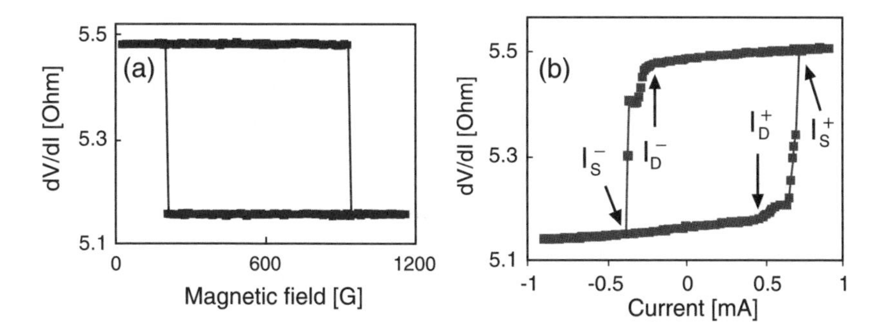

FIG. 5.5. (a) Switching of the magnetic layers in a nanopillar device between parallel and antiparallel orientations, driven by a magnetic field at 4.2 K. (b) Switching in the same device, driven by spin-transfer torques from an applied current at 4.2 K. The current can excite a precessional excitation at the current I_D, prior to full magnetic reversal at I_S.

wide-bandwidth electrodes and detecting the microwave signals that are emitted as the magnets precess under a DC current bias [19]. A microwave signal is generated naturally, because the relative angle of the two magnetic layers changes as the free layer undergoes elliptical precession in a thin-film multilayer device (the circular precession pictured in Fig. 5.3 corresponds to the unphysical case of pure uniaxial anisotropy), so that by the GMR effect the resistance oscillates at microwave frequencies. In the presence of a DC current, these resistance oscillations produce an oscillating voltage at microwave signals that can be detected. The average (low-frequency) resistances dV/dI measured in dynamical states are found to be intermediate between the P and AP resistances.

Figure 5.5 displays the hysteretic switching of the free layer in a Py (2 nm)/Cu (6 nm)/Py (20 nm) spin-valve nanopillar as driven by a magnetic field applied in the plane of the sample along the easy axis (Fig. 5.5a) and by a current (Fig. 5.5b). In both cases the background temperature is 4.2 K. The resistance trace in Fig. 5.5(a) shows a simple switching of the free-layer moment between P and AP alignment with the fixed layer, and provides a determination of the resistance values corresponding to these two states. The hysteresis loop is offset from zero field in Fig. 5.5(a) because of a magnetic dipole coupling between the layers. In Fig. 5.5(b) a magnetic field is applied to cancel this dipole coupling between the magnetic layers, leaving the free-layer subject to zero net magnetic field. Figure 5.5(b) shows that the current can also be used to switch the magnets between the same two P and AP states. Starting in the lower-resistance P state, a sufficiently large positive current causes a jump to the higher-resistance AP state, achieving the same value of dV/dI as does magnetic field switching once the current is reduced back to zero. By our sign convention, positive currents correspond to electron flow from the free layer to the fixed layer, the sign of current predicted within the spin-transfer theory to destabilize the P configuration and stabilize the AP one. Once the magnets are in the high-resistance AP configuration, they can be switched back to the P orientation by applying a sufficiently large negative current, again in agreement with the spin-transfer predictions. The switching can be identified with the mechanism of the spin-transfer torques, rather than current-induced magnetic fields, based both on the asymmetry with respect to the sign of the current and on the fact that in nm-scale devices the current levels are too small to produce magnetic fields of the magnitude needed to switch the magnets [11].

It should be noticed in Fig. 5.5(b) that the differential resistance exhibits small peaks or shoulders starting at the currents I_D^+ and I_D^-, prior to the large jumps in resistance [35]. Based on measurements of microwave-frequency signals emitted by the device, these features can be associated with the turn-on of a dynamical state in which the free layer undergoes small-angle precession. At low temperature, spin transfer excites the free layer into a precessional mode first, before eventually reversing the moment to reach a final static state antiparallel to the fixed layer moment. This sequence of transitions is in agreement with numerical [19, 26, 29] and analytical [43] solutions of the LLG equations that include a spin-transfer-torque term.

A variety of different dynamical modes are also observed at larger values of magnetic field, larger than the in-plane coercive field of the free layer [19, 32]. Figure 5.6 shows the progression seen in the resistance traces of a Co (3 nm)/Cu (10 nm)/Co (40 nm) spin-valve device at room temperature as a function of increasing in-plane magnetic field, along with the microwave spectra measured at selected values of current for $H = 2\,\mathrm{kOe}$. For sufficiently large values of H applied in plane, the resistance traces no longer display hysteretic switching; instead, there are non-hysteretic peaks and shoulders in dV/dI. These features can be associated with the turning-on and turning-off of dynamical modes as a function of current. The first dynamical mode to appear, in this sample near $I = 2.0\,\mathrm{mA}$, is small-angle precession, with a peak in the microwave signal at the ferromagnetic resonance frequency. Subsequently, the current can excite modes with larger tipping angles, corresponding to larger-amplitude microwave signals with reduced frequencies, before the microwave dynamics eventually cease in this sample beyond 7.5 mA. At even larger currents, additional microwave signals appear that can be associated with coupled dynamics of both the free and the fixed magnetic layers.

We have mentioned above that it is natural for a spin-transfer-driven nanomagnet to produce microwave signals when it is in the dynamically precessing regime, because of the GMR oscillations in the presence of a DC bias current. There are also other possible mechanisms that can lead to the generation of microwaves, for instance by inductive pickup of the oscillating magnetic field associated with the precessing magnet. One way to distinguish between mechanisms is to study the harmonic content of the microwave signals. In the very symmetric case that the applied magnetic field, the moment of the fixed layer, and the easy axis of the free layer are all aligned, then under the influence of spin transfer the

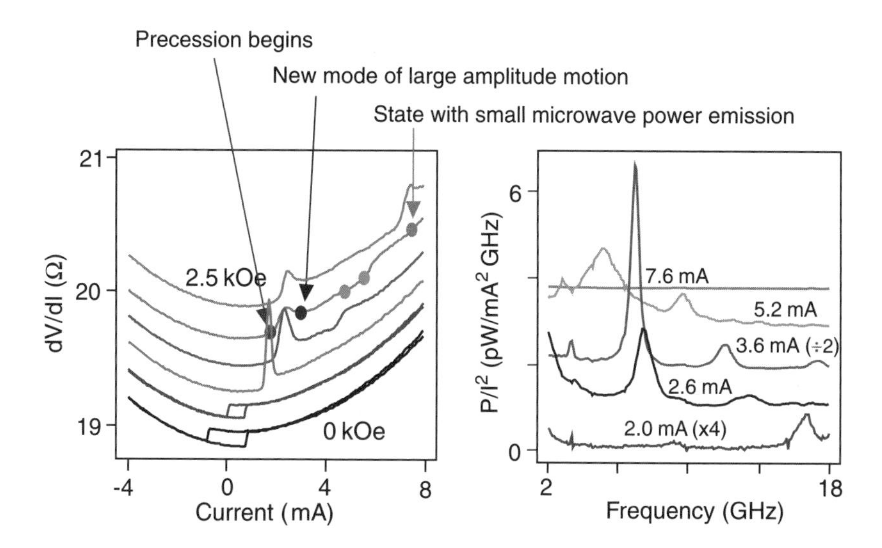

FIG. 5.6. Left: differential conductance of a nanopillar device for applied magnetic fields from 0 Oe to 2.5 kOe, at increments of 0.5 kOe. Right: microwave power spectra measured at 2 kOe, for selected values of current. The spectra are offset vertically. Different magnitudes of current excite different types of dynamical magnetic modes.

free layer should undergo symmetric elliptical precession about the direction of the fixed moment. For the GMR mechanism one would expect in this case to see a microwave signal only at integer multiples of *twice* the precession frequency, with no signal at the fundamental precession frequency, because the resistance signal would go through two cycles for each precession period. However, if the precession axis of the free layer is misaligned from the fixed layer, the GMR mechanism should also give signals at the fundamental precession frequency. If the precession orbit stays approximately in-plane, as expected for precession of a thin-film magnet, then within the GMR mechanism the microwave powers at the fundamental precession frequency P_f and at the second harmonic P_{2f} should be related to the misalignment angle $\theta_{\rm mis}$ and the maximum amplitude of the precession away from the precession axis $\theta_{\rm max}$ according to [19]

$$P_f = \frac{I^2 \Delta R_{\rm max}^2 \theta_{\rm mis}^2 \theta_{\rm max}^2}{32R}, \tag{5.1}$$

$$P_{2f} = \frac{I^2 \Delta R_{\rm max}^2 \theta_{\rm max}^4}{512R}. \tag{5.2}$$

Here $\Delta R_{\rm max}$ is the difference in resistance between P and AP configurations and R is the sample resistance. We have tested these relationships by applying a 700 Oe external field at 5, 10 and 15 degrees from the easy axis of the free

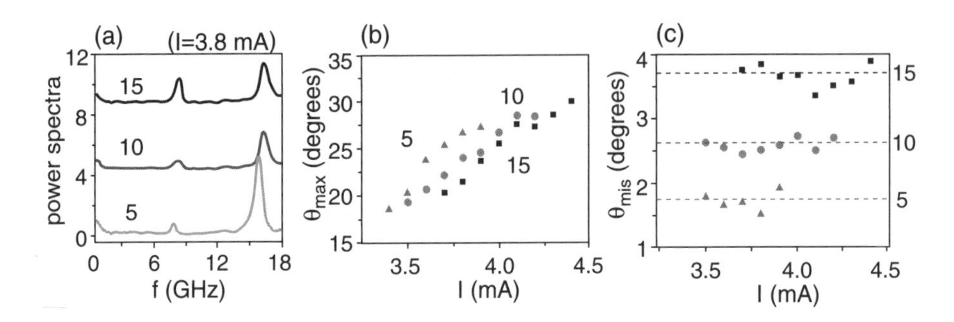

FIG. 5.7. Variation of microwave signals as a function of the angle of the applied magnetic field, for a field magnitude of 700 G, for a sample with an etched free layer and a continuous (un-etched) fixed layer. (a) The relative magnitudes of the first and second harmonics are different when the magnetic field is applied 5, 10, and 15 degrees from the easy axis of the free layer. (b and c) From the relative magnitude of the harmonics, we calculate the amplitude of the precessional motion ($\theta_{\max}$) and the average angle of misalignment between the fixed and free layer moments (θ_{mis}) using Eqs (5.1) and (5.2). These measurements demonstrate that the microwave signals originate from the GMR effect, not inductive signals.

layer in a nanopillar sample with an unpatterned fixed layer so as to induce a misalignment between the layers. We find good agreement with our expectations for the GMR mechanism (Fig. 5.7). At fixed current, the microwave power at the fundamental precession frequency grows monotonically as the misalignment angle is increased. As a function of increasing current, the formulas can be used to extract θ_{mis} and $\theta_{\max}$ from the measured powers. The results give values for θ_{mis} that are essentially independent of current, while $\theta_{\max}$ grows approximately linearly with current, as expected.

By combining an analysis of low-frequency resistance measurements and the microwave signals, it is possible to map out the dynamical phase diagram of the free layer as a function of magnetic field and current. Figure 5.8(a) shows the 4.2 K phase diagram for the Py (2 nm)/Cu (6n m)/Py (20 nm) device used in Fig. 5.5, and Fig. 5.8(b) displays the room-temperature phase diagram over a larger range of H and I for the Co (3 nm)/Cu (10 nm)/Co (40 nm) device of Fig. 5.6. In the Co/Cu/Co device, the fixed layer was not patterned, which reduced the strength of its dipole field acting on the free layer, as compared to the Py/Cu/Py device. These phase diagrams can be compared to the predictions of the spin-transfer theories, calculated by numerical solution of the LLG equation. It is important to take into account that the magnetic anisotropy in these samples is biaxial, with strong in-plane anisotropy and a weaker easy-axis anisotropy within the plane, rather than the simplified uniaxial anisotropy assumed in the introductory discussion near the beginning of this chapter. Figure 5.9 shows the calculated room-temperature phase diagram corresponding to the sample pa-

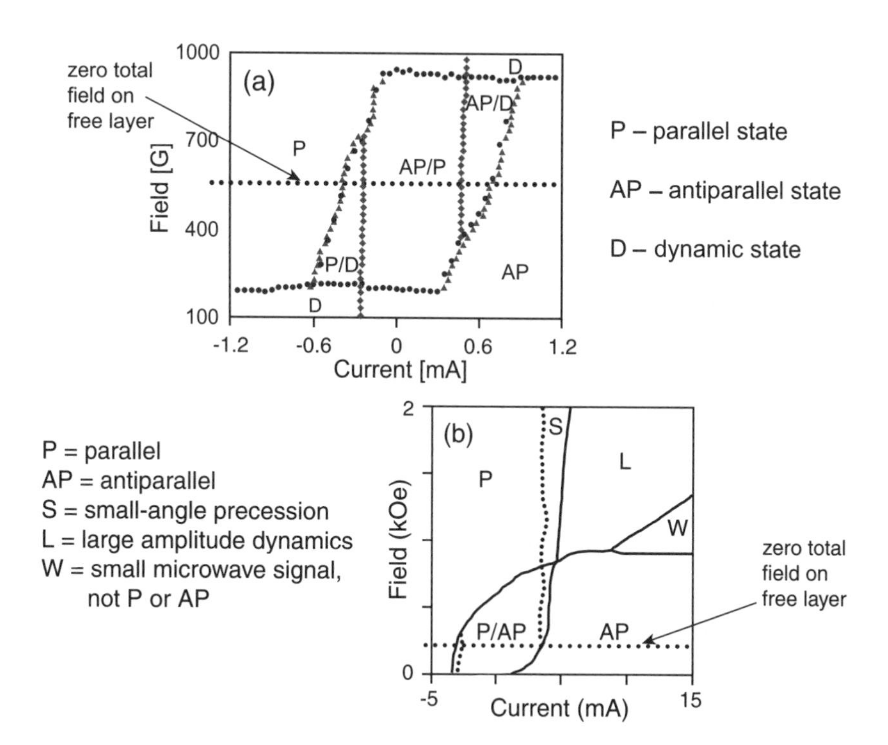

FIG. 5.8. (Top) phase diagram illustrating the different magnetic configurations as a function of current and applied magnetic field for a nanopillar with a permalloy free layer, at 4.2 K. Regions with two labels indicate hysteresis. (Bottom) phase diagram over a larger range of current and field for a nanopillar sample with a cobalt fixed layer, at room temperature.

rameters of the Co/Cu/Co device, for positive effective magnetic fields applied to the free layer. The calculated phase diagram is in excellent qualitative agreement with the measurements. The existence and relative positions of the P, AP, hysteretic, and small-angle precession states are all predicted well. There are differences, however, at large currents in the dynamical regime. The large-amplitude clamshell-type orbits are observed to extend to larger currents in the Co device than predicted by theory, while a predicted out-of plane orbit is not seen in Co. Instead, at large current the Co sample displays a mode "W" with resistance intermediate between the P and AP states but with no observable microwave signals. Interestingly, recent measurements of the dynamics of smaller-diameter Py free layers suggest somewhat better agreement with the simulations at high currents, with the presence of the out-of-plane dynamical mode and no observation of the peculiar "W" state [44].

When a magnetic field is applied perpendicular to the magnetic layers, rather than in the plane of the sample, a different set of dynamical modes can be excited.

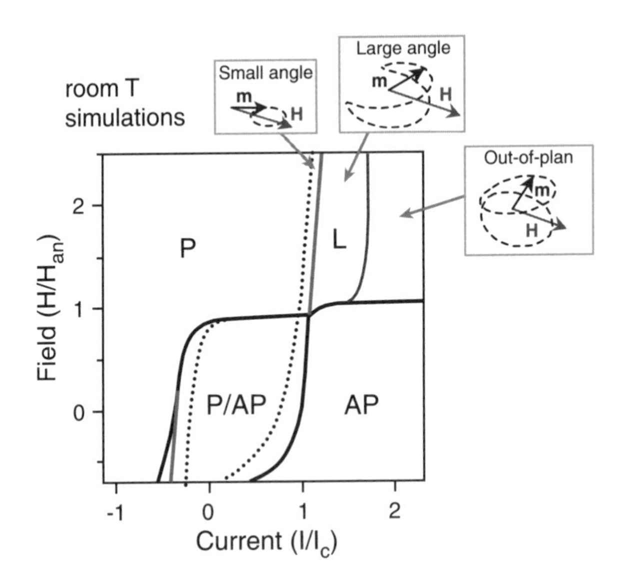

FIG. 5.9. Calculated dynamical phase diagram for parameters corresponding to a single-domain cobalt free layer at room temperature. (Compare to Fig. 5.8.) The Landau–Lifshitz–Gilbert simulation gives an accurate description of the existence and relative positions of the parallel, antiparallel, and dynamical states. However, there are differences at large currents which may indicate departures from single-domain behavior.

These modes have been analyzed in a level of detail similar to the dynamics for in-plane field and they have been compared to LLG simulations [32, 45]. The extent of agreement between the experimental and simulated dynamical phase diagram is again excellent at low currents, with some deviations at large currents.

The differences between experiments and predictions at large currents might be associated with the breakdown of the assumption of single-domain dynamics used in the calculations. The amplitude of the measured microwave signals supports this suggestion. Typically, the microwave power detected at large currents is 1/6 to 1/2 of the predicted values, which suggests that the free layers are precessing in a less-uniform fashion than in the single-domain simulations, or with smaller angles. More realistic simulations that go beyond the single-domain approximation are under development [29, 46–48]. Analytic calculations have suggested that short-wavelength (non-uniform) modes may be excited by spin transfer in addition to the uniform precessional mode [8,49,50]. Non-uniform excitations due to spin-transfer have in fact been observed clearly in devices containing just a single magnetic layer [11, 51, 52]. One of the unsolved questions for experimentalists in this field is how to distinguish unambiguously between single-domain motion and non-uniform excitations in spin-valve devices with two magnetic layers.

Thus far we have discussed the spin-transfer excitations from primarily a qualitative point of view, but we also wish to emphasize that quantitative agreement with predictions is quite satisfactory, as well (at least outside the regime of large currents). For instance, the calculated value of the $T = 0$ critical current required to destabilize the parallel orientation of the free layer in a nanopillar sample and initiate small-angle precession is, for an in-plane magnetic field and a fixed-layer moment both aligned with the easy axis of the free layer [12]

$$I_D^+ = \frac{\alpha e M V}{\hbar g(0)} \left[2\pi M_{\text{eff}} + H_{\text{eff}}\right], \tag{5.3}$$

where α is the Gilbert damping, e is the electron charge, M is the magnetization density of the free layer, V is the free-layer volume, H_{eff} is the effective field acting on the free layer, $4\pi M_{\text{eff}}$ is the effective demagnetization field (including uniaxial anisotropies as well as shape anisotropy), and $g(0)$ is an efficiency parameter $\approx .15$ [5]. Assuming the values $4\pi M = 4\pi M_{\text{eff}} \approx 7.5\,\text{kOe}$ for our Py samples, $V = 1.2 \times 10^{-17}\ \text{cm}^3$, and $\alpha = 0.01$, the equation for the critical current reduces to $I_D^+ = 0.3\,\text{mA} + (0.07\,\text{mA/kOe})H_{\text{eff}}$, with total uncertainties in the parameters amounting to a factor of 2 or 3. The measured low-temperature values for the corresponding Py free layers are in the range $I_D^+ = (0.35\text{–}0.5)\,\text{mA} + [(0.09\text{–}0.15)$ $\text{mA/kOe}]H_{\text{eff}}$, in very good agreement. Also in accord with Eq. (5.3), the critical current has been measured to scale with the total volume of the free layer [18], and it is larger for the large-M material Co than it is for smaller-M permalloy.

5.1.4.3 *Effects of temperature* The effects of temperature on the magnetic dynamics driven by spin-transfer torques merit additional discussion, because temperature can alter the dynamics in important ways both for the precessional regimes and for the process of hysteretic switching.

In the precessional regime, thermal fluctuations set a lower limit on the widths of the peaks in the microwave spectra (e.g., in Fig. 5.6b). These widths are a quite different physical quantity than the widths of resonance peaks measured in the ordinary way by applying a swept-frequency AC drive to an oscillator and monitoring the amplitude of the response as a function of drive frequency. In such an AC measurement, the width of the resonance peak is related to the quality factor of the oscillator. This is not the case for the spectral peaks driven by DC spin transfer currents; if the DC-drive spin-transfer measurement could be performed at $T = 0$ then each precessional cycle should take the same amount of time, and the spectral peaks should be delta functions with negligible frequency widths. Thermal fluctuations cause the precessional periods to differ from cycle to cycle, and give the peaks in the frequency spectra non-zero widths. A calculation that treats the consequences of the thermal fluctuations as a random walk perturbation of the precessional orbit predicts that the peak width should scale with temperature as $T^{1/2}$ and as $(MV)^{-1/2}$ where MV is the total magnetic moment of the free layer [53]. By decreasing T to 77 K and increasing MV using a thicker free layer (4 nm of Py), we have achieved spectral widths as narrow as 4 MHz for a 4.8 GHz mode. Even narrower peaks have been achieved by the NIST group [54], 2 MHz for a 35 GHz mode at room temperature, by using nanofabricated point-contact samples in which the free layer is part of an extended thin film. We believe that the coupling to the extended thin film helps to reduce the effects of thermal fluctuations beyond what is achieved with patterned free layers.

In the regime of hysteretic switching, thermal fluctuations allow the free layer to switch at lower average drive currents than are required near $T = 0$. As the temperature is raised, the current range over which hysteresis is observed becomes smaller, and it eventually disappears when the free-layer becomes superparamagnetic, fluctuating between P and AP alignment at rates faster than experimental time scales even when the drive current is zero [34]. At non-zero temperatures, if a sample is biased at a current less than the low-temperature switching current, thermal fluctuations will eventually allow the sample to switch. For a given bias current, the probability distribution of the switching times is to good accuracy a single exponential [17,20,34]. At fixed current, and for temperatures from above room temperature down to at least 20 K, the logarithm of the average switching rate depends approximately linearly on $1/T$, indicating that the switching process is thermally activated [34]. At fixed temperature, the average switching rate is a very strong function of current. Therefore, at non-zero temperatures, the way that one should ordinarily view the action of the spin-transfer torque from a spin-polarized current is that it modifies the process of thermally activated switching. Near room temperature, non-thermally activated switching processes can be achieved only by employing current pulses with sub-nanosecond-scale rise times and with amplitudes greater than the $T = 0$ switching currents [20].

The central role of thermal fluctuations in spin-transfer-induced magnetic switching may at first seem non-intuitive, given our description of the switching process in the first sections of this chapter. There, the condition for the current required to generate spin-transfer excitations at $T = 0$ was explained to be the current value at which the torque from the spin-transfer effect became stronger than the torque due to damping, so that the free-layer moment began to spiral away from the direction of the applied field, instead of relaxing back toward the low-energy configuration pointing in the field direction. There is no energy barrier that is mentioned explicitly in this condition, so it may not be obvious why thermal activation is relevant. However, if one thinks about the process of magnetic reversal at zero current, for instance reversal achieved by sweeping an applied magnetic field, in this case there is an energy barrier arising from magnetic anisotropy forces that can be overcome by thermal activation. These two pictures can be reconciled if one thinks about the role of the spin-transfer torque as being to modify the effective damping that acts on the free-layer moment. If the current bias is positive, but less than the threshold required at $T = 0$ to excite spin-transfer excitations, one can think of the effective damping torque as being reduced by the spin-transfer contribution to a value less than the natural $I = 0$ damping, but not all the way to zero. With less effective damping, the free layer moment can undergo fluctuations to larger excursion angles than it would at $I=0$, when driven by fluctuating torques of the same strength. Therefore, at a fixed value of temperature, the spin-transfer torque can enhance the consequences of thermal fluctuations, and eventually enable a sufficiently large fluctuation that the free layer can overcome the anisotropy barrier and switch. In the limit as the temperature goes to zero, this picture reduces to the simple $T = 0$ mechanism described near the beginning of this chapter, because in the absence of thermal fluctuations the condition for the free-layer to climb up and out of the anisotropy-energy barrier is that the spin-transfer torque must be greater than the damping torque.

A simple argument that provides quantitative predictions within this theoretical framework was presented by Koch, Katine, and Sun [20]. They considered the case of a free-layer subject to purely uniaxial anisotropy, with the easy axis, the magnetic field, and the free-layer moment all lying in the $\hat{z}$ direction. In the absence of a spin-transfer torque, the magnetic dynamics can be described by the LLG equation with a Langevin random field term added to model thermal fluctuations:

$$\frac{1}{\gamma}\frac{d\hat{m}}{dt} = \hat{m} \times \left[\boldsymbol{H}_{\text{eff}} - \frac{\alpha}{m\gamma}\frac{d\hat{m}}{dt} + \sqrt{\alpha k_{\text{B}} T}\boldsymbol{A}(t)\right]. \tag{5.4}$$

Here H_{eff} includes contributions from both the applied field and the anisotropy field, γ is the gyromagnetic ratio, and $\boldsymbol{A}(t)$ is a random vector that does not depend on the temperature T. The first term on the right causes precession, the second term is the intrinsic damping contribution, and the final term is the Langevin random field. The random-field contribution must have the given form in order to satisfy the fluctuation-dissipation theorem. The solution to this LLG equation for small magnetic fields is stochastic magnetic switching, with a rate for switching from the P to AP configuration that is given by

$$\Gamma_{\text{P}\rightarrow\text{AP}} \propto \exp[-U_0/(k_{\text{B}}T)], \tag{5.5}$$

where U_0 is an activation energy. Now, imagine that a current is applied so that a spin-transfer torque acts on the free layer with the sign so as to destabilize the P orientation. In the simple geometry that we have assumed, for small-angle precession the spin-transfer torque will always act in the opposite direction from the intrinsic damping torque, and will have a magnitude proportional to the current I, so we can include the effects of the spin-transfer torque by modifying the second term in the LLG equation:

$$\frac{1}{\gamma}\frac{d\hat{m}}{dt} = \hat{m} \times \left[\boldsymbol{H}_{\text{eff}} - \frac{\alpha(1 - I/I_D)}{m\gamma}\frac{d\hat{m}}{dt} + \sqrt{\alpha k_{\text{B}} T}\boldsymbol{A}(t)\right]. \tag{5.6}$$

I_D is the $T = 0$ critical current for spin-transfer excitations. Note that the third term in the equation is unchanged; the fluctuating fields that drive thermal fluctuations are not altered by the current. However, if we define $\alpha_{\text{eff}} = \alpha(1 - I/I_D)$, we can rewrite the final term in the LLG equation,

$$\frac{1}{\gamma}\frac{d\hat{m}}{dt} = \hat{m} \times \left[\boldsymbol{H}_{\text{eff}} - \frac{\alpha_{\text{eff}}}{m\gamma}\frac{d\hat{m}}{dt} + \sqrt{\frac{\alpha_{\text{eff}} k_{\text{B}} T}{1 - I/I_D}} \boldsymbol{A}(t) \right]. \tag{5.7}$$

This equation now has exactly the same form as the original LLG equation but with a rescaled temperature, so that we can write down immediately that the solution for the switching rate is the same as for the original equation but with this rescaled temperature.

$$\Gamma_{\text{P}\to\text{AP}}(I) \propto \exp[-U_0(1 - I/I_D)/(k_{\text{B}}T)]. \tag{5.8}$$

This solution can be understood in terms of either a current-dependent effective activation energy $U_{\text{eff}} = U_0(1 - I/I_D)$ [17, 29, 55] or as a current-dependent fictitious temperature $T_{\text{eff}} = T/(1 - I/I_D)$ [20, 55] – these are just different words to describe the same enhancement of the switching rate brought about by the spin-transfer torque. The ability to define an effective activation barrier does not imply that the spin-transfer torque can be represented by a real potential energy; in fact the Slonczewski expression for the spin-transfer torque is explicitly non-conservative. If one chooses instead to think in terms of a fictitious temperature, then it is important to note that T_{eff} remains proportional to the real sample temperature T. The spin-transfer current merely amplifies the effects of the existing thermal fluctuations at the true sample temperature. Therefore, within this theoretical model spin transfer should not be viewed as actually heating the sample. For a true heating mechanism, one would expect an elevated temperature that at large current levels should eventually depend only on the current, and not on the background temperature, so that in a normal heating process the elevated temperature would not be always proportional to the true temperature.

Although this simple derivation applies only to the case of samples with uniaxial anisotropy, analytical calculations and numerical modeling within this theoretical framework for samples with more realistic biaxial anisotropy also predict to good accuracy that the same result given in Eq. (5.8) should hold [29, 55].

An alternative model for the temperature dependence of spin-transfer switching was suggested by Urazhdin [35, 37] and the Lausanne group [36]. They proposed that when the spin polarization of the incident current is opposite to the moment of the free layer, then above a threshold current spin-flip scattering of electrons should excite non-uniform magnons, effectively raising the magnetic temperature of the free layer so as to accelerate switching. The difference from the model described above is that the dominant effect is assumed to be the generation of incoherent magnons, not a coherent motion of the free layer as a single domain. The result is therefore a real heating effect that should depend on the magnetic configuration; for positive currents in our convention the free layer should be heated in the parallel orientation but not the antiparallel one, thereby

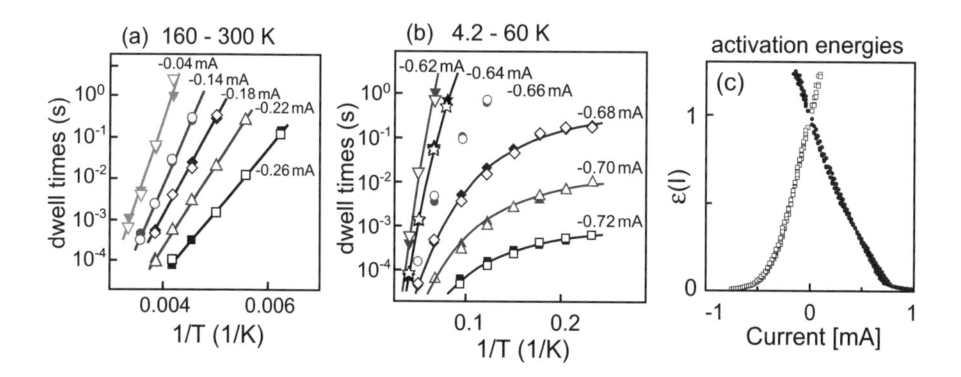

FIG. 5.10. (a) Plot of dwell times in the high resistance state (solid symbols) and the low resistance state (open symbols) as a function of inverse temperature. In this temperature range (160–300 K), the switching processes are thermally activated, and the spin-transfer current changes the effective activation energy. (b) Below approximately 15 K, the temperature dependence of the dwell times departs from simple activated behavior due to heating, but the heating is simply Ohmic and does not depend significantly on the magnetic configuration. (c) The switching rates as a function of current, temperature, and magnetic field can be collapsed onto one set of curves for the effective activation energies.

driving the free layer eventually to the antiparallel state. The degree of heating was argued to be very large, e.g. 400 K/mA above a threshold in Ref. [35] and 500 K to 1100 K in Ref. [36].

These two different physical pictures have been tested by making detailed measurements of spin-transfer-driven switching times as a function of temperature, current, and magnetic field [34]. The results are consistent with the predictions that treat the spin-transfer torque as modifying an effective activation barrier for switching, but they are inconsistent with a magnetic-configuration-dependent heating picture, at least for the low-switching-current samples that were studied. Figure 5.10(a) shows measured switching times plotted on a log scale as a function of $1/T$ for switching from the P to the AP configuration (open symbols) and AP to P (closed symbols) for a Py (2 nm)/Cu (6 nm)/Py (20 nm) spin-valve sample. The temperature range corresponds to 160–300 K. For each value of current bias, the magnetic field is adjusted until the two switching times are approximately equal and then the temperature is varied, keeping both the current and the magnetic field fixed. From the linear dependence of $\log(\tau)$ vs. $1/T$ it is clear that the switching is thermally activated to good accuracy and depends on the real sample temperature in this range of T. As the magnitude of the current is increased, the slopes of the lines in Fig. 5.10(a) decrease, indicating that the effective activation barrier is decreasing. There is no sign of a heating effect that depends on the magnetic configuration, because no significant difference is observed between the T dependence of the P $\rightarrow$ AP and AP $\rightarrow$ P times. Heating effects can be seen by going to lower temperatures, less than 20 K, and larger current magnitudes. Figure 5.10(b) displays the switching times measured between 4.2 and 25 K for the same sample, in a regime of large-enough currents that the transitions are between the parallel state and a dynamically processing state (D). Only below 20 K are heating effects evident, in the deviation from straight lines and the low-T saturation in the $\log(\tau)$ vs. $1/T$ plots. However, even when the heating is important, the P $\rightarrow$ D and D $\rightarrow$ P times show the same dependence on temperature – there is no evidence of heating that depends on the magnetic configuration. The temperature that is observed is consistent with the expectations for simple Ohmic heating in the devices. By extracting the dependence on magnetic field from the switching rates and taking into account the measured temperature dependence of the magnetic properties of the free layer, the dependence of the effective activation barrier on current can be determined throughout the temperature range from 20 K to room temperature. All of the measurements collapse nicely onto two lines representing the activation energies of the two switching processes (Fig. 5.10c) [34]. This demonstrates that the use of a current-dependent effective activation barrier provides an accurate description of the switching rates. In accord with calculations that model the spin-transfer torque as modifying the effective magnetic damping, the activation barriers depend linearly on current in the range of current where transitions originate from static magnetic states. When switching transitions begin from dynamical states, the dependence of the effective barrier on current appears to be slightly weaker than linear, but the transitions remain thermally activated and dependent on the real sample temperature.

5.1.5 *Applications of spin transfer torques*

5.1.5.1 *Magnetic random access memory* The ability of the spin-transfer effect to produce controllable switching of two magnetic layers back and forth between a high-resistance antiparallel state and a low-resistance parallel state suggests immediately that spin transfer might be a useful mechanism for writing information within non-volatile magnetic random access memories. Spin-transfer switching is already more efficient than the alternative of using current-generated magnetic fields to control magnetic bits (about 0.3 mA is required for spin transfer [56] vs. $\sim$ 10 mA for field switching). Furthermore, spin-transfer has a number of other potential advantages relative to field switching: no need to fabricate extra word and bit lines to generate the magnetic fields; reduced perturbation of neighboring magnetic elements when writing to one bit in an array; the possibility for continued scaling of magnetic bit sizes to the lithography limits of silicon processing, because spin-transfer switching currents can be minimized while independently maintaining the magnetic-anisotropy barriers required for thermal stability; and much less demanding device tolerances, because spin-transfer architectures will not require switching characteristics to be as uniform as field-switching. However, two main challenges have stood in the way of immediate applications of spin transfer in memory technologies. First, all-metal spin-valve devices have low resistances, 1–10 Ω, much less than the 1–10 kΩ needed to provide reasonable signal–to–noise when a silicon circuit is used to read the magnetic configuration. Second, while the critical currents required to produce magnetic switching have decreased steadily with improvements in device processing, from 5 mA in early Co devices [12] to approximately 0.3 mA in the best existing Py samples [56], even smaller critical currents would be desirable. Critical currents of 0.1 mA capable of switching magnets on the ns time scale would enable control of magnetic bits using minimum-area silicon CMOS transistors, and could make possible very dense memory circuits.

A solution to the problem of low resistance has recently been demonstrated. Spin-transfer switching has now been observed in devices in which the free and fixed magnetic layers are separated by a low-resistance tunnel barrier, rather than a metallic layer. Huai *et al.* achieved switching in tunnel junctions with R as large as 127 Ω with $\Delta R/R = 5\%$ at room temperature [57]. Fuchs *et al.* then demonstrated spin-transfer switching with devices fully in the desired resistance range, up to 1760 Ω with $\Delta R/R = 11\%$ at 77 K [58]. The magnitudes of the switching currents in the tunnel junctions are comparable to the values found in metallic spin valves with similar free layers. This indicates that the efficiency of a spin-transfer torque applied through a tunnel junction is not significantly reduced in comparison to metallic spin valves.

The question of how to make further reductions in the magnitude of the switching currents is the subject of active research by many groups. There exist several promising strategies that should enable the design goal of 0.1 mA to be reached while still maintaining the thermal stability of the free layer. Some of the strategies that have been proposed are: (i) increasing the spin-transfer torque by sandwiching the free magnetic layer between two oppositely directed magnetic fixed layers [59], (ii) properly controlling the device length-to-width ratio [60], (iii) controlling the resistivity and the degree of spin-relaxation in the electrodes [61–63], (iv) using spin transfer and heating together in devices with low-Curie-point magnets [64], and (v) optimizing materials to reduce the magnetization of the free layer and the anisotropy (recall Eq. 5.3). One class of potentially promising materials that may allow greatly-reduced switching currents are ferromagnetic semiconductors, which have magnetization densities greatly reduced from typical transition metals. Chiba *et al.* have recently demonstrated spin-transfer switching in (Ga,Mn)As devices at current densities of $(1.1\text{–}2.2) \times 10^5$ A/cm^2, nearly two orders of magnitude less than in transition-metal devices [65]. However, (Ga,Mn)As has a Curie point well below room temperature, so that this particular material is not yet suitable for commercial applications.

5.1.5.2 *Spin-transfer-driven microwave sources and oscillators* The microwave-frequency dynamical magnetic modes that can be excited by DC spin-transfer torques are under investigation for use in high-speed signal processing, as nanoscale microwave sources, oscillators, and amplifiers. The NIST group has suggested, for instance, that precessing spin-transfer devices may be applied as sources and detectors for wireless chip-to-chip communications [66]. Some of the important recent developments in developing high-frequency applications are the demonstration of frequency tunability over tens of GHz using either current or magnetic field [19, 31], the demonstration of precessional dynamics in the absence of any applied magnetic field with exchange-biased devices [67], and the measurement of precessional signals directly in the time-domain as well as the frequency domain [60]. Some of the key issues that remain to be resolved are similar to the ongoing work in the MRAM devices – e.g., to optimize the devices to achieve maximum signal levels. In this respect it is significant that spin-transfer-driven precession has been observed in magnetic tunnel junctions as well as metallic spin-valve structures [58].

Considerable effort is now also focused on achieving a very well-defined frequency response from the precessing magnets. This work requires the development of ways to reduce the frequency widths of the peaks in the microwave spectra generated by the magnetic precession, and also the elimination of broad backgrounds that often lie underneath the peaks in these spectra. We have already discussed strategies to achieve narrow spectral peaks, in the section on temperature dependence above. The narrowness of the peak widths is limited by thermal fluctuation effects, and can be improved by increasing the moment of the free layer or reducing the temperature. The narrowest power spectra reported to date are peaks 2 MHz wide in 35-GHz devices in which the free layer is exchange-coupled to an extended film [54]. We have not yet discussed the broad backgrounds that are often found in the microwave spectra of spin-transfer-driven precessing magnets, but such a background is visible in Fig. 5.6(b) in the 2.6 mA trace. Typically the backgrounds display maximum power near zero frequency, with shapes that correspond roughly to $1/f$ or Lorentzian distributions. Time-domain measurements have shown that the backgrounds are associated with telegraph-type switching between different resistance states, with switching rates ranging from less than 1 MHz to greater than 1 GHz [68]. Our group has found in some nanopillar samples that the magnitude of the low-frequency background is largest for a magnetic field aligned in plane along the easy axis of the free layer. When the field is tilted, the background is often reduced. We speculate that the tilted field may break close symmetries between energetically similar dynamical modes and reduce the likelihood that the system will undergo thermally activated switching between different orbits [29].

5.1.5.3 *Read heads* There is one other area of application in which the spin-transfer effect enters as a nuisance, not as a benefit. In current-perpendicular-to-the-plane magnetic read heads that are under development for high-density magnetic disk drives, spin-transfer torques can generate noise and reduce the effectiveness of the sensor [69–71]. A full understanding of spin-transfer excitations will be important in order to minimize these harmful effects. One will need to consider strategies such as increasing the thickness of the free layers to decrease their tendency to be excited by spin-polarized currents.

5.2 Electrons in micro- and nanomagnets

We will now change subjects, to address the question of how the transport properties of magnets change as one reduces their size from macroscopic dimensions down to the nanometer scale. One should realize, first, that there are many senses in which the properties of magnets are rather remarkably similar over this huge range in size. As current passes through a thin film of a metallic ferromagnet such as Co, only a couple of monolayers thickness can produce a similar degree of polarization in the resulting current as a magnet centimeters thick [72]. A large number of single molecules (such as Mn_{12}-acetate) can be viewed as ferromagnets in very much the same way as bulk magnets. Even individual electronic or nuclear spins exhibit precession and relaxation behaviors similar to ferromagnetic resonance in bulk magnets. However, there are differences. Several simple and important differences will *not* be emphasized in this chapter. For instance, there is the issue of magnetic domains. Most magnetic materials, when made larger than on the order of a few hundreds of nm, will become subdivided into domains in which their moments point in different directions, in order to minimize the total magnetostatic energy at the expense of some cost in exchange and anisotropy energy. The understanding of such processes is part of the field of micromagnetics. A very important phenomenon for applications is superparamagnetism. For a magnet that is sufficiently small, thermal fluctuations can overcome the magnetic anisotropy barriers holding the magnet's moment in place, so that its orientation may fluctuate on time scales shorter than typical experimental scales. The sample will then exhibit paramagnetic, rather than ferromagnetic properties. For transition-metal magnets at room temperature, the size scale at which superparamagnetic effects become important is typically tens of nm. For even smaller magnets, the increased surface-to-volume ratio together with changes in electronic bonding for atoms at surfaces as compared to embedded within the bulk can lead to significant changes in both the average magnetization of the nanomagnet and the magnetic anisotropy properties, compared to bulk systems.

All of these topics are interesting in their own right, but we will choose to focus here on properties of magnets that affect electrical transport on small length scales. We will concentrate primarily on devices in which a magnetic sample is contacted to two electrodes through tunnel barriers so that electrons can hop from one electrode, onto the magnet, and then to the second electrode to generate a current. By varying the applied voltage so as to change the electron energy, this type of measurement provides a means to probe the electronic density of states within the magnet. Typically, the devices will also have a gate electrode with which the energy of the electronic states in the magnetic island can be tuned relative to the Fermi energy in the source and drain electrodes (see Fig. 5.11), thus enabling transistor action. This half of the chapter will be organized to describe new phenomena that emerge on (1) the micron scale, (2) the 3–10 nm scale, and (3) the molecular scale. The emphasis will be primarily experimental,

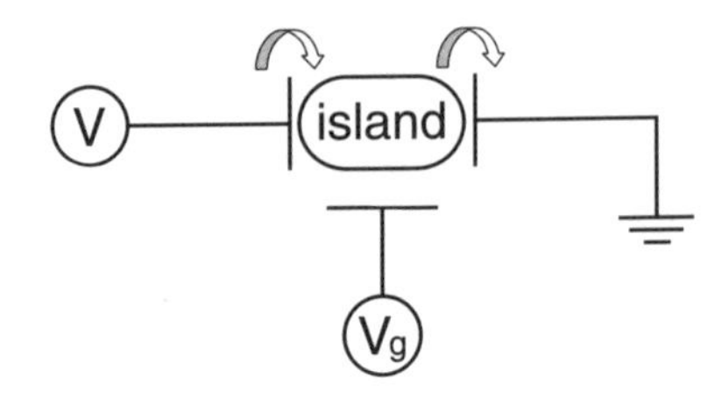

FIG. 5.11. Circuit schematic for a single-electron transistor.

but we also wish to note that there is increasingly active theoretical work related to this field, with several predictions that have not yet been tested experimentally [73–80].

5.2.1 *Micron-scale magnets and Coulomb blockade*

If a metal island within a tunneling device has a diameter of about 1 micron or smaller, the capacitance C of the island to the outside word is typically a few fF. This means that the scale of the charging energy required to move even a single electron onto or off the island, $e^2/(2C)$, is on the order of a few tenths of an meV, or in temperature units a few Kelvin. If the resistance of the tunnel junctions is sufficiently large to reduce quantum charge fluctuations ($R > h/e^2 = 25.8\ \mathrm{k}\Omega$), this means that at sufficiently low temperatures and low applied voltages, the electrons in the electrodes can have too little energy to enter the island. This means that to a good approximation no current can flow through the island; it is in the condition of "Coulomb blockade". However, if the device has a gate electrode, this can be used to shift the energy imbalance between states with n and n+1 electrons on the island. If the energies of these states are tuned to be equal, then there is no energy cost for changing the number of electrons on the island between these values. At such points, electrons can hop onto the island from one electrode and hop off to the other, creating a current. On account of the Coulomb energy, only one excess electron can hop onto the island at a time (for low temperatures and low source-drain voltages), so this device is known as a single electron transistor (SET) [81].

Figure 5.12(a) displays conductance oscillations of an SET with an Al island and Al electrodes at 4.2 K [82]. The darker represents higher values of conductance. As a function of the gate voltage V_g, the traces display a sequence of equally spaced conductance peaks, corresponding to the values of V_g where the n and n+1 electron states are degenerate, then where the n+1 and n+2 electron states are degenerate, etc. For a device made entirely out of non-magnetic materials such as Al, these conductance peaks show no dependence on the applied magnetic field. However, this absence of field dependence changes as soon as either the island or one or both leads are made from a magnetic material, as observed first by Ono, Shimada, and Ootuka [82]. Figure 5.12(b) shows the corresponding conductance curves for a magnetic Co island with Al electrodes. In this case, as the applied magnetic field is increased, the conductance peaks shift

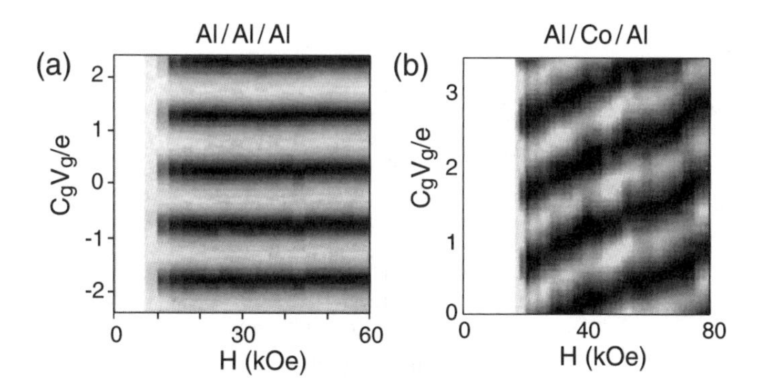

FIG. 5.12. Coulomb oscillations at 4.2 K (a) for a single-electron transistor with an aluminum island and aluminum source and drain electrodes and (b) for a cobalt island with aluminum electrodes. For the non-magnetic island, the Coulomb oscillations are independent of magnetic field, while for the magnetic cobalt island the oscillations shift with magnetic field. (From Ono, Shimada, and Ootuka.)

to higher values of gate voltage. This shift represents an increase in the Fermi energy for the electrons on the magnetic island. For higher magnetic fields, a greater energy is required to add an electron to the island, an energy that must be compensated by applying a more positive gate voltage in order to follow along one peak of the Coulomb-conductance curve.

The magnetic-field dependence of the Coulomb oscillations in the magnetic SET can be explained qualitatively, but not quantitatively, within a simple Stoner-model picture of magnetism on the cobalt island [82, 83]. In the Stoner model, it is assumed that strong exchange interactions in the magnet produce a splitting between spin-up and spin-down electronic states. In Co, the majority (spin-up) states have a lower density of states at the Fermi level than the minority states. As a magnetic field is applied, the energy of the majority states is decreased and the energy of the minority states is increased, due to the combined effects of the Zeeman energy and a contribution from exchange energy [83]. As a result, the system can lower its energy by allowing some fraction of minority spins to flip, enabling them to occupy lower-energy majority states (see Fig. 5.13). This process of flipping will continue until the majority and minority bands are filled up to the same energy, the new Fermi energy. In a non-magnetic material, for which the spin-up and spin-down electrons have the same density of states at the Fermi level, this process does not cause the Fermi energy to shift as a function of magnetic field. For every electron in a spin-down state whose energy is shifted above the original Fermi energy, there is an unoccupied spin-up state shifted below the original Fermi energy. After the electrons are redistributed among states, the Fermi energy remains unaltered. However, in a magnetic material like Co, there is a smaller density of states near the Fermi

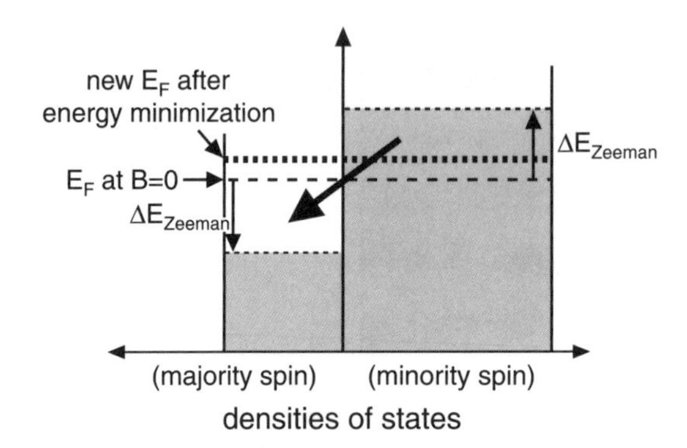

FIG. 5.13. Energy-level diagram illustrating why the Fermi energy will shift when a magnetic field is applied to a ferromagnetic metal sample.

level for majority electrons than for minority electrons. This means that there is not sufficient room in spin-up states lowered below the original Fermi energy to accommodate in all of the electrons in spin-down states that have been pushed up above the original Fermi level. When the electrons have flipped enough spins to reach a new equilibrium, the overall Fermi energy must therefore increase with the magnetic field. This is in agreement with the data shown in Fig. 5.12.

The reason that we say that the Stoner model provides only a qualitative, not a quantitative, description of the magnetic-field-induced changes in the Coulomb-blockade oscillations is that there appears to be an additional contribution due to a spin-dependence in the electronic wavefunctions at the magnetic interface, that can vary somewhat from sample to sample [84].

SETs can also be made in which the island and the electrodes are all made out of magnetic materials. These devices exhibit a number of interesting magnetoresistance effects, as the relative angle between the various magnetic moments change. In particular, small higher-order tunneling currents that can be observed using careful measurements within the Coulomb-blockade regime can exhibit much larger fractional magnetoresistance changes than the sequential-tunneling currents that flow once the Coulomb blockade is overcome [82]. If one makes the Coulomb island and the electrodes from different magnetic materials, the sign and magnitude of the field-induced shift in the Coulomb oscillations will depend on the relative polarizations of their densities of states at the Fermi level [82].

5.2.2 *Ferromagnetic nanoparticles*

If one studies the same sort of single-electron-transistor device, but shrinks the size of the metal island down to just a few nm, new sorts of physics become accessible. In this size range, it becomes possible to resolve experimentally the individual electrons-in-a-box quantum states inside a metal, and to measure the energy spectrum of such levels. Our research group has studied the electronic

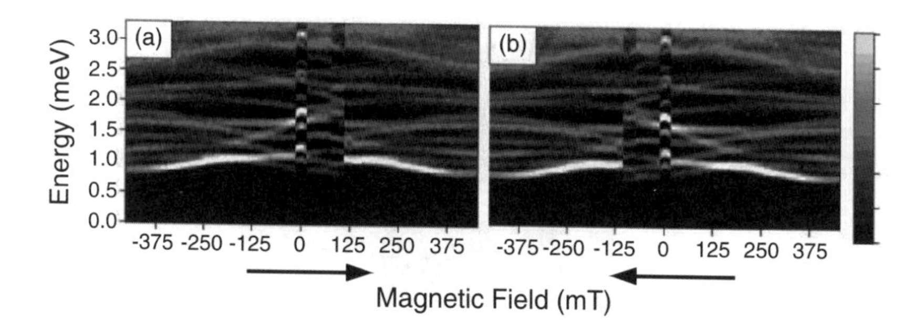

FIG. 5.14. Tunneling resonances for a cobalt nanoparticle as a function of magnetic field, at approximately 80 mK. The magnetic field is swept negative to positive on the left, and positive to negative on the right.

states in a variety of different types of metals in this way, and we have observed that the level spectra in general cannot be described by a simple non-interacting electron-in-a-box model. In fact, the energy levels are affected by all the different types of forces and interactions that influence the electronic structure in metals, and therefore they provide a means to understand these interactions at a very fundamental level [85].

In this section, we will explore the basic question, "What is the true nature of the quantum states inside a small ferromagnet?" We will see that the answer to this question is not straightforward, and a more sophisticated approach is required than the simple Stoner-type models that we considered above. In particular, it will be necessary to understand not only the effects of strong exchange interactions but also spin-orbit coupling – physics that is not taken into account in the Stoner model [86, 87].

The lowest-energy states in the tunneling spectrum from a Co nanoparticle attached to Al electrodes is shown in Fig. 5.14, for an applied magnetic field swept both (a) from negative to positive and (b) from positive to negative, at a temperature of $\sim$ 80 mK [87]. The intensity scale in the figure represents the measured value of dI/dV, due to tunneling via the discrete states on the particle. Black represents $dI/dV = 0$ and white a large conductance. The magnetic fields (H) shown in Fig. 5.14 are in the range for which the field acts to reorient and reverse the particle's magnetic moment. Consider first Fig. 5.14(a). At large negative fields, where the moment of the nanoparticle should be aligned with the direction of the magnetic field, a number of different resonances can be seen. As H is swept toward zero and the moment in the nanoparticle relaxes toward some easy-axis direction, the levels shift and can cross. Each level displays a different behavior as a function of H. Near zero field, there is a glitch in the data. This is due to a transition in the Al electrodes from normal to superconducting and back to normal again as the field is swept through zero. The tunneling resonances all simply shift in energy by an amount equal to the superconducting gap in the electrodes; there is no underlying discontinuity here for the energy levels inside

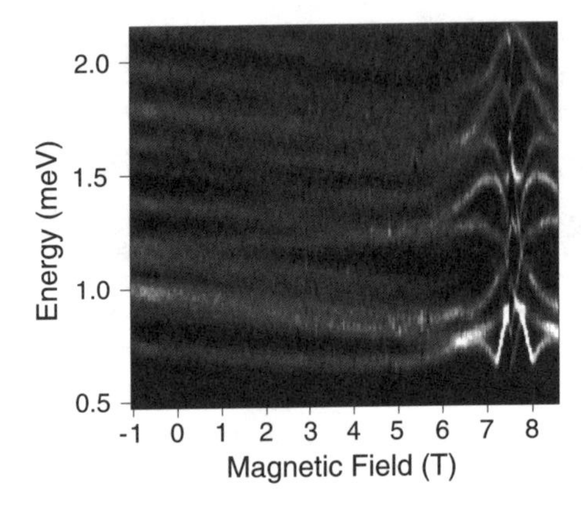

FIG. 5.15. Tunneling resonances for the same cobalt nanoparticle as in Fig. 5.14, for a larger range of magnetic field. The field is swept negative to positive.

the Co. However, at approximately $H = 120\,\mathrm{mT}$ in Fig. 5.14(a), there is another discontinuity in all the resonances. This discontinuity we can associate with the magnetic reversal of the Co particle, because if the field is swept in the opposite direction then it exhibits magnetic hysteresis – it is observed at the opposite sign of H (Fig. 5.14b). At this reversal transition, some levels jump up in energy and some jump down. For fields beyond the transition, the tunneling spectra are symmetric about $H = 0$.

The dependence of the tunneling spectra on larger values of magnetic fields is shown in Fig. 5.15, for H swept from negative to positive values. Between the reversal field at 120 mT and about 1 tesla, each of the levels continues to exhibit a strongly non-monotonic but continuous dependence on H. Beyond about 1 tesla, the character of the H dependence changes, with each resonance moving approximately linearly. All of the observed resonances move with a positive slope, and for all but one of the levels this slope has approximately the same value. Only the level with energy near 1.2 meV at large H shifts more slowly, with a slope that is positive but close to zero. The values of the slope correspond to g-factors for Zeeman splitting between 0.06 and 1.1. The fact that the g-factors are reduced below 2 indicates that the resonances are not purely spin up or spin down, but are mixed by spin-orbit scattering [85].

Some features of these observations can be understood within simple models, while some cannot. Within any model of magnetism, one expects the existence of a strong exchange field, so that the energies of spin-up and spin-down states should be different even in the absence of an applied magnetic field. This is true in the data – there is no indication of a degeneracy between spin-up and spin-down levels at $H = 0$ that undergoes Zeeman splitting when H is applied, as is seen universally in non-magnetic nanoparticles [85]. It is also natural that the levels in the nanoparticle might change energy as the magnetic moment of the nanoparticle reorients and reverses. However, the mechanism behind this behavior is not what one might expect. Within a Stoner-type model, one might anticipate that the rotation of the nanoparticle's moment could lead to a change in the effective magnetic field acting on the electrons, because the demagnetization field will change as the moment rotates within a non-spherical nanoparticle. If this were the mechanism, then the same effective field would act on every electronic state and there should be at most two classes of energy-level trajectories as a function of H – one for spin-up and one for spin-down. Furthermore, any non-monotonic dependence on H should cease soon after the reversal field (120 mT), as the moment becomes aligned with H. Instead, what is observed in the data is that every resonance appears to have a different dependence on H at low field, and the non-monotonic shifts continue until H is approximately eight times the switching field. The magnitude of the changes in energy at low H is also surprisingly large. The maximum change in demagnetization field should be no greater than the intrinsic magnetization of Co – about 1.6 T. Instead, the energy shifts at low H can be considerably larger than the changes induced by an external field of over 8 T (Fig. 5.15). One must conclude for all these reasons that a mechanism other than a changing demagnetization field is required to explain the low-field changes in the energy levels as a function of H.

The high-field data exhibit additional unexpected phenomena. Within the Stoner model, level shifts at large H should be entirely due to the Zeeman spin energy and there should be two classes of states observed, again for spin-up and spin-down. Depending on whether the tunneling spectrum corresponds to an electron being added to or subtracted from the nanoparticle (something that is not known independently for this ungated sample), the spin-down (minority spin) electrons that are expected to be more numerous near the Fermi energy could move either up or down in energy as a function of H. However, the expected polarization of the density of states near the Fermi level in Co is only $\approx 40\%$, so that there should be at most about 2.3 spin-down states for every spin-up. It is therefore odd within the Stoner model how every one of the energy levels in Fig. 5.15 has a positive slope as a function of H, and twelve of the thirteen levels visible exhibit approximately the same value for this slope.

If one simply counts the levels in either Fig. 5.14 or Fig. 5.15, there is one more surprise. The mean energy spacing between resonances is less than 0.2 meV, a value which is similar for all the Co nanoparticles that we have measured. However, from measurements on ensembles of nanoparticles made by our fabrication recipe, we expect that the particle diameters should range from 1–4 nm, and consequently that the spacing between the electron-in-a-box energy levels should be much larger, between 1 and 40 meV. The most likely explanation for the very small energy scale that we observe appears to be that magnetic particles have low-energy collective electronic excitations – spin waves – in addition to simple electron-hole excitations. These collective excitations appear to be mixing into the tunneling measurement. One mechanism by which this might happen is a non-equilibrium effect [87]. If spin waves are generated by the process of current flow, and they have lifetimes longer than the time between electron tunneling events, then the presence of a spin-wave excitation on the island can shift the energy required for an electron to tunnel into the lowest-energy electronic state. If it is possible to excite an ensemble of different spin-wave states, then each might produce a different shift, so that a single electronic orbital level might appear in the tunneling measurement at an ensemble of different energies. This mechanism has the capability to explain both the multiplicity of tunneling resonances observed at low energy and their similar dependence on H at large fields, because the different tunneling resonances observed might all originate from the same orbital state. Measurements of tunneling spectra as a function of gate voltage in different samples have confirmed that such non-equilibrium processes are active in Co nanoparticles [87].

Two main theoretical approaches have been pursued in trying to explain these experiments. Both extend beyond a simple Stoner model by incorporating effects of spin-orbit interactions that give rise to magnetic anisotropy, in addition to the exchange interactions that enable ferromagnetism. With spin-orbit interactions included, one can easily obtain larger changes in the level energies as the magnetic moment of the nanoparticle rotates than are expected purely from the changing demagnetization field of the nanoparticle. One theoretical approach is to use an effective spin Hamiltonian [88–90]. This picture disposes of the single-electron point of view at the heart of the Stoner model, and assumes from the beginning that one should calculate the tunneling energies as the energy difference between the true many-electron quantum states before and after the tunneling event. In the presence of strong exchange interactions, the energy required to produce excitations that change the total spin from its preferred value should be very large [88], so for purposes of calculating low-energy excitations only the states in the lowest-energy total-spin multiplets need be considered. After including in the Hamiltonian terms for the kinetic, exchange, and Zeeman energies, and a phenomenological term to include spin-orbit-induced anisotropy in a simple way, it is possible to diagonalize the model Hamiltonian within the lowest-energy spin-multiplet for both n and $n+1$ electrons, and then take the difference to model the tunneling energies. If the same strength of anisotropy energy per spin is assumed for both the n and $n+1$-electron states, then the results do not look much like the data. The predicted energies do not have a strongly non-monotonic dependence on H and the jumps in energy at the magnetic-reversal field are too small [87, 90]. However, if the anisotropy energy per unit spin is allowed to be different for the different charge states, by on order 1–3%, qualitatively much better agreement with the measured H dependence of the energy levels can be found [87, 90]. This degree of fluctuation is rather surprising – a 4-nm-diameter Co nanoparticle contains approximately 1000 conduction electrons, and we are saying that adding one electron might change the total anisotropy energy of the system by 3%.

The second theoretical approach is more numerical – to calculate the quantum states of a magnetic nanoparticle using a tight-binding Hamiltonian that includes exchange interactions and a realistic microscopic accounting of spin-orbit interactions [91]. This approach is able to provide a qualitatively good explanation for the measured H dependence of the ferromagnetic energy levels, as well. The calculation finds that each electron orbital experiences a different contribution to its energy from spin-orbit interactions. The contribution may be either positive or negative, so that the average shift is smaller by generally a factor of 100 than the absolute value of the typical shift for an individual orbital. Under these conditions, the change in occupation of one orbital can produce a sizable change in the total anisotropy energy of the entire magnetic nanoparticle. The conclusion is that changes on the order of a few percent that are needed to explain the measurements within the effective spin Hamiltonian are quite reasonable.

In summary, at this time many features of the experimental tunneling spectra for the ferromagnetic nanoparticles can be explained qualitatively by considering the combined effects of magnetic exchange and anisotropy forces, but there is no overall theory that provides a good account of all the experimental observations in one framework. What appears to be needed is to develop theoretical techniques that combine the ability of the effective-spin models to include effects due to collective excitations and non-equilibrium effects together with the capacity of the numerical models to account realistically for spin-orbit interactions. Cehovin, Canali, and MacDonald have performed recent work in this direction by considering the quasiparticle and collective excitations in a more unified framework, and have found that these excitations can be strongly coupled in small ferromagnetic particles, so that the distinction between them blurs [92]. However, additional work will still be necessary to achieve a fully satisfactory understanding of the quantum-mechanical states inside small metallic ferromagnets.

5.2.3 *Magnetic molecules and the Kondo effect*

The ultimate limit for exploring small-scale magnetism is to conduct experiments with single spin moments. This limit is now becoming realizable in experiments on unpaired electrons in both semiconductor quantum dots [93, 94] and single molecules [95, 96]. We will focus in this last section on recent results from molecules. Part of the interest in this subject comes from demonstrations that by controlling the form of chemical bonds in single molecules, one can control the magnitude of the total spin in the ground state, from $S = 0$ and 1/2, to greater than S=10. Such molecules may therefore serve as a simpler model system than nanoparticles for probing the formation of the ferromagnetic state and the microscopic basis of magnetic anisotropy. (In the magnetic nanoparticles discussed above, the total spin is approximately 1000.) The primary way in which single-molecule magnets have been studied before now is to assemble them into crystals and then apply bulk methods such as SQUID susceptibility, neutron scattering, and ESR. Electrical-transport techniques for probing the energy levels in single molecules have been developed very recently, and at this stage they are still quite crude. Thus far, the first measurements have been published only for spin-0 and spin-1/2 molecules [95, 96, 98], and preliminary experiments are underway on Mn_{12}-acetate, with spin 10 [99]. However, if the techniques continue to develop, they should provide a valuable new form of spectroscopy. Already they allow tests of some theoretical predictions (e.g., the properties of quantum levels coupled to ferromagnetic electrodes) that are not easily explored using conventional quantum dots.

The primary challenge in making electrical measurements on single molecules is that most molecules are, of course, small. A way must be found to position metallic source and drain electrodes just 1–2 nm apart, with a molecule spanning the gap. This small size scale is not within the resolution of existing lithographic techniques, so a trick is required. Thus far, three experimental techniques have provided useful solutions to this problem: scanning probe microscopy [100], mechanically controlled break junctions [101], and junctions formed by electromigration [102]. We will emphasize results from the electromigration technique, because it has been the primary method used to probe spin properties of molecules, and of the three techniques it is the only one so far to allow the use of a gate electrode for tuning the molecular energy levels.

To make single-molecule devices by the electromigration technique, one starts by depositing on a substrate a thin wire of metal, about 100 nm or less in width. The first devices used Au, but Pt, Ni, and other metals have now also been employed successfully. The process of electromigration consists simply in ramping an applied voltage across the ends of the wire. At sufficiently large voltages, atoms can be excited to move about on the surface of the wire or along grain boundaries, forming weak spots. There, the current becomes more concentrated, exciting more atoms to move, and the wire can be made to narrow progressively until, eventually, it breaks completely (typically near 500 mV bias). If this is done carefully at low temperatures, the width of the gap between the two broken ends of the wire can be made reproducibly in the range of 1-2 nm, as determined from the residual tunneling conductance measured across the gap after the breaking process is done. A gate electrode can be included in the device simply by using a conducting substrate (Si, or even better Al) separated from the broken electrodes by an insulating oxide.

A molecule can be incorporated into the gap in either one of two ways. The conceptually simpler way is to introduce a dilute drop of solution containing the desired molecules after the electromigration is done, allow some time for the molecules to become attached, and then blow dry the solution. This works. However, we find that we achieve a better percentage yield of devices with molecules spanning the gap (about a 5–10% success rate) if we deposit the molecules on the unbroken metal wire *before* electromigration, and then break the wire with the molecules present. The reason why this gives an improved yield is not known. Perhaps molecules are attracted to the high-field-gradient region in the gap while electromigration is taking place, or perhaps the molecules attach themselves firmly to the electrodes before electromigration and then the wire breaks underneath them, leaving them in a position to bridge the opening. Of course, there is a worry that if the molecules are present during electromigration, then they might be damaged by this process. For this reason, it is important to perform

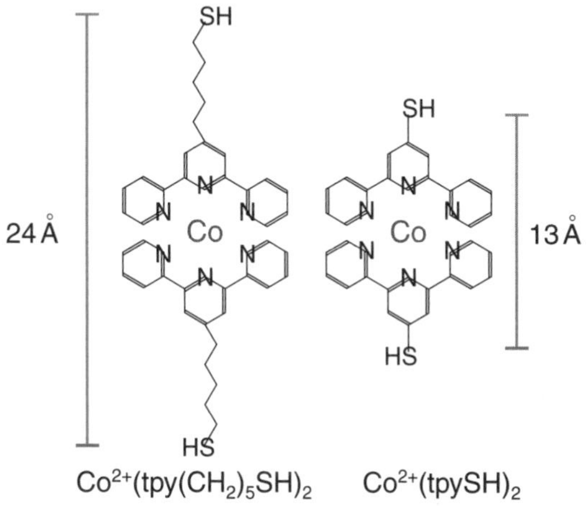

FIG. 5.16. Diagrams of cobalt coordination complexes used for making single-molecule transistors.

careful control experiments, to make sure that one is not making measurements on random chunks of carbonaceous residue created by the electromigration at the bottom of the cryostat. One of the most important controls is that different types of molecules exhibit different transport properties, as described below.

The first molecular-scale transistors achieved with the electromigration technique were C_{60} transistors, made by Hongkun Park and collaborators in the group of Paul McEuen [103]. However, we first wish to discuss devices made with a different type of molecule, a transition metal complex consisting of one Co atom bound to terpyridine-derived organic "linker arms" terminating in a sulfur atom that can bind to gold (Fig. 5.16) [95]. This molecule has the advantage that the organic linkers can be synthesized with different lengths, thereby providing control over the electrical resistance of the molecule. This particular coordination complex was selected because it is known through electrochemical measurements in solution to require a small energy cost (on the order of 0.1–0.2 eV) to change the charge state of the Co atom from 2+ to 3+. This provides hope that the same transition might be accessible at low temperatures when the molecule is bound to electrodes, so that the molecule can act like a single-electron transistor. In the case of the molecule, the "island" containing low-energy states through which the electron can tunnel is a single atom, rather than 1000 or more atoms as in the larger devices discussed above, but the principle of operation should be the same.

Figure 5.17 displays the *I*-*V* curves for a device made with the longer Co molecule, in which an alkyl chain with five carbon atoms is placed between the terpyridine group and the sulfur atom in each linker arm [95]. The device exhibits Coulomb-blockade characteristics. At fixed gate voltage, the current is essentially zero at low voltage, increasing to non-negligible values only above

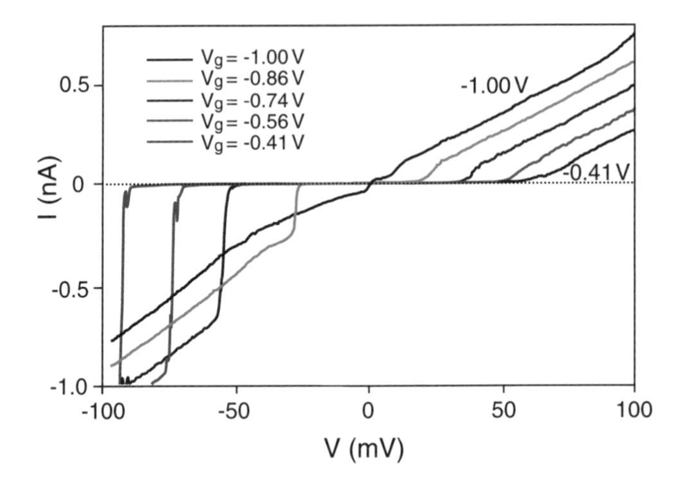

FIG. 5.17. Coulomb-blockade curves at approximately 100 mK, measured at different values of gate voltage, for a transistor made from a single molecule of the longer Co-containing molecule.

certain threshold voltages for positive and negative bias. These thresholds can be changed by varying the gate voltage, and for a particular value of V_g the conductance thresholds can be tuned to zero, meaning that the charge states involved in tunneling are degenerate. The primary way in which these curves differ from the Coulomb-blockade curves for metal nanoparticles is that the maximum Coulomb blockade energy is much larger. Because the devices become unstable at large source-drain voltages, we cannot actually measure the maximum Coulomb blockade energy, but we can determine a lower bound – that it is greater than 150 meV. The true charging scale is probably on the order of atomic energies, in the eV range. We find that every device requires a different gate voltage to tune the Coulomb blockade to the degeneracy point even when they are made from the same molecules, presumably because each molecule experiences a different electrostatic environment. However these differences mean that as long as measurements are taken at gate voltages near the degeneracy point, then current flow can occur only via a single quantum state on one molecule, even if more than one molecule bridges between the two electrodes. A close examination of the conductance through the molecule at low temperature shows tunneling through excited quantum states which produce resonances at energies offset by a fixed amount from the Coulomb-blockade threshold. These resonances are most likely due to vibration-assisted tunneling involving normal modes of the molecule, as first discussed in connection with C_{60} [103] and demonstrated persuasively with the mass-spring-mass molecule C_{140} [104].

We mentioned previously that it is crucial to conduct control experiments to ensure that the measured properties are due to the molecule under study, and not some other process. We have conducted controls using the same fabrication

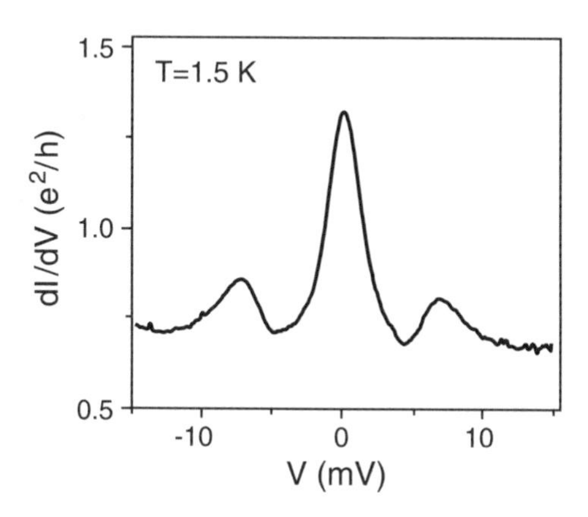

FIG. 5.18. Differential conductance vs. source-drain voltage at 1.5 K, for a molecular device made from the shorter Co-containing molecule. Instead of Coulomb blockade, the sample exhibits a peak at V=0 that can be identified with the Kondo effect.

technique, but with no molecules, with just the Co ions but no linker arms, and with just the organic linker arms but with no Co atoms. In none of these cases did we observe Coulomb-blockade characteristics like the one shown in Fig. 5.17. The excitation energies that we associate with molecular vibrations can also in some cases be identified with the particular molecule under study (for example, the 11-meV stretching mode of C_{140} [104]).

Another very important test is to change the structure of the molecule and investigate how this modifies the measured transport properties. Figure 5.18 shows dI/dV vs. V for the shorter Co-containing molecule, in which the high-resistance alkly chains connecting the terpyridine groups to the sulfur bonding atom have been removed [95]. Instead of exhibiting zero current near $V = 0$ due to the Coulomb blockade, the conductance of this device contains a peak at $V = 0$. The amplitude of the peak can be nearly as large as $2e^2/h$, the maximum conductance possible through one quantum state. To explain the physics behind this peak, it is important to note that the Co^{2+} electronic ground state contains an unpaired spin-1/2 electron and is therefore magnetic. When this state is coupled strongly, through low-resistance molecular bonds, to the gold electrodes, the result is the Kondo effect – a correlated electronic state forms involving the unpaired spin on the molecule and the electrons in the electrodes that enables a coherent transport of charge transfer through the molecule [105]. The consequences of the Kondo effect have been studied in detail recently using semiconductor quantum dots [106, 107] and carbon nanotubes [108, 109]. The signature of the effect is that this coherence requires low applied voltages, low temperatures, and low magnetic fields, or else it is disrupted. The requirement for low V is the reason that the Kondo effect produces a sharp peak in the conductance centered at low V. The dependence on temperature and magnetic field for this

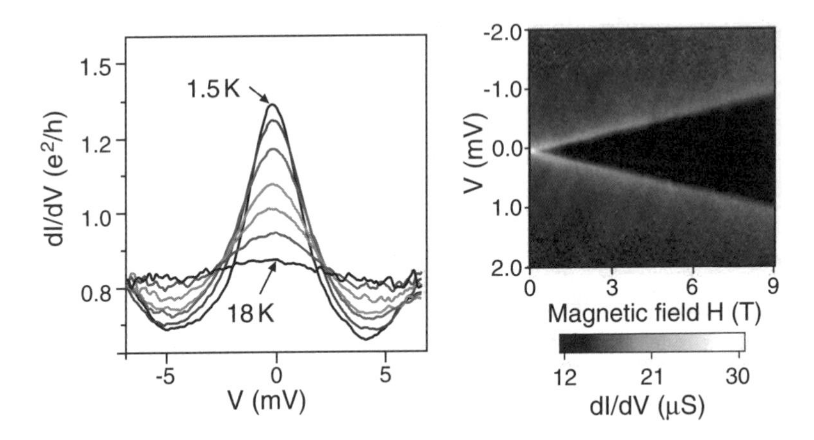

FIG. 5.19. Temperature and magnetic-field dependence of the peak in dI/dV for a device made from the shorter Co-containing molecule. The data exhibit the signatures of the Kondo effect.

sample are shown in Fig. 5.19. We observe characteristic Kondo temperatures for this molecule that range from less than 1 K to greater than 50 K, presumably depending on the precise nature of the coupling between the molecule and the electrodes. The form of the magnetic-field dependence is a splitting that can be explained by the Zeeman splitting of the spin-up and spin-down quantum states associated with the unpaired electron. Similar observations of the Kondo effect have been made in devices containing a di-vanadium molecule by the group of Hongkun Park [96].

We close with one example of a magnetic effect that can be studied more easily using a molecular system than by using more conventional quantum dots – the coupling of a Kondo center to magnetic electrodes [97]. With semiconductor quantum dots, it has not yet proved possible to achieve strong coupling between the dot and a ferromagnetic electrode. However, with a molecular system, it is straightforward to use nickel electrodes, in place of gold, and to observe the consequences on the Kondo effect. Figure 5.20 shows dI/dV as a function of both V and the applied magnetic field H for a C_{60} molecule with Ni electrodes [97]. C_{60}, if it happens to be sufficiently strongly coupled to magnetic electrodes, can exhibit the Kondo effect in the same way as the Co-containing molecule. The Kondo effect with non-magnetic electrodes takes the same form shown in Fig. 5.18, a single peak in dI/dV centered at $V = 0$. However, in the presence of Ni electrodes, the Kondo peak is split, as if there were a magnetic field. Moreover, the degree of splitting depends sensitively on the relative orientation of the magnetic moment in the two Ni electrodes. As a magnetic field is swept to turn the magnets from a parallel alignment to approximately an antiparallel alignment, back to a parallel alignment again, the splitting shown in Fig. 5.20 changes from 16 mV, to a minimum of 7.6 mV, and back to 16 mV

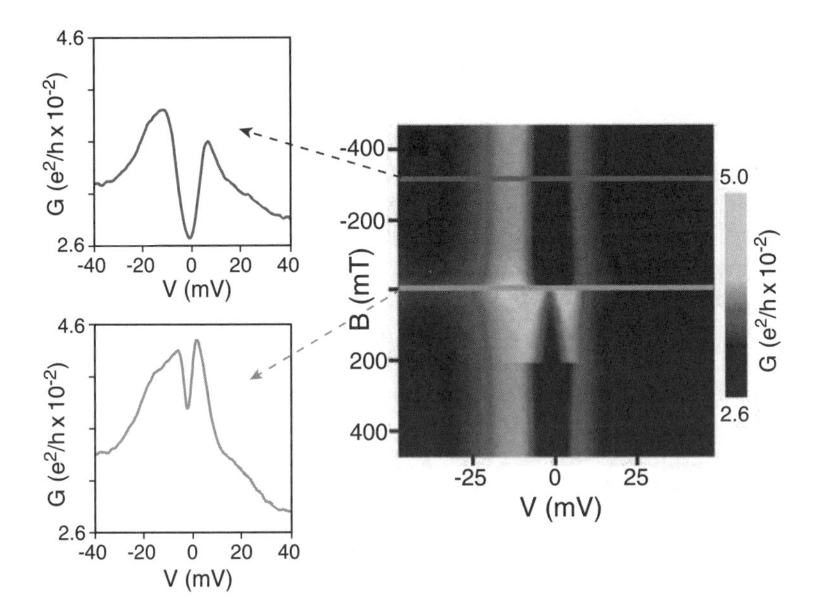

FIG. 5.20. (In gray-scale) differential conductance vs. source-drain voltage and magnetic field for a single-molecule device consisting of a C_{60} molecule between magnetic Ni electrodes, at 1.5 K. The field is swept from negative to positive, and the relative orientation of the electrodes switches from parallel to antiparallel and then back to parallel again. (Line graphs) differential conductance versus source-drain voltage when the magnetic electrodes are parallel (top) and approximately antiparallel (bottom), showing how the magnitude of the splitting depends on the magnetic configuration.

again. The splitting is much too large to be explained by a real magnetic field. A splitting of 16 mV would correspond to a magnetic field of approximately 70 tesla, much larger than the maximum local filed of about 0.6 tesla that could be generated by the intrinsic magnetization of Ni. However, the splitting is in very good accord with predictions of an effective exchange interaction between the quantum levels on the dot and the polarized density of electronic states in the electrodes [110, 111]. In this theory, charge fluctuations between the quantum dot and the electrodes renormalize differently the energies of the spin-up and spin-down states on the molecule, so they become split as they would in a magnetic field. In agreement with the theory, the size of the splitting is on the scale of the lifetime broadening of the molecular levels, and it is greater for parallel electrodes than for antiparallel because in the parallel case the effective exchange interaction with the two electrodes add together, while in the antiparallel case they have opposite signs.

References

[1] L. Berger, J. Appl. Phys. **55**, 1954 (1984).
[2] P. P. Freitas and L. Berger, J. Appl. Phys. **57**, 1266 (1985).
[3] C.-Y. Hung and L. Berger, J. Appl. Phys. **63**, 4276 (1988).
[4] J. C. Slonczewski, Phys. Rev. B **39**, 6995 (1989).
[5] J. C. Slonczewski, J. Magn. Magn. Mater. **159**, L1 (1996).
[6] L. Berger, Phys. Rev. B **54**, 9353 (1996).
[7] R. N. Louie, Ph.D. dissertation, Cornell University (1997).
[8] Y. B. Bazaliy, B. A. Jones, and S.-C. Zhang, Phys. Rev. B **57**, R3213 (1998).
[9] M. Tsoi, A. G. M. Jansen, J. Bass, W.-C. Chiang, M. Seck, V. Tsoi, and P. Wyder, Phys. Rev. Lett. **80**, 4281 (1998); erratum, ibid. **81**, 493 (1998).
[10] J. Z. Sun, J. Magn. Magn. Mater. **202**, 157 (1999).
[11] E. B. Myers, D. C. Ralph, J. A. Katine, R. N. Louie, and R. A. Buhrman, Science **285**, 867 (1999).
[12] J. A. Katine, F. J. Albert, R. A. Buhrman, E. B. Myers, and D. C. Ralph, Phys. Rev. Lett. **84**, 3149 (2000).
[13] J. C. Slonczewski, J. Magn. Magn. Mater. **247**, 324 (2002).
[14] M. D. Stiles and A. Zangwill, Phys. Rev. B **66**, 014407 (2002).
[15] X. Waintal, E. B. Myers, P. W. Brouwer, and D. C. Ralph, Phys. Rev. B **62**, 12317 (2000).
[16] K. Xia, P. J. Kelly, G. E. W. Bauer, A. Brataas, and I. Turek, Phys. Rev. B **65**, 220401 (2002).
[17] E. B. Myers, F. J. Albert, J. C. Sankey, E. Bonet, R. A. Buhrman, and D. C. Ralph, Phys. Rev. Lett. **89**, 196801 (2002)
[18] F. J. Albert, N. C. Emley, E. B. Myers, D. C. Ralph, and R. A. Buhrman, Phys. Rev. Lett. **89**, 226802 (2002).
[19] S. I. Kiselev, J. C. Sankey, I. N. Krivorotov, N. C. Emley, R. J. Schoelkopf, R. A. Buhrman, and D. C. Ralph, Nature **425**, 380 (2003).
[20] R. H. Koch, J. A. Katine, and J. Z. Sun, Phys. Rev. Lett. **92**, 088302 (2004).
[21] M. A. Zimmler, B. Özyilmaz, W. Chen, and A. D. Kent, J. Z. Sun, M. J. Rooks, and R. H. Koch, Phys. Rev. B **70**, 184438 (2004).
[22] M. AlHajDarwish, H. Kurt, S. Urazhdin, A. Fert, R. Loloee, W. P. Pratt, Jr., and J. Bass, Phys. Rev. Lett. **93**, 157203 (2004)
[23] A. Brataas, Y. V. Nazarov, and G. E. W. Bauer, Phys. Rev. Lett. **84**, 2481 (2000).
[24] Y. Tserkovnyak, A. Brataas, and G. E. W. Bauer, Phys. Rev. B **66**, 224403 (2002).
[25] A. A. Kovalev, A. Brataas, and G. E. W. Bauer, Phys. Rev. B **66**, 224424 (2002).
[26] J. Z. Sun, Phys. Rev B **62**, 570 (2000).
[27] Y. B. Bazaliy, B. A. Jones, and S.-C. Zhang, J. Appl. Phys. **89**, 6793 (2001).
[28] J. Grollier, V. Cros, H. Jaffrès, A. Hamzic, J. M. George, G. Faini, J. B. Youssef, H. Le Gall, and A. Fert, Phys. Rev. B **67**, 174402 (2003).
[29] Z. Li and S. Zhang, Phys. Rev. B **68**, 024404 (2003).

[30] M. Tsoi, A. G. M. Jansen, J. Bass, W.-C. Chiang, V. Tsoi, P. Wyder, Nature **406**, 46 (2000).
[31] W. H. Rippard, M. R. Pufall, S. Kaka, S. E. Russek, and T. J. Silva, Phys. Rev. Lett. **92**, 027201 (2004).
[32] S. I. Kiselev, J. C. Sankey, I. N. Krivorotov, N. C. Emley, M. Rinkoski, C. Perez, R. A. Buhrman, and D. C. Ralph, Phys. Rev. Lett. **93**, 036601 (2004).
[33] Y. Tserkovnyak, A. Brataas, and G. E. W. Bauer, Phys. Rev. B **67**, 140404 (2003).
[34] I. N. Krivorotov, N. C. Emley, A. G. F. Garcia, J. C. Sankey, S. I. Kiselev, D. C. Ralph, and R. A. Buhrman, Phys. Rev. Lett. **93**, 166603 (2004).
[35] S. Urazhdin, N. O. Birge, W. P. Pratt, Jr., and J. Bass, Phys. Rev. Lett. **91**, 146803 (2003).
[36] A. Fábián, C. Terrier, S. Serrano Guisan, X. Hoffer, M. Dubey, L. Gravier, and J.-Ph. Ansermet, Phys. Rev. Lett. **91**, 257209 (2003).
[37] S. Urazhdin, Phys. Rev. B **69**, 134430 (2004).
[38] J. A. Katine, F. J. Albert, and R. A. Buhrman, Appl. Phys. Lett. **76**, 354 (2000).
[39] J. Grollier, V. Cros, A. Hamzic, J. M. George, H. Jaffrès, A. Fert, G. Faini, J. B. Youssef, and H. Legall, Appl. Phys. Lett. **78**, 3663 (2001).
[40] J. Z. Sun , D. J. Monsma, D. W. Abraham, M. J. Rooks, and R. H. Koch, Appl. Phys. Lett. **81**, 2202 (2002).
[41] J. E. Wegrowe *et al.*, J.-E. Wegrowe, A. Fabian, Ph. Guittienne, X. Hoffer, D. Kelly, J.-Ph. Ansermet, and E. Olive, Appl. Phys. Lett. **80**, 3775 (2002).
[42] For a review of GMR, see the articles in IBM J. Res. Dev. **42** (1998).
[43] T. Valet, preprint.
[44] S. I. Kiselev, J. C. Sankey, I. N. Krivorotov, N. C. Emley, A. G. F. Garcia, R. A. Buhrman, D. C. Ralph, cond-mat/0504402.
[45] B. Özyilmaz, A. D. Kent, D. Monsma, J. Z. Sun, M. J. Rooks, and R. H. Koch, Phys. Rev. Lett. **91**, 067203 (2003).
[46] J. Miltat, G. Albuquerque, A. Thiaville, and C. Vouille, J. Appl. Phys. **89**, 6982 (2001).
[47] X. Zhu, J.-G. Zhu, and R. M. White, J. Appl. Phys. **95**, 6630 (2004).
[48] K. J. Lee, A. Deac, O. Redon, J. P. Nozieres, and B. Dieny, Nature Materials **3**, 877 (2004).
[49] M. L. Polianski and P. W. Brouwer, Phys. Rev. Lett. **92**, 026602 (2004).
[50] M. D. Stiles, J. Xiao, and A. Zangwill, Phys. Rev. B **69**, 054408 (2004).
[51] Y. Ji, C. L. Chien, and M. D. Stiles, Phys. Rev. Lett. **90**, 106601 (2003).
[52] B. Özyilmaz, A. D. Kent, J. Z. Sun, M. J. Rooks, and R. H. Koch, Phys. Rev. Lett. **93**, 176604 (2004).
[53] J. C. Sankey, I. N. Krivorotov, S. I. Kiselev, P. M. Braganca, N. C. Emley, R. A. Buhrman, D. C. Ralph, cond-mat/0505733.
[54] W. H. Rippard, M. R. Pufall, S. Kaka, T. J. Silva, and S. E. Russek, Phys. Rev. B **70**, 100406(R) (2004)
[55] Z. Li and S. Zhang, Phys. Rev. B **69**, 134416 (2004).

[56] E. M. Ryan *et al.*, unpublished.
[57] Y. Huai, F. Albert, P. Nguyen, M. Pakala, and T. Valet, Appl. Phys. Lett. **84**, 3118 (2004).
[58] G. D. Fuchs, N. C. Emley, I. N. Krivorotov, P. M. Braganca, E. M. Ryan, S. I. Kiselev, J. C. Sankey, D. C. Ralph, R. A. Buhrman, and J. A. Katine, Appl. Phys. Lett. **85**, 1205 (2004).
[59] L. Berger, J. Appl. Phys. **93**, 7693 (2003).
[60] I. N. Krivorotov, N. C. Emley, J. C. Sankey, S. I. Kiselev, D. C. Ralph, R. A. Buhrman, Science **307**, 228 (2005).
[61] S. Urazhdin, Norman O. Birge, W. P. Pratt, Jr., and J. Bass, Appl. Phys. Lett. **84**, 1516 (2004).
[62] J. Manschot, A. Brataas, and G. E. W. Bauer, Appl. Phys. Lett. **85**, 3250 (2004).
[63] Y. Jiang, S. Abe, T. Ochiai, T. Nozaki, A. Hirohata, N. Tezuka, and K. Inomata, Phys. Rev. Lett. **92**, 167204 (2004).
[64] J. M. Daughton, unpublished.
[65] D. Chiba, Y. Sato, T. Kita, F. Matsukura, and H. Ohno, Phys. Rev. Lett. **93**, 216602 (2004).
[66] T. J. Silva, W. H Rippard, and S. E. Russek, unpublished.
[67] W. H. Rippard, M. R. Pufall, and T. J. Silva, Appl. Phys. Lett. **82**, 1260 (2003).
[68] M. R. Pufall, W. H. Rippard, S. Kaka, S. E. Russek, and T. J. Silva, Phys. Rev. B **69**, 214409 (2004).
[69] A. V. Nazarov, H. S. Cho, J. Nowak, S. Stokes, and N. Tabat, Appl. Phys. Lett. **81**, 4559 (2002).
[70] J. G. Zhu and X. C. Zhu, IEEE Trans. Magn. **40**, 182 (2004).
[71] M. Covington, A. Rebei, G. J. Parker, and M. A. Seigler, Appl. Phys. Lett. **84**, 3103 (2004).
[72] S. K. Upadhyay, R. N. Louie, and R. A. Buhrman, Appl. Phys. Lett. **74**, 3881 (1999).
[73] J. Barnaś and A. Fert, Phys. Rev. Lett. **80**, 1058 (1998).
[74] S. Takahashi and S. Maekawa, Phys. Rev. Lett. **80**, 1758 (1998).
[75] A. Brataas, Y. V. Nazarov, J. Inoue, and G. E. W. Bauer, Phys. Rev. B **59**, 93 (1999).
[76] G. Usaj and H. U. Baranger, Phys. Rev. B **63**, 184418 (2001).
[77] J. Martinek, J. Barnaś, S. Maekawa, H. Schoeller, and G. Schön, Phys. Rev. B **66**, 014402 (2002).
[78] N. Sergueev, Q.-f. Sun, H. Guo, B. G. Wang, and J. Wang, Phys. Rev. B **65**, 165303 (2002).
[79] J. König and J. Martinek, Phys. Rev. Lett. **90**, 166602 (2003); M. Barun, J. König, and J. Martinek, Phys. Rev. B **70**, 195345 (2004).
[80] X. Waintal and P. W. Brouwer, Phys. Rev. Lett. **91**, 247201 (2003).
[81] M. Tinkham, *Introduction to Superconductivity, 2nd Edition*, (McGraw Hill, New York, 1996).

[82] K. Ono, H. Shimada, and Y. Ootuka, J. Phys. Soc. Jpn. **66**, 1261 (1997); **67**, 2852 (1998).
[83] A. H. MacDonald, in *Proceedings of the XVI Sitges Conference on Statistical Mechanics*, (Springer-Verlag, Berlin, 2000), p. 211.
[84] M. M. Deshmukh and D. C. Ralph, Phys. Rev. Lett. **89**, 266803 (2002).
[85] J. von Delft and D. C. Ralph, Phys. Rep. **345**, 61 (2001).
[86] S. Guéron, M. M. Deshmukh, E. B. Myers, and D. C. Ralph, Phys. Rev. Lett. **83**, 4148 (1999).
[87] M. M. Deshmukh, S. Kleff, S. Gueron, E. Bonet, A. N. Pasupathy, J. von Delft, and D. C. Ralph, Phys. Rev. Lett. **87**, 226801 (2001).
[88] C. M. Canali and A. H. MacDonald, Phys. Rev. Lett. **85**, 5623 (2000).
[89] C. M. Canali and A. H. MacDonald, Solid State Comm. **119**, 253 (2001).
[90] S. Kleff, J. von Delft, M. M. Deshmukh, and D. C. Ralph, Phys. Rev. B **64**, 220401 (2001).
[91] A. Cehovin, C. M. Canali, and A. H. MacDonald, Phys. Rev. B **66**, 094430 (2002).
[92] A. Cehovin, C. M. Canali, and A. H. MacDonald, Phys. Rev. B **68**, 014423 (2003).
[93] L. P. Kouwenhoven, D. G. Austing, and S. Tarucha, Rep. Prog. Phys. **64**, 701 (2001).
[94] R. Hanson, B. Witkamp, L. M. K. Vandersypen, L. H. Willems van Beveren, J. M. Elzerman, and L. P. Kouwenhoven, Phys. Rev. Lett. **91**, 196802 (2003).
[95] J. Park, A. N. Pasupathy, J. I. Goldsmith, C. Chang, Y. Yaish, J. R. Petta, M. Rinkoski, J. P. Sethna, H. D. Abruna, P. L. McEuen, D. C. Ralph, Nature **417**, 722 (2002).
[96] W. Liang, M. P. Shores, M. Bockrath, J. R. Long, H. Park, Nature **417**, 725 (2002).
[97] A. N. Pasupathy, R. C. Bialczak, J. Martinek, J. E. Grose, L. A. K. Donev, P. L. McEuen, and D. C. Ralph, Science **306**, 86 (2004).
[98] L. H. Yu and D. Natelson, Nano Lett. **4**, 79 (2004).
[99] M.-H. Jo, H. Park, *et al.*, communication.
[100] L. A. Bumm, J. J. Arnold, M. T. Cygan, T. D. Dunbar, T. P. Burgin, L. Jones, II, D. L. Allara, J. M. Tour, and P. S. Weiss, Science **271**, 1705 (1996).
[101] M. A. Reed, C. Zhou, C. J. Muller, T. P. Burgin, and J. M. Tour, Science **278**, 252 (1997).
[102] H. Park, A. K. L. Lim, and A. P. Alivisatos, J. Park, and P. L. McEuen Appl. Phys. Lett. **75**, 301 (1999).
[103] H. Park, J. Park, A. K. L. Lim, E. H. Anderson, A. P. Alivisatos, and P. L. McEuen Nature **407**, 57 (2000).
[104] A. N. Pasupathy, J. Park, C. Chang, A. V. Soldatov, S. Lebedkin, R. C. Bialczak, J. E. Grose, L. A. K. Donev, J. P. Sethna, D. C. Ralph, and P. L. McEuen, Nano Lett. **5**, 203 (2005).
[105] A. C. Hewson, *The Kondo Problem to Heavy Fermions* (Cambridge Univ. Press, Cambridge, 1993).

[106] D. Goldhaber-Gordon, H. Shtrikman, D. Mahalu, D. Abusch-Magder, U. Meirav, and M. A. Kastner, Nature **391**, 156 (1998).
[107] S. M. Cronenwett, T. H. Oosterkamp, and L. P. Kouwenhoven, Science **281**, 540 (1998).
[108] J. Nygard, D. H. Cobden, and P. E. Lindelof, Nature **408**, 342 (2000).
[109] M. R. Buitelaar, T. Nussbaumer, and C. Schonenberger, Phys. Rev. Lett. **89**, 256801 (2002).
[110] J. Martinek, Y. Utsumi, H. Imamura, J. Barnaś, S. Maekawa, J. König, and G. Schön, Phys. Rev. Lett. **91**, 127203 (2003).
[111] J. Martinek, M. Sindel, L. Borda, J. Barnaś, J. König, G. Schön, and J. von Delft, Phys. Rev. Lett. **91**, 247202 (2003).

6 Tunnel spin injectors

Xin Jiang and Stuart Parkin

6.1 Introduction

The manipulation of spin polarized current in metallic magnetic nanostructures, which gives rise to the phenomenon of giant magnetoresistance [1–8], has had a huge impact on the storage of digital information in recent years [9, 10]. The magnetic engineering of magnetic nanostructures, atomic layer by atomic layer, allowed for the development of the spin-valve sensor which is a highly sensitive detector of small magnetic fields, operating at room temperature and above [9]. Such sensors can detect much tinier quantities of magnetic material, in the form of magnetized regions within a magnetic film, than was previously possible and so enabled the storage capacity of hard disk drives to increase by more than four hundred-fold in a six year time frame after the introduction of spin-valve magnetic recording read head sensors into the marketplace in late 1997 by IBM. Notwithstanding the remarkable impact of the spin-valve sensor, its sensitivity as measured by its magnetoresistance has hardly changed in the past decade and a half since its invention (see Fig. 6.1).

The most important ingredients to build useful spin-valve devices, namely the use of ultrathin ferromagnetic layers comprised of cobalt or cobalt alloys sandwiched around non-magnetic spacer layers of copper [11], the importance of spin-dependent interface scattering and the consequent engineering of interfaces in spin-valve structures [12], and the use of oscillatory interlayer coupling to create artificial antiferromagnetic structures [13, 14], were realized more than fifteen years ago [9]. The limited magnetoresistance [12, 15, 16] displayed by spin-valve structures arises from the dilution of the polarized current in metallic structures by unpolarized electrons, for example, those in the non-magnetic spacer layers. By contrast, much more highly spin-polarized currents can be generated by extracting electrons from the surface of ferromagnetic or ferrimagnetic materials through vacuum, for example by applying electric fields in scanning tunneling microscopes [17] or by photo-exciting electrons in photoemission studies [18], or through tunnel barriers comprised of ultrathin insulating layers in solid state tunneling junction devices. The latter subject is the primary focus of this chapter.

In recent years there has been renewed interest in the phenomenon of spin-dependent tunneling particularly with regard to the property of tunneling magnetoresistance (TMR) exhibited by magnetic tunnel junctions (MTJs). As will be discussed in this chapter tunneling magnetoresistance is related to the de-

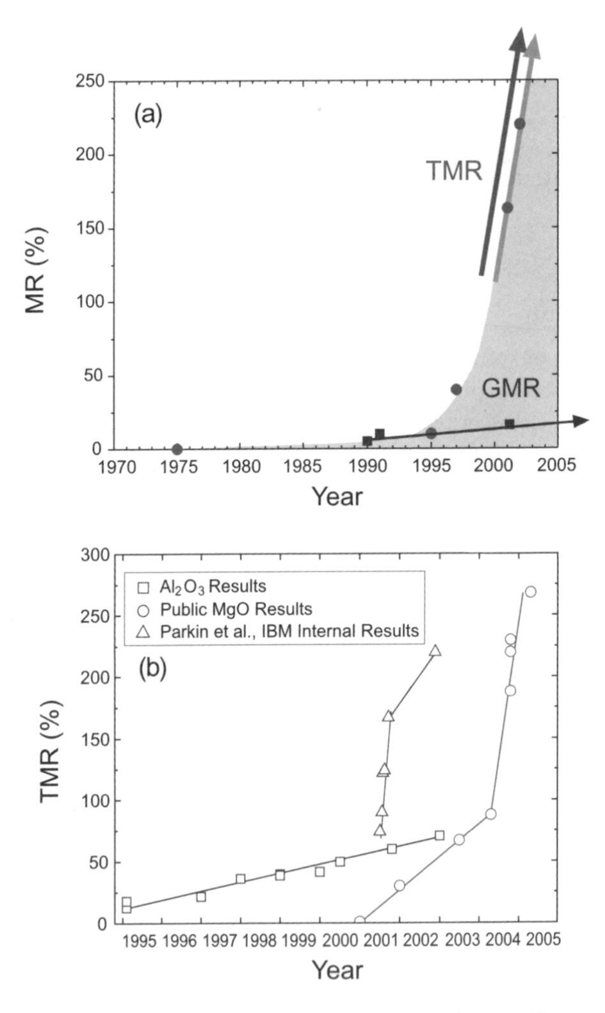

FIG. 6.1. (a) Evolution of the magnetoresistance of metallic exchange biased spin-valve sandwiches and magnetic tunnel junctions. (b) Evolution of the magnetoresistance of magnetic tunnel junctions with alumina and MgO tunnel barriers showing both publicly announced results and the first giant TMR results obtained for MTJs with MgO tunnel barriers by Parkin *et al.* at the IBM Almaden Research Center beginning in July 2001 which were only published several years later in December 2004.

gree of spin polarization of the tunneling current flowing through such devices. Although, for many years, tunneling spin polarization (TSP) was regarded as a unique property of a particular magnetic metal, it has more recently been appreciated that the spin polarization of the tunneling current is strongly influenced by chemical bond- and spin-dependent tunneling matrix elements and spin-filtering effects and thus is dependent on the detailed structure and morphology of the tunneling barrier and its interface with the magnetic material. This understanding has resulted in a rapid evolution of the magnitude of the room temperature tunneling spin polarization and tunneling magnetoresistance exhibited by devices with conventional $3d$ transition metal ferromagnetic electrodes in the past decade (see Fig. 6.1).

6.2 Magnetic tunnel junctions

A magnetic tunneling junction (MTJ) is comprised of a sandwich of two thin ferromagnetic or ferrimagnetic (F) layers separated by an ultrathin insulating layer [9, 19–23]. The most useful structures for studying the physics of tunneling or for technological applications are those in which the direction of the magnetization of each of the magnetic electrodes can be independently set. This is most simply accomplished using the magnetic engineering techniques developed for spin-valves [9]. In particular, one of the magnetic electrodes is coupled to a layer of an antiferromagnetic (AF) metal: for spin-valve structures, where the current flows in the plane of the device, it is preferred that the antiferromagnetic layer be non-conducting to reduce shunting of the electrical sense current through this layer, but for tunneling structures the AF layer must usually be conducting. The AF layer provides an exchange bias field [24, 25] at the interface with the ferromagnetic layer, most likely resulting from uncompensated magnetic moments at the AF/F interface [26–28]. The direction of the exchange bias field is set by cooling the AF/F bilayer in a magnetic field, sufficiently large to orient the F layer's moment, from above the blocking temperature of the AF layer. The blocking temperature is the temperature at which the AF domains are free to rotate and is typically lower than the AF Néel temperature.

Typical resistance versus field curves for an exchange biased MTJ are shown in Fig. 6.2. This MTJ has a tunnel barrier formed from amorphous alumina and magnetic electrodes formed from a crystalline $Co_{70}Fe_{30}$ layer, exchange biased with $Ir_{22}Mn_{78}$ (lower electrode-moment direction shown by the gray arrow) and an amorphous layer of $(Co_{70}Fe_{30})_{85}Zr_{15}$ (upper counter-electrode-moment direction shown by the dark arrow). The CoFeZr layer is magnetically very soft so that its moment can be switched between a direction parallel to the reference electrode, when the resistance of the device R_P is low, to a direction antiparallel to the reference electrode when the resistance R_{AP} is high, in small magnetic fields as low as a few Oersted. Indeed the TMR exhibited by this device is $\sim 66\%$ at room temperature and about 120% at 5 K. The TMR is defined as $TMR = (R_{AP} - R_P)/R_P$. The highest TMR values at room temperature using alumina tunnel barriers have, to date, been obtained using CoFeB electrodes [29]. These structures exhibit TMR values of $\sim 70\%$.

Co-Fe alloys can be made amorphous by adding various glass-forming elements, typically significantly smaller or larger atoms, such as B, Si, Hf and Zr or the rare-earth elements [30]. Interestingly, and perhaps surprisingly, the amorphization of ferromagnetic electrodes formed from Co-Fe results in higher tunnel

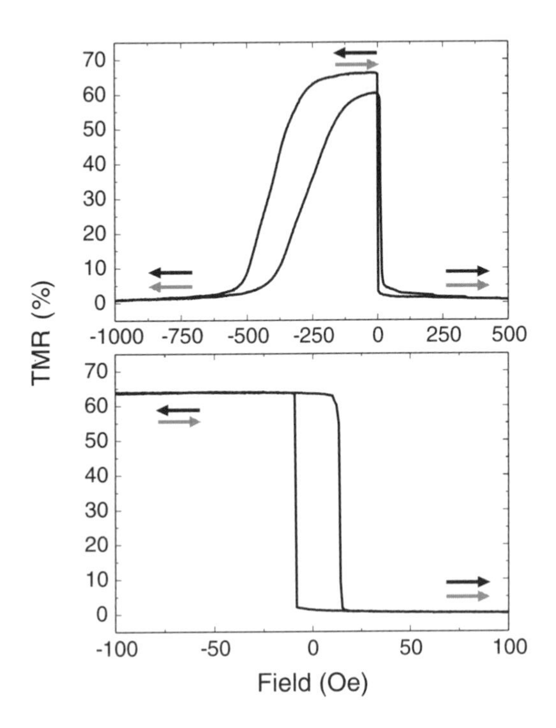

FIG. 6.2. Resistance versus field curves of a typical magnetic tunnel junction exhibiting ~66% tunneling magnetoresistance at room temperature.

magnetoresistance than the corresponding crystalline electrode even though the magnetization of the alloy is reduced (by dilution due to the addition of the glass-forming element). This is most likely a result of an increased density of states in the amorphous alloy due to energy band narrowing [31]. Indeed x-ray emission spectroscopy on MTJs with ultrathin CoFe interface layers sandwiched between amorphous alumina tunnel barriers and amorphous CoFeB layers shows that the CoFe interface layers, when amorphous, exhibit an increased density of states (apparently derived from the Fe constituent of the alloy). The CoFe interface layers are amorphous when sufficiently thin (less than ~20 Å thick), as seen by cross-section transmission electron microscopy [32], and it is when the CoFe layer is amorphous that higher TMR values are observed.

The evolution of the room-temperature TMR of MTJs is shown in Fig. 6.1. The first demonstration of spin-dependent tunneling in an MTJ was made by Julliere in 1975 [33]. Julliere observed a small TMR of ~14% at low temperature and very low junction bias in junctions with Co and Fe electrodes, and a tunnel barrier formed from Ge which was oxidized after deposition. Subsequently, other tunnel barriers including NiO [34] and GdO_x [35] were studied but only small TMR effects were found and only at low temperatures. It took 20 years after Julliere's work until significant room-temperature TMR (~10%) was observed in 1995 by two groups, simultaneously. These groups both used amorphous Al_2O_3 barriers but with different ferromagnetic electrodes composed of Fe, and CoFe and Co, respectively [22,36]. In between these results and the first work of Julliere there was prolonged work on magnetic tunnel junctions with alumina barriers and permalloy electrodes which had resulted in TMR values at room temperature of a few percent, several years earlier [19]. There had also been extensive studies of the magnitude of the spin polarization of tunneling current from ferromagnetic electrodes in tunnel junctions with superconducting and ferromagnetic counter-electrodes (FIS) using superconducting tunneling spectroscopy (STS) [37]. These studies had shown that the tunneling current was significantly spin-polarized for a wide range of transition metal and rare-earth metal ferromagnets.

As shown in Fig. 6.1(a) the TMR value of $\sim 10\%$ was similar to the giant magnetoresistance [38] exhibited by metallic spin-valve sandwich structures [2] which was perhaps one of the reasons why the observation of TMR of $\sim 10\%$ in 1995 exceeded a certain threshold value that energized renewed interest in MTJs. Moreover, there had long been interest in the United States in the possibility of building a high performance, non-volatile magnetic random access memory (MRAM) for military applications by the Defense Advanced Research Projects Agency (DARPA). In 1995 this agency decided to fund a program to develop devices using spintronic materials and concepts and in 1996 DARPA funded two independent programs at Motorola and Honeywell to build MRAM based on metallic spin-valve devices, and one program at IBM to research and develop an MRAM based on magnetic tunnel junctions.

In 1995 successful prototype magnetic recording read heads for high-density disk drive storage devices using spin-valve sensors had already been demonstrated [9, 39]. The MTJ structures at that time had very high resistances and it was widely believed that the resistance of MTJs would always be too high for use in magnetic field sensors in disk drives where the most important metric is the signal to noise ratio at high frequencies (up to ~ 1 GHz today). Thus, MTJs with resistance-area products in the range from 0.1 to a few $\Omega\,\mu m^2$ would likely be needed for magnetic bit areal densities in the 100 to 500 Gbit/in^2 range. At these densities it is considered that spin-valve sensors will likely have insufficient signal to noise to read the magnetic bits at the required data rates. However, the evolution of magnetic recording from conventional longitudinal (in-plane) recording to perpendicular recording may prolong the use of spin-valve sensors because of the greater flux from perpendicularly magnetized magnetic media [40, 41].

By contrast, for magnetic random access memory applications, much higher resistance devices are needed. In 1996 it was recognized that MTJs had significant potential for applications as storage elements in magnetic random access memories [23] but only if their TMR values could be significantly increased from what was then $\sim 10\%$ to more than at least $\sim 20\%$, and their resistance-area product decreased from what was then $\sim 10^9\,\Omega\,\mu m^2$ to $\sim 10^2$ to $10^3\,\Omega\,\mu m^2$ i.e. by a factor of by about 10,000,000 [42]. A number of significant milestones in

Table 6.1 History of developments in MTJs useful for applications.

Year	Development	Ref.
1960	Giaever (GE) – Tunneling probe of superconducting state	[164]
1971	Meservey and Tedrow (MIT) – Spin polarized tunneling between superconductors and ferromagnets	[165]
1974	Slonczewski (IBM) – MTJ concept proposed (internally in IBM)	
1975	Julliere (CNR-France) – First MTJ demonstration: Fe/Ge/Co, $\Delta R/R \sim 14\%$ at 4.2 K	[33]
1982	Maekawa and Gafvert (IBM) – MTJ demonstration: Ni/NiO/Ni, Fe, Co, $\Delta R/R \sim 0.4$–2% at 4.2 K	[34]
1990-1993	Miyazaki *et al.* (Tohoku University) – MTJ demonstration: NiFe/Al-Al_2O_3/Co, $\Delta R/R \sim 2.7\%$ at room temperature	[19,166]
1995	Miyazaki *et al.* (Tohoku University) – Large room-temperature TMR: Fe/Al-Al_2O_3/Co, $\Delta R/R \sim 18\%$ at room temperature Moodera *et al.* (MIT) – Large room-temperature TMR: Co-Fe/Al-Al_2O_3/Co, $\Delta R/R \sim 10\%$ at room temperature	[22, 36]
1996	Parkin *et al.* (IBM) – Large room temperature TMR in exchange biased MTJs with Al_2O_3 and CoFe electrodes: $\Delta R/R >$ 25% in shadow masked and patterned junctions; reproducible	[23,167]
1998	Parkin *et al.* (IBM) – Extraordinarily large room temperature TMR in exchange biased MTJs with Al_2O_3 and CoFe electrodes; high thermal stability: $\Delta R/R > 35\%$ in sub-micron junctions; $\Delta R/R > 47\%$ in shadow masked junctions with specific resistances ~ 60 to $> 10^9\,\Omega\,\mu m^2$; thermal stability ($>$ 300 °C)	[23]
1999-2000	Scheuerlein *et al.* (IBM) – First MTJ MRAM demonstration: < 3 ns read and write	[168]
2001-2002	Parkin *et al.* (IBM) – Giant tunneling magnetoresistance in highly textured (100) oriented MgO based MTJs: $\Delta R/R$ $\sim 220\%$ at room temperature, $\sim 300\%$ at 4K (IBM internal results)	
2002	Sony – Increased TMR at room temperature using CoFeB alloy electrodes and alumina tunnel barriers: $\Delta R/R \sim 70\%$ (Presented at Intermag 2002)	[169]
2002	Durlam *et al.* (Motorola) – 1 Mbit MRAM in 0.6 μm technology	
2003	Sitaram *et al.*; Bette *et al.* (IBM & Infineon) – 128 kbit MRAM core in 0.18 μm technology	[170]
2003	Durlam *et al.* (Motorola) – 4 Mbit MRAM in 0.18 μm technology using the toggle write mode	
2004	Debrosse *et al.* (IBM & Infineon) – 16 Mb MRAM in 0.18 μm technology	[171]
2004	Parkin *et al.* (IBM) – Giant TMR in MgO based MTJs published: $\Delta R/R \sim 220\%$ at room temperature, highly thermally stable	[63]
2004	Seagate – First MTJ read heads using leaky TiO_2 tunnel barriers shipped	
2005	Anelva / AIST – TMR $\sim 271\%$ at room temperature in MgO based MTJs (Presented at Intermag 2005 (FB05))	
2003-2005	Cypress and Freescale (Motorola) – First MRAM products sampled at 256 kbit and 4 Mbit, respectively, using alumina tunnel barriers	

¶ All MRAM demonstrations and product samples, to date, have used alumina tunnel barriers.

the development of magnetic tunnel junctions useful for applications are listed in Table 6.1.

6.2.1 *Tunneling spin polarization*

Julliere's work was predated by work by Tedrow and Merservey [43] in 1971 that showed that the tunneling current in tunneling junctions with very thin superconducting aluminum films and ferromagnetic nickel films was spin polarized. Meservey and Tedrow subsequently used this technique of superconducting tunneling spectroscopy (STS) to explore the relationship between the polarization of the tunneling current (at 0.25 K) and the magnetization of numerous ferromagnetic metals and alloys [37]. Julliere proposed that the TMR which he observed was related to the spin polarization $P_{1,2}$ of the two ferromagnetic electrodes of the MTJ according to the relationship TMR $= 2P_1P_2/(1 - P_1P_2)$, where $P_{1,2}$ are defined as

$$P_{1,2} = \frac{|M_\uparrow|^2 N_\uparrow - |M_\downarrow|^2 N_\downarrow}{|M_\uparrow|^2 N_\uparrow + |M_\downarrow|^2 N_\downarrow},$$

where $\uparrow\downarrow$ represent the up and down spin electrons in each of the ferromagnetic electrodes (see Fig. 6.3). The tunneling matrix elements for the spin-up and spin-down electrons $|M_{\uparrow,\downarrow}|^2$ represent their respective tunneling probabilities and $N_{\uparrow,\downarrow}$ their respective density of states at the Fermi energy in each of the ferromagnetic layers. This relationship is schematically shown in Fig. 6.3. This relationship only applies at low bias voltage and low temperature so that $P_{1,2}$ represent the polarization of the electrons at the Fermi energy.

The spin polarization of a ferromagnetic material can be measured in a variety of ways. The spin-dependent density of states can be directly probed using spin-polarized photoemission [18]. Using spin as well as angle and energy resolved electron detection the density of filled electron states in crystalline metals can be probed within each energy band and as a function of momentum within the Brillouin zone. Since the escape depth of the electrons is quite small this means that this technique is limited to exploring the electronic structure of the surface of magnetic metals. Moreover, the connection between the electronic structure measured from photoemission and the sign and magnitude of the tunneling spin polarization measured using STS is not direct. The TSP of all of the 3*d* transition metal ferromagnets is measured to be positive (majority spin polarized) yet the density of states at the Fermi energy is dominated by either minority spin polarized d states in hcp Co and fcc Ni (which are "strong" ferromagnets with largely filled majority bands) or by majority spin polarized d states in bcc Fe [44]. It was realized that the tunneling probability of the electrons would be related to the degree of localization of the electron wavefunctions so that the tunneling current would likely be dominated by the more itinerant *sp* electrons [45]. Thus there could be a very weak relationship between the density of electron states and the polarization of the tunneling current.

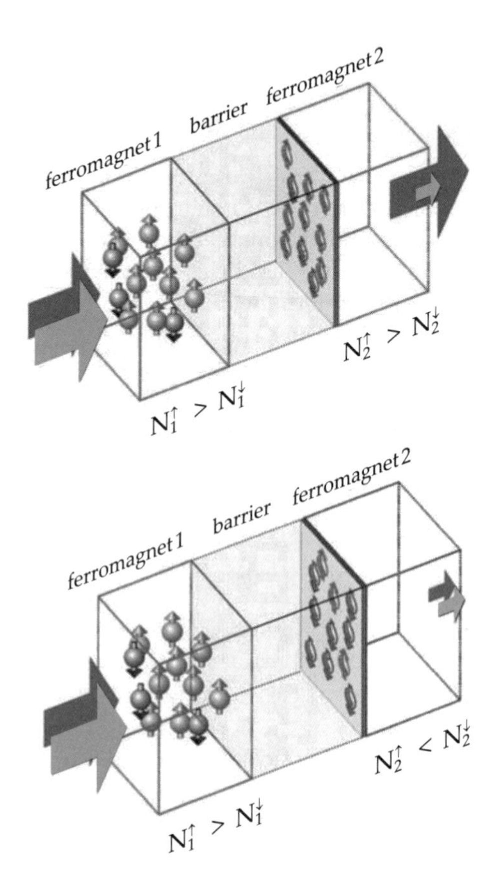

FIG. 6.3. Schematic cartoon showing the operation of a magnetic tunnel junction device.

The polarization of a ferromagnet can also be measured with superconducting point contacts using Andreev reflection spectroscopy [46–50]. This technique is discussed elsewhere in this book in Chapter 9 so will not be discussed further here. This method is far from measuring the spin polarization of the current in tunnel junctions since it relies on diffusive transport at a superconductor/ferromagnet contact. One further drawback of this technique is that it does not give the sign of the polarization. Perhaps the most useful technique for the measurement of the polarization of ferromagnets which is most relevant to spin-dependent tunneling is that of superconducting tunneling spectroscopy which is described in detail in the review by Merservey and Tedrow [37]. Thus only a very brief description will be given here.

In STS a tunneling junction is formed between the ferromagnet of interest and a superconducting layer of aluminum. In the presence of a large in-plane magnetic field the quasiparticle density of states in the superconductor is Zeeman-split providing for spin polarized states into which the electrons tunnel. Thus the superconducting electrode serves as an analyzer of the spin polarization of the tunneling current. By fitting the conductance versus bias voltage curves using the Maki density of states for BCS superconductors the TSP can be inferred with high precision [9, 37, 51, 52]. Spin-orbit mixing, magnetic field depairing and the superconducting energy gap are fitting parameters. The calculated conductance curves also depend on the measured field and temperature as well as the derived spin polarization of the tunneling electrons.

An important conclusion from the work of Tedrow and Meservey was that the magnitude of the TSP was linearly related to the magnetization of the ferromagnetic electrode for various series of $3d$ transition metal ferromagnetic alloys [53]. This conclusion was rationalized by arguments concerning the degree of itinerancy of the tunneling electrons and their magnetization [45]. However, the rekindled interest in magnetic tunneling junctions in the past decade has led to further studies of the TSP of $3d$ transition metal ferromagnets using both STS [9,54] and Andreev point contact spectroscopy [50]. These new studies have led to a very different and perhaps surprising conclusion. The more recent STS studies find that the spin polarization for a wide range of Ni-Fe and Co-Fe alloys varies little with the alloy composition and is about 45–55% [9, 54]. Andreev reflection spectroscopy measurements on Ni-Fe alloys also find very little variation of the polarization of the ferromagnet with the composition of the alloy [50]. The latter work by Nadgorny *et al.* finds support for their results from theoretical calculations of the density of states and the Fermi velocity of the electronic bands at the Fermi energy: variations of one compensate for variations of the other. The difference between these newer studies and the older STS studies suggests that the latter may have been affected, in some cases, by the quality of the tunnel junctions used. The polarization of the tunneling current is strongly influenced by the structure and composition of the interface between the tunnel barrier and the ferromagnet. Intermixing or reaction of the ferromagnet and the tunnel barrier material would very likely result in diminished polarization of the tunneling current. Nevertheless, the STS technique is a powerful method for probing the spin polarization of tunneling electrons and the quality of the interface between the ferromagnet and tunnel barrier.

The maximum TSP recorded for MTJs with alumina tunnel barriers is about 55% [9]. Although many other tunnel barriers have been studied the TSP values are typically much smaller with the sole exception, at room temperature, of crystalline MgO tunnel barriers which will be discussed in the following section.

6.2.2 *Giant tunneling using MgO tunnel barriers*

The magnitude of the TMR is directly related to the spin-polarization of the tunneling electrons [37], which itself is determined by the spin dependence of the density of states near the Fermi energy of each of the F electrodes, and the tunneling matrix elements for these electrons [51]. Thus, the higher the tunneling spin polarization (TSP), the higher the TMR, so that there has been

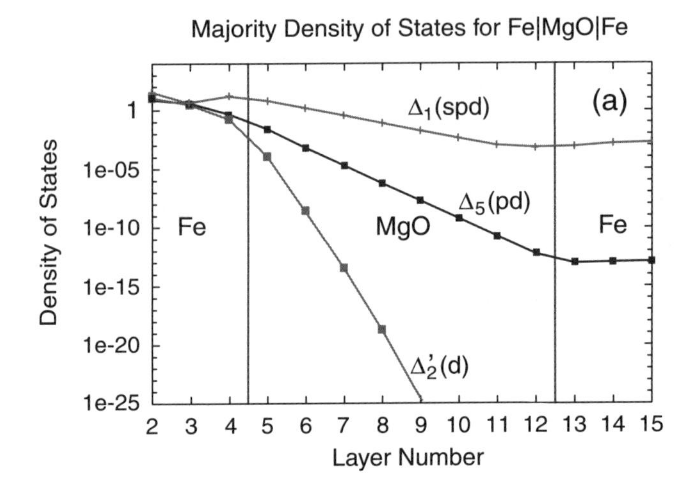

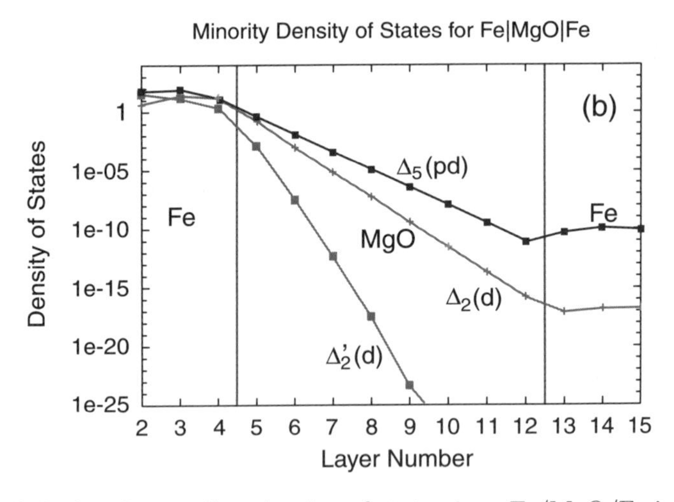

FIG. 6.4. Calculated tunneling density of states in a Fe/MgO/Fe junction for the majority (a) and minority (b) bands according to Butler *et al.* [61].

considerable interest in nominally half-metallic ferromagnetic electrodes, such as the manganite perovskites [55–57] and CrO_2 [58]. However, whilst TMR values of $\sim 1000\%$ [55, 59] and TSP values of $\sim 70\%$ [52] have been measured at low temperatures in tunnel junctions with electrodes formed from the perovskite manganites, very small effects are found at room temperature, only partly accounted for by the low Curie temperature of these ferromagnets. By contrast, conventional ferromagnetic metals formed from Fe, Co and Ni have much higher Curie temperatures, well above room temperature. Just as important, by using the phenomenon of oscillatory interlayer coupling in transition metal multilayers [13], useful magnetic structures for technological applications can be readily magnetically engineered from thin layers of these ferromagnets [9]. However, as discussed above, after several decades of work, the highest TSP found in tunnel junctions incorporating these metals is $\sim 55\%$ for alumina tunnel barriers [9,37], with corresponding TMR values in MTJs of up to $\sim 70\%$ at room temperature.

Alumina tunnel barriers are amorphous. It has recently been speculated that crystalline tunnel barriers may give rise to much higher TSP and TMR values because of a highly spin-dependent evanescent decay of certain wavefunctions, with particular transverse momentum values, across the tunnel barrier [60]. In particular, calculations for perfectly ordered (100) oriented Fe/MgO/Fe MTJs, suggest TMR values of hundreds or even thousands of percent, for sufficiently thick MgO tunnel barriers [61,62] (see Fig. 6.4). Recently, support for these predictions has been found in experimental observations, first by Parkin *et al.* [63] and subsequently by Yuasa *et al.* [64] of giant values of tunneling magnetoresistance at room temperature in MTJs with MgO tunnel barriers. Parkin *et al.* [63] report values of TMR at room temperature of more than 220% in highly textured polycrystalline MTJs prepared on amorphous SiO_2 layers by magnetron sputter deposition, while Yuasa *et al.* [64] report TMR values of up to 180% at room temperature in epitaxial Fe/MgO/Fe sandwiches prepared by molecular beam epitaxy.

Detailed results will be discussed for sputter-deposited highly textured MTJs with (100) oriented MgO tunnel barriers from the work by Parkin *et al.* [63] which was the first observation of large TMR at room temperature exceeding 100% (see Fig. 6.1 b). Typical resistance versus field curves are shown in Fig. 6.5 for two similar sputter deposited MTJs with lower ferromagnetic electrodes formed from Fe and $Co_{70}Fe_{30}$, patterned by in-situ shadow-masks. Very large TMR values of more than 165% at room temperature are found. The MTJs were prepared, nominally at ambient temperature, by sputter deposition using a combination of ion-beam and magnetron sputtering. The lower ferromagnetic electrode is formed by first depositing an underlayer of 100 Å TaN on an amorphous layer of SiO_2 formed on a Si(100) substrate. An antiferromagnetic layer of 250 Å $Ir_{22}Mn_{78}$, which is used to exchange bias the lower F layer [9], is then deposited by ion beam sputtering, followed by the F layer, which is formed from a bilayer of 8 Å $Co_{84}Fe_{16}$ and either 18 Å Fe (a and b) or 30 Å $Co_{70}Fe_{30}$ (c and d). The highest TMR values were found for MTJs with Co-rich Co-Fe electrodes. MTJs with Fe electrodes systematically gave lower TMR values.

The MgO tunnel barrier is prepared by first depositing a thin layer of Mg metal, followed by an MgO layer formed by reactive sputtering of Mg in an Ar-O_2 plasma. The Mg layer is sufficiently thick to prevent oxidation of the underlying F layer but thin enough that it is fully oxidized during the MgO layer deposition. Figure 6.6 shows a cartoon comparing the formation of alumina and MgO layers by oxidation of respective metallic layers of Al and Mg. This latter method, which is the preferred method of forming alumina, is not suitable for the formation of pin-hole-free tunnel barriers of MgO for the reason shown in the figure. Whereas the volume of a metallic layer of aluminium increases when fully oxidized by

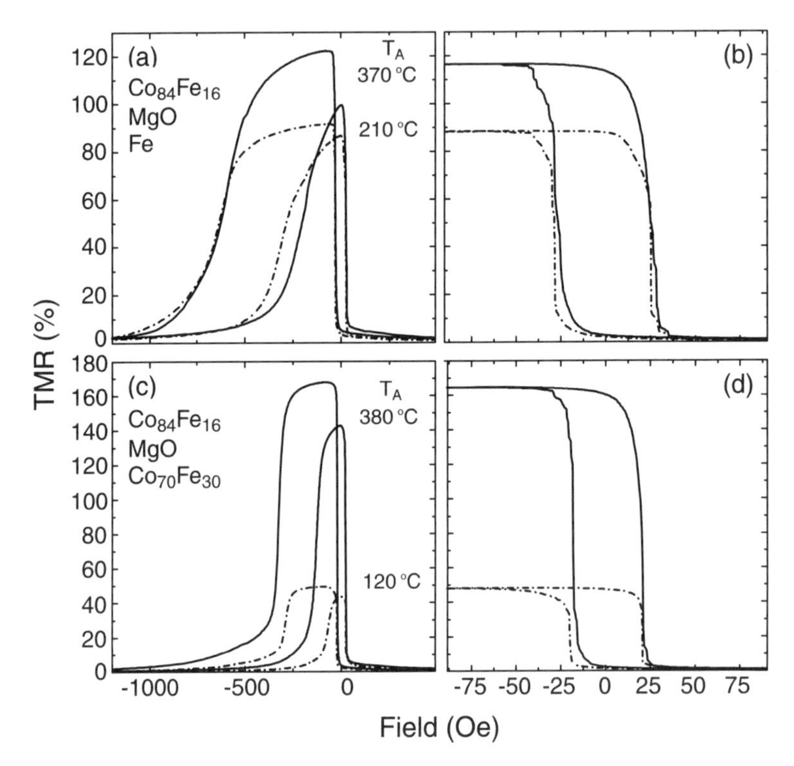

FIG. 6.5. Room-temperature tunneling magnetoresistance (TMR) versus field plots for magnetic tunnel junctions with structures as follows: (a) and (b) 100 TaN/250 IrMn/8 $Co_{84}Fe_{16}$/18 Fe/5 Mg/22 MgO/100 $Co_{84}Fe_{16}$/100 TaN, (c) and (d) 100 TaN/250 IrMn/8 $Co_{84}Fe_{16}$/30 $Co_{70}Fe_{30}$/5 Mg/24 MgO/150 $Co_{84}Fe_{16}$/100 Mg (all thickness in Å). The corresponding final anneal temperature T_A, after which the data is measured at room temperature, is shown in the figure. The field range in (a) and (c) is sufficiently broad that both the switching of the exchanged biased lower ferromagnetic electrode and the upper ferromagnetic electrodes are seen, whereas data in (b) and (d) correspond to a minor hysteresis loop where the field range is limited so that only the upper electrode switches. The exchange bias field is larger in (a) than in (c) because of the thinner F electrode in (a).

about 27% the volume of a layer of Mg decreases by about 20% so leading to the likely possibility of pin-holes in the MgO barrier.

The upper F electrode of the MTJs of Fig. 6.5 are formed from a layer of $Co_{84}Fe_{16}$ with capping layers formed from TaN or Mg. The TMR of the MTJs, as deposited, is modest but is dramatically increased by thermal annealing at temperatures of up to ~ 400 °C. Typically, the TMR increases monotonically and almost linearly with increasing anneal temperature, whereas the resistance of the junction changes little up to a critical temperature, in the range ~ 350–425 °C.

Forming Al_2O_3 by oxidation of metallic Al layer → high quality barrier

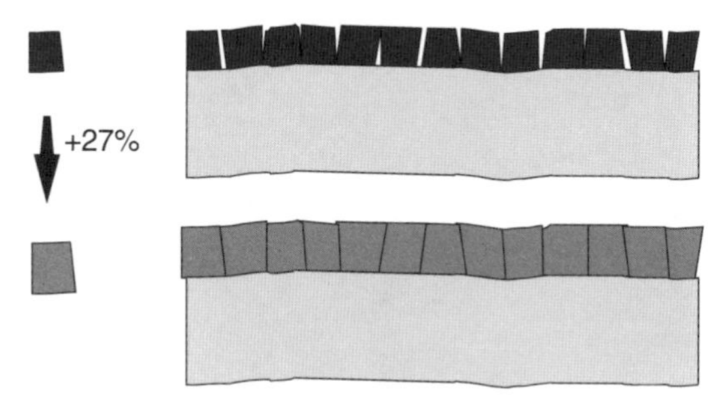

Al_2O_3 unit cell is 27% larger than that of metallic Al, so a metallic layer with pinholes between the grains can still yield a continuous oxide layer with post-deposition oxidation.

Forming MgO by oxidation of metallic Mg layer → poor quality barrier

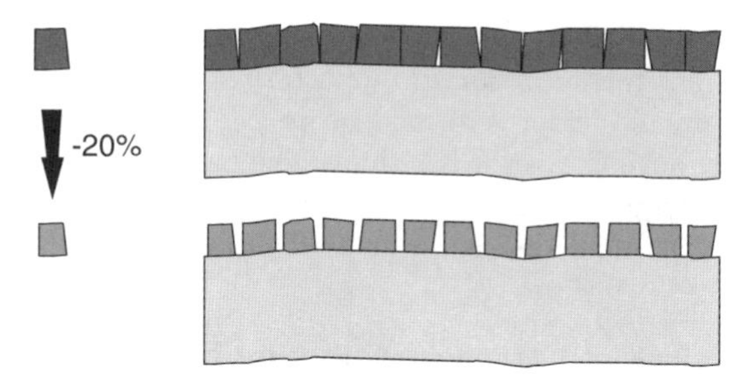

MgO unit cell is 20% smaller than that of metallic Mg, so even a conformal, nearly perfectly continuous metallic layer can yield a discontinuous MgO layer with post-deposition oxidation.

FIG. 6.6. Schematic illustration of the formation of aluminium oxide (upper) and a MgO (lower) layers by the oxidation of thin metal layers of aluminium and magnesium, respectively. The volume of an oxidized layer of Al composed of Al_2O_3 is increased by $\sim$27% whereas the volume of a fully oxidized layer of Mg is actually reduced in volume by $\sim$20%. This means that the alumina layer is much less likely to contain "cracks" or pin-holes than the MgO layer. Thus it is very difficult to form high-quality MgO tunnel barriers by the oxidation of Mg layers whereas the same method produces high-quality pin-hole-free tunnel barriers of alumina.

At this temperature the junction breaks down with a loss of both resistance and TMR. The two junctions shown in Fig. 6.5 survived to anneal temperatures of $\sim$370 °C and 380 °C, respectively, with respective TMR values of 123% and 168%.

The dependence of TMR on MgO barrier thickness was explored using both

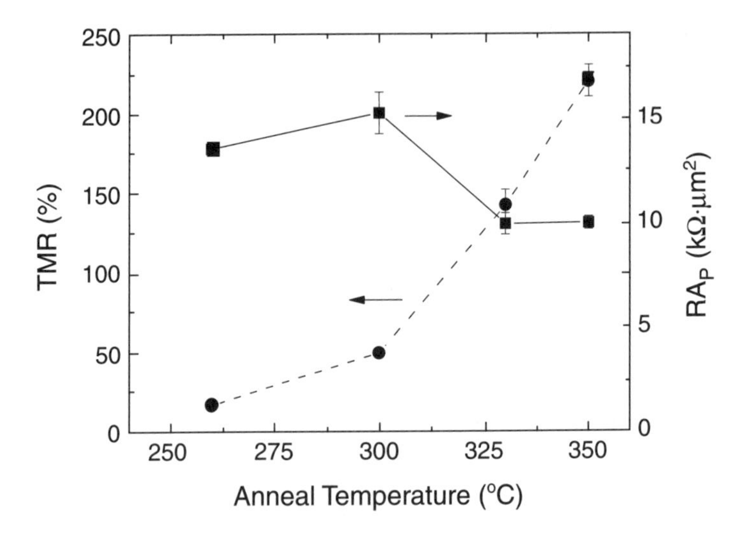

FIG. 6.7. Dependence on anneal temperature of TMR and RA in the parallel state, RA_{P}, measured at room temperature using the current-in-plane tunneling measurement technique on an unpatterned magnetic tunnel junction film. The MgO tunnel barrier is formed by first depositing a Mg layer ~ 12 Å thick, followed by a MgO layer ~ 8 Å thick. The parallel state resistance-area product is $\sim 10^4\,\Omega\,\mu\mathrm{m}^2$, which is about 20 times smaller than the sample of Fig. 6.5(c) and (d). After annealing at 350 °C the TMR attains a value of $220 \pm 10\%$.

shadow-masked junctions, and, at room temperature, by the technique of current-in-plane tunneling (CIPT) [65] on un-patterned films where the MgO thickness was linearly increased across the wafer. The CIPT technique allows the study of junctions with much lower resistance-area (RA) products (thinner MgO barriers). Both techniques give similar TMR values for MTJs with similar RA values ($\sim 160\%$ for $RA \sim 10^5\,\Omega\,\mu\mathrm{m}^2$) but the CIPT results reveal high TMR values even for RA values decreased to as little as $\sim 100\,\Omega\,\mu\mathrm{m}^2$. TMR values as high as $220 \pm 10\%$ are found for MTJ stacks with $RA \sim 10^4\,\Omega\,\mu\mathrm{m}^2$ after an anneal at 350 °C, as shown in Fig. 6.7.

Figure 6.8 shows the TMR and RA of the junction of Fig. 6.5(c) and (d) as a function of temperature. The TMR increases as the temperature is reduced attaining a value of $\sim 300\%$ at 4 K. The increase in TMR results from an increase of the resistance of the AP state of the junction whereas the P state resistance hardly changes at all on cooling. With increasing temperature, increasing magnetic disorder will decrease R_{AP} but should increase R_{P}, whereas, by contrast, thermal excitations across the barrier will decrease both R_{AP} and R_{P}. Thus, the weak dependence of R_{P} is likely a coincidental cancellation of these two effects.

The high TMR values indicate, within the Julliere model [33, 37, 51], high TSP values. The magnitude of the spin polarization of the tunneling current was

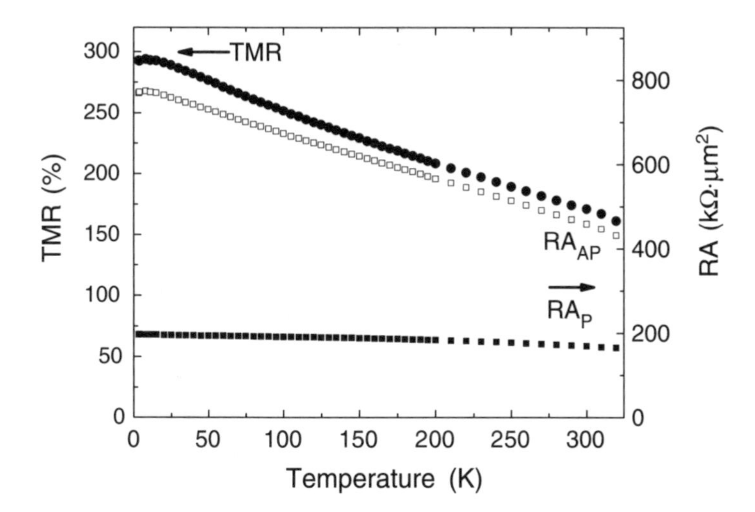

FIG. 6.8. TMR (•) and RA (■□) versus temperature for the same magnetic tunnel junction shown in Fig. 6.5(c) and (d) (annealed at 380 °C) in magnetic fields of −100 and +1000 Oe to set the state of the junction in the antiparallel (AP) and parallel (P) states, respectively. The TMR values are extracted from measuring the resistance versus field loop at each temperature.

measured using the superconducting tunneling spectroscopy technique (STS) [37]. Results for Fe and $Co_{70}Fe_{30}$ electrodes are shown in Fig. 6.9. As deposited, these electrodes have TSP values of ~57% and 52%, but after annealing, these values increase dramatically, attaining values of ~74% and 85% at anneal temperatures of 380 and 410 °C, respectively. These values are almost twice as high as those found for the same ferromagnets with alumina tunnel barriers [9, 37]: these results unambiguously demonstrate that a ferromagnetic metal does not have a unique spin-polarization value, as has long been assumed. Similar conclusions have previously been inferred from changes in the sign of TMR for junctions containing $SrTiO_3$ barriers and Co and manganite ferromagnetic electrodes [66].

Transmission electron micrographs show an excellent morphology of the MTJ structures with extremely smooth and flat layers as shown in the micrograph in Fig. 6.10(a). However, a significant numbers of defects along the (111) planes in the IrMn layer, most likely stacking faults, can be seen in the micrograph. Diffractograms (not shown) from selected areas of the digital micrograph reveal that the IrMn layer is fcc with a (100) texture. High-resolution micrographs show a high degree of epitaxy of the MgO and upper and lower Co-Fe layers as shown in Fig. 6.10(b). Both the Co-Fe layers are bcc with a (100) texture and the MgO is cubic (NaCl structure) and also (100) textured. The texturing of the MgO layer is good but is not perfect as shown by the rotated grain in the micrograph in Fig. 6.10(c).

The texture of the MTJ is extremely sensitive to the underlayers on which

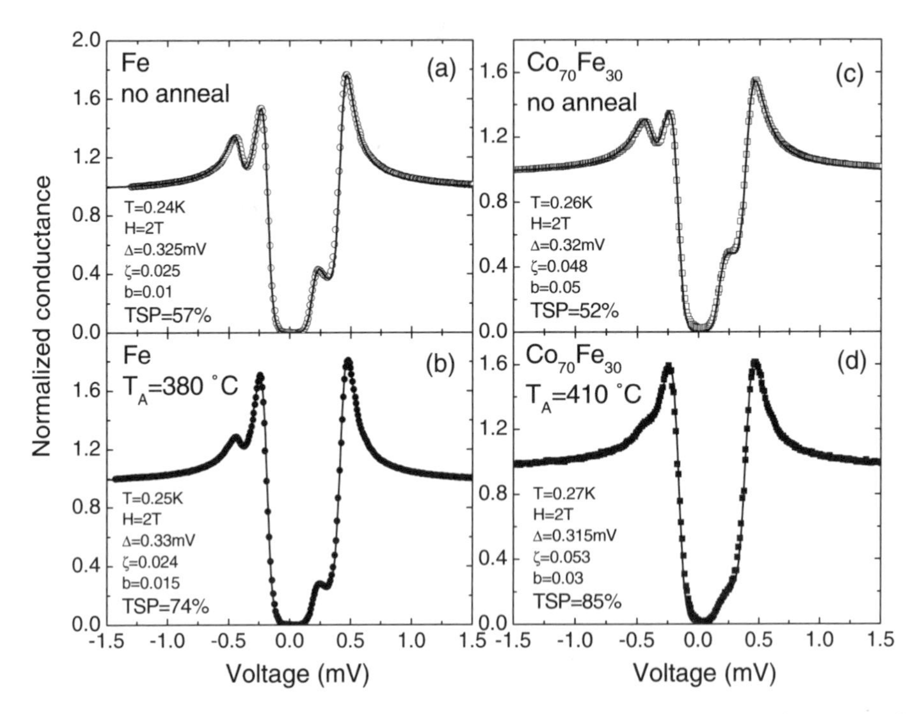

FIG. 6.9. Conductance versus bias voltage curves (symbols) and fits (solid lines) for STS junctions with (a) Fe and (b) $Co_{70}Fe_{30}$ ferromagnetic electrodes and superconducting counter-electrodes of $Al_{96}Si_4$. (a) and (b) correspond to the as-deposited junctions (no anneal) and (c) and (d) to junctions annealed at 380 °C and 410 °C, respectively. The measurements were taken at ~ 0.25 K in a field of 2 T applied in the plane of the films. The values for the tunneling spin polarization (TSP) were extracted by fitting the data curves with the following fitting parameters [37, 51, 52] indicated in the figure: superconducting gap Δ, depairing parameter ζ, and spin-orbit parameter b.

the IrMn layer is grown. The orientation of the IrMn layer is (100) for underlayers formed from Ta, TaN or TaN/Ta bilayers but can be changed to (111) for underlayers such as Ta/Pt or Ti/Pd. In the latter case, bcc CoFe layers grow (110) oriented and the MgO layer is also (110) oriented. CoFe layers form a bcc structure for Fe contents ranging from about 14 to 100 atomic % Fe. For these (110) oriented structures much lower TMR values are found, comparable in magnitude to those found for amorphous alumina tunnel barriers. Similarly, when the ferromagnet is not bcc structured, as for Co and Co-Fe alloys with low Fe content ($\sim$10 atomic %), similarly low TMR values are found. Thus, the magnitude of the TMR is strongly affected by the crystal orientation and epitaxy of the CoFe/MgO/CoFe sandwich, consistent with theoretical predictions for Fe/MgO/Fe [61]. However, the highest TMR (and TSP) values are found, not

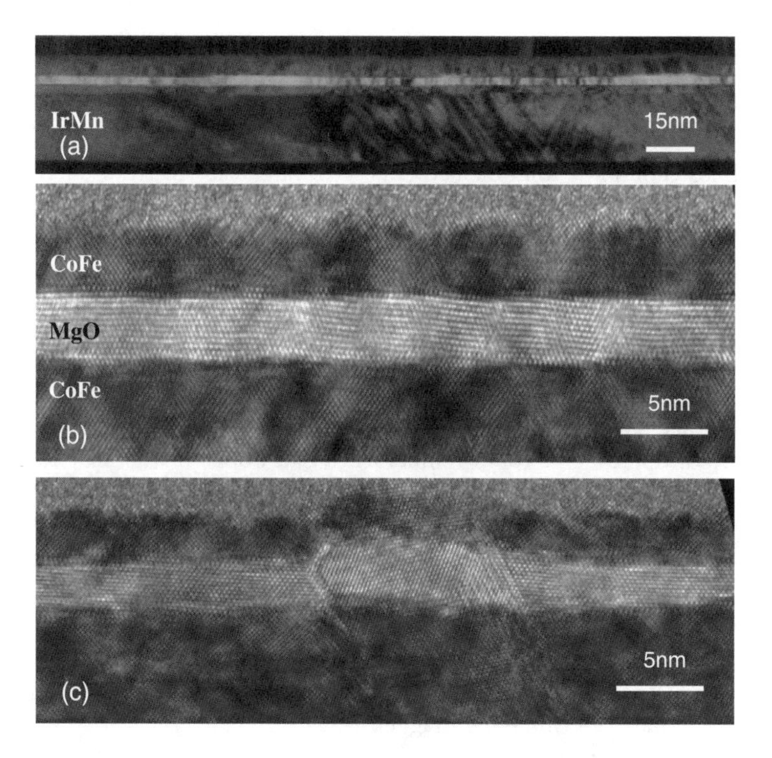

FIG. 6.10. Transmission electron micrographs of magnetic tunnel junctions showing a highly oriented (100) MgO tunnel barrier. (a) Low magnification image showing the growth of ultrasmooth underlayers formed from TaN, Ta, IrMn and CoFe, each readily distinguishable, which form a template for the growth of the (100) oriented MgO tunnel barrier (lightest layer); (b) and (c) high-resolution images along the [110] zone axes showing atomically resolved lattice planes with (100) planes perpendicular to the growth direction. The (100) planes in the grain in the center of (c) are rotated by $\sim -15°$.

for Fe electrodes, but for Co-rich Co-Fe electrodes with Fe contents in the range of 14–40 atomic%. This appears to be because of lower thermal stability for Fe and Fe-rich Co-Fe electrodes: the method we have outlined for the formation of high-quality MgO tunnel barriers requires high-temperature anneals, presumably to fully oxidize the Mg underlayer and to increase the crystalline perfection and orientation of the MgO barrier. Interestingly, even small amounts of disorder within the tunnel barrier or at its interfaces is theorized to significantly reduce both TSP and TMR [67, 68].

Using the Julliere model [33, 37, 51] a TSP of 85% would give rise to a TMR of $\sim$520% at low temperatures. Furthermore, if we assume a similar temperature dependence to that shown in Fig. 6.8, TMR values of more than 260% at room temperature are predicted from the TSP values. Such TMR values are very close to the highest TMR values which have been measured in MTJs. In any case STS measurements will likely give rise to higher TSP values than those inferred from TMR, since the AP alignment of the ferromagnetic electrodes in the latter is unlikely to be perfect, whereas the large magnetic fields used in the TSP studies ensure almost complete alignment of the ferromagnetic electrode moment with the magnetic field.

The results discussed here for sputter deposited (100) oriented CoFe/MgO/Co Fe magnetic tunneling junctions with MgO tunnel barriers are quite remarkable with giant tunneling magnetoresistance values of up to $\sim 220\%$ at room temperature. Improved crystal perfection and orientation is very likely to lead to even higher TMR values. The simplicity and ease of fabrication of such junctions, together with the ability to build complex magnetically engineered structures from these materials, suggests that these materials will have a major impact on spintronic devices operable at room temperature and above in the near future. For example, higher signal strengths from higher TMR values might make easier the implementation of advanced MRAM architectures such as ultradense cross-point random access memory [69].

6.3 Magnetic tunnel transistor

The magneto-transport properties of the magnetic tunneling junction discussed in the previous section depend strongly on the bias voltage applied across the device. In particular, the tunneling magnetoresistance can vary strongly with bias voltage, typically decreasing with increasing bias. The bias voltage dependence is influenced by the spin-dependent character and density of the electronic states in the ferromagnetic electrodes as well as by the spin-dependent tunneling matrix elements, each of which will vary as a function of increasing energy of the electrons above the Fermi energy. Effectively, the spin polarization of the tunneling electrons varies with bias voltage. Electrons with energies above the Fermi energy are described as "hot" electrons since they occupy non-thermalized states at an effective temperature higher than the actual equilibrium temperature. The spin-dependent transport properties of such electrons have been extensively studied using techniques such as ballistic electron emission microscopy (BEEM) and spin polarized photoemission (SPPE) [70–84].

The scattering of hot electrons can be quite different from that of electrons at or close to the Fermi energy since the scattering processes can involve the emission of energetic phonons or magnons. In ferromagnetic metals it is well known that the mean free path of the electrons at the Fermi energy is spin dependent [85]. The minority electrons usually have a smaller mean free path than the majority electrons, i.e., the minority electrons are more easily scattered than the majority electrons.

6.3.1 *Hot electron devices*

Spin-dependent hot electron transport in ferromagnetic metals has been extensively investigated particularly as it may be important for the functional-

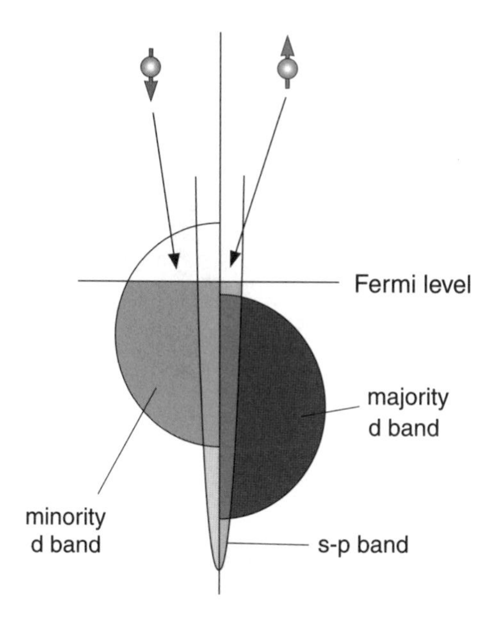

FIG. 6.11. Illustration of spin-dependent electron scattering in $3d$ transition metals. The horizontal and vertical axes represent density of states and electron energy, respectively. The filled region represents occupied states. Only the minority electrons (spin-down) can be effectively scattered into the empty minority d-band.

ity of many magneto-electronic devices [70–84]. The spin dependence of hot-electron scattering in ferromagnetic metals is often considered as a consequence of electron-electron scattering. In ferromagnetic $3d$ transition metals, the electron d-band is split into the majority and minority bands. A portion of the minority electron band is empty, whilst the majority electron band is almost fully occupied. As a result, minority electron scattering into the empty d-band is very effective, giving rise to a shorter mean free path. In contrast, the majority electrons can only be scattered into the less abundant sp band and thus have a larger mean free path. This simple picture is illustrated in Fig. 6.11.

Various techniques have been used to study hot-electron transport in ferromagnetic metals. These techniques usually require a source of hot electrons with certain energy, a medium in which the electrons travel, and an analyzer of the transmitted electrons. Commonly used electron sources include a semiconductor photocathode, a scanning tunneling microscopy (STM) tip, a metal/semiconductor Schottky barrier, or a metal/insulator tunnel barrier. The transmission medium normally contains a single ferromagnetic metal layer or a ferromagnetic multilayered structure. The electron analyzer may be a detector of the electron spin polarization and intensity, such as the Mott detector, or simply an energy analyzer, such as a metal/semiconductor Schottky barrier.

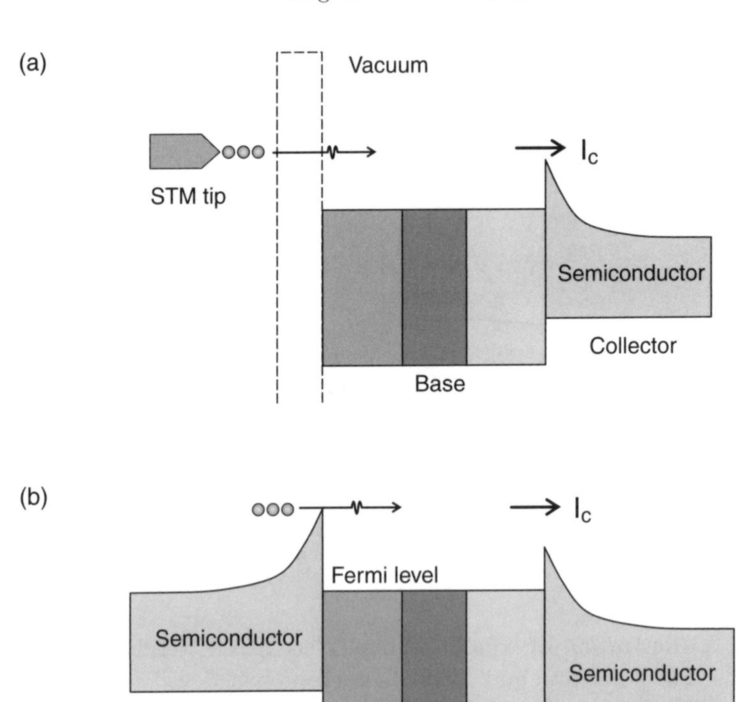

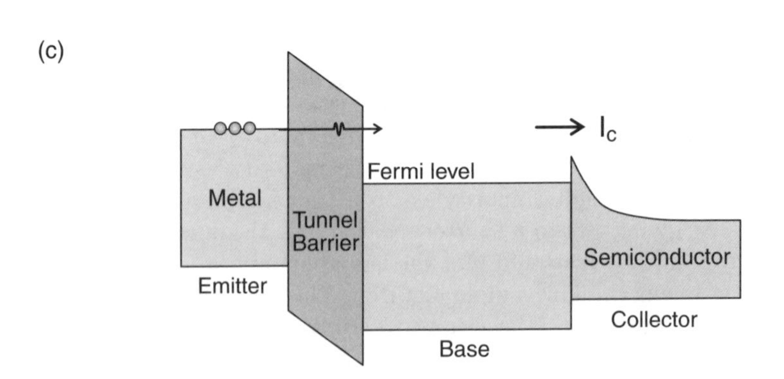

FIG. 6.12. Schematic drawings of the BEEM (a), the spin-valve transistor (b), and the magnetic tunnel transistor (c).

One such technique is ballistic electron emission microscopy (BEEM) [86]. In BEEM, an STM tip is positioned above a metal/semiconductor Schottky barrier, as shown in Fig. 6.12(a). Hot electrons are generated by tunneling from the STM tip into the metal base through the vacuum barrier. If these electrons have energy larger than the Schottky barrier height at the metal/semiconductor interface, they can ballistically traverse the base layer and be collected into the semiconductor collector. By analyzing the collector current (I_C), one can extract useful information about hot-electron transport in the metal base and at the metal/semiconductor interface. BEEM was first introduced by Kaiser and Bell as a means of investigating subsurface interface electronic structures in metal/semiconductor heterostructures. The metal base often consists of noble metals, such as Au and Cu, which have very large electron mean free paths. Later on, Rippard and Buhrman developed a variation of BEEM, where they replaced the noble metal base with a ferromagnetic single layer or multilayers [87]. This variation is sometimes called the ballistic electron magnetic microscopy (BEMM). Since the collector current is very sensitive to hot-electron transport properties in the base, BEEM can be used to study spin-polarized hot-electron transport in ferromagnetic metals [80]. Figure 6.13 shows BEEM images of Co/Cu multilayers grown on Si. It can be seen that BEEM has excellent spatial resolution of a few nanometers, thus allowing the imaging of local magnetic structures and hot-electron transport. In addition, the hot-electron energy can be easily controlled by varying the bias voltage between the STM tip and the metal base.

Although BEEM is a very useful technique, the use of an STM tip as the electron source makes it impractical for device applications. In 1995, Monsma *et al.* introduced a three-terminal solid state hot electron device – the spin-valve transistor [82] (Fig. 6.12b). Similar to BEEM, the spin-valve transistor has a metal base and a semiconductor collector. But instead of an STM tip, a semiconductor emitter is used to create hot electrons by thermionic emission from the emitter Schottky barrier. These electrons subsequently travel across the base layers and are collected by the collector Schottky barrier. To operate the spin-valve transistor, the emitter Schottky barrier height must be larger than the collector Schottky barrier height. This can be realized by inserting different metal seed layers at the emitter/base and base/collector interfaces. The metal base is comprised of a ferromagnetic metal/normal-metal/ferromagnetic metal spin-valve. Very large magnetic field sensitivity has been demonstrated for spin-valve transistors at room temperature [88, 89]. A detailed discussion of the spin-valve transistor can be found in the review article by Jansen [90].

In the spin-valve transistor, the hot-electron energy is determined by the emitter Schottky barrier height. Thus it is not possible to vary the electron energy continuously in one single device. In addition, the difference between the emitter and collector Schottky barrier heights is typically small. Therefore, the collection efficiency of the spin-valve transistor is limited. In order to overcome these problems, a magnetic tunnel transistor is developed to study spin-dependent hot-electron transport in ferromagnetic metals [91–93]. One form of magnetic tunnel transistor consists of a ferromagnetic metal emitter, a ferromagnetic metal base, and a semiconductor collector (Fig. 6.12c). The emitter and the base are separated by a thin insulating tunnel barrier. A Schottky barrier is formed between the base and the collector with a barrier height Φ_S. Hot electrons are injected

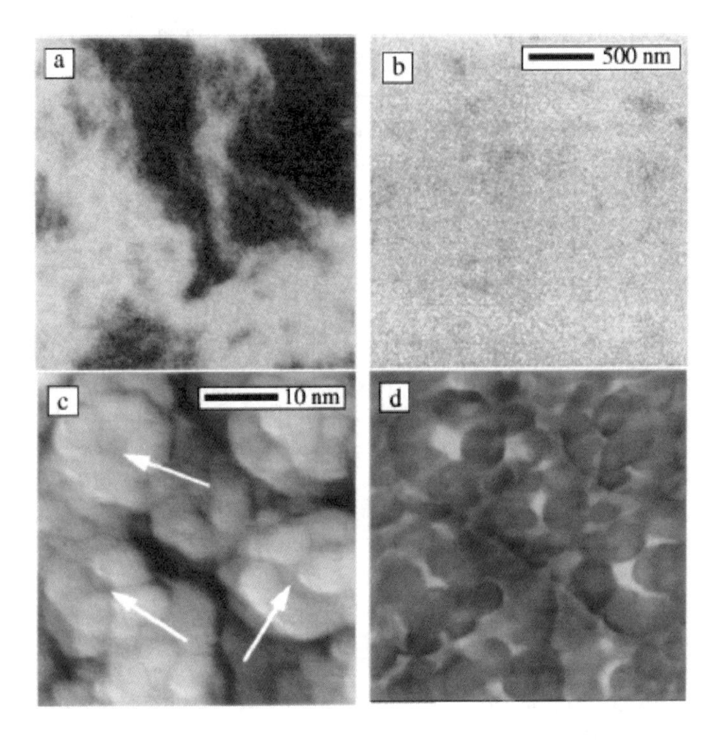

FIG. 6.13. BEEM images taken by Rippard and Buhrman[80] before (a) and after (b) applying a 60 Oe magnetic field. The sample is Co (30-Å)/Cu (45-Å)/Co (30-Å)/Cu (9-Å)/Au (75-Å) grown on Si. (c) and (d) are the height image and BEMM image of a Co (3-Å)/Cu (30-Å)/Au (100-Å) film grown on Si, respectively.

from the emitter into the base when an emitter/base bias voltage (V_{EB}) is applied across the tunnel barrier, forming the emitter current (I_{E}). These electrons traverse the ferromagnetic base layer and get scattered. The transmitted electron beam is subsequently analyzed by the semiconductor collector: when the electrons have energy higher than the Schottky barrier height, they can be collected into the semiconductor collector and contribute to the collector current; otherwise they will be reflected by the Schottky barrier. Since the minority electrons have a short mean free path, they are easily scattered in the base layer. As a result, they lose energy and cannot be collected. In contrast, the majority are more likely to maintain their energy due to their larger mean free path. Therefore, they are more likely to overcome the Schottky barrier and be collected. This spin-dependent transmission of hot electrons is called spin filtering since the minority electrons are selectively "filtered" out from the hot-electron current after transmission through the ferromagnetic base layer.

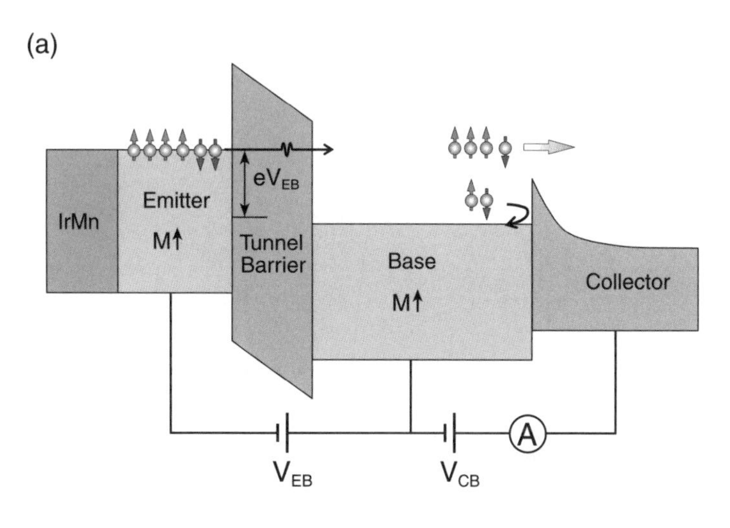

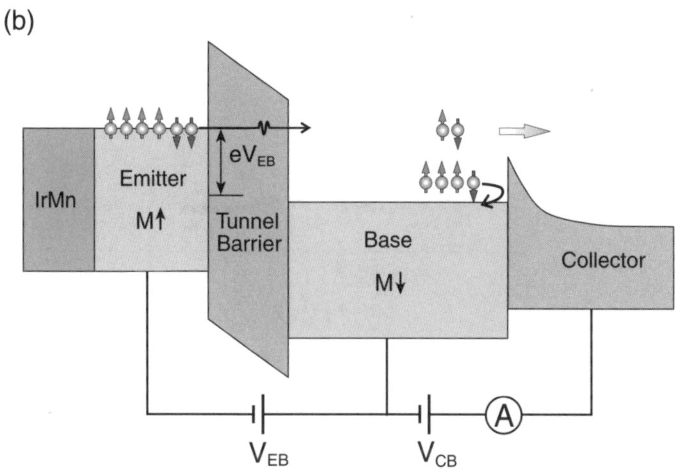

FIG. 6.14. Schematic band diagrams of magnetic tunnel transistor in parallel (a) and antiparallel (b) alignments of the emitter and base magnetic moments.

A direct consequence of the spin filtering effect in the magnetic tunnel transistor is that the collector current is very sensitive to the relative alignment of the magnetic moments of the emitter and the base. In the parallel alignment, most of the electrons injected into the base are majority electrons and are scattered less in the base, giving rise to a large collector current (Fig. 6.14a). In the antiparallel alignment, most of the injected electrons are minority electrons and are more easily scattered, leading to a small collector current (Fig. 6.14b). The switching between the parallel and antiparallel alignment of the magnetic moments can be realized by growing an anti-ferromagnetic IrMn layer on top of the

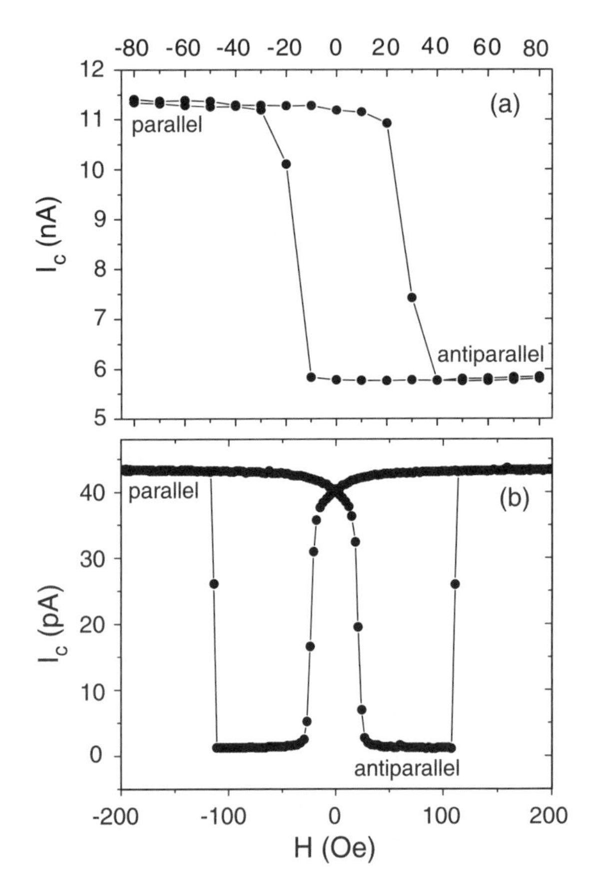

FIG. 6.15. Collector current as a function of the magnetic field for a magnetic tunnel transistor with a single layer (a) and spin-valve base (b) at 77 K. The bias voltage is $V_{EB} = 1.0\,V$ for (a) and 0.8 V for (b).

emitter layer. This IrMn layer acts as an exchange coupling layer to the emitter and pins its magnetic moments to a fixed direction. In this case, by applying an appropriate magnetic field, the magnetic moment of the base can be aligned to be either parallel or antiparallel to that of the emitter. The relative change in the collector current when the alignment is switched from antiparallel to parallel can be quantified by the magneto-current (MC), defined as:

$$\mathrm{MC} = \frac{I_{C,P} - I_{C,AP}}{I_{C,AP}}$$

where $I_{C,P}$ and $I_{C,AP}$ are the collector current for parallel (P) and antiparallel (AP) alignment of the emitter and base magnetic moments, respectively.

Figure 6.15(a) shows the collector current as a function of magnetic field at 77 K for a magnetic tunnel transistor with the following structure:

$$GaAs/30\text{-}Å\ Co_{84}Fe_{16}/25\text{-}Å\ Al_2O_3/50\text{-}Å\ Co_{84}Fe_{16}/300\text{-}Å\ Ir_{22}Mn_{78}/50\text{-}Å\ Ta.$$

As the magnetic field is swept between ±80 Oe, the base magnetic moment switches at ∼ ±20 Oe, whilst the emitter magnetic moment remains the same orientation in the applied field range. At $V_{EB} = 1.0$ V, $I_{C,P}$ (∼11.5 nA) is almost twice as large as $I_{C,AP}$, giving rise to an MC value of 97%. The magnetic tunnel transistor in Fig. 6.15 is fabricated using shadow masks and has a large base region area (∼1 × 8 mm^2) [93]. Thus, the measurement is taken at 77 K in order to reduce the leakage current from the Schottky barrier [93]. For the rest of the section, all the transport measurements of the magnetic tunnel transistor are taken at 77 K unless otherwise stated.

The moderate MC value in Fig. 6.15(a) is a consequence of the modest tunneling spin polarization of the CoFe/Al_2O_3 emitter, which is typically ∼45%. Even if the spin filtering in the base is perfect, the maximum MC will be on the order of 100%. To enhance the MC effect, a different type of magnetic tunnel transistor can be formed using a non-magnetic metal emitter and ferromagnetic spin-valve base. In such a device, the initially unpolarized electrons are spin filtered by the two ferromagnetic layers in the spin-valve base. Since the spin filtering can create spin polarization of more than 90%, the MC values can be much larger [94]. As demonstrated in Fig. 6.15(b), the MC value exceeds 3400% in a magnetic tunnel transistor with the following structure:

$$GaAs/50\text{-}Å\ Co_{70}Fe_{30}/40\text{-}Å\ Cu/50\text{-}Å\ Ni_{81}Fe_{19}/25\text{-}Å\ Al_2O_3/300\text{-}Å\ Cu.$$

Note that the collector currents shown in Fig. 6.15 are quite small. This is because most of the hot electrons are lost due to scattering in the base and at the base/collector interface. Increasing the hot-electron energy can give rise to larger collector currents up to several μA. By optimizing film growth and improving interface properties, the magnitude of the collector current may be further improved.

6.3.2 *Energy-dependent electron transport in the magnetic tunnel transistor*

6.3.2.1 *Hot-electron attenuation lengths* The magnetic tunnel transistor is a very useful tool to study spin-dependent hot-electron transport and to probe electronic structures in metals and semiconductors. One such example is to measure the spin-dependent hot-electron attenuation length in ferromagnetic thin films [84].

In a simple model ignoring spin-flip processes, the collector current of a magnetic tunnel transistor is carried by the majority and minority electrons independently. The attenuation of the hot-electron current in each channel is described by the corresponding bulk attenuation length and interface collection efficiency. The collector current for parallel and antiparallel alignments of the emitter and base magnetic moments can be modeled by the following formula:

$$I_{C,P(AP)} = I_E \frac{1+P_E}{2} e^{-t/\lambda_{\uparrow(\downarrow)}} \alpha_{C\uparrow(\downarrow)} + I_E \frac{1-P_E}{2} e^{-t/\lambda_{\downarrow(\uparrow)}} \alpha_{C\downarrow(\uparrow)} \qquad (6.1)$$

where I_E is the tunnel current, P_E is the spin polarization of the electrons injected from the emitter into the base (the emitter spin polarization), t is the base layer thickness, $\lambda_{\uparrow(\downarrow)}$ is the attenuation length for majority (minority) electrons within the ferromagnetic base layer, and $\alpha_{C\uparrow(\downarrow)}$ is the electron collection efficiency at the base/collector interface.

Note that the electron attenuation length $\lambda_{\uparrow(\downarrow)}$ describes the exponential decay of the hot-electron current with increasing base layer thickness. Thus $\lambda_{\uparrow(\downarrow)}$ is correlated, but not equivalent, to the electron mean free path. For example, GaAs has a lowest energy conduction band near the Γ point in the Brillouin zone and two sets of higher energy conduction bands near the L and X points. When GaAs is used as a collector, at electron energy barely above the collector Schottky barrier height, only a small portion of the Γ conduction band is available for electron collection. Therefore, only electrons with very small lateral wave vectors can be collected. Since a scattering event in the base region reduces the energy and/or changes the lateral wave vector of the electron, it almost certainly removes the scattered electron from the collector current. This is, however, no longer true when the electron energy is well above the Schottky barrier height. The scattered electron might still retain enough energy to overcome the Schottky barrier even after losing a small amount of energy. Moreover, the number of GaAs conduction-band states increases rapidly at elevated energy. Additional conduction-band states in the L and X bands also become available for electron collection at high energy. Thus, electrons with large lateral wave vectors can be collected as well and, as a consequence, the contribution of scattered electrons to the collector current is increased.

In Eq. (6.1), multiple passages of hot electrons through the base layer are not considered. Since the hot-electron attenuation length in ferromagnetic $3d$ transition metals is small [70–74], the error introduced by such a simplification is negligible. For hot-electron transport in noble metals, such as Au, the multiple passages of electrons can be important [95].

It is possible to extract the hot-electron attenuation length using Eq. (6.1). To this end, one can fabricate a series of magnetic tunnel transistors with different base layer thickness and measure the base layer thickness dependence of the collector current in the parallel and antiparallel alignments at a given emitter/base bias voltage V_{EB} (i.e. hot-electron energy). By fitting the data to Eq. (6.1), the hot-electron attenuation length at this given electron energy can be obtained. Finally, the same measurements can be repeated at various electron energies to obtain the energy dependence of the electron attenuation length.

Figure 6.16 shows the hot-electron attenuation length in $Co_{84}Fe_{16}$ thin films measured using a set of magnetic tunnel transistors with the following structure:

$$\text{GaAs}\,(001)/t\,\text{Co}_{84}\text{Fe}_{16}/25\text{-Å Al}_2\text{O}_3/50\text{-Å Co}_{84}\text{Fe}_{16}/300\text{-Å Ir}_{22}\text{Mn}_{78}/50\text{-Å Ta}.$$

Here t is the base layer thickness. Over the energy range of $1-1.9\,\text{eV}$, the majority electron attenuation length is about 5–6 times larger than the minority electron attenuation length. Moreover, the majority attenuation length decreases with

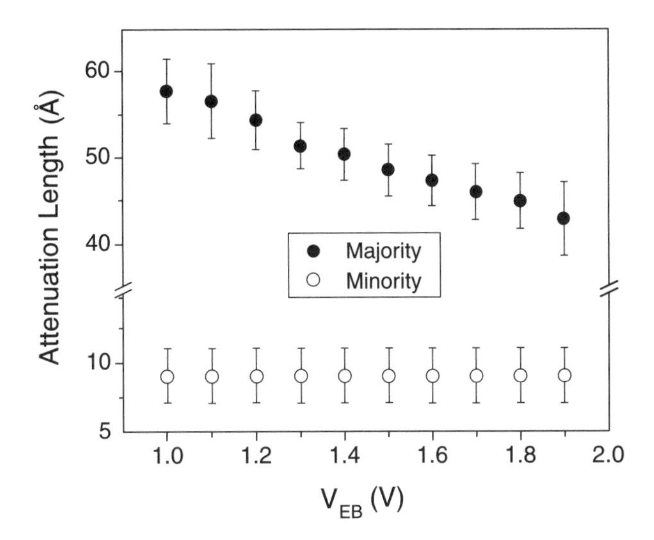

FIG. 6.16. Majority and minority electron attenuation lengths in $Co_{84}Fe_{16}$ thin films as a function of the electron energy.

the electron energy, whilst the minority electron attenuation length hardly shows any energy dependence. The decrease of the majority electron attenuation length at elevated energies is due mainly to a larger electron-electron scattering rate at high energies, which is the most important scattering mechanism for the majority electrons [96–98]. On the other hand, the minority electrons are subject to more efficient scattering because of the abundant available states to scatter into in the minority *d*-band near the Fermi level and additional scattering mechanisms such as spontaneous spin wave scattering [99]. As a result, the minority electron attenuation length is much smaller. The large spin asymmetry in the attenuation lengths implies that the hot electron current in the magnetic tunnel transistor, after spin-filtering in the base region, can be highly spin-polarized [84]. Using the attenuation lengths plotted in Fig. 6.16, the spin polarization of the hot-electron current at the base/collector interface is more than 95% when the CoFe base layer thickness exceeds ~ 35 Å.

6.3.2.2 *Bias voltage dependence of the magneto-current* The collector current in a magnetic tunnel transistor is determined not only by the electron transmission in the ferromagnetic base layer, but also by the conduction band structure of the semiconductor collector. For example, the bias dependence of the MC can be very different depending on whether the collector is GaAs or Si [100, 101]. This is illustrated in Fig. 6.17, where the MC is measured as a function of V_{EB} for a set of magnetic tunnel transistors with the following structures:

$$\text{semiconductor/base/23-Å } Al_2O_3\text{/50-Å } Co_{84}Fe_{16} \text{ or } Co_{70}Fe_{30}\text{/300-Å } Ir_{22}Mn_{78}\text{/50-Å Ta.}$$

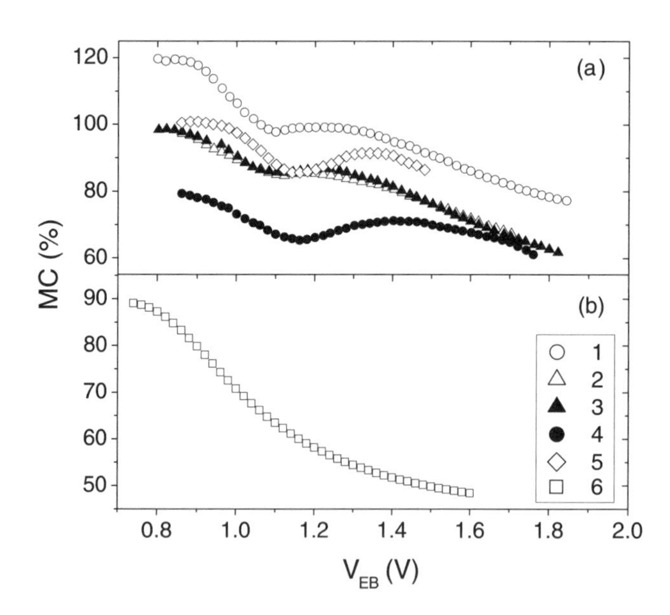

FIG. 6.17. Bias dependence of the MC in magnetic tunnel transistors with a GaAs (a) or Si (b) collector.

Here the base is $Co_{84}Fe_{16}$, $Co_{70}Fe_{30}$ or $Ni_{81}Fe_{19}$, and the semiconductor is n-type GaAs (001), GaAs (111) or Si (001). An overview of the sample structures is given in Table 6.2. For the magnetic tunnel transistor with a GaAs collector, a pronounced non-monotonic bias dependence of the MC is observed (Fig. 6.17a). When the bias voltage V_{EB} exceeds the Schottky barrier height ($\sim 0.78\,\mathrm{V}$), a large MC is obtained. The MC decreases with V_{EB} up to about 1.1 V, then increases slightly, and finally decreases gradually. The quantitative difference between these samples is most likely due to variations in film growth and/or tunnel barrier formation. For instance, the growth of magnetic tunnel junctions depends critically on the materials used and the growth of the base layers can be quite different on GaAs (001) and GaAs (111) substrates. This likely accounts for the different tunneling magnetoresistance (TMR) values measured for samples 1–5 (Table 6.2). In contrast, for the magnetic tunnel transistor with a Si collector (sample 6), such a non-monotonic bias dependence of the MC is absent: the MC decreases continuously as a function of the bias voltage (Fig. 6.17b).

The bias dependence of the MC can be understood using a simple model taking into account inelastic electron scattering in the base, electron scattering at the base/collector interface, and the semiconductor collector band structure. In this model, the collector current is calculated by the following equation:

$$
\begin{aligned}
I_{\mathrm{C,P(AP)}}\left(V_{\mathrm{EB}}\right) &= I_{\mathrm{E}} \frac{1+P_{\mathrm{E}}}{2} \exp\left(-t/\lambda_{\uparrow(\downarrow)}\right) \\
&\times \int_{\Phi_S}^{eV_{\mathrm{EB}}} f_{\uparrow(\downarrow)}(E) \int D_{\uparrow(\downarrow)}(E, \boldsymbol{k}_{\parallel}) T(E, \boldsymbol{k}_{\parallel}) d\boldsymbol{k}_{\parallel} dE
\end{aligned}
$$

Table 6.2 The base, collector structures and the TMR values of the MTT samples.

Sample number	Base	Collector	TMR (%)
1	30-Å $Co_{84}Fe_{16}$	GaAs (001)	46.4
2	45-Å $Co_{84}Fe_{16}$	GaAs (001)	40.7
3	100-Å $Co_{84}Fe_{16}$	GaAs (001)	31.7
4	74-Å $Ni_{81}Fe_{19}$	GaAs (001)	14.7
5	30-Å $Co_{84}Fe_{16}$	GaAs (111)	29.0
6	3-Å Fe / 50-Å $Co_{70}Fe_{30}$	Si (001)	33.8

$$+ I_{\mathrm{E}} \frac{1 - P_{\mathrm{E}}}{2} \exp\left(-t/\lambda_{\downarrow(\uparrow)}\right)$$
$$\times \int_{\Phi_S}^{eV_{\mathrm{EB}}} f_{\downarrow(\uparrow)}(E) \int D_{\downarrow(\uparrow)}(E, \boldsymbol{k}_{\parallel}) T(E, \boldsymbol{k}_{\parallel}) d\boldsymbol{k}_{\parallel} dE, \qquad (6.2)$$

where I_{E} is the emitter current, P_{E} is the emitter spin polarization, t is the base layer thickness, $\lambda_{\uparrow(\downarrow)}$ is the spin-dependent attenuation length for the quasi-elastic majority (minority) electrons, E is the hot-electron energy, $f_{\uparrow(\downarrow)}$ and $D_{\uparrow(\downarrow)}$ are the energy and angular distribution functions of the hot electrons at the base/collector interface, T is the transmission coefficient across the Schottky barrier, and $\boldsymbol{k}_{\parallel}$ is the component of the electron wave vector parallel to the layers.

P_{E} and $\lambda_{\uparrow(\downarrow)}$ used in Eq. (6.2) are subtly different from those defined in Eq. (6.1). In Eq. (6.1), the collector electronic structure is not explicitly considered. Rather, the influence of the semiconductor conduction-band structure on the collector current is included in the emitter spin polarization, and the electron attenuation length characterizes the base layer thickness dependence of the collector current attenuation. In contrast, Eq. (6.2) explicitly takes into account the collector electronic structure effects by including the energy and angular distribution of the hot electrons and the calculated electron transmission coefficients. P_{E} is a constant in Eq. (6.2) and $\lambda_{\uparrow(\downarrow)}$ describes the attenuation of the quasi-elastic majority (minority) electron current in the base.

The angular distribution $D_{\uparrow(\downarrow)}$ is assumed to be a two-dimensional Gaussian distribution centered at $\boldsymbol{k}_{\parallel} = 0$ with a width of $\sigma_{\uparrow(\downarrow)}$, where σ is a fraction of k_{B}, the maximum hot-electron wave vector in the base layer:

$$k_{\mathrm{B}} = \sqrt{2m(eV_{\mathrm{EB}} + E_{\mathrm{F}})/\hbar^2},$$

where m is the electron mass in the base and E_{F} is the Fermi energy. The energy distribution $f_{\uparrow(\downarrow)}$ is assumed to be a half-Gaussian centered at eV_{EB} with a width of $\varepsilon_{\uparrow(\downarrow)}$. The transmission coefficient T can be calculated using quantum mechanics [102].

In the magnetic tunnel transistor, the spin-polarized hot electrons that are injected from the emitter into the base initially have very narrow energy and angular distributions because the tunneling process is highly sensitive to the height and thickness of the tunnel barrier. Specifically, the electrons are injected with energy close to the emitter Fermi level and with small parallel wave vector components. As these electrons traverse the base region, they experience inelastic scattering and lose energy. As a consequence, the energy distribution of the hot electrons broadens. Since the scattering rate is normally higher for the minority electrons than for the majority electrons, the minority electrons are more likely to lose energy and thus have a broader energy distribution. Additional electron scattering occurs at the base/collector interface, after which a fraction of the incident electrons are collected by the semiconductor. This interface scattering could broaden the angular distribution of the hot electrons.

The GaAs conduction band has lowest energy at the center of the Brillouin zone (Γ valley). At an energy of $\sim 0.29\,\mathrm{eV}$ above the top of the Schottky barrier, there are eight local minima along the $\langle 111 \rangle$ axis (L valleys). At an even higher energy, $\sim 0.48\,\mathrm{eV}$ above the Schottky barrier height, there are six local minima along the $\langle 001 \rangle$ axis (X valleys) [103]. When the bias voltage exceeds the Schottky barrier height by a small margin, a hot-electron current is collected through the central Γ valley. Because of their narrow energy distribution, a relatively large portion of the majority electrons is able to surmount the Schottky barrier and hence contributes to the collector current. On the other hand, only a small portion of the minority electrons has enough energy to be collected. This is schematically illustrated in Fig. 6.18. The large spin asymmetry results in a large MC. At elevated emitter/base bias voltages, increasingly more of the scattered minority electrons are able to surmount the Schottky barrier. The increase of the minority electron current gives rise to a smaller MC. If all the collector conduction bands open up at the same energy level, a monotonic decrease of the MC with bias is expected, as observed in the magnetic tunnel transistor with a Si collector. However, for GaAs, the L valleys and the X valleys open up at higher energies than the Γ valley. When these valleys just become available for hot-electron injection, they favor the collection of the majority electrons and thus tend to increase the MC. This leads to the observed non-monotonic bias dependence of the MC when the collector is GaAs.

Figure 6.19(a) depicts the measured MC for sample 1 together with the calculated MC using the simple model. The agreement between the experimental data and the calculated results is very good. In Fig. 6.19(b), the measured transfer ratios (defined as $I_{\mathrm{C}}/I_{\mathrm{E}}$) for parallel and antiparallel alignment of the base and emitter magnetic moments are plotted together with the calculated results. Again, excellent agreement is obtained. In Fig. 6.19(c), the contributions to the collector current from the different conduction band valleys are calculated separately. It can be seen that at low bias voltages, all the electrons are injected into the Γ valley. At $V_{\mathrm{EB}} \sim 1.1\,\mathrm{V}$, the L valleys become available for electron collection. Initially, many more majority electrons are collected than minority

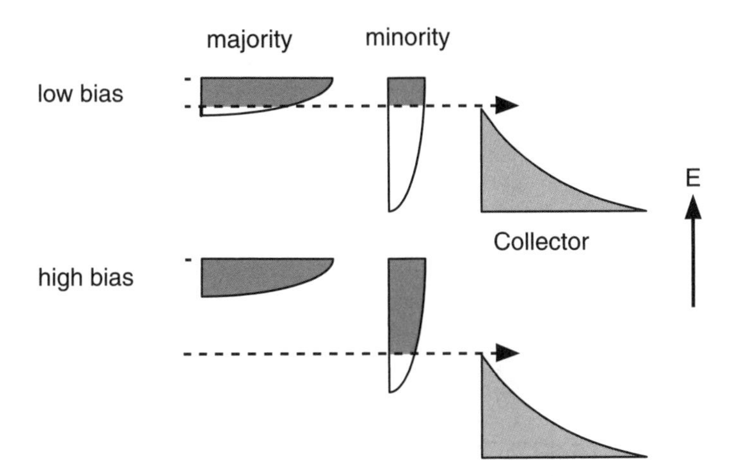

FIG. 6.18. Schematic illustration of the energy distribution for the majority and the minority electrons at the base/collector interface. The shaded areas indicate electrons that have enough energy to surmount the Schottky barrier. At low bias (upper half), most of the majority electrons but only a small fraction of the minority electrons have enough energy to overcome the Schottky barrier. At high bias (lower half), this asymmetry is reduced.

electrons, and an increase of the MC results. The longitudinal electron effective mass of the L valleys is much larger than that of the Γ valley. Consequently, the number of available energy states in the L valleys increases very rapidly with bias. The contribution to the collector current from the L valleys is therefore significant, as reflected by a small kink in the transfer ratio at $V_{\mathrm{EB}} \sim 1.1\,\mathrm{V}$ (Fig. 6.19b). As the bias voltage increases further, more minority electrons are injected into the L valleys and the MC starts to decrease again. The energy states in the X valleys become available for electron collection at $V_{\mathrm{EB}} \sim 1.3\,\mathrm{V}$. However, the current collected through the Γ and L valleys is already very large and hence no significant change in the MC or the transfer ratio is caused by the small additional current collected through the X valleys.

For the calculations shown in Fig. 6.19, a broad angular distribution is assumed for both the majority and the minority electrons. This assumption is essential to reproduce the non-monotonic bias voltage dependence of the MC and can be rationalized by spin-independent electron scattering at the base/collector interface [104]. Since the L valleys are located far from the center of the Brillouin zone, a large parallel wave vector is required for the electrons to access the conduction band states in these valleys. If the angular distribution is narrow, only few electrons have large parallel wave vectors and thus the L valleys hardly contribute to the collection of hot electrons. This would result in a monotonic decrease of the MC with bias, as illustrated in Fig. 6.20(a). In this calculation the same fitting parameters are assumed as those used in Fig. 6.19, except for

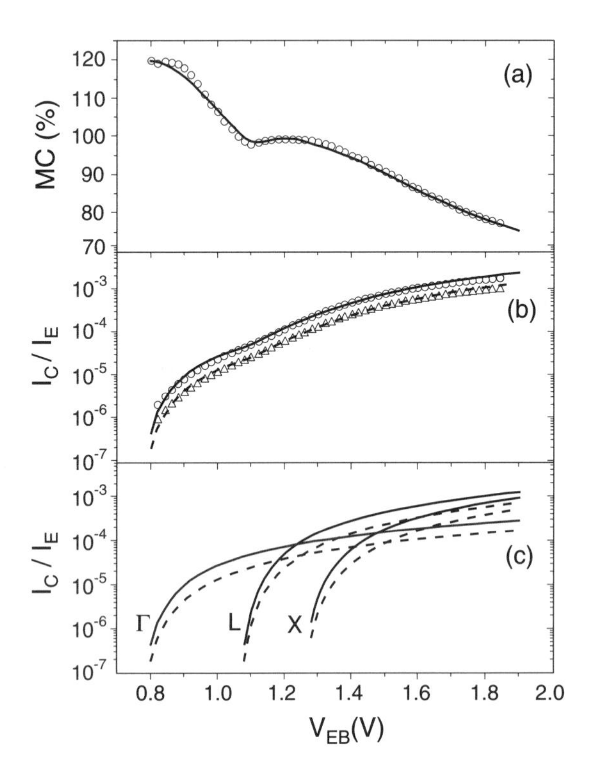

FIG. 6.19. The measured (symbols) and calculated (lines) bias voltage dependence of the MC (a) and I_C/I_E (b) for both parallel (circles/solid line) and antiparallel (triangles/dashed line) alignments of the emitter and base magnetic moments in the magnetic tunnel transistor with a GaAs (001) collector. (c) The calculated I_C/I_E for the Γ, L, and the X conduction-band valleys separately. The solid and dashed lines in this calculation are for the parallel and antiparallel configuration, respectively. The angular distribution is assumed to be broad and identical for the majority and the minority electrons.

the assumption of a much narrower angular distribution. As discussed above, the narrow angular distribution results in a small contribution of the L valleys to the collector current (Fig. 6.20c). Although the opening of the L valleys still tends to increase the MC, the collector current is dominated by electron collection through the central Γ valley and therefore the MC varies monotonically with the bias voltage.

The assumption of a spin-dependent hot-electron energy distribution is also important in order to understand the experimental results. Figure 6.20(c)-(f) show the calculated results assuming the same energy distribution for the majority and the minority electrons. In addition, the majority and the minority electrons are assumed to have a narrow and broad angular distribution, respectively. At low bias, the calculated MC stays approximately constant since the

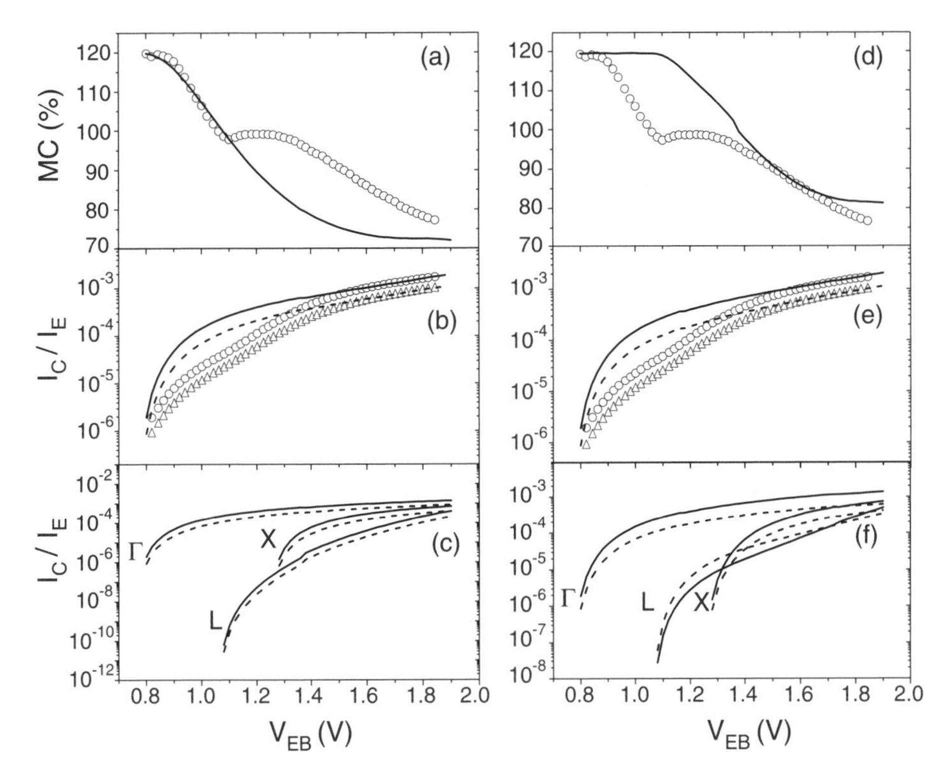

FIG. 6.20. Similar to Fig. 6.19, but assuming narrow and identical angular (a)-(c) or energy (d)-(f) distributions for the majority and the minority electrons.

same energy distribution is assumed for both the majority and the minority electrons. When the L valleys open up, more minority electrons are collected than majority electrons because they have larger parallel wave vectors and therefore can access the L bands more easily. As a result, the MC decreases. On the other hand, if a broad angular distribution is assumed for the majority electrons and a narrow angular distribution is assumed for the minority electrons, the MC will become negative (not shown), which is opposite to the experimental data. Note that a very similar bias dependence of the MC is measured for both the GaAs (001) and the GaAs (111) collectors, although the projection of conduction bands onto the interface plane is very different for the two substrate orientations. This is a strong indication that the measured bias voltage dependence of the MC cannot be explained by different angular distributions of the majority and the minority electrons.

The simple model can account for the bias dependence of the MC in the magnetic tunnel transistor with a Si collector as well. Si has six conduction valleys located along the ⟨001⟩ axes and they all have the same energy minimum. Hence, a monotonic decrease of the MC with bias is expected. In Fig. 6.21, the

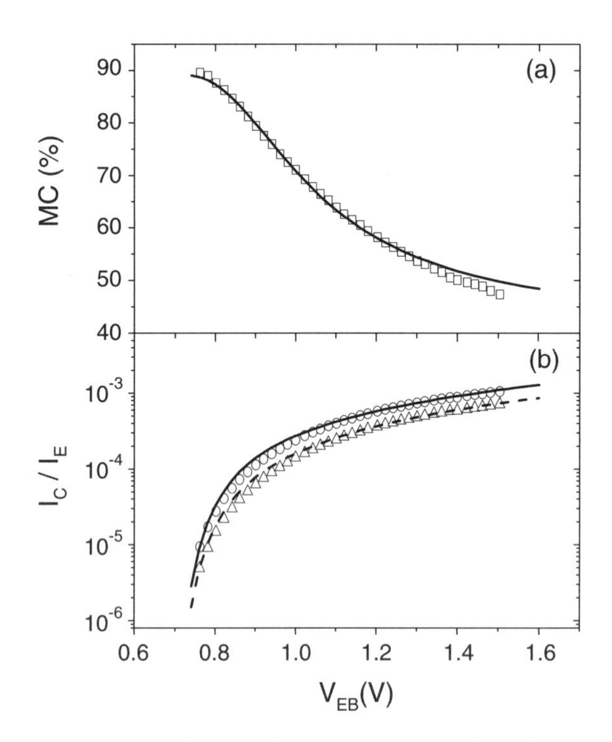

FIG. 6.21. The measured (symbols) and calculated (lines) bias voltage dependence of the MC (a) and I_C/I_E (b) for both parallel (circles/solid line) and antiparallel (triangles/dashed line) alignments of the emitter and base magnetic moments in the magnetic tunnel transistor with a Si collector. The angular distribution is assumed to be broad and identical for majority and minority electrons.

measured and calculated MC and I_C/I_E is depicted. The agreement between the experimental data and the calculated results is satisfactory.

6.4 Tunnel-based spin injectors

6.4.1 *Spin injection*

The development of modern electronic devices has followed Moore's law for several decades [105]. As the device size continues to shrink towards fundamental physical limits, there is an increasing interest in exploring alternate technologies. Spin-based electronics, or spintronics, is a promising technology, where the spin states of carriers are utilized as an additional degree of freedom for information processing and storage. Spins have played an important role in conventional magneto-electronic devices, such as spin-valves and magnetic tunnel junctions. The discovery of the giant magnetoresistance and the tunneling magnetoresistance effects has had a profound impact on the storage and recording

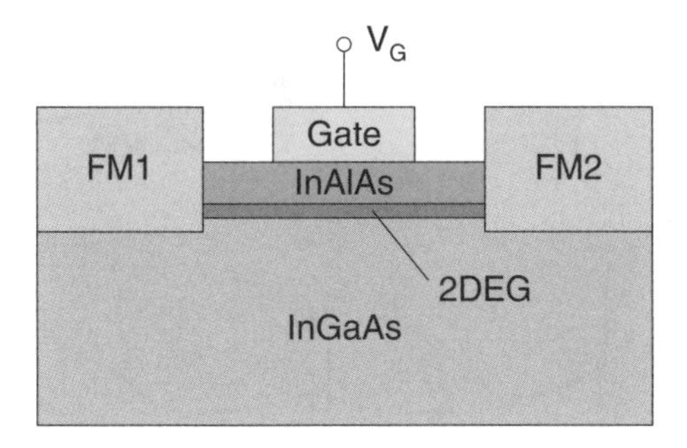

FIG. 6.22. Schematic drawing of the Datta–Das transistor. FM1 and FM2 are two ferromagnetic contacts to the 2DEG.

industry [106–113]. In semiconductor electronics, however, the role of spins is rather passive. It is interesting to note that semiconductors actually have many desirable properties when spins are concerned. The electron spin relaxation time in semiconductors can be several orders of magnitude longer than the electron momentum and energy relaxation times [114]. Using an electric field, electrons in GaAs can be dragged over a distance of $100\,\mu$m without losing spin coherence [115]. In addition to the long spin lifetimes and large spin diffusion lengths, semiconductors offer the flexibility of varying carrier doping profiles and spin relaxation rates, which could be very useful for building spintronic devices. For example, Ohno *et al.* showed that it is possible to control the ferromagnetism of InMnAs thin films by modulating the hole concentration [116]. Karimov *et al.* demonstrated that electron spin relaxation rates in GaAs heterostructures can be varied by a factor of ten by applying a gate voltage [117]. Murakami *et al.* predicts that a dissipationless spin current can flow in GaAs in the presence of an electric field [118]. These studies suggest that spin-based semiconductor electronics has the potential to bring an entirely new generation of devices with fast speed, high density, low power consumption and non-volatility [119].

The first semiconductor spintronic device was proposed by Datta and Das in 1990 (Fig. 6.22) [120], often referred to as the Datta–Das "transistor". This device is comprised of two ferromagnets in contact with a semiconductor two-dimensional electron gas (2DEG) formed at a InAlAs/InGaAs interface. The two ferromagnetic contacts are used to create and detect spin polarized electrons, respectively. The 2DEG provides a channel for electron transport between the contacts. Due to the Rashba effect [121, 122], the mobile electrons in the 2DEG sense an effective magnetic field and precess around the field. The strength of the magnetic field can be controlled by applying a gate voltage across the 2DEG [123]. Therefore, it is possible to modulate the electron spin precession inside the 2DEG and consequently the magnitude of the current. Although a simple concept, the Datta–Das transistor contains all the essential components of a semiconductor spintronic device: the creation, transport, manipulation, and detection of spin-polarized electrons. The first step, the creation of spin polarized

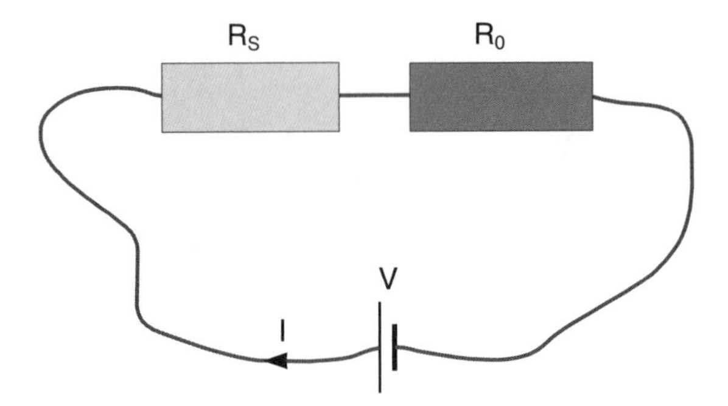

FIG. 6.23. A simplistic picture to illustrate the conductivity mismatch problem in spin injection.

electrons inside semiconductors, is often called spin injection. It is a prerequisite for semiconductor spintronics.

It has long been known that spin-polarized electrons can be generated in direct band gap semiconductors by optical pumping with circularly polarized light [124]. For device applications, however, electrical means of spin injection is much more desirable. The first attempts of electrical spin injection were carried out using Ohmic contacts formed by ferromagnetic metals [125–127]. Since the electrons in the ferromagnetic metals are spin polarized, it was expected that the electrons injected into the semiconductors would maintain their spin orientation and thus give rise to successful spin injection. Despite significant effort, however, unambiguous spin injection was not demonstrated. Later on, it was realized that the conductivity mismatch between the metal Ohmic contact and the semiconductor might present a fundamental obstacle for spin injection [128]. The concept of the conductivity mismatch can be understood in a simple picture shown in Fig. 6.23. In the spin injection experiment, the resistance of the device can be divided into a spin-dependent part R_S and a spin-independent part R_0. The semiconductor resistance is normally spin independent, whilst the ferromagnet/semiconductor contact resistance is spin-dependent. When an Ohmic contact is used, R_S is much smaller than R_0. Therefore, the electron transport will be dominated by the spin-independent semiconductor resistance. As a result, the electron current will be largely unpolarized. In order to achieve efficient spin injection, the spin-dependent conductivity needs to be smaller than its spin-independent counterpart. Note that Fig. 6.23 is a rather simplistic picture of viewing the conductivity mismatch problem. A strict treatment of this subject can be found in the paper by Schmidt *et al.* [128]

The first evidence of electrical spin injection was reported by Hammar *et al.* [129]. They used NiFe contacts to inject electrons into an InAs 2DEG, with a thick AlGaSb insulating barrier inserted between the NiFe and the 2DEG. Due to the Rashba effect, the spin degeneracy of the electron density of states in the 2DEG is lifted. As a result, the electron transport in the 2DEG channel is spin dependent. When the magnetization of the NiFe contacts is varied by a magnetic

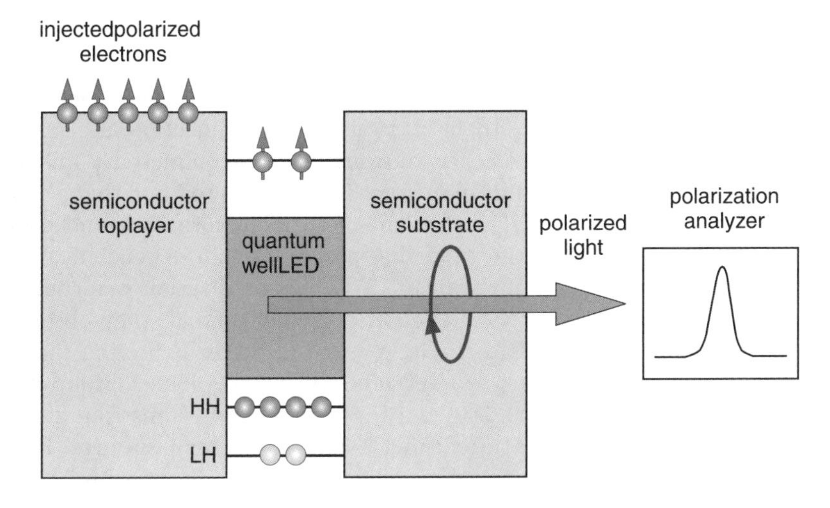

FIG. 6.24. Schematic drawing of a quantum-well LED detector used for measuring spin polarization of injected electrons. HH and LH represent the heavy and light holes in the quantum well.

field, the resistance of the device is expected to change [130]. Indeed, Hammar *et al.* observed a resistance change of $\sim 0.9\%$. This observation, although a small effect, was encouraging progress in spin injection. Later on, Hu *et al.* also reported spin injection from NiFe into an InAs 2DEG [131], with a smaller resistance change of $\sim 0.2\%$.

A much larger spin injection effect was obtained using diluted magnetic semiconductors, which have conductivity matched to that of normal semiconductors. The use of magnetic semiconductors was proposed by Oestreich *et al.* [132]. Using time-resolved photoluminescence, they showed that magnetic semiconductor $Cd_{0.98}Mn_{0.02}Te$ could serve as a very good spin aligner in a magnetic field of 2.5 tesla. Moreover, the spin polarized electrons generated in the $Cd_{0.98}Mn_{0.02}Te$ layer could be efficiently transported into an adjacent CdTe layer. Shortly afterwards, Fiederling *et al.* [133], Ohno *et al.* [134], and Jonker *et al.* [135] demonstrated electrical spin injection using BeMnZnSe, GaMnAs, and ZnMnSe as the spin injectors, respectively. In their experiments, a quantum-well light-emitting diode (LED) was used as an optical detector to measure the injected electron spin polarization (Fig. 6.24). The injected electrons recombine with holes inside the quantum well and emit light. The circular polarization of the light is correlated to the spin polarization of the electrons prior to recombination through the optical selection rules [124]. Therefore, the spin polarization can be inferred from the light polarization. Very large spin polarization, more than 80%, has been reported [136]. Such a high spin polarization is very desirable for spintronic applications. However, magnetic semiconductors also have their limitations: these materials, to date, show good magnetic properties only at temperatures well below room temperature and/or in a large magnetic field, thereby limiting their usefulness.

Ferromagnetic 3*d* transition metals have Curie temperatures much higher than room temperature, making them attractive for spin injection into semiconductors. However, care must be taken to overcome the conductivity mismatch between the metals and the semiconductors. Rashba pointed out that the mismatch problem can be resolved if the ferromagnetic metal forms a tunnel contact with the semiconductor since the tunneling process is spin dependent and the tunnel contact can have high impedance [137]. This predication was confirmed by Zhu *et al.* [138]. Using a Fe/GaAs Schottky tunnel contact, they observed a spin polarization of $\sim 2\%$ in GaAs/InGaAs quantum wells. Following the same route, higher spin polarization was later reported by the Jonker group at the Naval Research Laboratory [139, 140] and the Crowell and Palmstrøm group at the University of Minnesota [141], reaching $\sim 30\%$ at low temperatures. Besides Schottky tunnel contacts, oxide tunnel barriers were utilized for spin injection as well. Manago and Akinaga used ferromagnetic metals in conjunction with Al_2O_3 tunnel barriers as the spin injector and observed a signal of $\sim 1\%$ at room temperature [142]. van 't Erve *et al.* used a Fe/Al_2O_3 tunnel contact and observed a spin polarization of 40% at 5 K [143]. Motsnyi *et al.* [144, 145] and Van Dorpe *et al.* [146] used the oblique Hanle effect [124] to investigate spin injection from a ferromagnetic metal/Al_2O_3 tunnel barrier injector into a GaAs LED. They applied a magnetic field at an angle to the ferromagnetic metal film plane and measured the circular polarization of the light emitted from the LED. By model fitting to the luminescence polarization, the spin injection efficiency can be extracted. Motsnyi *et al.* [145] measured a light polarization of $\sim 4\%$ at 80 K and $\sim 1\%$ at room temperature and concluded that the actual spin injection efficiency is about 21% and 16%, respectively.

When a Schottky or Al_2O_3 tunnel contact is used for spin injection, the maximum spin polarization that can be achieved may be limited by the tunneling spin polarization from the ferromagnetic metal. For instance, for 3*d* transition metals and their alloys, the tunneling spin polarization is normally no more than 50% when an Al_2O_3 tunnel barrier is used [54]. To overcome this limit, two approaches can be adopted. One is to take advantage of the spin filtering effect of hot electrons in ferromagnetic metals to obtain high spin polarization. The other is to develop new materials which give rise to high tunneling spin polarization. These two methods will be discussed in the next subsection.

6.4.2 *Spin injection using tunnel injectors*

6.4.2.1 *Spin injection into GaAs from a magnetic tunnel transistor* In the previous section, it has been shown that the spin polarization of the hot-electron current in a magnetic tunnel transistor can be as high as 95% at the base/collector interface due to efficient spin filtering in the base. Moreover, the presence of a tunnel barrier in the magnetic tunnel transistor solves the conductivity mismatch problem. These properties make the magnetic tunnel transistor very intriguing as

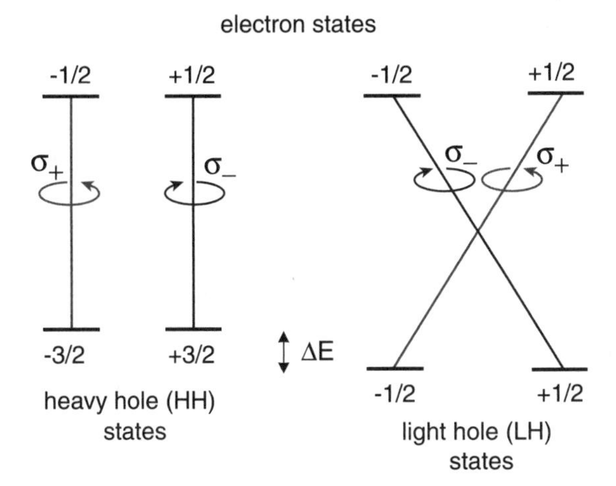

FIG. 6.25. Optical selection rules for electron-hole recombination in the Faraday geometry.

a spin injector. To detect the injected electron spin polarization, a quantum-well LED structure can be incorporated as the collector of the magnetic tunnel transistor. In this scheme, the injected electrons recombine with unpolarized holes from the substrate in the quantum well and emit light. By analyzing the circular polarization of the light, the spin polarization of the electrons can be determined using optical selection rules [124]. As shown in Fig. 6.25, there are two types of holes existing in the quantum well: heavy holes (HH) and light holes (LH). They can both recombine with the electrons and emit photons with opposite helicity. In general, careful analysis has to be taken in order to extract the spin polarization from the luminescence spectra. However, in the quantum well, the energy degeneracy of the heavy and light hole states is lifted due to confinement and/or strain effects. If the energy splitting (ΔE) between the two hole states is larger than the spectral resolution, it is possible to measure the circular polarization of the heavy hole emission only. In this case, the luminescence polarization is simply equal to the electron spin polarization. Note that the selection rules in Fig. 6.25 are only valid in the Faraday geometry, i.e., with the spin and light propagation directions both perpendicular to the quantum-well plane. Experimentally, this implies that a large magnetic field is required to rotate the electron spins out of the film plane.

In the spin injection experiment, the quantum well detector is buried inside the semiconductor heterostructure. The injected electrons are first transported into the quantum-well region, where they spend a certain amount of time (described by the recombination time) before recombining with the holes and emitting light. The electroluminescence (EL) polarization does not include any spin relaxation effects prior to recombination and, therefore, sets a lower bound on the spin polarization of the injected electrons. To properly interpret the experimental data, it is necessary to take into account various spin relaxation processes in the semiconductor. Three spin relaxation mechanisms have been found to be important: the Elliott–Yafet (EY), D'yakonov–Perel' (DP), and Bir–Aronov–Pikus (BAP) mechanisms (see discussion in Chapter 2). The EY process derives from the mixing of electron wave functions with opposite spin states due to spin-orbit coupling [147, 148]. As a result, electron momentum scattering leads to spin relaxation, with a rate proportional to the momentum scattering rate. The DP process is present in semiconductors without inversion symmetry [149, 150]. The mobile electrons see an effective magnetic field whose magnitude and orientation depend on the electron momentum. Spin precession around this magnetic field gives rise to spin relaxation. Momentum scattering randomizes the direction of the effective magnetic field and reduces the average precession effect. The DP spin relaxation rate is therefore inversely proportional to the momentum scattering rate, which is opposite to the EY process. The BAP process is due to electron-hole exchange and annihilation interactions [151]. The relative importance of the three processes depends on the details of the sample structure and the experimental conditions, such as the semiconductor doping profile, experiment temperature, etc.

Figure 6.26 shows the electroluminescence spectra taken at 1.4 K when a magnetic tunnel transistor is used as the spin injector. The emitter of this device is 50 Å $Co_{84}Fe_{16}$. The base consists of 35 Å $Ni_{81}Fe_{19}$ and 15 Å $Co_{84}Fe_{16}$ with the NiFe layer adjacent to the collector. The tunnel barrier is Al_2O_3 with a thickness of $\sim$ 22 Å. Three GaAs/$In_{0.2}Ga_{0.8}As$ quantum wells are incorporated in the collector as the optical detector. The electron energy in these measurements is set to be $\sim$2 eV. The thin and thick lines in Fig. 6.26 represent the left (σ^+) and right (σ^-) hand circular polarization components, respectively. Note that the width of the luminescence peaks is only $\sim$25 Å which is limited by the spectrometer resolution for the given signal level. According to absorption studies, the separation in wavelength between the heavy and light hole emissions is $\sim$400 Å in these GaAs/$In_{0.2}Ga_{0.8}As$ quantum wells. Therefore, the narrow luminescence linewidth enables the unambiguous detection of electron-heavy hole recombination. As discussed before, the circular polarization of the electroluminescence is equal to the spin polarization of the electrons just before recombination.

The electroluminescence polarization in Fig. 6.26 clearly depends on the magnetic field. At zero field, the intensities of the σ^+ (I^+) and σ^- (I^-) components are the same. At high fields, there is a significant difference between I^+ and I^-. Here, the intensities are calculated by integrating the areas under the peaks. The electroluminescence polarization is defined as:

$$P_{\text{EL}} = \frac{I^+ - I^-}{I^+ + I^-}.$$

P_{EL} is $\sim$13% at 2.5 T and ~ -13% at -2.5,T. The sign of P_{EL} indicates injection of the majority electron spins ($-1/2$ spin state) into the quantum wells. This result is consistent with the sign of the collector current polarization observed

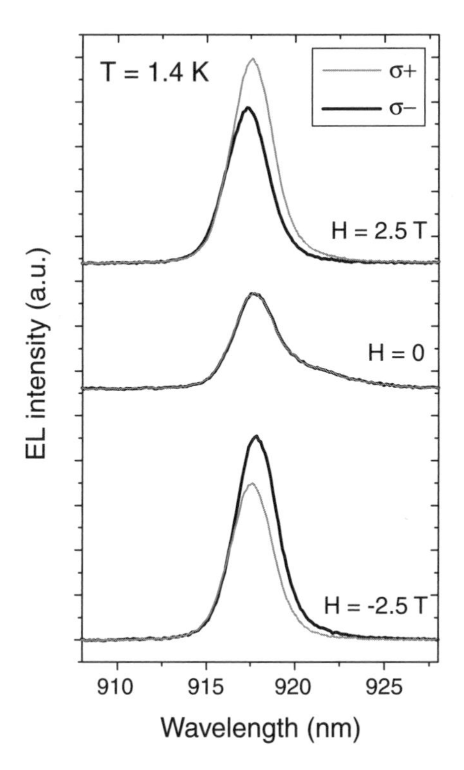

FIG. 6.26. Electroluminescence spectra measured at magnetic fields of 0 and $\pm 2.5\,\mathrm{T}$ at $1.4\,\mathrm{K}$ for a magnetic tunnel transistor injector. The thin and thick lines represent the σ^+ and σ^- circular polarization components, respectively.

in electrical transport measurements in magnetic tunnel transistors. Excitons in $In_{0.2}Ga_{0.8}As$ have a large g-factor, leading to a large Zeeman splitting energy in the quantum wells, which is shown by the shift of the luminescence peak center positions for the σ^+ and σ^- components at high fields.

Figure 6.27(a) shows P_{EL} measured as a function of the magnetic field. P_{EL} increases rapidly with the field up to $\sim 2\,\mathrm{T}$, where the magnetic moments of the ferromagnetic emitter and base layers are expected to be completely rotated out of plane by the field. Above $2\,\mathrm{T}$, P_{EL} continues to increase with the field roughly linearly. This background polarization at high fields may be due to thermalization of the excitons in the quantum wells. Field-dependent spin relaxation and/or recombination rates may also contribute to this background. The polarization P_{C}, obtained after subtracting this linear background polarization from P_{EL}, is shown in Fig. 6.27(b). A P_{C} value of $\sim 10\%$ is obtained at $2.5\,\mathrm{T}$. Note that polarization-dependent reflection at the ferromagnetic base/GaAs interface may give rise to a contribution to the electroluminescence polarization. However, it is found that this effect is very small ($< 1\%$) by passing linearly polarized light through the backside of the wafer and measuring the polarization of

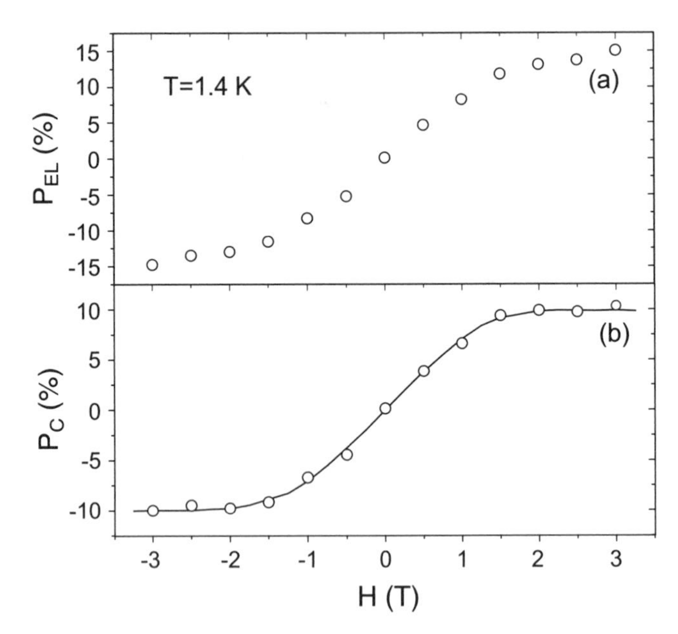

FIG. 6.27. Electroluminescence polarization before (a) and after (b) subtraction of the linear background as a function of the magnetic field for a magnetic tunnel transistor injector. The solid line in (b) is the magnetization of the ferromagnetic layers measured with a SQUID magnetometer at 10 K, which is scaled to compare with the polarization data.

light reflected from the ferromagnetic layers. The solid line in Fig. 6.27(b) shows the magnetization of the ferromagnetic layers measured with a superconducting quantum interference device (SQUID) magnetometer at 10 K for magnetic fields oriented perpendicular to the sample. The field dependence of the sample magnetization is in excellent agreement with the field dependence of P_C, confirming that P_C is related to the injection of spin-polarized hot electrons from the magnetic tunnel transistor.

The DP spin relaxation rate in GaAs is approximately proportional to the third power of the electron energy. As a result, DP spin relaxation becomes very effective at elevated electron energies [124]. The injected hot electrons likely lose a significant amount of spin polarization during the process of thermalization to the bottom of the conduction band. After the hot electrons enter the quantum well region, further spin relaxation can occur before they recombine with the holes [152]. It would be interesting to use other semiconductor collectors (e.g. Si or GaN), where the DP mechanism is not so important [153, 154]. Under these circumstances, the measured spin polarization should be much larger.

6.4.2.2 *Spin injection into GaAs from a CoFe/MgO tunnel injector* The small collector current of the magnetic tunnel transistor requires the device to operate

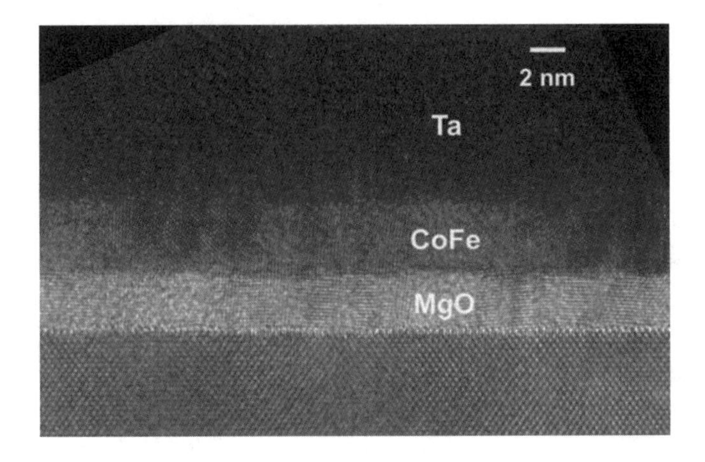

FIG. 6.28. HRTEM image of a CoFe/MgO tunnel spin injector.

at large electron energy in order to obtain enough signal in the spin injection experiments. Electron spin relaxation, however, is very efficient at high energy, leading to a moderate electroluminescence polarization of $\sim 10\%$. To increase the electron current at lower energy while maintaining a high spin polarization, a crystalline CoFe/MgO tunnel injector is formed. As discussed in Section 6.1, the tunneling spin polarization of a CoFe/MgO(100) structure is predicted to be very high using a first-principles calculation [61,62,155]. Experimentally, tunneling spin polarization as high as 85% has been observed by Parkin *et al.* [156]. Therefore, a CoFe/MgO tunnel injector should be able to inject highly spin polarized electrons into the semiconductor

Figure 6.28 shows a high-resolution transmission electron microscopy (HRTEM) image of a CoFe/MgO spin injector grown on a GaAs heterostructure. Both the MgO and CoFe layers are very smooth and are polycrystalline with a strong (100) texture along the growth direction. Such crystallographic orientations are consistent with theoretically predicted orientations which should give rise to high tunneling spin polarization [61, 62, 155].

Figure 6.29 shows the electroluminescence spectra using CoFe/MgO tunnel injectors for spin injection. (a) and (b) represent two samples with a GaAs/$Al_{0.08}$ $Ga_{0.92}As$ (sample I) and a GaAs/$Al_{0.16}Ga_{0.84}As$ (sample II) quantum-well detector, respectively. For both samples, the injector consists of 50 Å $Co_{70}Fe_{30}$ and 30 Å MgO capped with 100 Å Ta. The spectra in (a) are taken at 100 K with a bias voltage (V_T) of 1.8 V applied across the semiconductor heterostructure. The spectra in (b) are taken at 290 K with $V_T = 2.0$ V. The electroluminescence peaks at longer wavelengths are due to the recombination of electrons with the heavy holes in the quantum well, whilst the peaks at shorted wavelengths are due to the recombination of electrons with light holes [157]. For both samples, the electroluminescence intensities of the left and right circular polarization components are magnetic field dependent, giving rise to a significant polarization as

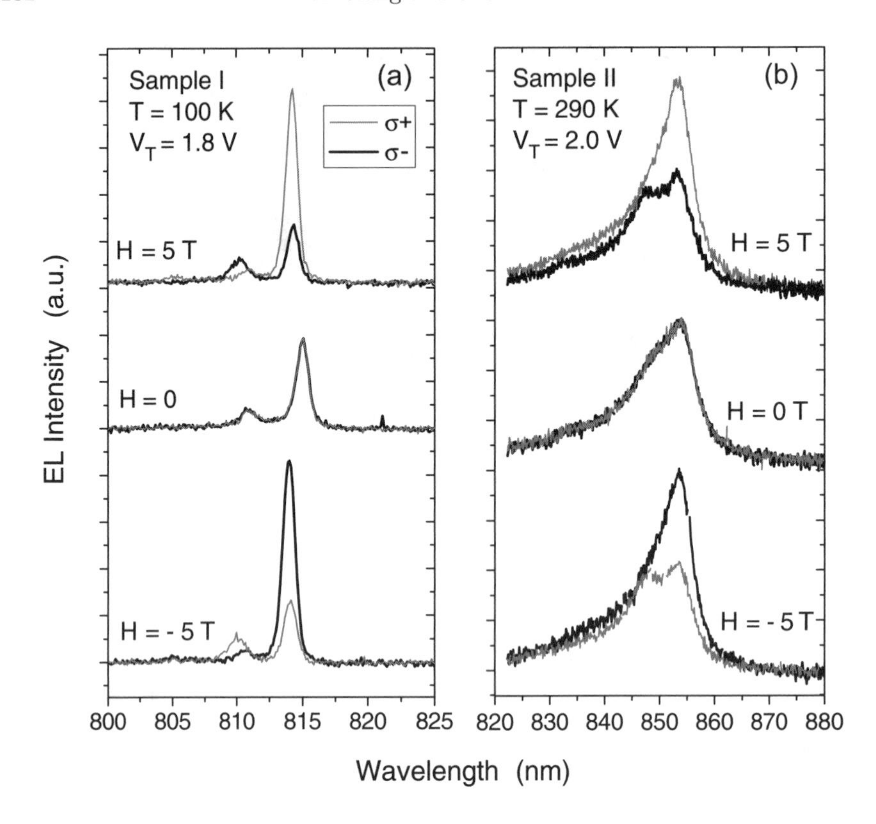

FIG. 6.29. Electroluminescence spectra of sample I (a) and II (b) with CoFe/MgO tunnel spin injectors. The thin and thick lines in (a) and (b) represent the σ^+ and σ^- circular polarization components of the luminescence, respectively.

the CoFe moment is rotated out of the film plane. Again, the sign of the luminescence polarization indicates majority spin injection from CoFe. Since the circular polarization of the heavy-hole emission has a simple relationship with the spin polarization of the electrons just prior to recombination, henceforth, only the heavy-hole luminescence polarization is discussed and it is referred to as $P_{\rm EL}$. For sample I, the heavy-hole emission is well resolved from the light hole emission due to its narrow linewidth ($\sim 10\,$Å). Therefore, it is rather straightforward to determine $P_{\rm EL}$. In contrast, the heavy and light hole peaks for sample II are broad at 290 K and are thus less well resolved. In order to extract $P_{\rm EL}$ for this sample, the luminescence spectrum is fit with two Lorentzians and $P_{\rm EL}$ is calculate from the fit.

The magnetic field dependences of $P_{\rm EL}$ for sample I at 100 K and sample II at 290 K are depicted in Fig. 6.30(a) and (b). In each case the polarization increases rapidly with field up to $\sim 2\,$T, when the CoFe moment is rotated completely out

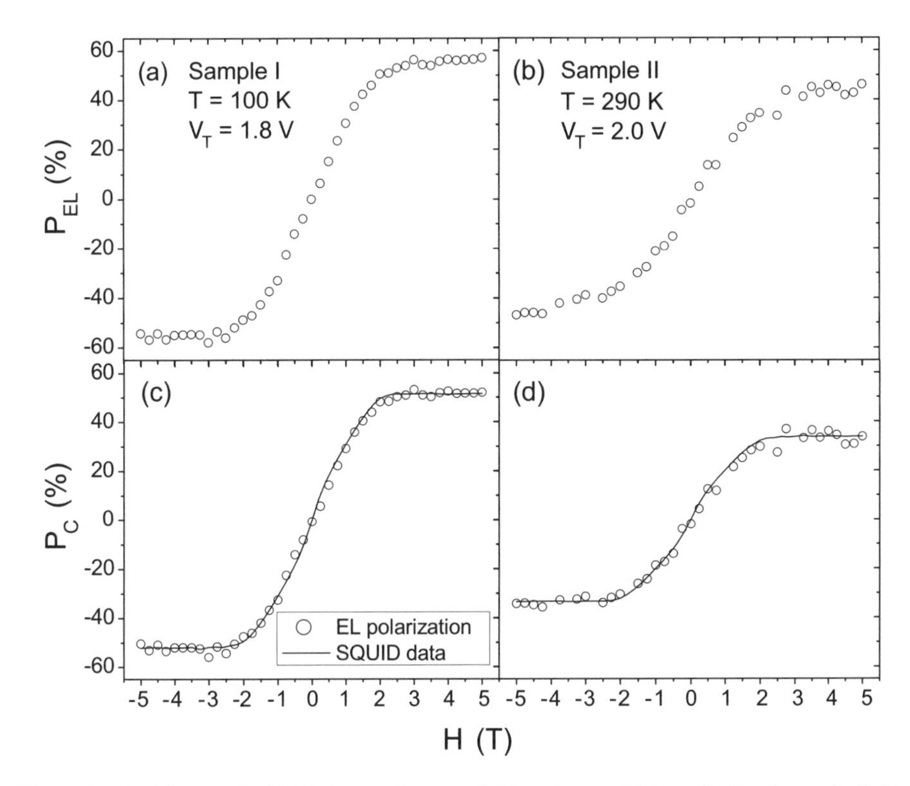

FIG. 6.30. Magnetic field dependence of P_{EL} (a and b) and P_{C} (c and d) for sample I at 100 K and sample II at 290 K (open circles). The solid lines in (c and d) show the field dependence of the CoFe moment measured with a SQUID magnetometer at 20 K, which has been scaled to allow comparison with P_{C}.

of plane. Above 2 T, P_{EL} continues to increase with field approximately linearly, but at a much lower rate, reaching 57% and 47% at 5 T for sample I and II, respectively. A linear variation of P_{EL} with field above 2 T is observed for both samples over a wide temperature range. The slope of this background polarization usually varies gradually from a negative value at low temperatures to a positive value at high temperatures, crossing zero at $\sim$40–50 K. Several factors may contribute to the background polarization. At low temperatures, thermalization of electron spins in the quantum well due to Zeeman splitting could give rise to a negative background since GaAs has a negative g-factor. At high temperatures, however, the Zeeman energy is negligible compared to $k_{\mathrm{B}}T$, and therefore cannot explain the observed background polarization. It is likely due to field dependent spin relaxation and/or electron-hole recombination times. It is well known that a perpendicular magnetic field can suppress DP spin relaxation in GaAs [124], which would therefore give rise to a positive background. Moreover, it is found

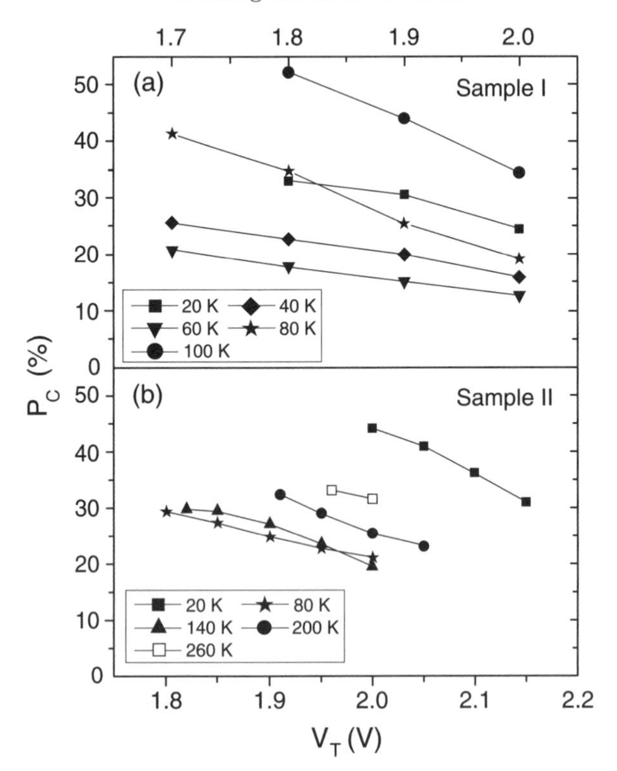

FIG. 6.31. Bias and temperature dependence of P_C of sample I (a) and II (b). Note the different bias ranges for (a) and (b).

that the luminescence intensity from the quantum well increases with increasing fields, implying a shorter recombination time at higher fields which would also give rise to a positive background.

The electroluminescence polarization after subtraction of the linear background is shown in Fig. 6.30(c) and 6.30(d), which is a measure of spin polarization when the magnetic field influence on the polarization is excluded. P_C values as high as 52% and 32% were obtained at 100 K and 290 K, respectively. The solid lines in Fig. 6.30(c) and 6.30(d) show the field dependence of the CoFe moment measured in a perpendicular magnetic field with a SQUID magnetometer at 20 K. The excellent agreement between the polarization data and SQUID data confirms that the large polarization originates from spin injection.

The bias and temperature dependence of P_C is shown in Fig. 6.31 for the two samples. The relatively small confinement potential of the GaAs/$Al_{0.08}Ga_{0.92}As$ quantum well results in weak luminescence signals at high temperatures and consequently limits the measurements on sample I to below 100 K. In contrast, measurements on sample II are possible up to room temperature due to the use of a deeper GaAs/$Al_{0.16}Ga_{0.84}As$ quantum well. For both samples, P_C decreases with increasing bias at a given temperature. A similar bias dependence was

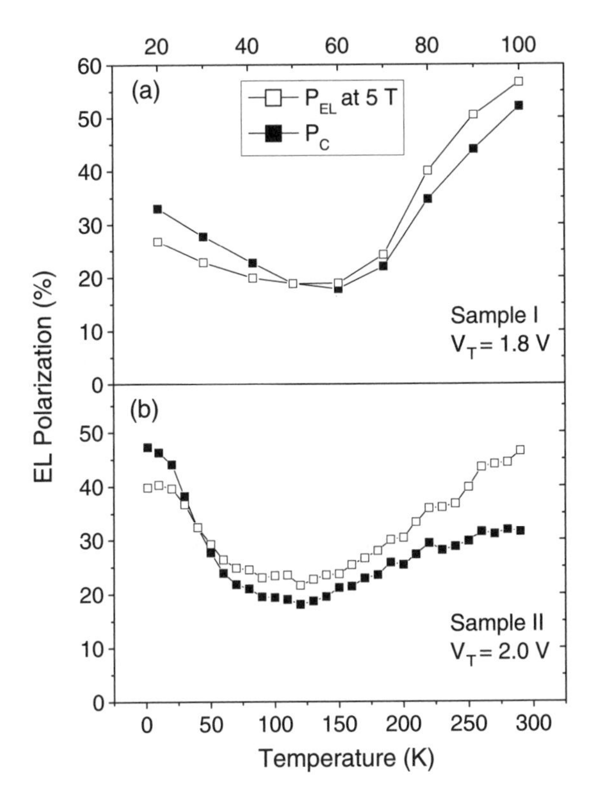

FIG. 6.32. Temperature dependence of the electroluminescence polarization of sample I (a) and II (b). The open and closed squares correspond to values of P_{EL} at 5 T and of P_{C} respectively. Note the different temperature ranges for (a) and (b).

observed in optical experiments and was attributed to spin relaxation through the DP mechanism before photo-excited electrons reached the quantum well [158, 159]. In semiconductors lacking inversion symmetry, DP spin relaxation occurs due to spin precession about an effective magnetic field whose orientation and magnitude depends on the electron momentum. Larger electron momentum at higher bias results in a bigger effective field and consequently more rapid spin relaxation [124].

A non-monotonic temperature dependence of the electroluminescence polarization is found for both samples, which is plotted in Fig. 6.32. The bias voltage is $V_{\mathrm{T}} = 1.8\,\mathrm{V}$ and $2.0\,\mathrm{V}$ for sample I and II, respectively. In the spin injection experiment, the electroluminescence polarization depends on the spin relaxation rate and the electron recombination time in the quantum well detector. The measured luminescence polarization P in a steady state is given by [124]

$$P = \frac{\tau_s}{\tau_s + \tau_R} P_0,$$

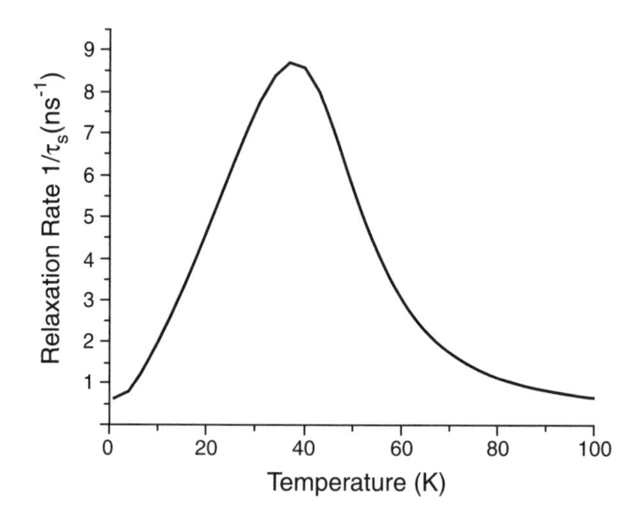

FIG. 6.33. DP spin relaxation rate in a quantum well calculated by Puller *et al.* [160].

where P_0 is the initial spin polarization of the electrons injected into the quantum well, τ_s and τ_R are the spin and electron lifetimes, respectively. The DP spin relaxation rate in a quantum well is given by

$$\tau_s^{-1} \propto \tau_p T,$$

where τ_p is the momentum relaxation time and T is the temperature [150]. At very low temperatures, τ_p is dominated by ionized impurity scattering which has a weak temperature dependence, so that $\tau_p T$ and, consequently, the spin relaxation rate increase with temperature. At higher temperatures, when polar optical phonon scattering dominates the momentum scattering, $\tau_p T$ and, therefore, the spin relaxation rate decrease with increasing temperature [160]. As a result, the luminescence polarization tends to increase with temperature.

Figure 6.33 shows the DP spin relaxation rate in a quantum well calculated by Puller *et al.* [160] The spin relaxation rate has a maximum in the intermediate temperature range, which is qualitatively consistent with the experimental data shown in Fig. 6.32. The electron recombination time also varies with temperature [161, 162] and contributes to the temperature dependence of the electroluminescence polarization. It has been found that the radiative electron recombination time increases with temperature whilst the non-radiative recombination time decreases with temperature at high temperatures [162]. This interplay gives rise to a peak in the temperature dependence of the recombination time which could account for the temperature dependence of the electroluminescence polarization. Both the spin relaxation rate and the electron recombination time are dependent on the details of the quantum well detectors, which likely accounts for the quantitative differences between samples I and II.

The observation of efficient spin injection up to 290 K using a CoFe/MgO tunnel injector is consistent with the high Curie temperature of CoFe and the weak temperature dependence of spin-dependent tunneling. The actual spin injection efficiency will be higher than that inferred from the polarization of the quantum well electroluminescence because of spin relaxation in the quantum-well detector. Moreover, the spin relaxation time and electron recombination time are both strongly temperature dependent so giving rise to a non-monotonic temperature dependence of the luminescence polarization. The MgO-based spin injector can readily be fabricated by sputter deposition. In addition, the MgO barrier prevents intermixing of the ferromagnetic metal and the semiconductor, leading to improved device thermal stability [163]. These desirable features make MgO-based tunnel spin injectors attractive for future semiconductor spintronic applications.

References

[1] P. M. Levy, in *Solid State Physics*, edited by H. Ehrenreich and D. Turnbull (Academic Press, New York, 1994), Vol. **47**, p. 367.
[2] S. S. P. Parkin, in *Annual Review of Materials Science*, edited by B. W. Wessels (Annual Reviews Inc., Palo Alto, 1995), Vol. **25**, p. 357.
[3] S. S. P. Parkin, in *Ultrathin Magnetic Structures*, edited by B. Heinrich and J. A. C. Bland (Spinger-Verlag, Berlin, 1994), Vol. **II**, p. 148.
[4] A. Fert and P. Bruno, in *Ultrathin Magnetic Structures*, edited by B. Heinrich and J. A. C. Bland (Springer-Verlag, Berlin, 1994), Vol. **II**, p. 82.
[5] R. E. Camley and R. L. Stamps, J. Phys. C. Condens. Matter 5, 3727 (1993).
[6] E. Y. Tsymbal and D. G. Pettifor, in *Solid State Physics*, edited by H. Ehrenreich and F. Spaepen (Academic Press, 2001), Vol. **56**.
[7] M. A. M. Gijs and G. E. W. Bauer, Adv. Phys. **46**, 285 (1997).
[8] J. Mathon, Contemp. Phys. **32**, 143 (1991).
[9] S. S. P. Parkin, X. Jiang, C. Kaiser, *et al.*, Proc. IEEE **91**, 661 (2003).
[10] S. S. P. Parkin, in *Spin dependent transport in magnetic nanostructures*, edited by S. Maekawa and T. Shinjo (Taylor & Francis, London, 2002).
[11] S. S. P. Parkin, Z. G. Li, and D. J. Smith, Appl. Phys. Lett. **58**, 2710 (1991).
[12] S. S. P. Parkin, Phys. Rev. Lett. **71**, 1641 (1993).
[13] S. S. P. Parkin, N. More, and K. P. Roche, Phys. Rev. Lett. **64**, 2304 (1990).
[14] S. S. P. Parkin and D. Mauri, Phys. Rev. B **44**, 7131 (1991).
[15] W. F. Egelhoff, T. Ha, R. D. K. Misra, *et al.*, J. Appl. Phys. **78**, 273 (1995).
[16] W. F. Egelhoff, P. J. Chen, C. J. Powell, *et al.*, J. Appl. Phys. **79**, 5277 (1996).
[17] M. Bode, Rep. Prog. Phys. **66**, 523 (2003).
[18] P. D. Johnson, Rep. Prog. Phys. **60**, 1217 (1997).
[19] T. Miyazaki, T. Yaoi, and S. Ishio, J. Magn. Magn. Mater. **98**, L7 (1991).
[20] E. Y. Tsymbal, O. N. Mryasov, and P. R. LeClair, J. Phys. C: Condens. Matter **15**, R109 (2003).
[21] T. Miyazaki and N. Tezuka, J. Magn. Magn. Mater. **151**, 403 (1995).

[22] J. S. Moodera, L. R. Kinder, T. M. Wong, *et al.*, Phys. Rev. Lett. **74**, 3273 (1995).
[23] S. S. P. Parkin, K. P. Roche, M. G. Samant, *et al.*, J. Appl. Phys. **85**, 5828 (1999).
[24] W. H. Meiklejohn and C. P. Bean, Phys. Rev. **102**, 1413 (1956).
[25] A. Yelon, in *Physics of Thin Films: Advances in Research and Development*, edited by M. Francombe and R. Hoffman (Academic Press, New York, 1971), Vol. **6**, p. 205.
[26] F. Nolting, A. Scholl, J. Stöhr, *et al.*, Nature **405**, 767 (2000).
[27] H. Ohldag, A. Scholl, F. Nolting, *et al.*, Phys. Rev. Lett. **91**, 017203 (2003).
[28] A. Scholl, H. Ohldag, F. Nolting, *et al.*, Rev. Sci. Instrum. **73**, 1362 (2002).
[29] D. Wang, C. Nordman, J. M. Daughton, *et al.*, IEEE Trans. Magn. **40**, 2269 (2004).
[30] A. R. Ferchmin and S. Kobe, *Amorphous Magnetism and Metallic Magnetic Materials Digest* (North Holland Publishing Company, New York, 1983).
[31] A. P. Malozemoff, A. R. Williams, V. L. Moruzzi, *et al.*, Phys. Rev. B **30**, 6565 (1984).
[32] S. S. P. Parkin, C. Kaiser, P. Rice, *et al.*, (unpublished).
[33] M. Julliere, Phys. Lett. **54A**, 225 (1975).
[34] S. Maekawa and U. Gäfvert, IEEE Trans. Magn. **MAG-18**, 707 (1982).
[35] J. Nowak and J. Rauluszkiewicz, J. Magn. Magn. Mater. **109**, 79 (1992).
[36] T. Miyazaki and N. Tezuka, J. Magn. Magn. Mater. **139**, L231 (1995).
[37] R. Meservey and P. M. Tedrow, Phys. Rep. **238**, 173 (1994).
[38] A. Barthélémy, A. Fert, and F. Petroff, in *Handbook of Magnetic Materials*, edited by K. Buschow (Elsevier, 1999), p. 1.
[39] J. C. S. Kools, IEEE Trans. Magn. **32**, 3165 (1996).
[40] A. Moser, K. Takano, D. T. Margulies, *et al.*, J. Phys. D: Appl. Phys. **35**, R157 (2002).
[41] R. Wood, Y. Sonobe, Z. Jin, *et al.*, J. Magn. Magn. Mater. **235**, 1 (2001).
[42] R. E. Scheuerlein, in *NCE Seventh Biennial IEEE International Nonvolatile MemoryTechnology Conference* (Proc. IEEE Inter. Nonvolatile Memory Technology Conf., Albuquerque, NM, USA, 1998), p. 47.
[43] P. M. Tedrow and R. Meservey, Phys. Rev. Lett. **27**, 919 (1971).
[44] D. A. Papaconstantopoulos, *Handbook of the Band Structure of Elemental Solids* (Plenum Press, New York, 1986).
[45] M. B. Stearns, J. Magn. Magn. Mater. **5**, 167 (1977).
[46] R. J. Soulen, J. M. Byers, M. S. Osofsky, *et al.*, Science **282**, 85 (1998).
[47] I. I. Mazin, A. A. Golubov, and B. Nadgorny, J. Appl. Phys. **89**, 7576 (2001).
[48] S. K. Upadhyay, A. Palanisami, R. N. Louie, *et al.*, Phys. Rev. Lett. **81**, 3247 (1998).
[49] G. T. Woods, R. J. Soulen, I. Mazin, *et al.*, Phys. Rev. B **70**, 054416 (2004).
[50] B. Nadgorny, R. J. Soulen, M. S. Osofsky, *et al.*, Phys. Rev. B **61**, R3788 (2000).
[51] S. Maekawa and T. Shinjo, (Taylor & Francis, London, 2002).
[52] D. C. Worledge and T. H. Geballe, Phys. Rev. B **62**, 447 (2000).

[53] R. Meservey, D. Paraskevopolous, and P. M. Tedrow, Phys. Rev. Lett. **37**, 858 (1976).
[54] D. J. Monsma and S. S. P. Parkin, Appl. Phys. Lett. **77**, 720 (2000).
[55] J. Z. Sun, D. W. Abraham, K. Roche, *et al.*, Appl. Phys. Lett. **73**, 1008 (1998).
[56] M.-H. Jo, N. D. Mathur, N. K. Todd, *et al.*, Phys. Rev. B **61**, R14905 (2000).
[57] J. O'Donnell, A. E. Andrus, S. Oh, *et al.*, Appl. Phys. Lett. **76**, 1914 (2000).
[58] J. S. Parker, S. M. Watts, P. G. Ivanov, *et al.*, Phys. Rev. Lett. **88**, 196601 (2002).
[59] M. Bowen, M. Bibes, A. Barthélémy, *et al.*, Appl. Phys. Lett. **82**, 233 (2003).
[60] P. Mavropoulos, N. Papanikolaou, and P. H. Dederichs, Phys. Rev. Lett. **85**, 1088 (2000).
[61] W. H. Butler, X.-G. Zhang, T. C. Schulthess, *et al.*, Phys. Rev. B **63**, 054416 (2001).
[62] J. Mathon and A. Umerski, Phys. Rev. B **63**, 220403 (2001).
[63] S. S. P. Parkin, C. Kaiser, A. Panchula, *et al.*, Nature Mater. **3**, 862 (2004).
[64] S. Yuasa, T. Nagahama, A. Fukushima, *et al.*, Nature Mater. **3**, 868 (2004).
[65] D. C. Worledge and P. L. Trouilloud, Appl. Phys. Lett. **83**, 84 (2003).
[66] J. M. De Teresa, A. Barthélémy, A. Fert, *et al.*, Science **286**, 507 (1999).
[67] E. Y. Tsymbal and D. G. Pettifor, Phys. Rev. B **58**, 432 (1998).
[68] M. Zwierzycki, K. Xia, P. J. Kelly, *et al.*, Phys. Rev. B **67**, 092401 (2003).
[69] W. Reohr, H. Hönigschmid, R. Robertazzi, *et al.*, IEEE Circuits & Devices Magazine **18**, 17 (2002).
[70] D. P. Pappas, K. P. Kämper, B. P. Miller, *et al.*, Phys. Rev. Lett. **66**, 504 (1991).
[71] G. Schönhense and H. C. Siegmann, Annalen der Physik **2**, 465 (1993).
[72] M. Getzlaff, J. Bansmann, and G. Schönhense, Solid State Commun. **87**, 467 (1993).
[73] J. C. Gröbli, A. Kündig, F. Meier, *et al.*, Physica B **204**, 359 (1995).
[74] E. Vescovo, C. Carbone, U. Alkemper, *et al.*, Phys. Rev. B **52**, 13497 (1995).
[75] H.-J. Drouhin, A. J. v. d. Sluijs, Y. Lassailly, *et al.*, J. Appl. Phys. **79**, 4734 (1996).
[76] A. Filipe, H.-J. Drouhin, G. Lampel, *et al.*, Phys. Rev. Lett. **80**, 2425 (1998).
[77] D. Oberli, R. Burgermeister, S. Riesen, *et al.*, Phys. Rev. Lett. **81**, 4228 (1998).
[78] M. Aeschlimann, M. Bauer, S. Pawlik, *et al.*, Phys. Rev. Lett. **79**, 5158 (1997).
[79] R. Knorren, K. H. Bennemann, R. Burgermeister, *et al.*, Phys. Rev. B **61**, 9427 (2000).
[80] W. H. Rippard and R. A. Buhrman, Phys. Rev. Lett. **84**, 971 (2000).
[81] W. Weber, S. Riesen, and H. C. Siegmann, Science **291**, 1015 (2001).
[82] D. J. Monsma, J. C. Lodder, T. J. A. Popma, *et al.*, Phys. Rev. Lett. **74**, 5260 (1995).
[83] D. J. Monsma, R. Vlutters, and J. C. Lodder, Science **281**, 407 (1998).
[84] S. van Dijken, X. Jiang, and S. S. P. Parkin, Phys. Rev. B **66**, 094417 (2002).
[85] N. Mott, Adv. Phys. **13**, 325 (1964).
[86] W. J. Kaiser and L. D. Bell, Phys. Rev. Lett. **60**, 1406 (1988).

[87] W. H. Rippard and R. A. Buhrman, Appl. Phys. Lett. **75**, 1001 (1999).
[88] R. Jansen, P. S. A. Kumar, O. M. J. v. t. Erve, *et al.*, Phys. Rev. Lett. **85**, 3277 (2000).
[89] R. Jansen, O. M. J. van 't Erve, S. D. Kim, *et al.*, J. Appl. Phys. **89**, 7431 (2001).
[90] R. Jansen, J. Phys. D: Appl. Phys. **36**, R289 (2003).
[91] K. Mizushima, T. Kinno, T. Yamauchi, *et al.*, IEEE Trans. Magn. **33**, 3500 (1997).
[92] R. Sato and K. Mizushima, Appl. Phys. Lett. **79**, 1157 (2001).
[93] S. van Dijken, X. Jiang, and S. S. P. Parkin, Appl. Phys. Lett. **80**, 3364 (2002).
[94] X. Jiang, R. Wang, S. van Dijken, *et al.*, Phys. Rev. Lett. **90**, 256603 (2003).
[95] L. D. Bell, Phys. Rev. Lett. **77**, 3893 (1996).
[96] J. J. Quinn, Phys. Rev. **126**, 1453 (1962).
[97] E. Zarate, P. Apell, and P. M. Echenique, Phys. Rev. B **60**, 2326 (1999).
[98] H.-J. Drouhin, J. Appl. Phys. **89**, 6805 (2001).
[99] R. Vlutters, O. M. J. van 't Erve, S. D. Kim, *et al.*, Phys. Rev. Lett. **88**, 027202 (2002).
[100] S. van Dijken, X. Jiang, and S. S. P. Parkin, Phys. Rev. Lett. **90**, 197203 (2003).
[101] X. Jiang, S. van Dijken, R. Wang, *et al.*, Phys. Rev. B **69**, 014413 (2004).
[102] D. L. Smith, M. Kozhevnikov, E. Y. Lee, *et al.*, Phys. Rev. B **61**, 13914 (2000).
[103] J. S. Blakemore, J. Appl. Phys. **53**, R123 (1982).
[104] D. L. Smith, E. Y. Lee, and V. Narayanamurti, Phys. Rev. Lett. **80**, 2433 (1998).
[105] G. E. Moore, Electronics **38**, 114 (1965).
[106] M. N. Baibich, J. M. Broto, A. Fert, *et al.*, Phys. Rev. Lett. **61**, 2472 (1988).
[107] G. Binasch, P. Grünberg, F. Saurenbach, *et al.*, Phys. Rev. B **39**, 4828 (1989).
[108] S. S. P. Parkin, N. More, and K. P. Roche, Phys. Rev. Lett. **64**, 2304 (1990).
[109] S. S. P. Parkin, R. Bhadra, and K. P. Roche, Phys. Rev. Lett. **66**, 2152 (1991).
[110] M. Julliere, Phys. Lett. **54A**, 225 (1975).
[111] T. Miyazaki and N. Tezuka, J. Magn. Magn. Mater. **139**, L231 (1995).
[112] J. S. Moodera, L. R. Kinder, T. M. Wong, *et al.*, Phys. Rev. Lett. **74**, 3273 (1995).
[113] S. Parkin, X. Jiang, C. Kaiser, *et al.*, Proc. IEEE **91**, 661 (2003).
[114] J. M. Kikkawa and D. D. Awschalom, Phys. Rev. Lett. **80**, 4313 (1998).
[115] J. M. Kikkawa and D. D. Awschalom, Nature (London) **397**, 139 (1999).
[116] H. Ohno, D. Chiba, F. Matsukura, *et al.*, Nature **408**, 944 (2000).
[117] O. Z. Karimov, G. H. John, R. T. Harley, *et al.*, Phys. Rev. Lett. **91**, 246601 (2003).
[118] S. Murakami, N. Nagaosa, and S.-C. Zhang, Science **301**, 1348 (2003).
[119] S. A. Wolf, D. D. Awschalom, R. A. Buhrman, *et al.*, Science **294**, 1488 (2001).
[120] S. Datta and B. Das, Appl. Phys. Lett. **56**, 665 (1990).
[121] E. I. Rashba, Sov. Phys. Solid State **2**, 1109 (1960).
[122] Y. A. Bychkov and E. I. Rashba, JETP Lett. **39**, 78 (1984).
[123] J. Nitta, T. Akazaki, H. Takayanagi, *et al.*, Phys. Rev. Lett. **78**, 1335 (1997).

[124] F. Meier and B. P. Zakharchenya, *Optical Orientation* (NorthHolland, Amsterdam, 1984).
[125] F. G. Monzon and M. L. Roukes, J. Magn. Magn. Mater. **198-199**, 632 (1999).
[126] S. Gardelis, C. G. Smith, C. H. W. Barnes, *et al.*, Phys. Rev. B **60**, 7764 (1999).
[127] A. T. Filip, B. H. Hoving, F. J. Jedema, *et al.*, Phys. Rev. B **62**, 9996 (2000).
[128] G. Schmidt, D. Ferrand, L. W. Molenkamp, *et al.*, Phys. Rev. B **62**, R4790 (2000).
[129] P. R. Hammar, B. R. Bennett, M. J. Yang, *et al.*, Phys. Rev. Lett. **83**, 203 (1999).
[130] M. Johnson, Phys. Rev. B **58**, 9635 (1998).
[131] C.-M. Hu, J. Nitta, A. Jensen, *et al.*, Phys. Rev. B **63**, 125333 (2001).
[132] M. Oestreich, J. Hübner, D. Hägele, *et al.*, Appl. Phys. Lett. **74**, 1251 (1999).
[133] R. Fiederling, M. Keim, G. Reuscher, *et al.*, Nature **402**, 787 (1999).
[134] Y. Ohno, D. K. Young, B. Beschoten, *et al.*, Nature **402**, 790 (1999).
[135] B. T. Jonker, Y. D. Park, B. R. Bennett, *et al.*, Phys. Rev. B **62**, 8180 (2000).
[136] B. T. Jonker, A. T. Hanbicki, Y. D. Park, *et al.*, Appl. Phys. Lett. **79**, 3098 (2001).
[137] E. I. Rashba, Phys. Rev. B **62**, R16267 (2000).
[138] H. J. Zhu, M. Ramsteiner, H. Kostial, *et al.*, Phys. Rev. Lett. **87**, 016601 (2001).
[139] A. T. Hanbicki, B. T. Jonker, G. Itskos, *et al.*, Appl. Phys. Lett. **80**, 1240 (2002).
[140] A. T. Hanbicki, O. M. J. van 't Erve, R. Magno, *et al.*, Appl. Phys. Lett. **82**, 4092 (2003).
[141] C. Adelmann, X. Lou, J. Strand, *et al.*, Phys. Rev. B **71**, R121301 (2005).
[142] T. Manago and H. Akinaga, Appl. Phys. Lett. **81**, 694 (2002).
[143] O. M. J. van 't Erve, G. Kioseoglou, A. T. Hanbicki, *et al.*, Appl. Phys. Lett. **84**, 4334 (2004).
[144] V. F. Motsnyi, J. D. Boeck, J. Das, *et al.*, Appl. Phys. Lett. **81**, 265 (2002).
[145] V. F. Motsnyi, P. V. Dorpe, W. V. Roy, *et al.*, Phys. Rev. B **68**, 245319 (2003).
[146] P. Van Dorpe, V. F. Motsnyi, M. Nijboer, *et al.*, Jpn. J. Appl. Phys. **42**, L502 (2003).
[147] R. J. Elliott, Phys. Rev. **96**, 266 (1954).
[148] Y. Yafet, in *Solid State Physics*, edited by F. Seitz and D. Turnball (Academic, New York, 1963), Vol. **14**, p. 1.
[149] M. I. D'yakonov and V. I. Perel', Sov. Phys. JETP **33**, 1053 (1971).
[150] M. I. D'yakonov and V. Y. Kachorovskii, Sov. Phys. Semicond. **20**, 110 (1986).
[151] G. L. Bir, A. G. Aronov, and G. E. Pikus, Sov. Phys. JETP **42**, 705 (1976).
[152] T. Amand, B. Dareys, B. Baylac, *et al.*, Phys. Rev. B **50**, 11624 (1994).
[153] J. Fabian and S. Das Sarma, J. Vac. Sci. Technol. B **17**, 1708 (1999).
[154] S. Krishnamurthy, M. van Schilfgaarde, and N. Newman, Appl. Phys. Lett. **83**, 1761 (2003).
[155] X.-G. Zhang and W. H. Butler, Phys. Rev. B **70**, 172407 (2004).
[156] S. S. P. Parkin, C. Kaiser, A. F. Panchula, *et al.*, Nature Materials **3**, 862 (2004).

[157] P. D. Johnson, in *Annu. Rev. Mater. Sci.*, edited by B. W. Wessels (Annual Reviews Inc., Palo Alto, 1995), Vol. **25**, p. 455.
[158] H. Sanada, I. Arata, Y. Ohno, *et al.*, Appl. Phys. Lett. **81**, 2788 (2002).
[159] E. A. Barry, A. A. Kiselev, and K. W. Kim, Appl. Phys. Lett. **82**, 3686 (2003).
[160] V. I. Puller, L. G. Mourokh, N. J. M. Horing, *et al.*, Phys. Rev. B **67**, 155309 (2003).
[161] J. Feldmann, G. Peter, E. O. Göbel, *et al.*, Phys. Rev. Lett. **59**, 2337 (1987).
[162] M. Gurioli, A. Vinattieri, M. Colocci, *et al.*, Phys. Rev. B **44**, 3115 (1991).
[163] R. Wang, X. Jiang, R. M. Shelby, *et al.*, Appl. Phys. Lett. **86**, 052901 (2005).
[164] I. Giaever, Phys. Rev. Lett. **5**, 147 (1960).
[165] P. M. Tedrow and R. Meservey, Phys. Rev. Lett. **26**, 192 (1971).
[166] T. Yaoi, S. Ishio, and T. Miyazaki, IEEE Trans. J. Magn. Jpn. **8**, 498 (1993).
[167] W. J. Gallagher, S. S. P. Parkin, Y. Lu, *et al.*, J. Appl. Phys. **81**, 3741 (1997).
[168] R. Scheuerlein, W. Gallagher, S. Parkin, *et al.*, 2000 IEEE Int. Solid State Circuits Conf. Digest Tech. Papers, 218 (2000).
[169] H. Kano, K. Bessho, Y. Higo, *et al.*, in *Intermag 2002*, (Amsterdam, The Netherlands, 2002), paper BB04.
[170] A. R. Sitaram, D. W. Abraham, C. Alof, *et al.*, Proc. 2003 Symposia on VLSI Technology and Circuits (2003).
[171] J. DeBrosse, D. Gogl, A. Bette, *et al.*, IEEE J. Solid State Circuits **39**, 678 (2004).

7 Theory of spin-transfer torque and domain wall motion in magnetic nanostructures

S. E. Barnes and S. Maekawa

7.1 Introduction

The use of magnetic materials to store information is an indispensable part of current technology. The amount of information stored on the active magnetic medium of hard drives denies superlatives. This information is invariably written by reversing micron-sized magnetic domains using magnetic fields. The use of magnet RAM, however, has long been abandoned in favor of bistable electronic memory and currently electronic "flash" memory is even making inroads as non-volatile storage. The switching, i.e., coercive field B_c of a small mono-domain is roughly independent of size while the magnetic field, at constant current density, of a wire decreases as the radius r. Making current switched memory using the magnetic field therefore becomes more difficult as the size decreases. In contrast, for thin films the total angular momentum of a magnetic domain decreases as r^2 as does the angular momentum current carried by the conduction electrons in a wire of the same dimensions. If this angular momentum can be transferred directly, thereby switching the magnetic domain, the scaling problems associated with magnetic field switching are avoided. This is one of the promises of the subject discussed in this chapter.

It is also a striking experimental fact [1–4], comparable to the existence of the Josephson effect, that a direct current (dc) can induce radio frequency (rf) oscillations of nanodomains. It does not need much imagination to find direct applications for this effect.

Further interest in the spintronic manipulation of magnetic moments which has received little or no attention arises in connection with quantum computing. Molecular magnets incorporated in molecular spintronic circuits can function as qubits to be manipulated by currents using the principles of angular momentum transfer to be described in this chapter.

The idea of angular momentum transfer between a current and a magnetic domain wall was suggested independently by Slonczewski [5] and Berger [6]. Despite the elapsed time since this seminal work, the precise nature of the process involved remains unclear although there is clear experimental proof [7–14] for both spin valves and domain walls [15–19, 21] that such a coupling exists. The reader will find a review of the historical development in Chapter 5. The purpose of this chapter is to introduce a coherent theory of angular momentum transfer. The reader will find detailed reviews [22, 23] and some key papers [24–29] elsewhere. Of conceptual importance for this development is the idea that the angular momentum transfer process is not adiabatic. With the transfer of each electron there is a certain probability of an adiabatic rotation which transfers the transverse component of the spin current as discussed in Chapter 5, but in addition, with some other probability, a transition is made to an excited state involving magnons. Emphasized is the idea that coupling is to the various "dynamical" modes of the interface where the transfer occurs. The stability or otherwise of these modes in the presence of a current determines the observed static and dynamic states. Of importance is a bottleneck in the relaxation at a ferromagnetic to non-magnetic metal interface. An important issue is the correctness of traditional Gilbert relaxation. In the context of the dynamics of a current driven domain, it will be argued that this needs to be modified. For a domain wall these modifications imply that the "particle derivative" appears in the place of the usual partial derivative. It will be shown there exist certain non-resonant solutions to the dynamical equations which are of considerable importance to the interpretation of experiment.

An interesting subject which will also be addressed here is the existence of a spin-motive-force (smf). Spintronics implies the use of the spin degrees of freedom of electrons in order to develop new functional materials and devices. It is invariably the case that the forces which drive the currents are of electronic origin, i.e., arise from the Lorenz force $\boldsymbol{f}_e = q(\boldsymbol{E} + \boldsymbol{v} \times \boldsymbol{B})$ where $\boldsymbol{E}$ and $\boldsymbol{B}$ are true electric and magnetic fields. This ignores the force

$$\boldsymbol{f}_s = \frac{\partial}{\partial \boldsymbol{r}}(\boldsymbol{\mu} \cdot \boldsymbol{B}) \tag{7.1}$$

and which occurs, for example, in the famous Stern–Gerlach experiment [30]. Each element of $\boldsymbol{f} \equiv \boldsymbol{f}_e + \boldsymbol{f}_s$ can convert magnetic to electrical (potential) energy, i.e., can produce a non-conservative force which leads to an emf $\mathcal{E} = \oint (\boldsymbol{f}/e) \cdot d\boldsymbol{r}$. It should be evident that a time dependent $\boldsymbol{f}_s$ can produce non-conservative forces which couple to the spin degrees of freedom and which is therefore an smf. What is implied is that a current flows whenever the magnetic energy of a domain wall changes. This provides an alternative to the giant magnetoresistance effect (GMR) as a means by which to measure the state of a spintronics device and opens up the possibility of power amplification [31].

7.2 Landau–Lifshitz equations

The conservation of charge is embodied in the continuity equation

$$\frac{\partial \rho}{\partial t} + \boldsymbol{\nabla} \cdot \boldsymbol{J} = 0 \tag{7.2}$$

where ρ is the charge density and $\boldsymbol{J}$ the charge current density. At a fundamental level charge conservation reflects an electromagnetic gauge invariance in Schrödinger's equation via Neother's theorem. In the steady state $(\partial \rho / \partial t) = 0$ and the resulting $\boldsymbol{\nabla} \cdot \boldsymbol{J} = 0$ is just a version of Kirchhoff's node rule. This rule is very extensively used in the description of regular electronics. However, there is no equivalent gauge invariance which would imply the conservation of spin currents and no Kirchhoff's node rule for such currents. Also instead of a scalar charge, $\boldsymbol{M}$, a vector quantity (magnetization or angular momentum), is under consideration. Specifically, $\boldsymbol{M}$ denotes a local moment which is equivalent to the magnetization: $-g\mu_{\mathrm{B}}\boldsymbol{M}$ with the g-factor g, μ_{B} the Bohr magneton, and also to the angular momentum per unit volume: $\hbar\boldsymbol{M}$, with $\hbar$ the Planck constant divided by 2π. Note that for an electron the magnetization opposes the angular momentum due to its negative charge.

Even in the absence of spin currents $\boldsymbol{j}_s$, as reflected by the Landau–Lifshitz equations,

$$\frac{\partial \boldsymbol{M}}{\partial t} = -\frac{g\mu_{\mathrm{B}}}{\hbar}\boldsymbol{M} \times \boldsymbol{B}, \tag{7.3}$$

the magnetization $\boldsymbol{M}$ has a finite rate of change. The magnetic field $\boldsymbol{B}$ is given by the variation $-(\delta F/\delta \boldsymbol{M})$ since the free energy F must contain a term $-\boldsymbol{M} \cdot \boldsymbol{B}$ reflecting both the real applied magnetic fields plus those which arise from internal and anisotropy fields, for example. Now $\boldsymbol{M}$ is the *order parameter* for the broken symmetry of a ferromagnet. As such it is difficult to find processes by which $\boldsymbol{M}$ might relax. While $\boldsymbol{M}$ is not conserved, i.e., in quantum mechanical terms it does not commute with the Hamiltonian $\mathcal{H}$, its motion does trace out constant energy surfaces when relaxation is absent. Putting $\boldsymbol{M}$ in motion using an external radio frequency (rf) field in a ferromagnetic resonance experiment (FMR) shows that relaxation of $\boldsymbol{M}$ does indeed occur. Schematically then the Landau–Lifshitz equations become,

$$\frac{\partial \boldsymbol{M}}{\partial t} = -\frac{g\mu_{\mathrm{B}}}{\hbar}\boldsymbol{M} \times \boldsymbol{B} + \boldsymbol{\mathcal{R}}. \tag{7.4}$$

The precise form of the relaxation term $\boldsymbol{\mathcal{R}}$ will be discussed in the next section.

The spin current is a tensor quantity. There is a vector current, for example, $\boldsymbol{j}_z$ which is the current in the M_z component of $\boldsymbol{M}$ and the vector made of these is the dyadic $\mathcal{J} \equiv (\boldsymbol{j}_x, \boldsymbol{j}_y, \boldsymbol{j}_z)$ so that adding the appropriate divergence terms,

$$\frac{\partial \boldsymbol{M}}{\partial t} + \boldsymbol{\nabla} \cdot \mathcal{J} = -\frac{g\mu_{\mathrm{B}}}{\hbar}\boldsymbol{M} \times \boldsymbol{B} + \boldsymbol{\mathcal{R}} \tag{7.5}$$

which illustrates that spin currents are only conserved in the steady state when the right-hand side is zero. That this is rarely the case is what adds so many possibilities to spintronics which do not exist in electronics. When the spin current is due to spin diffusion, for example, $\boldsymbol{j}_z = -D\boldsymbol{\nabla} M_z$ where D is the diffusion constant proportional to the momentum relaxation time. With this

$$\boldsymbol{\nabla} \cdot \mathcal{J} = -D\nabla^2 \boldsymbol{M}, \tag{7.6}$$

the standard diffusion term. However it is the case that certain "gauge fields" generate a *spin-motive-force* (smf), i.e., the equivalent of an emf but which drives

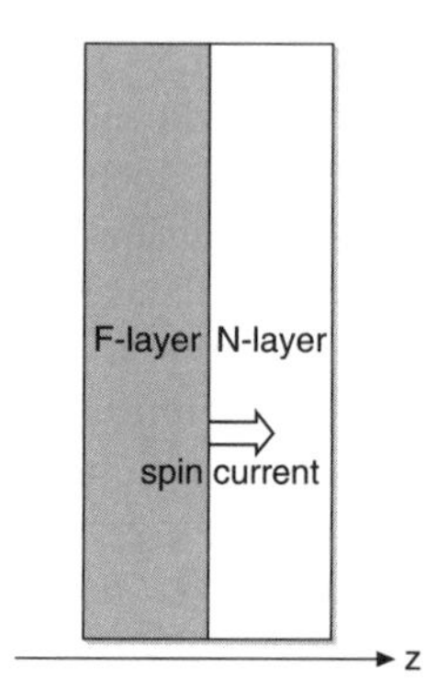

Fig. 7.1. The relaxation induced by contact with the N-layer may be viewed as being a spin current across the boundary.

a spin rather than charge current. The important spin currents in what follows are due to an emf or an smf and not simple diffusion.

When dealing with multilayers it is often the case that what might usually be thought of as relaxation might also be considered as a spin current [22]. Consider the FN sandwich of Fig. 7.1. Due to various interactions, the N-layer will induce some additional relaxation $S(\beta/M_F)\boldsymbol{M}_F \times (\partial \boldsymbol{M}_F/\partial t)$ due to the interface, where S is the surface area and β a constant which measures the size of the interaction. This angular momentum lost by the F-layer must be gained by the N-layer and so $S\beta(d\boldsymbol{M}_F/dt)$ might be added to the right-hand side of Eq. (7.5) for $\boldsymbol{M}_N$ as a "scattering in" term. Such a transfer of angular momentum might instead be viewed as being a spin current density,

$$\boldsymbol{j}_z = \frac{\beta}{M_F}\boldsymbol{M}_F \times \frac{\partial \boldsymbol{M}_F}{\partial t} \tag{7.7}$$

which crosses the surface in the xz-plane, since when the total $\boldsymbol{M}_N$ is calculated, the surface term $\int \mathcal{J} \cdot d\boldsymbol{S} = \int \boldsymbol{j}_z dS = S(\beta/M_F)\boldsymbol{M}_F \times (\partial \boldsymbol{M}_F/\partial t)$, which arises from the divergence, has the same effect on the equation for $\boldsymbol{M}_N$. Such currents are distinct from those which arise either from diffusion or an smf/emf.

7.2.1 *Relaxation and the Landau–Lifshitz equations*

Traditionally it is usual to assume Gilbert relaxation for which

$$\boldsymbol{\mathcal{R}}_G = \frac{\alpha}{M}\boldsymbol{M} \times \frac{\partial \boldsymbol{M}}{\partial t}. \tag{7.8}$$

The relaxation parameter α reflects the fraction of the magnetic (Zeeman) energy which is lost in a spin reversal.[13] The original Landau–Lifshitz equations were written with,

[13]The Gilbert term is the simplest relaxation term which preserves the length of $\boldsymbol{M}$, i.e., it is the case that $(d/dt)M^2 = 2\boldsymbol{M} \cdot (d\boldsymbol{M}/dt) = 0$. This follows since both $\boldsymbol{M} \cdot \boldsymbol{M} \times \boldsymbol{B}$ and $\boldsymbol{M} \cdot \boldsymbol{M} \times (\partial \boldsymbol{M}/\partial t)$ are trivially zero.

$$\mathcal{R}_L = -\frac{g\mu_B}{\hbar}\frac{\lambda}{M}\boldsymbol{M} \times (\boldsymbol{M} \times \boldsymbol{B}). \tag{7.9}$$

In fact, *in the absence of the divergence term*, for a single domain ferromagnet, these two forms of relaxation are mathematically equivalent, even when α is large. In passing from Gilbert to Landau–Lifshitz, $g \to g/(1+\alpha^2)$ and $\lambda = \alpha/(1+\alpha^2)$.

Now, by definition, the change in the free energy $\delta F = -\boldsymbol{B} \cdot \delta M$, and since $\boldsymbol{M}$ is assumed to have a fixed length, the condition for a turning point, i.e., $\delta F = 0$ is satisfied when $\boldsymbol{B}$ and $\boldsymbol{M}$ are either parallel or antiparallel, this corresponding to $\boldsymbol{M} \times \boldsymbol{B} = 0$. This is the case even when $\boldsymbol{B}(t)$ is a function of time. Further it is clearly a *sufficient* condition that the second law of thermodynamic be satisfied that relaxation ceases at such a turning point in the energy and in particular when the system is in an energy minimum. For this reason, the Landau–Lifshitz relaxation term is always consistent with the second law while Gilbert relaxation[14] is not. If however the *total derivative* is defined as

$$\frac{D\boldsymbol{M}}{Dt} \equiv \frac{\partial \boldsymbol{M}}{\partial t} + \boldsymbol{\nabla} \cdot \mathcal{J} \tag{7.10}$$

then it is again the case that the Landau–Lifshitz–Gilbert equations of the form

$$\frac{D\boldsymbol{M}}{Dt} = -\frac{g\mu_B}{\hbar}\boldsymbol{M} \times \boldsymbol{B} + \frac{\alpha}{M}\boldsymbol{M} \times \frac{D\boldsymbol{M}}{Dt} \tag{7.11}$$

are equivalent to those with Landau–Lifshitz relaxation.

Both of these options omit longitudinal relaxation. Imagine $\boldsymbol{M}$ lies along the easy axis *but* does not have its equilibrium value $M_0\hat{z}$. The relevant relaxation equation is often $dM_z/dt = -(1/T_1)[M_z - M_0]$. This T_1 time brings the order parameter back to its equilibrium value.[15] This longitudinal relaxation term is proportional to the deviation from local equilibrium and this alternative, adapted to the relaxation the transverse component of $\boldsymbol{M}$ discussed above, has proved useful in connection with other (magnetic resonance) problems where the mathematical equivalent of a divergent term arises[32]. The relaxation term is now of the very simple form:

$$\mathcal{R}_{\text{loc}} = -\frac{1}{\tau_e}\delta\boldsymbol{M}; \qquad \delta\boldsymbol{M} = [\boldsymbol{M} - \boldsymbol{M}_0(t)] \tag{7.12}$$

[14] The second law can be violated with $\mathcal{R}_G$, given a suitable choice of the divergence term. Assume $\boldsymbol{B} = B_0\hat{z}$, the torque term does no work, i.e., $\boldsymbol{M}$ processes with a constant θ. In the absence of a divergence term, the effect of Gilbert relaxation is to reduce θ, thereby reducing the energy. It is always possible to orientate adjacent domains at some small angles and which produces an arbitrary divergence with the adiabatic passage of electrons. Imagine $\boldsymbol{\nabla} \cdot \mathcal{J} + (g\mu_B/\hbar)\boldsymbol{M} \times \boldsymbol{B} = -(g\mu_B/\hbar)\boldsymbol{M} \times \boldsymbol{B}$ thereby precisely reversing the direction of procession. Since the induced motion still has constant θ, the electrons do no work, there is no energy inflow, and for perfect conductors no entropy production. Now $\mathcal{R}_G$ changes sign and sends the system to higher energies, in a process which can be made periodic, the heat bath *does* work on the magnetization and this is a clear violation of the second law.

[15] It is to be noted that $|\boldsymbol{M}| \neq M_0$ does *not always* cost an energy $\sim J$, the exchange constant. The states created by M^+ are higher in energy than the ground state by energies of the order of those associated with the anisotropy.

where $\boldsymbol{M}_0(t)$ is the equilibrium value that $\boldsymbol{M}$ would have if the Landau–Lifshitz defined internal field $\boldsymbol{B}(t)$ was frozen in time. An appropriate approximation for this *destination vector* might be $\boldsymbol{M}_0(t) \approx M^2\boldsymbol{B}(t)/\boldsymbol{M}\cdot\boldsymbol{B}(t)$, see below. Microscopic derivations, performed in another related context [32], using linear response theory show, for conduction electrons subject to relaxation via spin orbit scattering, the relaxation term *is* towards local equilibrium. Linear response will remain a good approximation since it only requires the various fields to be small compared with E_{F} the Fermi energy. Adapting this microscopic considerations to the present problem, i.e., when the conduction electrons become coupled to the ferromagnetic order parameter, the relevant Landau–Lifshitz equations become

$$\frac{\partial \boldsymbol{M}}{\partial t} + \boldsymbol{\nabla}\cdot\mathcal{J} = -\frac{g\mu_{\mathrm{B}}}{\hbar}\boldsymbol{M}\times\boldsymbol{B} - \frac{M_e}{M}\frac{1}{\tau_{\mathrm{so}}}\delta\boldsymbol{M}; \quad \text{with } \delta\boldsymbol{M} = [\boldsymbol{M} - \boldsymbol{M}_0(t)], \tag{7.13}$$

where $1/\tau_{\mathrm{so}}$ is the conduction electron spin-orbit time identified in the microscopic theory and where M_e and M are the magnitude of the conduction electron and total magnetizations. The modified Gilbert, Eq. (7.11), and local equilibrium, Eq. (7.11), relaxation terms have the same form with the above for $\boldsymbol{M}_0(t)$ when it is assumed that the order parameter has a fixed length.[16]

The discussion of this section implicitly assumes that the $\boldsymbol{M}$ is the order parameter of a single domain ferromagnet. The divergence term thereby reflects the effect of spin currents which enter and leave the volume in which reside the spins which make up this order parameter. Each more complicated situation requires a re-examination of the appropriate form for $\boldsymbol{R}$. In particular, the relaxation of a domain wall is such a situation. This will be treated near the end of this chapter.

7.3 Models for itinerant ferromagnets

The Stoner model is considered by many as the basic model for a ferromagnet. The Hamiltonian

$$\mathcal{H} = \sum_{\boldsymbol{k}\sigma} \epsilon_{\boldsymbol{k}\sigma} c^{\dagger}_{\boldsymbol{k}\sigma} c_{\boldsymbol{k}\sigma} - \mu\hat{N} \tag{7.15}$$

where $\epsilon_{\boldsymbol{k}\sigma} = \epsilon_{\boldsymbol{k}} + \sigma(Jm/2)$, reflects the basic version of this model in which the internal magnetic field Jm is to be determined self-consistently and where $\epsilon_{\boldsymbol{k}\sigma}$ is the energy of an itinerant electron with wave-vector $\boldsymbol{k}$ and spin σ. First-principles LDA and LDA+U calculations might well be considered as extensions

[16] Given that $\mathcal{R}_G$ can be reduced to $\mathcal{R}_L$, the latter has the form

$$-\frac{\alpha g\mu_{\mathrm{B}}}{\hbar M(1+\alpha^2)}\boldsymbol{M}\times(\boldsymbol{M}\times\boldsymbol{B}) = -\frac{\alpha g\mu_{\mathrm{B}}}{\hbar M(1+\alpha^2)}[(\boldsymbol{M}\cdot\boldsymbol{B})\boldsymbol{M} - M^2\boldsymbol{B}]. \tag{7.14}$$

With the destination vector $\boldsymbol{M}_0(t) \approx M^2\boldsymbol{B}(t)/\boldsymbol{M}\cdot\boldsymbol{B}(t)$, and if the order parameter is rigid, only the transverse part of $\mathcal{R}_{\mathrm{loc}}$ has an effect, and this is also proportional to $[(\boldsymbol{M}\cdot\boldsymbol{B})\boldsymbol{M} - M^2\boldsymbol{B}]$.

of this time honored model. The principal difficulty from the current perspective, is that the anisotropy energies present in real ferromagnetic materials do not appear in a natural fashion in such a model. These can be added in a somewhat ad hoc fashion by attaching these energies, and the associated dynamics of central interest here, to the axes of quantization. To avoid these difficulties, here attention will be focused on the s–d-exchange model

$$\mathcal{H} = \sum_{\boldsymbol{k}\sigma} \epsilon_{\boldsymbol{k}\sigma} c^{\dagger}_{\boldsymbol{k}\sigma} c_{\boldsymbol{k}\sigma} - \mu \hat{N} - \sum_i (J \boldsymbol{S}_i \cdot \boldsymbol{s}_i + A^0 {S_{iz}}^2 - K^0_{\perp} {S_{iy}}^2) - J^0_s \sum_{<ij>} \boldsymbol{S}_i \cdot \boldsymbol{S}_j \tag{7.16}$$

where A^0 is the easy and $K^0_{\perp}$ the hard axis anisotropy parameters,[17] while $\boldsymbol{S}_i$ and $\boldsymbol{s}_i$ are the spins of the local and conduction, i.e., itinerant, electrons. Now anisotropy fields appear in $\mathcal{H}$ as energies associated with the local moments. For spintronic applications very soft magnetic material such as Permalloy are of greatest interest and for the relevant geometries, i.e., thin films and wires, this anisotropy energy comes from the demagnetization field rather than from an intrinsic anisotropy energy, i.e., both effects are lumped into the two parameters A^0 and $K^0_{\perp}$.

Often, when dealing with non-elemental $3d$ magnets the crystal field levels of the magnetic ion are treated in different fashions. In cubic symmetry the triplet t_{2g} levels are much more localized than the doublet e_g orbitals which are typically orientated in such a fashion that they mutually hybridize more effectively. Such a situation is often described by the double exchange (or Zener) model,

$$\begin{aligned}\mathcal{H} = &- t \sum_{<ij>\sigma} \left(c^{\dagger}_{i\sigma} c_{j\sigma} + \text{h.c.} \right) + U \sum_i n_{i\uparrow} n_{i\downarrow} \\ &- \sum_i \left(J \boldsymbol{S}_i \cdot \boldsymbol{s}_i + A^0 {S_{iz}}^2 - K^0_{\perp} {S_{iy}}^2 \right) - J^0_s \sum_{<ij>} \boldsymbol{S}_i \cdot \boldsymbol{S}_j\end{aligned} \tag{7.17}$$

where U is the Coulomb energy which induces strong correlation effects. In this model the local spin $\boldsymbol{S}_i$ reflects the spin angular momentum of the t_{2g} levels. These have a direct exchange J^0_s with neighboring spins and an intra-atomic exchange J with the conduction electrons occupying a tight binding conduction band formed from the e_g orbitals. The orbital degeneracy of the e_g is ignored so that only U is left to induce the strong correlation effects expected for $3d$-ions. If they represent ionic parameters, it should be that $U > J$.

Often calculations can be simplified by taking the *half-metal limit* of any of these models. This is the limit where the exchange splitting is sufficiently large that only the majority spin band is filled, i.e., that $J \gg t$. With $U > J$ this also

[17] It is usual to obtain the angular dependence of the anisotropy energies by, for example, the obvious substitution $S_{iz} \to S\cos\theta$, where θ is the angle to the z-axis. With this, e .g., $A^0 {S_{iz}}^2 \to A^0 S^2 \cos^2\theta$. The result with quantum corrections is rather $AS^2 \cos^2\theta$, where $A = [(2S-1)/2S]A^0$.

implies the strong correlated limit, i.e., $U \gg t$, in which the double occupation of a given site by electrons of opposite spin is energetically forbidden. In fact, in the limit $J \gg t$ the value of U becomes irrelevant since, with no orbital degeneracy, a site which is doubly occupied is also a spin singlet and has an energy which is higher by an amount $\sim J$, as compared to placing the same two electrons on different sites and in the maximum spin state. Thus most of the essential complications exist in a tight binding s–d-exchange model of the form:

$$\mathcal{H} = - \sum_{<ij>\sigma\sigma'} \left(c^{\dagger}_{i\sigma} t_{ij\sigma\sigma'} c_{j\sigma'} + \text{h.c.} \right) - \mu \hat{N} \\ - \sum_{i} \left(J \boldsymbol{S}_i \cdot \boldsymbol{s}_i + A^0 {S_{iz}}^2 - K^0_{\perp} {S_{iy}}^2 \right) - J^0_s \sum_{<ij>} \boldsymbol{S}_i \cdot \boldsymbol{S}_j \tag{7.18}$$

in which the Coulomb energy U is set equal to zero. This is the model considered here. It will also be used to describe ferromagnets which are *not* in the half-metal limit.

It is important to have an idea about the order of magnitude of the parameters. The largest parameters are t, U and J depending upon the material and the bands, which are being modeled and are of the order of an electron volt. Of particular importance to the dynamics of magnets of interest here are the magnitudes of the effective anisotropy fields A^0 and $K^0_{\perp}$. For a thin film of Permalloy the largest contribution to A^0 is from the demagnetization field usually written as $4\pi M$ (in the cgs system) and which has a value of $\sim 1\,\text{T}$ (MKSA units). The in-plane anisotropy is much smaller and can be engineered in various ways. For a small element of a magnetic device the coercive field, B_c, i.e., that required to reverse the magnetization as a single large spin, is $\approx SA^0$, although experimentally it is smaller for materials which are of sufficient dimensions that the magnetization reverses through domain wall propagation. Such coercive fields vary but are, for example, of $\sim 10^{-2}\,\text{T}$ for the devices discussed in Chapter 5.

7.3.1 *Description of the ferromagnetic spin*

Fundamental to a ferromagnet is the existence of an order parameter, $\boldsymbol{M}$. This is equal to the total angular momentum,

$$\hbar \boldsymbol{M} = \hbar \sum_i \boldsymbol{M}_i, \tag{7.19}$$

where $\hbar \boldsymbol{M}_i = (\hbar \boldsymbol{S}_i + \hbar \boldsymbol{s}_i)$ is the angular momentum of a given site and which comprises the sum of that of the local moment, $\boldsymbol{S}_i$, and conduction electron, $\boldsymbol{s}_i$. In what follows it will be assumed that the individual elements of a device are mono-domains, or, at the most, that there is a single domain wall.[18] It is

[18] As discussed in Chapter 5, in order to explain the experiments, there are repeated suggestions that this is not adequate when relatively large currents are involved. Such extensions of the model are beyond the scope of the present development, although some of the experimental facts which prompt such discussion will here find alternative mono-domain explanations.

intuitively evident for a single-domain ferromagnet that (usually) not only are each of the $\boldsymbol{M}_i$ parallel to $\boldsymbol{M}$ but that this applies to each of $\boldsymbol{S}_i$ and $\boldsymbol{s}_i$. If this was not an itinerant magnet it would be possible to insist that $\boldsymbol{M}$ is the maximum spin state made out of the $\boldsymbol{S}_i$ and that deviations from this simply correspond to magnons, i.e., when no magnons are excited $(\boldsymbol{S}_i/S) = (\boldsymbol{M}/M)$ expressing the classical idea that all the moments are aligned and therefore parallel. In an itinerant magnet the conduction electrons see an internal field $-s_z SJ$ where s_z is the z-component of the magnetization of these electrons. The electrons with wave-vector $\boldsymbol{k}$ have an energy $\epsilon_{\boldsymbol{k}} + s_z(g_e\mu_{\mathrm{B}}H - JS)$ where H is an applied field and $\epsilon_{\boldsymbol{k}}$ is the kinetic energy. For a given $\boldsymbol{k}$ there is a large shift $\pm JS$ as the electron belongs to the majority (minority) spin direction with $s_z = +\frac{1}{2}(-\frac{1}{2})$, and implies a very large energy cost for flipping an electron when the kinetic energy is unchanged. The Fermi sea comprises a series of spin singlets, i.e., values of $\boldsymbol{k}$ for which there are both $s_z = \pm\frac{1}{2}$ electrons and other majority electron states for which only $s_z = +\frac{1}{2}$ is occupied. The spin state is therefore $|M = m_e + m_s, M_z = m_e + m_s\rangle \equiv |m_e\rangle|m_s\rangle|s\rangle$ where $|m_e\rangle$ is the maximal spin state for the unpaired electrons while $|s\rangle$ is the product state of all of the paired electron singlets, and where $|m_s\rangle$ is the maximal spin state of the local moments, to form the spin state $|m_e + m_s, m_e + m_s\rangle$. Adding a majority electron at the Fermi surface produces $|m_e + m_s + 1, m_e + m_s + 1\rangle$ while adding a minority electron results in $|m_e + m_s - 1, m_e + m_s - 1\rangle$. This description is not quite precise since the interaction is really $-\sum_i J\boldsymbol{S}_i \cdot \boldsymbol{s}_i$ and is rotationally invariant. Consider lowering the state $|m_e + m_s + 1, m_e + m_s + 1\rangle$ using the total spin lowering operator $M^- = S^- + s^-$ to give $|m_e + m_s + 1, m_e + m_s\rangle \propto (S^- + s^-)|m_e + m_s + 1, m_e + m_s + 1\rangle$, where $s^- = \sum_i s_i^-$. Due to the rotational invariance this has the same energy as $|m_e + m_s + 1, m_e + m_s + 1\rangle$ and equally important, there are matrix elements of $c^\dagger_{\boldsymbol{k}\downarrow}$ between $|m_e + m_s, m_e + m_s\rangle$ and $|m_e + m_s + 1, m_e + m_s - 1\rangle$. It is evident that these statements remain true when J is sufficiently large that there are no minority electrons, i.e., for a half-metal. There is *not* a similar process in which an up electron is put in the minority band since the state $(S^+ + s^+)|m_e + m_s - 1, m_e + m_s - 1\rangle = 0$. There are thus two independent conduction channels in parallel and for the majority band it is possible to add an opposite spin electron with a lowering of the total spin state. It is this latter possibility that will lead to angular momentum transfer. Neglected in this are states higher in energy by $\sim J$. In what follows, provided the energies involved are small compared with J,[19] it will suffice to deal with a half-metal which is shunted by a resistance reflecting the minority conduction channel.

The transfer of angular momentum is an intrinsically quantum process and requires, at least formally, the quantization of the local moment degree of freedom. The Holstein-Primakoff representation is used for this purpose. It consists of writing the spin operators as:

[19]This does *not* imply the adiabatic transfer of electrons as will be seen below.

$$S_{iz} = \hbar\left(s - b_i^\dagger b_i\right); \quad S^- = \hbar\left(2s - b_i^\dagger b_i\right)^{1/2} b_i^\dagger; \quad S^+ = \hbar\left(S^-\right)^\dagger \tag{7.20}$$

where $S^- = S_x - iS_y$ is the spin lowering operator and where s is the total spin quantum number for the local moments. It is left as an exercise for the reader to confirm that these obey the angular momentum commutation rules: $[S_x, S_y] = iS_z$ plus cyclic permutations. It is often the case, during the dynamics, that the vector $\boldsymbol{S}_i$ rotates by only a small amount and in this case the occupation number $\hat{n}_i = b_i^\dagger b_i$ is very close to zero. In this case the approximation $S^- = (2s)^{1/2} b_i^\dagger$ is reasonable.

7.3.2 *Quantum effects*

The most basic quantum effect is to make microwave transitions between the different energy levels, i.e., to perform a FMR experiment. As will be expanded upon below, the FMR resonance frequency is $\omega_0 = M\sqrt{(K_\perp + A)A}/\hbar$ which is typically in the GHz region but can be made much smaller. These modes of, for example, NF systems are indeed detected [33, 34] .

Typical domains used in spin valves have dimensions of, say, $10\times100\times100\text{nm}^3$ which corresponds to a total spin $M \sim 10^6$ which is enormous compared to values $M \sim 10$ for which the effects of quantum tunneling are well documented [35]. However the demagnetization field of a thin film is very large compared with the in-plane anisotropy which reduces by a certain amount the number of quantum levels involved. Since FMR resonance frequency is ω_0 while the barrier height for spin reversal is M^2A, there are only a number

$$N_\ell \sim \frac{M^2 A}{M\sqrt{(K_\perp + A)A}} \sim M\sqrt{\frac{A}{K_\perp}} \sim 10^4 \tag{7.21}$$

of levels below the barrier. Since the Zeeman energy is proportional to $g\mu_B M$, the separation between levels on the magnetic field axis is $\sim \hbar\omega_0/(g\mu_B M)$ which is in the mG region and is distinct from the effect of a field on ω_0 and which involves the difference between these energy levels. A reduction of the dimensions to $2\times10\times10\text{nm}^3$ would give a spin $M \sim 10^3$. For Permalloy it is not impossible to make $K_\perp/A \sim 10^4$ whence quantum effects might be observable even with traditional pillar designs. Clearly there are many more exotic possibilities with molecular magnets and molecular (spin-)electronics. Of particular interest is the fact that the (thermally assisted) tunneling rate is periodic with a period $\sim S/\omega_0$. In fact, with the current spin Hamiltonian it is known [36] that all levels cross simultaneously which strongly enhances this effect. It is implied that there would be some field periodic modulation in, for example, the critical current for spin reversal.

When a large magnetic film is applied *perpendicular* to a thin field it might also be possible to observe effects which are periodic in the field and drive current. Ignoring the smaller anisotropy A the effective spin Hamiltonian is then

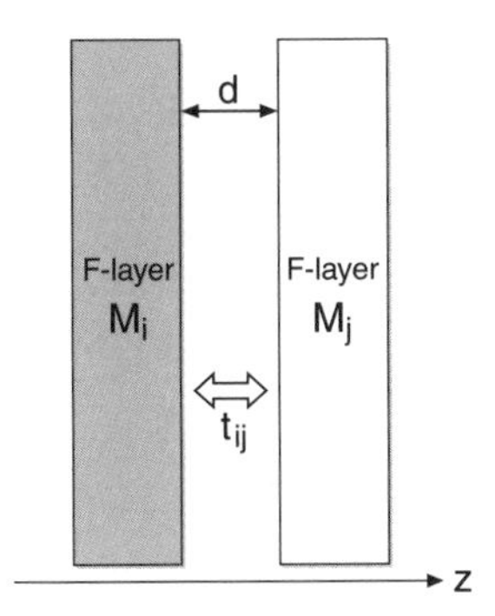

FIG. 7.2. Simple model for direct coupling between two ferromagnetic layers.

$$-S_z g\mu_B H + K_\perp S_z{}^2 = K_\perp (S_z - S_0)^2 - K_\perp S_0{}^2; \quad S_0 = \frac{g\mu_B H}{2K_\perp}. \tag{7.22}$$

For $S_0 < S$ there are many low-lying excitations which, when A is accounted for, have the usual the FMR energy $\hbar\omega_0 \approx S\sqrt{(K_\perp + A)A}$ for small S_0 but which drops to $S^{1/2}\sqrt{(K_\perp + A)A}$ when $S_0 \to S$. When S_0 is a half-integer many levels are doubly degenerate. When $S_0 < S$ the low-lying states are singlets and the splitting between adjacent levels increases dramatically. If $H = H_c$ when $S_0 = S$ then the transition energy between the n and $n+1$ levels is

$$\Delta E \approx \left(1 - \frac{H}{H_c} - n\right) g\mu_B H \tag{7.23}$$

and which is larger by more than a factor of S as compared with the values discussed in the previous paragraph. A typical value of $SK_\perp$ corresponds to a demagnetization field $\sim 1\,\mathrm{T}$. In an experiment with a field of say $5\,\mathrm{T}$ the voltage scale is $\sim 1/2\,\mathrm{mV}$ (or $1\,\mathrm{mA}$ for a device with a resistance of $\sim 2\,\Omega$). This might have some bearing on the fine structure seen in some recent experiments [37].

7.4 Angular momentum transfer for bi-layers

Consider the simple model of a bi-layer shown in Fig. 7.2. The conduction electrons hop from side to side via a hopping matrix element t_{ij}. As shown in Fig. 7.3(a), the electron before it hops is parallel to the direction of the total angular momentum $\hbar\boldsymbol{M}_i$; this has the total angular momentum quantum number and that for the z-component both equal to $m_i + 1/2$. It hops conserving its direction so that after the hoping event, the situation is as shown in Fig. 7.3 (b). The electron does not point along the direction of $\boldsymbol{M}_j$, i.e. this final state comprises a mixture of states with not only smaller values of the z-component, i.e., $m_j - 1/2$ but also with the same smaller value for the total angular momentum quantum number. These latter states are higher by energies $\sim J$, the intra-atomic exchange constant. The corrections due to the coupling

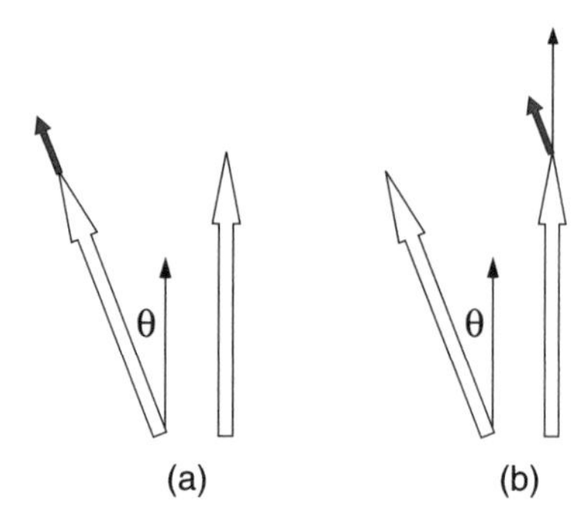

FIG. 7.3. (a) The electron is parallel to the local moment before it hops. (b) As it hops the electron keeps its direction fixed. The new state must be decomposed into eigenstates of the total angular momentum.

to these states is a small perturbation. However it remains the case that when projected only onto the states with the maximal value of the angular momentum $m_j + 1/2$, the result is *not* in the same direction as $\boldsymbol{M}_j$. In order to understand this consider the quantum mechanical state

$$|S\rangle = \left[\cos\frac{\theta}{2}|\uparrow\rangle + \sin\frac{\theta}{2}|\downarrow\rangle\right]. \tag{7.24}$$

Corresponding to a well known property of spin-one-half particles, this involves one-half of the usual Euler angles as defined by Fig. 7.4. By assumption that $\boldsymbol{M}_i$ is fixed, the state $|J = m, J_z = m\rangle$ of the left-hand side does not change and hence is not relevant in the hopping process. The *relevant initial* state before hopping is therefore,

$$|I\rangle \propto \left[\cos\frac{\theta}{2}|\uparrow\rangle + \sin\frac{\theta}{2}|\downarrow\rangle\right]|m\rangle \tag{7.25}$$

where $|m\rangle$ is the similar maximal spin state of the right-hand side. The only part of this wave function which is transmitted is the projection onto the states with total angular momentum $m + 1/2$. The state

$$|\uparrow\rangle|m\rangle = |m + 1/2, M_z = m + 1/2\rangle \tag{7.26}$$

already corresponds to the maximal total angular momentum. The second term however comprises a linear combination of

$$|m + 1/2, M_z = m - 1/2\rangle = \frac{1}{(2m+1)^{1/2}}[(2m)^{1/2}|\uparrow\rangle|m-1\rangle + |\downarrow\rangle|m\rangle] \tag{7.27}$$

and

$$|m - 1/2, M_z = m - 1/2\rangle = \frac{1}{(2m+1)^{1/2}}[|\uparrow\rangle|m-1\rangle - (2m)^{1/2}|\downarrow\rangle|m\rangle]. \tag{7.28}$$

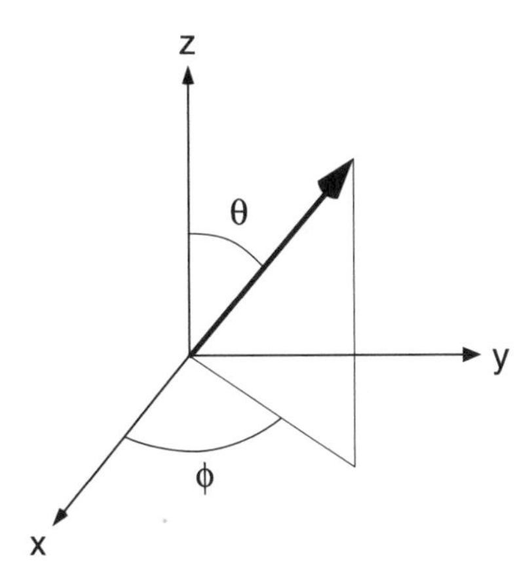

FIG. 7.4. The definition of Euler angles used here.

The required result is

$$|\downarrow\rangle|m\rangle = \frac{1}{(2m+1)^{1/2}}[|m+1/2, m-1/2\rangle - (2m)^{1/2}|m-1/2, m-1/2\rangle]. \tag{7.29}$$

The coupling to the high-energy state $|m-1/2, m-1/2\rangle$ represents a perturbation and can be dropped. It follows that the final state for the right-hand side is

$$|F\rangle \propto \cos\frac{\theta}{2}|m+1/2, m+1/2\rangle + \sin\frac{\theta}{2}\frac{1}{(2m+1)^{1/2}}|m+1/2, m-1/2\rangle. \tag{7.30}$$

The interpretation of this state is not immediately obvious. It might, for example, be considered as $|m+1/2, m+1/2\rangle$ rotated, or as indicating that the electron is transmitted in the state $|m+1/2, m+1/2\rangle$ with probability $\cos^2\theta/2$ and $|m+1/2, m-1/2\rangle$ with probability $\sin^2(\theta/2)/(2m+1)$ along with an infinite number of other probabilities. This later state is proportional to $S^-|m+1/2, m+1/2\rangle$, where S^- is the total spin lowering operator for the surface layer, i.e., corresponds to a magnon being created at the free layer surface. Such a magnon carries a spin current which is in addition to that associated with the conduction electron charge current.

The appropriate projection of $|F\rangle$ reflects the nature of an implicit "measurement" associated with the observation of a classical magnetization $\boldsymbol{M}_j$. Assume the free layer is thick compared to the spin diffusion length. In this case the magnon part of the spin current will have decayed before it can leave this layer. The modified Landau–Liftshitz–Gilbert equations are integrated over the volume of the free layer. Each of the terms is uniform apart from the divergence term. The latter reduces to the difference between the spin currents which pass through the two free layer surfaces, and given that the magnons decay in the body, this is simply due to the different direction of the polarization of these currents. The angle between these polarizations is θ and it follows that $\boldsymbol{M}_j$ will rotate by an angle

$$\beta = \frac{1}{2m+1} \sin\theta, \tag{7.31}$$

with the passage of one electron. When written in this rotated axis,

$$|F\rangle \approx \cos\frac{\theta}{2}|F_j\rangle_a + \sin^3\frac{\theta}{2}\frac{1}{(2m+1)^{1/2}}|F\rangle_t \tag{7.32}$$

where $|F\rangle_a = e^{i\beta S_y}|m+1/2, m+1/2\rangle$ corresponds to $|m+1/2, m+1/2\rangle$ rotated by β, while $|F\rangle_t$ is the state $|m+1/2, m-1/2\rangle$ in the new frame. Thus $\boldsymbol{M}_j$ is rotated by the angle β and the matrix elements that the conduction electron simply hops onto this layer is $\cos\theta/2$. With matrix element $\sin^3(\theta/2)/(2m+1)^{1/2}$ the electron similarly hops but with the production of a magnon at the surface.

This illustrates that the transmission of an electron is *not* adiabatic for finite angles θ. The part $|F\rangle_a$ corresponds to the adiabatic rotation while $|F\rangle_t$ reflects the correction to that approximation. The experienced reader might find the result β for the rotation angle a surprise. For a half-metal the result is usually [5]: $(2/2m+1)\tan\theta/2$. In the past the equivalent of the result for $|F\rangle$ has always been interpreted in terms of a simple adiabatic rotation. With the current result for β the transverse part of the incident spin current is transmitted.

A finite current, I, implies a finite rate of rotation $(\partial\theta/\partial t) \propto I\sin\theta$. In a rotating frame in which ϕ rotations have been eliminated, this amounts to a certain transverse torque which is identical, to within at most a sign, to the effect of adding relaxation to the Landau–Lifshitz equations. When electrons pass from the fixed to free domain, the sense of the rotation is such that this term adds to the thermodynamic relaxation rate, an effect which needs to be verified by experiment. From the present point of view the only effect of reversing the direction of current flow is to cause down *holes* to be incident on the interface. The effective charge of the carriers is irrelevant to the above argument and the only effect is to cause the current term to change sign and now reduce the effective rate of relaxation. It follows that there exists a critical current for which small-angle modes actually grow in magnitude. For currents greater than this value the free domain either is dynamic or the domain rotates until it finds a new equilibrium state.

The effective hopping matrix element is $t\cos\theta/2$ except that θ is very close to π whence it takes the very small value $t/(2m+1)$. This is an artefact of the half-metal limit with its single channel for electrical conduction. In realistic cases there are at least four such relevant channels. This implies that the angular dependence of the barrier conductance is much less dramatic. Consider, for example, the channel which exists when the left (right) domain polarizes its conduction electrons in the up (down) direction. The maximum conduction now occurs when the $\boldsymbol{M}_i$ and $\boldsymbol{M}_j$ are antiparallel. With a current coming from the left, rotation is still in the sense required to bring these to the parallel orientation, i.e., the only difference between this channel and that considered above is that the conductivity is large rather than small in the antiparallel configuration. When both domains comprise similar magnets with parallel majority electrons, then this channel corresponds to one of two which couple the majority to minority electrons and which are responsible for maintaining the conductance at a larger value when the magnetizations are antiparallel. Of course most experiments are performed on FNF systems in which the coupling between the two domains is via an intermediate non-magnetic metal. This requires a study of the transmission of spin currents at an FN interface. This adds a number of new interesting ingredients and will be taken up below when the gauge theory has been introduced.

7.4.1 *Gauge theory*

The rotation of one domain relative to the other modifies the hopping matrix elements between the domains in a manner which is equivalent to the introduction of fictive SU(2) gauge fields [24]. The situation which is envisaged is again that shown in Fig. 7.3, in which the left-hand domain has its magnetization in the z-direction while that of the right-hand side has its $\boldsymbol{M}_j$ defined by the Euler angles θ, ϕ. In the formal gauge theory approach, based upon the s–d-exchange model, the diagonalization of the $J\boldsymbol{S}_i \cdot \boldsymbol{s}_i$ interaction is achieved by a rotation of the axes of quantization. Since it is easier to deal with the single angle θ, the fields which are responsible for a finite $d\phi/dt$ are eliminated by the use of a suitable rotating frame, i.e., the instantaneous axes are such that $\phi = 0$, unless stated otherwise.

Thus initially in the rotating frame in which $\phi_j = 0$, when acting upon the vector,

$$\begin{pmatrix} c_{i\uparrow}^{\dagger} \\ c_{i\downarrow}^{\dagger} \end{pmatrix} \tag{7.33}$$

the relevant matrix for a rotation about the y axis is

$$r(\theta) = e^{i\theta s_y} = \cos\frac{\theta}{2} + i\sin\frac{\theta}{2}\sigma_y \tag{7.34}$$

where $s_y = \frac{1}{2}\sigma_y$ when written in terms of the usual Pauli matrix σ_y. (The result on the right-hand side uses the fact that ${\sigma_y}^2 = 1$.) When $r(\theta)$ is applied to the right-hand domain it makes an angle θ with that to the left. If it is assumed that the direct exchange J_s^0 coupling the two sides is zero, the only coupling is via the interdomain hopping matrix elements. Both mathematically and physically the effect of such a rotation on the Hamiltonian for both components of the system is null. For those matrix elements coupling these domains, the matrix element

$$t_{ij} = r_{\sigma\sigma'}(\theta)t = t\left(\cos\frac{\theta}{2} + i\sin\frac{\theta}{2}\sigma_y\right) \tag{7.35}$$

where, as usual, it is understood that in the absence of an explicit Pauli matrix there is a unit matrix . The spin off-diagonal parts of t_{ij} imply a mixing of spin up and down electrons but still leaves a single-particle effective Hamiltonian. It is simplest to solve for the single-particle states and then construct the many-particle states. The presence of σ_y in t_{ij} leads the spin off-diagonal terms $c^\dagger_{j\downarrow}c_{i\uparrow}$ + h.c. $= s^-_j c^\dagger_{j\uparrow}c_{i\uparrow}$ + h.c. using the fact that this is a single-particle Hamiltonian. Similarly $c^\dagger_{j\uparrow}c_{i\downarrow}$ + h.c. $= c^\dagger_{j\uparrow}c_{i\uparrow}s^-_i$ + h.c.

The next and key step is to replace s^- by $(M^-/2M)$. This is certainly a good approximation for a half-metal since excitations for which $\boldsymbol{s}$ and $\boldsymbol{M}$ are *not* parallel cost an energy $\sim J$. However as already discussed in Sec. 7.3.1, even for ferromagnets with partially polarized conduction electrons, there exist hopping matrix elements which are off-diagonal in spin space. In fact all that is necessary is that there be a well-defined total magnetization $\boldsymbol{M} = \sum_i(\boldsymbol{s}_i + \boldsymbol{S}_i)$. If this is in, for example, the maximum spin quantum state $|M, M\rangle$ then there are matrix elements of s^- which connect it to $|M, M-1\rangle$. These are given by the Wigner-Eckart theorem, which in the present context is simply an elegant fashion by which to insist that the matrix elements of s^- between these states are equal to those of $(M^-/2M)$. There may well be other matrix elements to other states but if interest is in the coupling between the current and $\boldsymbol{M}$ it will be permissible to replace $\boldsymbol{s}$ with $\boldsymbol{M}$. This is the case *even* for a non-magnetic metal with a magnetization, $\boldsymbol{M}_e$, induced by contact with a ferromagnet, i.e., the interaction derived below is valid for a FN interface.

With this, the interesting part of the transfer Hamiltonian becomes:

$$\mathcal{H}_t = -t\sum_{<ij>}\left\{c^\dagger_{i\uparrow}c_{j\uparrow}\left[\cos\frac{\theta}{2} + \sin\frac{\theta}{2}\left(\frac{M^+_j}{2M_j} + \frac{M^-_i}{2M_i}\right)\right] + \text{h.c.}\right\}. \tag{7.36}$$

The interactions between the spin and charge sectors is thus reduced to the spin-dependent part of this interaction which is small because of the factors of $1/M$. The effective field seen by the order parameter $\boldsymbol{M}_i$ at a mean field level is obtained by replacing the conduction electron operators by their averages, i.e.,

$$\mathcal{H}_t \to -t\sum_{<ij>}\left\{\langle c^\dagger_{i\uparrow}c_{j\uparrow}\rangle\left[\cos\frac{\theta}{2} + \sin\frac{\theta}{2}\left(\frac{M^+_j}{2M_j} + \frac{M^-_i}{2M_i}\right)\right] + \text{h.c.}\right\}. \tag{7.37}$$

Writing $M^+_i = M_{ix} + iM_{iy}$ it is observed that only that part proportional to M_{iy} couples to the expectation value of the spin current density,

$$j_s = -i\frac{et}{\hbar\mathcal{A}}\cos\frac{\theta}{2}\sum_{<ij>}\left(\langle c^\dagger_{i\uparrow}c_{j\uparrow}\rangle - \text{c.c.}\right) \tag{7.38}$$

where $\mathcal{A}$ is the cross-sectional area and where j_s is defined such that it is equal to the z-component of the up-spin charge current. The interaction between the current and the order parameter is

$$\mathcal{H}_i = \frac{j_s \hbar \mathcal{A}}{e} \tan\frac{\theta}{2}\left[\frac{M_{jy}}{2M_j} - \frac{M_{iy}}{2M_i}\right]. \tag{7.39}$$

It is *wrong* to simply write the M_{iy} in terms of spin deviation operators. As pointed out in the previous section, the axis of quantization, reflecting here the orientation of the expectation value of the magnetization, must be rotated with the passage of each electron. The Landau–Lifshitz Eqs. (7.14) are the equations of motion for this classical magnetization.[20] The decomposition of $\mathcal{H}_i$ in terms of rotations and excitations is determined by comparison with the divergence term in Eqs. (7.14). The net spin current leaving the region of the interface is determined by the charge current (and for other than half-metals, the material determined polarization p introduced earlier). Any spin wave excitations which carry a spin current will die out within the spin diffusion length. Using the Gauss theorem for a region which encloses the spin diffusion length, the rate of rotation is then determined by the change in the transverse magnetization. With one domain fixed, the rate of rotation of the free domain is therefore,

$$\left(\frac{\partial\theta}{\partial t}\right)_{\text{current}} = \frac{j_s \mathcal{A}}{4eM}\sin\theta, \tag{7.40}$$

which reduces the magnon interaction to

$$\mathcal{H}_i = \frac{j_s \hbar \mathcal{A}}{2ei}\sin^3\frac{\theta}{2}\left[\frac{b_j - b_j^\dagger}{(2M_j)^{1/2}} - \frac{b_i - b_i^\dagger}{(2M_i)^{1/2}}\right]. \tag{7.41}$$

Thus the effect of a spin current is to induce the rotation given by Eq. (7.40) and the creation of magnons via $\mathcal{H}_i$. These magnons will relax restoring the spin current. Observe that the rotation is about the *instantaneous* y-direction and that the interaction with the current involves the *difference* in magnetizations of the two domains.The current-induced rotations are never "propeller like" for a simple F-F interface. Evidently when one domain is fixed the corresponding operators are absent.

The passage to a rotating frame of reference used above involves the time dependent rotation $R(\theta) = e^{i\theta M_y}$. In a formal Schrödinger's equation, $i\hbar(\partial/\partial t)\psi = \mathcal{H}\psi$ not only does $\mathcal{H} \rightarrow R\mathcal{H}R^{-1}$ but also $i\hbar(\partial/\partial t) \rightarrow Ri\hbar(\partial/\partial t)R^{-1} = i\hbar(\partial/\partial t) + (\partial\theta/\partial t)M_y$ so that the effective equation in the rotating frame is $i\hbar(\partial/\partial t)\psi = \mathcal{H}'\psi$ where $\mathcal{H}' = R\mathcal{H}R^{-1} - (\partial\theta/\partial t)M_y$, i.e., added to the rotated Hamiltonian is a term which again looks like a magnetic field in the y direction. The fields $\boldsymbol{B}$ which occur in the Landau–Lifshitz equations are eliminated by taking the axis of quantization to be along the direction of this field and setting $\hbar\omega = g\mu_{\text{B}}B$. With this $\phi = 0$. What are left are the rotations $(\partial\theta/\partial t)_{\text{current}}$ due to the current and

[20] A derivation of the particular Landau–Lifshitz equations (7.14) is not given here. The derivation of similar spin transport equations can be found in the literature, see, for example, Ref. [32].

similar rotations induced by relaxation. It is required that a finite $\partial\theta/\partial t$ from one source, at least on average, eliminates that from the other. This procedure is not simple especially when large-angle modes are of importance.

7.4.2 *Spin-motiveforces (smf) in bi-layers*

The interaction between the current and the magnetization must be consistent with both the conservation of angular momentum and energy. Since the electrons must do work (positive or negative) in order to change the anisotropy and Zeeman energies this must appear as an emf/smf in the effective circuit. Equally that part of the interaction which creates magnons and which eventually relax must take the relevant energy from the battery in the form of a back-emf/smf [31]. When the magnetization makes an abrupt change of value so does the emf/smf and implies voltage jumps. The idea of a single terminal "spin battery" has been suggested by others; see in particular the review [22].

It is useful to incorporate $\mathcal{H}_i$ into

$$t_{ij} = e^{\frac{i}{\hbar}\int \boldsymbol{A}\cdot d\boldsymbol{s}} t \cos\frac{\theta}{2} \tag{7.42}$$

where the "gauge field" $\boldsymbol{A}$ is such that,

$$\frac{1}{\hbar}\int \boldsymbol{A}\cdot d\boldsymbol{s} = \frac{1}{\hbar}A_z d = \tan\frac{\theta}{2}\frac{M_y}{2M} \tag{7.43}$$

i.e., the only finite component of the gauge field is A_z and d is the separation between the elements of the domain. While the description of this interaction in terms of an effective vector potential might seem rather esoteric, it does serve two useful purposes. First it puts an emphasis on the fact that the associated term in the effective Hamiltonian is $\boldsymbol{A}\cdot\boldsymbol{j}$ and not $-\boldsymbol{M}\cdot\boldsymbol{B}$ and second that the time derivative of $\boldsymbol{A}$ must correspond to a force acting on the conduction electrons. In fact with the current set of units this force is

$$\boldsymbol{f}_s = -\frac{\partial \boldsymbol{A}}{\partial t}. \tag{7.44}$$

This leads to a back-emf of magnitude $\mathcal{E} = f_s d/e$ which opposes that which drives the current. This really must be called an smf since it couples to the spin of the electron and not its charge.

The force $\boldsymbol{f}_s$ must also be divided into that part which arises from the rotations of $\boldsymbol{M}$ and that which arises from the production of magnons. The former part corresponding to the adiabatic motion is

$$\frac{1}{\hbar}\int \boldsymbol{A}_a\cdot d\boldsymbol{s} = \frac{1}{\hbar}A_{az} d = \sin\theta\frac{M_y}{4M}. \tag{7.45}$$

While M_y is off-diagonal in spin space, its derivative is not. This is rather trivial to calculate although the presence of a $K_\perp$ anisotropy energy, which strongly squeezes the states, complicates matters. There are three directions which count: first the orientation of the fixed domain second the direction of the internal field, i.e., the axis of quantization, and third that of the free domain magnetization. Define γ to be the angle between the axis of quantization and the direction of $\boldsymbol{M}$. Now $M_y = M\sin(\theta+\gamma)\sin\phi$, so that for $\phi = 0$,

$$\frac{dM_y}{dt} = \frac{dM_y}{d\phi}\frac{d\phi}{dt} = M\sin(\theta+\gamma)\frac{d\phi}{dt} \tag{7.46}$$

and so the emf/smf is

$$\mathcal{E}_a = \frac{f_s d}{e} = -\frac{d}{e}\frac{\partial A}{\partial t} = -\frac{\hbar}{2e}\sin\theta\sin(\theta+\gamma)\frac{d\phi_j}{dt} \tag{7.47}$$

which has an easy interpretation. The derivative $(d\phi/dt) = -g\mu_{\mathrm{B}}B/\hbar$ so that the magnetic energy is $\hbar(d\phi/dt)m\cos(\theta+\gamma)$ while when an electron hops it changes the angle θ by $\beta = (1/2m)\sin\theta$ and the work done per electron corresponds to the above.

In general this emf/smf is time dependent. It is easy to show that $\mathcal{E}_a$ is time *independent* when $K_\perp = 0$. It becomes time dependent through the time dependence of the angle $\gamma(t)$ and must, at least in part, be responsible for the observed rf signals.

The derivation of the other, i.e., longitudinal part of the emf/smf is less than general because of the simple nature of model. It is better to derive the appropriate expression using the relevant physical principles. When an electron leaves a domain it causes the order parameter to change quantum number from $m+\frac{1}{2}$ to m and this implies a change in energy for the fixed domain of

$$\frac{\hbar}{2}\frac{d\phi_i}{dt} \tag{7.48}$$

since again $d\phi_i/dt$ reflects the field seen by the order parameter. However a domain has two surfaces and so the work done in adding the electron at one surface is compensated by that gained when it leaves. There is only net work done when the order parameter magnetization associated with the electron relaxes between the two surfaces. Thus the work done on the fixed domain is

$$\frac{2e}{j_s\mathcal{A}}\frac{\partial M_i}{\partial t}\frac{\hbar}{2}\frac{d\phi_i}{dt} \tag{7.49}$$

where $(2e/j_s\mathcal{A})(\partial M_i/\partial t)$ is the fraction of electrons which reverse their direction and where $(\partial M_i/\partial t)$ is the rate of relaxation of the order parameter. The equivalent longitudinal emf/smf is insensitive to the direction of the free domain and hence the net longitudinal emf/smf is

$$\mathcal{E}_t = \frac{\hbar e}{j_s\mathcal{A}}\left[\frac{\partial M_i}{\partial t}\frac{d\phi_i}{dt} + \frac{\partial M_j}{\partial t}\frac{d\phi_j}{dt}\right]. \tag{7.50}$$

This is an important effect for FNF systems in the antiparallel configuration, since, for example, for half-metal F-layers all electrons must relax in order to pass through the system and

$$e\mathcal{E}_t \sim g\mu_{\mathrm{B}}B. \tag{7.51}$$

Voltage jumps of this magnitude are observed when the system switches between the (near) parallel to (near) antiparallel configuration [38].

There are experimental manifestations of both contributions to the smf/emf. Often large peaks (see Chapter 5) are observed in (dV/dI) and find a natural explanation in terms of the change in $\mathcal{E}$ which occurs when the system jumps from one mode to another. When a device is current feed, a back-emf/smf appears as a magnetoresistance. This needs to be accounted for in any analysis of this resistance.

7.5 Magnetic dynamics of bi-layers

In thin films which make up a pillar device, because of the demagnetization field, the magnetization usually lies in the plane of the film, i.e., the axis perpendicular to the film is hard. (In some experiments a very large external field (see, for example, Refs. [37, 38]) is applied in order that this is also not the case and in ferromagnetic semiconductors this is not always the case (see, for example, Ref. [18]).) Either because the layer is oval or by accident there is an easy axis in the plane of the film, i.e., in the Hamiltonian, Eq. (7.18), both A and $K_\perp$ are positive and usually $A \ll K_\perp$ which implies that the motion lies close to the xz-plane. The natural, i.e., FMR, frequency for small oscillations of such a system is $\omega \approx M\sqrt{K_\perp A}/\hbar$. This follows from considering the traditional Landau–Lifshitz–Gilbert equations

$$\frac{\partial M_x}{\partial t} = -kM_y - \alpha\frac{\partial M_y}{\partial t}; \qquad \frac{\partial M_y}{\partial t} = aM_x + \alpha\frac{\partial M_x}{\partial t}, \tag{7.52}$$

where $k = [(K_\perp + A)M + g\mu_{\mathrm{B}}B]/\hbar$ and $a = [AM + g\mu_{\mathrm{B}}B]/\hbar$. Ignoring terms of order α^2, these become

$$\frac{\partial M_x}{\partial t} + \alpha a M_x = -kM_y; \qquad \frac{\partial M_y}{\partial t} + \alpha a M_y = aM_x, \tag{7.53}$$

and clearly, with α small, the natural frequency and width are, respectively, given by

$$\omega = \sqrt{ka} = \frac{1}{\hbar}\sqrt{[(K_\perp + A)M + g\mu_{\mathrm{B}}B][AM + g\mu_{\mathrm{B}}B]}, \tag{7.54}$$

and

$$\Delta = \frac{\alpha}{2}(k + a) = \frac{\alpha}{\hbar}\left(\frac{1}{2}K_\perp + A + g\mu_{\mathrm{B}}B\right). \tag{7.55}$$

The motion in the xy-plane is far from being a circle. It is strongly "squeezed" in the y-direction, i.e., the motion is an oval which extends the farthest along the easier x-direction. Without damping and with $x = M_x/M$ and $p_x = M_y/M$, the problem reduces to a Harmonic oscillator $\mathcal{H} = ({p_x}^2/2m) + (k_s/2)x^2$ where $1/m = k$ and $k_s = a$ so that $\omega = \sqrt{k_s/m}$, which reduces to the above result.

The Gilbert damping included in the above is *not* equivalent to the simple velocity damping of a harmonic oscillator since, with the above mapping, the term $-\alpha k M_y$ would be absent. (Although it is to be noted that the effect of this term is often small.) It is a commonplace observation that the usual velocity damping term changes the frequency of a harmonic oscillator. This is a real physical effect in, for example, an LCR circuit. In the absence of squeezing, i.e., with $K_\perp = 0$, the Gilbert term is precisely of such a form that it causes a width *without* any such frequency shift. Now at a fundamental level, in the quantum theory of the relaxation, this corresponds to the imaginary part of some self-energy and as such leads to a width without a shift (or rather a, usually small, shift given by the real part of the same self-energy). Such relaxation might correspond to a "squeezed" Gilbert term and the equations of motion become

$$\frac{\partial M_x}{\partial t} + \alpha\omega M_x = K_\perp M M_y; \qquad \frac{\partial M_y}{\partial t} - \alpha\omega M_y = -AMM_x, \tag{7.56}$$

which reduces to the problem without relaxation for M_{0x} and M_{0y} following the substitutions $M_x = e^{+\alpha\omega t}M_{0x}$ and $M_y = e^{-\alpha\omega t}M_{0y}$. That the usual Gilbert term *correctly* predicts the width of small-angle modes has to do with the fact that, in general, relaxation corresponds to a matrix with off-diagonal elements which couple different modes. The *two* roots to Eqs. (7.52) correspond to these two modes. Given that the z-axis is easy, these modes corresponding to clockwise and counter-clockwise rotations about this axis of quantization. The effective longitudinal field is that part of the spin Hamiltonian which commutes with S_z and here corresponds to $(K_\perp + A)M + g\mu_\mathrm{B}B$. The clockwise and counter-clockwise modes have frequencies $\pm(K_\perp + A)M + g\mu_\mathrm{B}B$ *but* with a coupling equal to $K_\perp M$ to give frequencies $\pm\sqrt{ka}$ displayed above. The symmetry is such that if the modes remained uncoupled, Gilbert damping would lead to a width $(\alpha/\hbar)[(K_\perp + A)M + g\mu_\mathrm{B}B]$ *without* a shift, however the coupling $K_\perp M$ causes the spin wave functions to be modified and relaxation results finally in both a width and a shift.

The generator of rotations by the Euler angle ϕ is S_z. If this commutes with the effective spin Hamiltonian, then $d\phi/dt$ is a *constant* of motion, independent of ϕ, and the angular frequency of the two modes described above are $\pm d\phi/dt$. In this special case Gilbert damping causes a width without shifts, and in particular there is no contribution to $d\phi/dt$ which arises from relaxation. This observation will be of particular importance when the relaxation of domain walls is introduced.

7.5.1 *Relaxation of dynamical modes*

The correct form of relaxation has experimental consequences far beyond this question of what are after all usually small frequency shifts. As seen above Gilbert relaxation gives a net width $\frac{1}{2}\alpha(k+a) \approx \frac{1}{2}\alpha k$, since for the thin films of interest $k \gg a$. On the other hand, squeezed Gilbert damping gives simply $\alpha\omega = \alpha\sqrt{ka}$ which has a completely different dependence on the physical parameters.

The usual Gilbert term damping with a characteristic α embodies the idea that the rate of relaxation is proportional to $(\partial \boldsymbol{M}/\partial t)$. In a bulk sample where relaxation is due to eddy currents, i.e., occurs via electromagnetic induction, this might well be the case. Simple estimates suggest that this is *not* the process of prime importance for the nanometric systems of interest here.

Experiments on thin films suggest that the FMR relaxation rate is roughly proportional to the electrical resistivity [39]. A similar result is true for the conduction electron resonance (CESR) of non-magnetic metals [40] where this has been reliably interpreted in terms of spin-orbit scattering, see also Ref. [23]. This lends support to the idea that relaxation in the present itinerant ferromagnets is due to spin-orbit scattering of the conduction electron component of the total magnetization. If this is the case the existence of a material determined α is questionable, for example, in the preliminary discussion of conduction electron spin-orbit scattering, Sec. 7.2.1, it was pointed out that this is characterized by a relaxation time $1/\tau_e$ and *not* by a parameter α.

It is also the case that the N-layer in FNF systems causes relaxation [34]. This has been investigated by FMR and these experiments will be discussed below in connection with the possible excitation modes of such sandwiches [41, 42]. (This might also be interpreted in terms of rotations due to the particle fluxes which impinge on the FN interfaces, as discussed earlier [22].) This relaxation is also proportional to the resonance frequency, i.e., this process *does* define an α. In fact α is proportional to the surface conductivity σ_s defined by $J = \sigma_s E$ where J is the charge current density when the electric field at the interface is E. For FNF systems and films of the thickness of typical free layers, this rate might well be comparable to that due to spin-orbit scattering inside the magnetic layer [41].

Lastly if relaxation due to what might be called pseudo-eddy currents. As outline at the end of the last section, that $\boldsymbol{M}$ is in motion implies an smf which drives a spin current through the interface. By Lenz's law this spin current opposes that implicit in the driving charge current and for the FMR modes being discussed here, reduces the spin drive. This constitutes an effective relaxation rate. The smf/emf will never overcome the driving emf for a current-driven system; however large internal magnetic fields, via Eq. (7.47), can produce an smf large enough to essentially cancel the spin current. Notice that the smf is proportional to the internal field B and depends on the angles θ and β. Thus for large fields the amplitude of the oscillations can be limited by this effect. It is suggested that a number of different effects linear in the external field owe their existence to the self-generated smf. Such relaxation is not of Gilbert form.

7.6 Description of the dynamical modes

In order to make a connection with the equations of motion for the Euler angles θ and ϕ, it is useful to develop the less familiar Lagrangian formalism of spin dynamics which uses the function $\mathcal{V}(\theta, \phi)$ as defined above to determine the dynamics. The appropriate Lagrangian is

$$\mathcal{L} = \mathcal{T} - \mathcal{V}; \quad \mathcal{T} = \hbar M(\cos\theta - 1)\frac{\partial \phi}{\partial t}. \tag{7.57}$$

This form for the kinetic energy $\mathcal{T}$ is known to reproduce the Landau–Lifshitz equations. The integral of $\mathcal{T}$ during one cycle is proportional to the solid angle subtended by the orbit and in this way reflects the Berry phase. Different choices for the z-axis have different uses and provide complementary pictures of the motion of the magnetization. The evident choice is to retain the definitions used up to here. With this $\mathcal{V} = -g\mu_{\mathrm{B}}BM\cos\theta - AM^2\cos^2\theta + K_\perp M^2\sin^2\theta\sin^2\phi$ and for small-angle FMR modes this formalism reproduces the last set of equations for M_x and M_y. In a certain sense this corresponds to the most physical description since the total field $\boldsymbol{B}$ causes the magnetization $\boldsymbol{M}$ to precess about the z-axis with a ϕ which increases by 2π per cycle. The system of equations is, however, complicated since both $\boldsymbol{M}(t)$ and $\boldsymbol{B}(t)$ need to be calculated at each instant.

7.6.1 *Effective particle description of the dynamical modes*

A simpler scheme maps the problem onto one of a massive particle in a potential well, provided $K_\perp$ is large. This provides a useful and quite accurate overall picture of the dynamics. For this purpose z is taken to be the hard direction with the x-axis being easy. Then $\mathcal{V} = K_\perp M^2\cos^2\theta - AM^2\sin^2\theta\cos^2\phi - g\mu_{\mathrm{B}}BM\sin\theta\cos\phi$ is the appropriate potential and it will be the case that $\theta \approx \pi/2$. The equations of motion are then

$$\frac{\partial \theta}{\partial t} = -\frac{\partial \mathcal{V}'}{\partial \phi} \tag{7.58}$$

where the effective potential is $\mathcal{V}'(\phi) = -(AM\cos^2\phi + g\mu_{\mathrm{B}}B\cos\phi)/\hbar$, and

$$\hbar\frac{\partial \phi}{\partial t} = 2K_\perp M\cos\theta - AM\cos\theta\cos^2\phi - g\mu_{\mathrm{B}}B\cot\theta\cos\phi. \tag{7.59}$$

If $K_\perp$ is sufficiently large, the first term on the right-hand side dominates, except when θ is too close to zero. With this the momentum is

$$p_\phi = \frac{\hbar}{2K_\perp M}\frac{\partial \phi}{\partial t}, \tag{7.60}$$

where[21] $p_\phi \approx \theta - \pi/2$. The equation of motion is then,

$$\frac{\partial p_\phi}{\partial t} = -\frac{\partial \mathcal{V}'}{\partial \phi}, \tag{7.61}$$

with the above for the momentum. The effective mass is $\hbar/2MK_\perp$, i.e., proportional to the hard axis anisotropy. As stated, in this formalism the dynamics

[21] The formal momentum conjugate to ϕ is $-(1-\cos\theta) \approx -1+\theta-\pi/2$; however the constant corresponds to an exact differential in $\mathcal{T}$ and is therefore arbitrary.

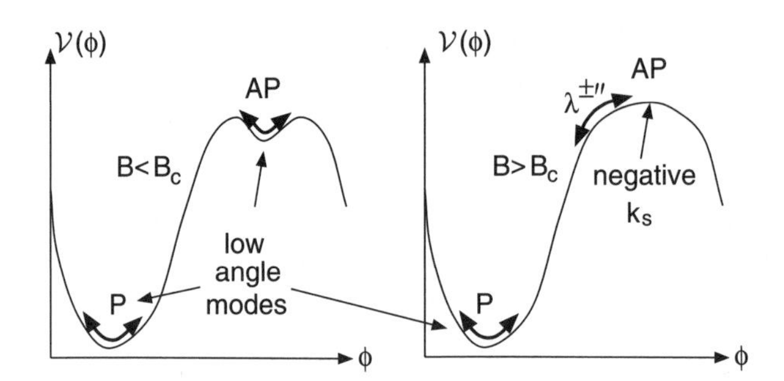

FIG. 7.5. The effective potential for both $B < B_c$ and $B > B_c$. The possible low-angle modes for $B < B_c$ are indicated. With $B > B_c$ there is a single small-angle mode possible, as indicated. The frequency and other properties of the large-angle mode for $B > B_c$ are controlled by the roots $\lambda^{\pm\prime\prime}$, as explained in the text. The potential extremes corresponding to parallel "P" and antiparallel "AP" configurations are also indicated.

of the free domain are mapped to those of a particle moving in the potential $\mathcal{V}'(\phi)$, i.e., the coordinate of the particle is ϕ and the momentum p_ϕ given above. The kinetic energy is actually that associated with the hard axis anisotropy. In a quantum description there are discrete energy levels when the particle energy is less than the barrier height. At the simplest level, for low-angle oscillations about the bottom of the well the modes correspond to those seen in FMR, with frequency ω and the separation between the quantum levels is given by $\hbar\omega$.

There are two overall situations reflecting the magnitude of the external field B. For $B < B_c$, the nominal coercive field, there are two energy minima in $\mathcal{V}'(\phi)$, see Fig. 7.5. Since tunneling is usually negligible, there are equilibrium points with the magnetization along either the positive or negative z-direction, which are stabilized by the anisotropy energy A. One well corresponds to a free domain being parallel and the other antiparallel to the fixed magnetization. Any relaxation process must cause the system to relax towards equilibrium, i.e., towards the bottom of one of the two potential wells. On the other hand, the current will cause spin pumping in one well but will add to the relaxation in the other. Consider the former situation. For small enough currents the pumping effect is overcome by relaxation and the system remains at the bottom of its well (at zero temperature). There exists a critical current density j_D at which the converse is true and beyond which either the particle is pumped upwards until it passes over the barrier or until it reaches a dynamic limit cycle for which the two effects cancel.

Such classical oscillations require coherence. Coherent states are never eigenstates of the quantum mechanical problem, since by definition these are stationary. Even when the relevant fields are quite small, the coupling to the external electromagnetic fields tends to induce coherence, making the classical description the correct one. However relaxation can lead to the destruction of coherence. In

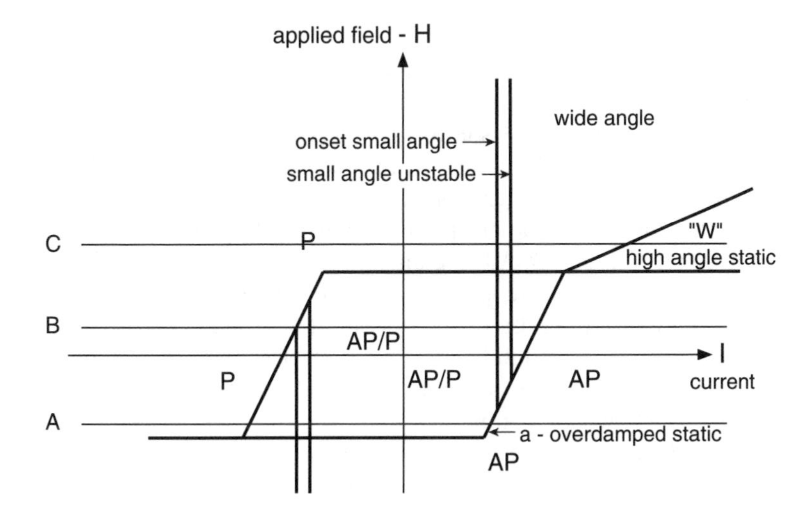

FIG. 7.6. Schematic phase diagram, see text.

particular, due to time reversal symmetry, the equations of motion generically have equal but opposite frequency roots corresponding to the magnetization rotating in counter-senses. Relaxation invariably couples these roots and when the coupling is too strong destroys coherent motion. Put simply, the harmonic oscillations about the bottom of the well will be over-damped under certain circumstances. Clearly when B is less than but very close to B_c the well of interest becomes very shallow and inevitably the motion becomes over-damped. Current-induced pumping couples to the magnetization in a different fashion than relaxation and simply raises the amplitude, invariably of the lowest (complex) energy mode. When it is over-damped, the system is lifted out of the potential well without inducing oscillations, see the next section. This is a tentative explanations of the cut “A” indicated in the phase diagram in Fig. 7.6 and for which the domain is reversed below B_c without going into small-angle oscillations. On the other hand for cut “B” the oscillations at the bottom of the well are not over-damped and the system is driven over the potential barrier, for $j > j_B$, to the state with a reversed magnetization by ever increasing oscillations.

The second situation arises when $B > B_c$ which eliminates the second minimum in the potential, see Fig. 7.5, and corresponding to cut “C”. What happens depends delicately on the relaxation model, the angular dependence of the angular momentum transfer term, and the existence of an smf. When the amplitude of a large-angle mode is sufficient that the classical turning point approaches the maximum of the potential, the frequency drops to zero since the particle spends all its time near the turning point. This is what happens approaching the region “W” where a signal is absent. The possible existence of such stable large-angle modes is taken up again below.

7.6.2 *Static critical current*

Under suitable circumstances there are two equilibrium states with the magnetization of the domains either parallel or antiparallel. There are competing interactions between the domains. Usually there is a normal metal layer interposed between the magnetic domains and which is believed to essentially eliminate any direct exchange interaction. There remains, however, demagnetization fields and these prefer the domains to be antiparallel. This is the situation for the experiments described in Chapter 5. A finite magnetic field parallel to the fixed domain is required in order to make the parallel configuration stable. This has the effect of displacing the phase diagram of the system along the field axis.

For relaxation to local equilibrium, the rate of change of ϕ is proportional to $(1/\tau_e)\sin\beta$ where β is the angle between $\boldsymbol{M}$, and its instantaneous equilibrium value $\boldsymbol{M}_0(t)$ which is directed along the direction of the internal field $\boldsymbol{B}$ and where $(1/\tau_e)$ is the intrinsic relaxation rate. If $\boldsymbol{M}$ defines the angles θ and ϕ then, except very close to the classical turning points of the particle description, $\boldsymbol{M}_0(t)$ makes an angle $(K_\perp/A)\theta$ to the z-axis. This leads to a contribution $(K_\perp/A)(1/\tau_e)p_\phi$ in the equivalent particle equation of motion. The effective gauge field produced by the current is exactly perpendicular to the plane defined by $\boldsymbol{M}_i = M\hat{x}$ and $\boldsymbol{M}_j$ and proportional to sine of the angle between them, i.e., given by $(1/4eM)j_s\mathcal{A}\boldsymbol{M}_j \times \hat{x}$. With the present definitions this is $\approx (j_s\mathcal{A}/4eM)(p_\phi\hat{z} - \sin\phi\hat{y})$ and, including these terms in the equations of motion, they become

$$\frac{\partial p_\phi}{\partial t} - \frac{K_\perp}{A}\frac{1}{\tau_e}p_\phi + \frac{j_s\mathcal{A}}{4eM}p_\phi = -\frac{\partial V'}{\partial \phi}, \tag{7.62}$$

with

$$\frac{\partial \phi}{\partial t} = \frac{2K_\perp M}{\hbar}p_\phi + \frac{j_s\mathcal{A}}{4eM}\sin\phi. \tag{7.63}$$

For small-angle modes $\sin\phi \approx \phi$ and the pumping, or antirelaxation, due to the current is squeezed and simply adds $e^{+\lambda_p t}$ to any magnetization as compared with the $j_s = 0$ solution. The solution of the $j_s = 0$ damped harmonic oscillator problem is then determined by the roots

$$\omega^\pm = i\frac{1}{\tau_e} \pm \sqrt{\frac{k_s}{m} - \left(\frac{1}{\tau_e}\right)^2} \tag{7.64}$$

where here k_s is the effective spring constant. If the system is under-damped then the effective relaxation rate[22] for small-angle modes is

$$\frac{1}{\tau_e} = \frac{K_\perp}{2A}\frac{1}{\tau}. \tag{7.65}$$

[22]There is the same rate of relaxation between the positive and negative energy roots and involved in the destruction of coherence, as discussed in the last section, see below.

Adding the effects of the current this becomes,

$$\frac{1}{\tau'} = \frac{1}{\tau_e} - \frac{j_s \mathcal{A}}{4eM}. \tag{7.66}$$

Clearly the system is stable to small-angle oscillations only when this is positive and this therefore defines a critical number current

$$j_D = \frac{4eM}{\mathcal{A}} \frac{1}{\tau_e} \tag{7.67}$$

beyond which the system is in small-angle motion. This is essentially the same as the expression, Eq. (5.3) in Chapter 5 with $(1/\tau_e)$ replaced by $\frac{1}{2}\alpha k$, this reflecting the assumption of traditional Gilbert damping.

When relaxation arises from conduction electron spin-orbit coupling, $(1/\tau_e)$ is, to a first approximation field independent, and j_D defines a vertical line[23] in the H-I phase diagram, Fig. 7.6.

It is an experimental consequence of this model that the net relaxation rate

$$\frac{1}{\tau} = \frac{1}{\tau_e}\left(1 - \frac{j}{j_D}\right) \tag{7.68}$$

for *driven* small-angle oscillations goes to zero approaching j_D. For the antiparallel stable point the current contribution adds to the thermodynamic relaxation.

For fields less negative but close to the value B_c, k_s becomes small and the system will inevitably be over-damped. The roots now correspond to

$$\lambda^{\pm\prime} = \frac{1}{\tau_e} \pm \sqrt{\left(\frac{1}{\tau_e}\right)^2 - \frac{k_s}{m} - \frac{j_s \mathcal{A}}{4eM}}. \tag{7.69}$$

That $\lambda^{-\prime} = 0$ indicates the existence of a static solution with finite small-angles and this defines a critical current j_s at which rotation to the antiparallel orientation occurs without oscillations, i.e., along cut "A". Over-damped modes are a consequence of the non-squeezed nature of the relaxation.

7.6.3 *Stability of small-angle oscillations - dynamic critical points*

Relaxation couples to the effective particle momentum p while current pumping is squeezed. Because of the strong squeezing, the linear approximation, $\cos\theta \approx \theta - \pi/2 = p_\phi$, has a correction which is $\sim A/K_\perp$ times smaller that to the linear approximation $\sin\phi \approx \phi$. The dynamical critical current j_B beyond which small angle oscillations are no longer stable is estimated by making the approximation

[23]With traditional Gilbert damping this line has the slope implied by the field dependence of k and which is also relatively weak, see the comments in Chapter 5.

$\sin\phi \approx \phi(1 - \phi^2/6)$, and replacing ϕ^2 by its average over the motion. Equating the resulting relaxation rate to zero results in the equation,

$$0 = \frac{1}{\tau_e} - \frac{j_s \mathcal{A}}{4eM}\left(1 - \frac{1}{12}{\phi_m}^2\right), \tag{7.70}$$

in terms of j_D, this implies

$${\phi_m}^2 \approx \frac{12(j - j_D)}{j_D}. \tag{7.71}$$

With increasing j and ϕ_m, in the absence of a net external bias field, the equilibrium point will pass over the barrier between the parallel and antiparallel configurations when $\phi_m = \pi/2$. This gives the estimate,

$$j_B^0 - j_D \approx \frac{1}{5} j_D, \tag{7.72}$$

for this new dynamical critical current. This magnitude is not inconsistent with experiment; see the results discussed in Chapter 5. For a negative bias field, favoring the antiparallel configuration, the top of the barrier occurs for $\phi_m < \pi/2$ and $j_B(B) < j_B^0$ while for a positive field the reverse is true. This is reflected by a line in the phase diagram with a finite slope, Fig. 7.6. Sun [43] has used simulations based upon traditional Gilbert damping to also show the existence of a stable small angle oscillations.

When $B > B_c$, the coercive field, the antiparallel configuration is no longer *energetically* stable, see Fig. 7.5, and j_B is not immediately defined. The problem near to the antiparallel configuration is that of an inverted harmonic oscillator, i.e., one in which the "spring constant" k_s has changed sign. The motion for small deviations from this configuration is determined by the two roots which correspond to the relaxation rates

$$\lambda^{\pm\prime\prime} = \frac{1}{\tau_e} \pm \sqrt{\frac{k_s}{m} + \left(\frac{1}{\tau_e}\right)^2}. \tag{7.73}$$

These correspond to motion towards and away from the limiting situation.

7.6.4 *Stability of the negative temperature fixed point*

There exists the possibility of static solutions in the region just above the line where $B = B_c$, for which $\boldsymbol{M}$ is close to the antiparallel position.

The analysis of this situation parallels that for the existence of small-angle oscillations except now it is the existence of the large-angle modes which is important. Of the two roots displayed above it is the root

$$\frac{1}{\tau^-} = \sqrt{\frac{k_s}{m} + \left(\frac{1}{\tau_e}\right)^2} - \frac{1}{\tau_e} \tag{7.74}$$

which corresponds to the damped motion *away* from antiparallel. Unlike the situation for small-angle oscillations, the fact that this has no imaginary part does not imply the absence of large-angle modes. It is rather that this is *positive* implying, when $\boldsymbol{M}$ is near to the antiparallel configuration, it will move to larger angles and thereby eventually leave the small-angle region and undergo oscillations. In the region just above the line $B = B_c$ the "spring constant" k_s is small which implies $(1/\tau^-)$ is also small and relaxation strongly prolongs the net period. Now $(1/\tau^-)$ is the net torque due to relaxation and fields and this opposes $(1/4M)j_s\theta$ due to the current so that the net effective relaxation rate becomes

$$\frac{1}{\tau'} = \frac{1}{\tau^-} - \frac{j_s\mathcal{A}}{4eM} = \sqrt{\frac{k_s}{m} + \left(\frac{1}{\tau_e}\right)^2} - \frac{1}{\tau_e} - \frac{j_s\mathcal{A}}{4eM}. \tag{7.75}$$

This defines a critical current j_N given by

$$\frac{j_N\mathcal{A}}{4eM} = \sqrt{\frac{k_s}{m} + \left(\frac{1}{\tau_e}\right)^2} - \frac{1}{\tau_e}, \tag{7.76}$$

for which the large-angle oscillations approach the antiparallel configuration and the frequency approaches zero. This again shows, for larger currents, there exist static solutions near, but not at, the antiparallel configuration. Such static solutions will clearly have a magnetoresistance intermediate between that for the parallel and antiparallel situations. This offers a natural explanation of the region "W" already mention above. Such static solutions can exist because of the subtle difference between relaxation to local equilibrium, or of modified Gilbert form, for which these exist and traditional Gilbert damping, and the above-mentioned simulations, for which they do not.

This notwithstanding, the picture predicts that as j passes through j_B the dynamic mode changes from being of small to large-angle character in a *smooth* fashion. As the current increases further the large-angle mode becomes more and more dominated by the effects of relaxation driving the frequency to zero at j_N. Beyond this the dynamical modes disappear. The absence of a sharp transition between modes is at odds with experiment [1], but would appear to be in agreement with the work of Sun [43]. It contrasts with the finite-temperature results for Co domains to be found in Chapter 5. As is detailed there, the effects of a finite-temperature are simulated by adding a random magnetic field, $\boldsymbol{A}(t)$ proportional to $\sqrt{k_{\mathrm{B}}T}$. This is common to most finite-temperature simulations. Consider the effect of such a field when a spin is nearly parallel to a static field $\boldsymbol{B}$. The field evidently acts to confine the random walk generated by $\boldsymbol{A}(t)$. The vector $\boldsymbol{M}$ is effectively repelled from the direction defined by $\boldsymbol{B}$ but only by a finite amount. In the opposite situation when $\boldsymbol{M}$ begins in a region where is almost antiparallel to the field, the random walk generated by $\boldsymbol{A}(t)$ will cause $\boldsymbol{M}$ to steadily move away from this direction. In the case of interest, the antiparallel configurations is an energetic saddle point and the random walk is directed towards smaller θ (smaller p_ϕ in the particle picture) and smaller ϕ. The random walk flow will take the system to the lowest energy, i.e., the parallel configuration. This amounts to an additional relaxation process, separate from $(1/\tau_e)$. The fact that there is zero probability of finding $\boldsymbol{M}$ in the antiparallel direction causes there to be a divergence in this temperature-dependent supplementary effective relaxation. The net effect is that the antiparallel configuration is made unstable and beyond j_B the system will first jump to into large-angle and then out-of-plane modes as explained in Chapter 5. (Given suitable parameters it is possible to make the small to large-angle transition even for $B < B_c$.)

On physical grounds it is evident that at least one Lyapunov exponent turns positive near any critical current where $\boldsymbol{M}$ jumps and it may well be that the effects of $\boldsymbol{A}(t)$ mimic the chaotic dynamics near the antiparallel configuration which plays the role of a strange attractor.

Thus as a consequence of the random field $\boldsymbol{A}(t)$ the antiparallel configuration can never be stable at finite-temperature within such simulations and the system jumps from large-angle to out-of-plane modes with increasing current. Experiment on the Co system [1] for which these simulation were performed, in contrast, exhibits no out-of-plane modes. Roughly speaking, in the region of the phase space where the out-of-plane modes are expected from the simulations is found instead the region "W" for which no dynamic modes are evident. This was *not* identified with the antiparallel configuration since the magnetoresistance is smaller than would be expected. On the other hand, the frequency of the large-angle modes decreases strongly approaching this region which represents strong evidence that the magnetization *is* approaching antiparallel as it approaches region "W".

Accounting for the smf generated by the dynamical modes may help explain these observations.

7.6.5 *The role of the smf/emf due to dynamical modes*

The formal expression Eq. (7.47) for the transverse smf/emf is rather difficult to handle. When $\boldsymbol{M}$ approaches the antiparallel configuration both θ and γ are small and therefore so is this contribution independent of the value of $(d\phi/dt)$, i.e., the transverse part is not important to the above discussion of the area occupied by the "W" part of the phase diagram. This notwithstanding, the existence of this smf/emf does offer a natural explanation of the constant rf power lines when large-angle excitations are relevant. The frequency of such modes is controlled by the motion near the classical turning point of the equivalent particle. As already observed, this is the region where the current-magnetic coupling is the smallest. It follows that the matrix elements which determine the probability that an electron makes a transition between levels in a quantum picture is determined by the motion at smaller angles, i.e., in regions where the motion is independent of the frequency and the (small) amplitude measured relative to a fixed point. Given a fixed transition probability, the average energy *given* to an electron, to create the *rf* currents, must be proportional to the quantum of energy, i.e., $\hbar\omega_r \approx \hbar\omega_0 + zg\mu_B B$ where $z \approx 0.5$. Thus energy measured in a load is also proportional to $\hbar\omega_r$, i.e., contours of constant power obey approximately

$$j = C(\omega_0 + zg\mu_B B/\hbar) \tag{7.77}$$

where C is an undetermined constant.

More importantly the longitudinal smf/emf is large, determines the boundary of the "W" region, and explains why there is a jump between small and large-angle oscillations. Given that the demagnetization field in the two domains has the same magnitude and independent of the mechanism which couples them, the smf for antiparallel domains is $\mathcal{E}_t = g\mu_B B/e$. This is the work done[24] by an electron on the order parameters $\boldsymbol{M}_i$ and $\boldsymbol{M}_j$. By Lenz's law this is a back smf which opposes the driving spin current implicit in the charge current. Given an effective surface conductivity σ_s, the *spin*-current-induced by this back-smf is $\sim (\sigma_s g\mu_B/e)B$. The current-induced spin current is pj, where p is the effective polarization of the conduction electrons and j the charge current, and so the net spin current

$$j'_s = pj - \frac{\sigma_s g\mu_B}{e}B \tag{7.78}$$

which might be positive or negative. Given that j_B is rather small compared to the current at the boundary of the "W" region, it must be that this corresponds to a good approximation to $j'_s = 0$, i.e., the critical current

$$j_w = \frac{\sigma_s g\mu_B}{pe}B \tag{7.79}$$

is proportional to the applied field. It is a remarkable fact that the boundary of the "W" region is indeed a straight line which passes through the origin [1].

Consider modest currents but for large enough fields to correspond to a point above the boundary to the "W" region. Following the above discussion, the smf will cause the spin current to change sign if $\boldsymbol{M}$ finds itself near to the antiparallel direction. It is implied that the back-smf increases with the amplitude of large-angle modes and that the rise in this smf rather than relaxation will limit their amplitude. On the other hand, the longitudinal smf for small-angle modes is small since $\boldsymbol{M}_j$ is always close to the parallel configuration. Thus along a line of increasing current at constant applied field $B_0 > B_c$, there is a first a critical current j_D at which small-angle oscillations start. Since this is above the "W" region, and since $(1/\tau^-)$ is smaller than $(1/\tau_e)$ there should already exist a stable high energy state corresponding to large angle oscillations. There will be a critical current $\approx j_B$ at which the current drives the system over the barrier between these states and where the system switches from small to large-angle modes.

[24] It is to be emphasized that this is *not* directly the difference of Zeeman energy induced shifts in the effective potentials seen by the conduction electron.

7.6.6 *Interactions of ferro- and metallic layers*

A real experimental device usually has a metallic layer between the ferromagnets, i.e., is a FNF sandwich. The conduction electrons in the N-layer might be expected to interact with the magnetization in the F-layer via a contact effective exchange interaction at the interface. Through this interaction the effective potential for the up electrons is pushed, say, up while that for the down electrons moves in the opposite direction. In a metal, the carriers move to screen such local potential within the Fermi-Thomas screening length. However, because the electrons involved are near the Fermi surface this familiar RKKY interaction falls off as a power law and oscillates in space at $2k_{\mathrm{F}}$ where k_{F} is the Fermi wave-vector. Experiments suggest that this does not lead to an important interaction for N-layers of dimensions and disorder of interest here. This means that the direct exchange interaction which would exist for a direct F-F contact are, in fact, absent.

Of real interest for devices are the interactions which appear when there is a current present in the FNF system. For half-metal ferromagnets, the current in the F-layer injects up conduction electrons into the N-layer. In contrast the N-layer is weakly polarized and the same charge current only transfers a small part of these up conduction electrons to the second F-layer. From one point of view this situation is remedied by *spin accumulation.* It is envisaged that the spin-dependent chemical potential difference (sometimes called a branch in-balance) builds up in the N-layer until it is sufficiently large to drive the necessary spin current through this layer via the effects of the diffusion term.

From a different point of view it is to be observed that provided that the N-layer is thinner, $d < \ell_d$, than the spin diffusion length, ℓ_d, as is invariably the case, then the conduction electrons traverse this layer without a spin-flip and so although they might scatter many times they retain the memory of their initial spin direction. The current leaving the N-layer is only weakly polarized and cannot transmit the resulting incident angular momentum. It is necessary to also account for the *flux* of particles through the surface. Even in the absence of a charge current these correct for any spin imbalances between the two sides of the interface. Put differently these cause rapid spin relaxation. As will be discussed in detail below, provided this relaxation is rapid compared to the time by which the N-layer relaxes to the lattice, relaxation *does* efficiently transmit the angular moment given to the N-layer by the spin currents which flow into it.

This idea can be put in different language. The transmission of the angular momentum by the conduction electron can be thought of as the dynamic part of the RKKY interaction [44]. It has been shown that this part of the interaction has a much longer range than the static part of the same interaction discussed at the beginning of the subsection. This point of view has been extensively discussed elsewhere [22].

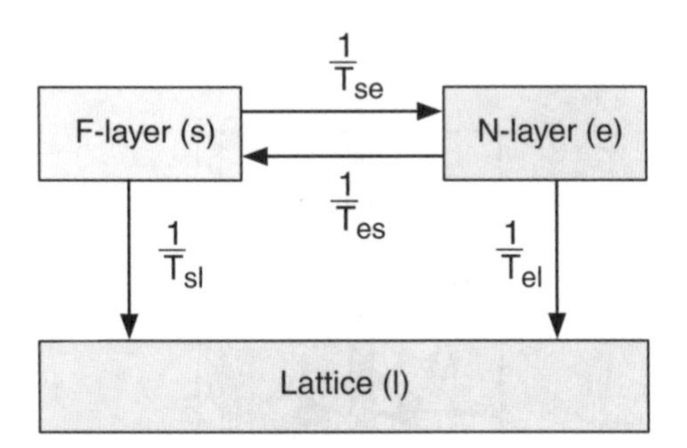

FIG. 7.7. The relaxation scheme relevant for a FN bottleneck.

7.6.7 *The relaxation bottleneck and dynamics of an FN-interface*

The efficient transfer of angular momentum between the two F-layers of an FNF device requires there to be a relaxation bottleneck. This means that any angular momentum given to the N-layer is lost via relaxation to one or other of the F-layers before it can be lost to the lattice, for example, via spin-orbit scattering of the conduction electrons in the N-layer. This bottleneck effect has been much discussed in the literature on magnetic resonance in dilute magnetic alloys [32]. The existence of this effect for the small-angle modes excited directly by FMR has been demonstrated experimentally [41], but described in different language.

The dynamical behavior of current-driven systems reflects the possible small and large-angle modes which the system is capable of exhibiting. It is therefore imperative to understand the nature of such modes in FN and FNF structures.

The basic effect will exist in a FN system. The conduction electron flux through the surface leads to cross-relaxation. The scheme is illustrated in Fig. 7.7. There are rates $(1/T_{se})$ and $(1/T_{es})$ which, respectively, transfer angular momentum between the spin, i.e., the order parameter, of the F-layer and the conduction electrons of the N-layer. If M_s (M_e) is the *uniform* component of the F(N)-layer magnetization, then a detailed balance condition $(M_s/T_{se}) = (M_e/T_{es})$ relates these relaxation rates. The magnetizations M_s and M_e lose angular momentum to the lattice via the rates $(1/T_{s\ell})$ and $(1/T_{e\ell})$. In general the two subsystems will have distinct resonance frequencies ω_s and ω_e since they see different demagnetization and anisotropy fields and may have significantly different g-factors. When $(1/T_{s\ell})$ and $(1/T_{e\ell})$ are both zero, the total magnetization $\boldsymbol{M}_T \propto \boldsymbol{M}_e + \boldsymbol{M}_s$ commutes with the interactions, and provided $\omega_s = \omega_e$, the total magnetization precesses about the internal field at this common frequency and the relaxation rate is zero. There is necessarily a second normal mode corresponding to the magnetization $\boldsymbol{M}_A = (\boldsymbol{M}_e/M_e) - (\boldsymbol{M}_s/M_s)$. This mode has the same resonance frequency but has a large relaxation rate $\approx (1/T_{s\ell}) + (1/T_{e\ell})$. It is easy to check that the transverse components of $\boldsymbol{M}_T$ and $\boldsymbol{M}_A$ commute. The second, asymmetric, mode is usually not important in resonance experiments since it does not couple to the uniform driving rf field. In the present context it is of prime

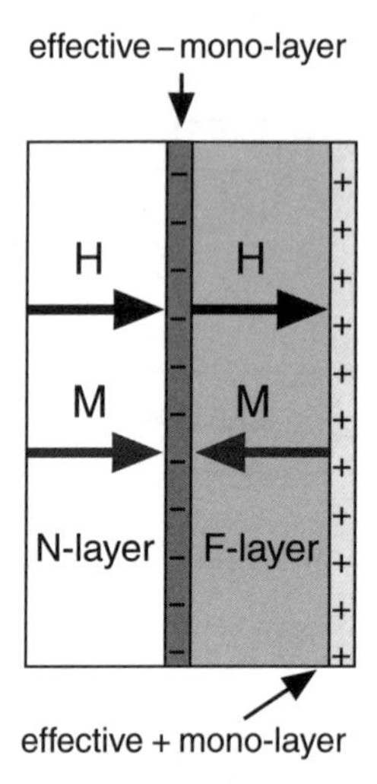

FIG. 7.8. The demagnetization field within the ferromagnetic (F)-layer can be pictured as being generated by a magnetic monolayer at the surfaces. It follows that the field in the non-magnetic N-layer actually adds to the external field.

importance since the angular momentum transfer term couples precisely to this degree of freedom. When $\omega_s \neq \omega_e$ the resonance of $\boldsymbol{M}_e + \boldsymbol{M}_s$ occurs at

$$\omega_s = \frac{\omega_s M_s + \omega_e M_e}{M_s + M_e} \tag{7.80}$$

and turning on the lattice relaxation, the width is

$$\frac{1}{\tau_s} = \frac{\frac{1}{T_{s\ell}} M_s + \frac{1}{T_{e\ell}} M_e}{M_s + M_e}. \tag{7.81}$$

Both are the magnetization weighted average of the relevant quantities. In this context it is worth noting that the dipole field which is responsible for the anisotropy field proportional to $K_\perp$ is a demagnetization field in the F-layer but has the opposite sign in the N-layer, see Fig. 7.8. This might lead to an appreciable correction to the effective value of this demagnetization field for an ultrathin F-layer in contact with more substantial N-layers.

For a large fixed F-layer, $M_s \ll m_e$ and the antisymmetric mode is to a good approximation just the resonance mode of the conduction electrons of the N-layer, with a correction $(M_e/M_s)\boldsymbol{M}_s$ just large enough to make this "orthogonal" to the symmetric mode $\boldsymbol{M}_e + \boldsymbol{M}_s$. Thus at a fixed FN contact, the current just drives the conduction electron magnetization of the N-layer.

In contrast, when the F-layer is free it is the symmetric mode which is of importance since, because of its small relaxation rate, it is easier to excite with the driving current. This will be taken up again in the next section.

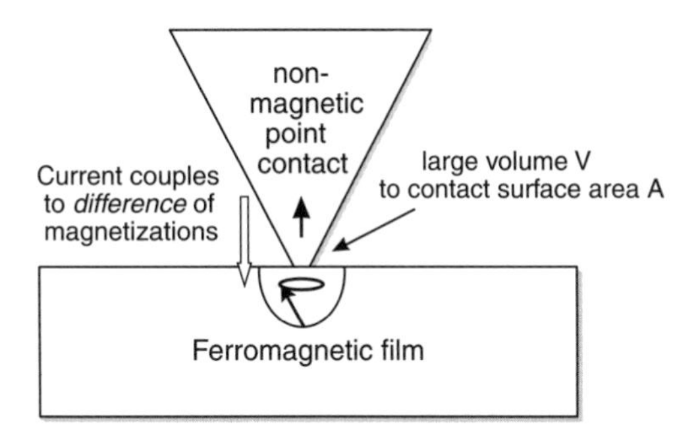

FIG. 7.9. The spin current couples to the difference of the magnetizations. In a point contact the relaxation coupling between the non-magnetic metal and ferromagnetic layer is weak compared to the lattice relaxation of the non-magnetic layer. This can break the relaxation bottleneck. A spin current exists because of the "filter" effect of the ferromagnet. The spin current therefore couples to the magnetization $\boldsymbol{M}_s$ of the ferromagnet.

For the reasons explained above, it can only be of interest to break the bottleneck when a small FN interface is involved, e.g., for a point contact. Such a geometry, Fig. 7.9, can have a small surface-to-volume ratio. The lattice rate $(1/T_{e\ell})$ is proportional to the volume within the spin diffusion length while $(1/T_{es})$ is proportional to the surface area leading to $(1/T_{e\ell}) > (1/T_{es})$ and the absence of a bottleneck. When the bottleneck is broken the resonance modes correspond to the excitation of $\boldsymbol{M}_e$ and $\boldsymbol{M}_s$ separately. These are both half-symmetric and half-antisymmetric and couple to the current. There is no need to have a fixed ferromagnetic layer to have such a coupling. Since $(1/T_{e\ell})$ is large the $\boldsymbol{M}_s$ modes are easiest to excite. When a large external static field is applied perpendicular to the F-layer magnetization the large-angle modes generate an equally large smf when θ is large leading to stable modes (rather than a smooth evolution with current as Landau–Lifshitz–Gilbert theory would predict). The large change in this smf when the system jumps from small to large-angle modes results in a large peak in dV/dt, as seen in experiment [45].

7.6.8 *Dynamics of FNF domains*

As already stated, a bottleneck is needed in order to have the efficient transfer of angular momentum between domains. Such a bottleneck involving both FN-surfaces can exist even when there are substantial differences between the resonance frequencies of the two domains. A similar situation has been analyzed in the past [32, 46]. Adapting this to the present situation, assuming that there is equal probability of the conduction electrons to cross-relax at each domain, it is found that the coupled Landau–Lifshitz equations are of the form

$$\frac{d\boldsymbol{\mathcal{M}}_i}{dt} = -\frac{g\mu_{\mathrm{B}}}{\hbar}\boldsymbol{\mathcal{M}}_i \times \boldsymbol{B}_i - \frac{1}{\tau_e}\delta\boldsymbol{\mathcal{M}}_i + \frac{1}{2T_{se}}\delta\boldsymbol{\mathcal{M}}_j$$

$$\frac{d\boldsymbol{\mathcal{M}}_j}{dt} = -\frac{g\mu_{\rm B}}{\hbar}\boldsymbol{\mathcal{M}}_i \times \boldsymbol{B}_j - \frac{1}{\tau_e}\delta\boldsymbol{\mathcal{M}}_j + \frac{1}{2T_{se}}\delta\boldsymbol{\mathcal{M}}_i \tag{7.82}$$

where, for example, $\boldsymbol{\mathcal{M}}_i = (\boldsymbol{M}_i + \frac{1}{2}\boldsymbol{M}_e)$. The fields $\boldsymbol{B}_i$ are given by the derivative of an appropriate energy function for each domain. The $\delta\boldsymbol{\mathcal{M}}_i$ are the deviations from local equilibrium defined in the usual fashion. The total symmetric bottleneck coordinate is $\boldsymbol{\mathcal{M}}_i + \boldsymbol{\mathcal{M}}_j = (\boldsymbol{M}_i + \boldsymbol{M}_j + \boldsymbol{M}_e)$ indicating why the factor of $\frac{1}{2}$ must appear in each of the $\boldsymbol{\mathcal{M}}_i$. The effective diagonal relaxation rate is

$$\frac{1}{\tau_e} = \frac{1}{T_{s\ell}} + \frac{M_e}{M_s}\frac{1}{2T_{e\ell}} + \frac{1}{2T_{se}}. \tag{7.83}$$

Equations of this structure are well known from the theory of motional narrowing of resonance lines [47] and experiment has shown that the appropriate mode locking and narrowing that these equations predict *do* occur when FMR experiments are performed on a system with two free domains [41]. Without a bottleneck such effects would not occur.

When the anisotropy is such that one domain is fixed, then for the free domain,

$$\frac{d\boldsymbol{\mathcal{M}}_i}{dt} = -\frac{g\mu_{\rm B}}{\hbar}\boldsymbol{\mathcal{M}}_i \times \boldsymbol{B}_i - \frac{1}{\tau_e}\delta\boldsymbol{\mathcal{M}}_i \tag{7.84}$$

which is just a simple Landau–Lifshitz equation with relaxation to local equilibrium but for the magnetization $\boldsymbol{\mathcal{M}}_i$ rather than just $\boldsymbol{M}_i$. This modification is all important since only the part $\boldsymbol{M}_e$ of $\boldsymbol{\mathcal{M}}_i$ couples to the current at the interface with the fixed domain.

This simplification ignores some important effects. The large-angle modes predicted by this simplified equation have a frequency which changes rapidly as a function of amplitude and thus with current in the experiments of interest here. In addition these modes have a very rich harmonic structure. In contrast the small-angle modes for the fixed domain have an almost fixed frequency. Whenever the harmonics of the large-angle modes cross those of the fixed domain, the coupling between the modes again becomes important and such mode locking is a plausible explanation for the frequency jumps seen in some experiments [48].

7.6.9 *Angular momentum transfer in FNF domains*

The presence of a bottleneck implies that angular momentum given to the conduction electrons is given back to the ferromagnetic domains before it can relax to the lattice. When this is not the case the angular momentum given to the conduction electrons by the angular momentum transfer term at one interface is not transmitted at the other, i.e., the two surfaces behave independently.

Assume that a bottleneck exists. In the frame of reference of the free domain the current at the interface with the fixed domain couples to M_{ey}. The dynamic modes of the free domain involve $\boldsymbol{\mathcal{M}}_i = (\boldsymbol{M}_i + \frac{1}{2}\boldsymbol{M}_e)$ rather than just $\boldsymbol{M}_i$. In the equation of motion for $\boldsymbol{\mathcal{M}}_i$ the coupling is to $\boldsymbol{\mathcal{M}}_{iy}$ but with an added factor of $\frac{1}{2}M_e/(M_s + (M_e/2))$. The net spin transfer is as in the treatment of a direct F-F contact *except* for an extra factor of one-half. This factor reflects the fact that the transverse angular momentum given to the N-layer is transferred via relaxation equally to the free and fixed domains.

With an extra factor of two the theory developed earlier is valid for a FNF system. The only new consideration is the relaxation coupling discussed in the previous subsection and which does not (automatically) exist for a direct interaction between F-layers. All expressions for critical currents obtained earlier can be taken over when multiplied by a factor of two.

7.6.10 *Angular momentum transfer in NFN systems*

It might reasonably be assumed that the relaxation is bottlenecked in NFN thin film systems with dimensions similar to the FNF spin valves. It is to be observed that the spin current at the two FN interfaces is in the same direction and, since the spin current coupling involves the difference of the magnetizations, the coupling to the F-layer cancels for a symmetric system. This observation is not new [49–51]. Almost any asymmetry will introduce such a coupling. Imagine that one N-layer is in contact on its other side with one which causes rapid spin relaxation. This adds a relaxation term

$$\frac{1}{\tau_\ell}\delta \boldsymbol{M}_{ej} \approx \frac{M_{ej}}{\tau_\ell}\left[\frac{\boldsymbol{M}_{ej}}{M_{ej}} - \frac{\boldsymbol{M}}{M}\right]. \tag{7.85}$$

This is relaxation to local equilibrium, which requires $\boldsymbol{M}_{ej}$ and $\boldsymbol{M}$ to be parallel, i.e., that $(\boldsymbol{M}_{ej}/M_{ej}) = (\boldsymbol{M}/M)$. Here $\boldsymbol{M}_{ej}$ is the magnetization of one of the N-layers and $\boldsymbol{M}$ that of the ferromagnet. Clearly this induces a coupling to the antisymmetric mode and acts to *partially* open the bottleneck. Put differently this adds to the symmetric mode a part of the antisymmetric one with a coefficient $\sim T_{es}/\tau_\ell$ and with this a coupling to the spin current.

Experiments [51] on such a system exhibit an extended "W" region with a straight line boundary which extrapolates to the origin as in Co FNF valves [1]. It is tempting to speculate that there are again non-dynamic states near an energy maximum.

7.7 Domain walls

Experimentally [16–18, 52] it has been shown that the critical current densities required to put in motion or displace a domain wall can be a couple of orders of magnitude smaller than that required to reverse the magnetization of a free domain. For this reason it is probable that the domain wall will find important applications in spintronics and it is therefore important to understand the dynamical properties of such entities. The theory presented here is based upon Refs. [31, 53].

A domain wall is a solitonic object which connects regions of a magnet in which the magnetization has a stable configuration. Thus, for example, for the wire shown in Fig. 7.10 the magnetization at the extreme ends is pointing along

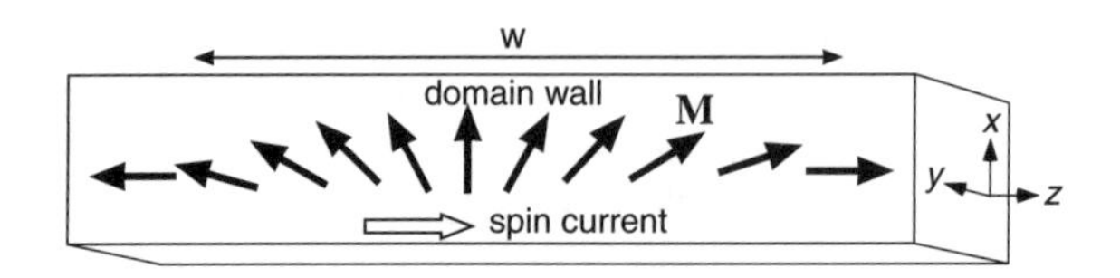

FIG. 7.10. The wire containing a domain wall extends down the z-direction. Due to the anisotropy field proportional to $K_\perp$, the magnetization lies in the xz-plane. The domain wall width is denoted by w.

the axis of the wire *but* in opposite directions.[25] In making a smooth connection between these two configurations there is a competition between the exchange energy which would like the spins to rotate slowly and the anisotropy energy which would have as few spins as possible deviate appreciably from the z-direction. In terms of the Euler angles introduced earlier, the direction of a given spin is specified by the angles θ and ϕ_i where i is a site index. Twisting the wall costs extra exchange energy and usually a wall will have a constant fixed value of $\phi_i = \phi$ with values of $0 < \theta < \pi$. When a second anisotropy term proportional to $K_\perp$ is present then the system has the lowest energy when $\phi = 0$ or π. The effective potential energy in such a wall, assuming $\phi = 0$, is in a continuum approximation,

$$V = \int \frac{dv}{v_c} \hbar^2 \left[\frac{1}{2} m^2 J \left(\frac{\partial \theta}{\partial z} \right)^2 a^2 - m^2 A \cos^2 \theta(z) \right] \equiv \int \frac{dv}{v_c} \mathcal{V}(z), \tag{7.86}$$

where m is the magnitude of spins per site, a is the lattice constant in the z-direction, and v_c is the unit volume. The Lagrange equation which minimizes $\mathcal{V}(z)$ with respect to variations in $\theta(z)$ is

$$-\frac{\partial}{\partial z} \left[a^2 J \frac{\partial \theta}{\partial z} \right] + 2A \cos\theta \sin\theta = 0. \tag{7.87}$$

This can be integrated by observing that it implies,

$$\frac{\partial}{\partial z} \left[\frac{1}{2} J \left(\frac{\partial \theta}{\partial z} \right)^2 a^2 - A \sin^2 \theta + C \right] = 0 \tag{7.88}$$

where C is an integration constant. Since the derivative $(\partial\theta/\partial z) \to 0$ for $z \to \pm\infty$, it follows that $C = 0$ and taking the appropriate root,

$$\left(\frac{\partial \theta}{\partial z} \right) a = \left(\frac{2A}{J} \right)^{1/2} \sin\theta. \tag{7.89}$$

[25] This is usually the case for Permalloy because this is the configuration in which the demagnetization field has the smallest energy. This energy is represented by the anisotropy energy A in the model, Eq. (7.18).

In turn integrating this using the appropriate boundary conditions gives as the result

$$\theta = 2\cot^{-1} e^{(z-z_0)/w} \equiv \theta_0(z - z_0), \tag{7.90}$$

where the wall has width $w = (J/2A)^{1/2}\, a$ and position z_0. Clearly the energy is independent of z_0 and directly implies that the translation operator commutes with the *many* states on the *ground* manifold of the problem.

It is an interesting exercise to check that when a static magnetic field, $\boldsymbol{B} = B\hat{z}$ is added, i.e., with $\mathcal{V}(z) \to \mathcal{V}(z) - g\mu_{\mathrm{B}} Bm\cos\theta$, this lifts the aforementioned degeneracy and leads to a problem without a static solution. Such a field applies a pressure on the wall, see below. Consider first the case with a magnetic field, but with $K_\perp = 0$. It is necessary to write the Lagrangian,

$$L = T - V; \;\; T = \hbar m \int \frac{dv}{v_c}(\cos\theta - 1)\frac{\partial\phi}{\partial t} \equiv \int \frac{dv}{v_c}\mathcal{T}(z). \tag{7.91}$$

In a rotating frame of reference obtained via $\phi \to \phi - \omega_{\mathrm{L}} t$, and it follows $\mathcal{T} \to \mathcal{T} - \hbar m(\cos\theta - 1)\omega_{\mathrm{L}}$, where $\omega_{\mathrm{L}} \equiv g\mu_{\mathrm{B}} B/\hbar$ is the Larmor frequency, and the field term cancels, to within a constant. The original field-free problem results in this frame and shows that the effect of a magnetic field is to cause uniform rotations about the z-axis. The wall does not move in the z-direction as might be expected, at least not without relaxation. This collective rotational motion of the wall can be extracted from the original kinetic energy. With $\phi(z,t) \to \phi_0 + \phi(z,t)$,

$$T = -2\hbar m \frac{\mathcal{A}}{v_c} z_0 \frac{\partial\phi_0}{\partial t} + \hbar m \int \frac{dv}{v_c}(\cos\theta - 1)\frac{\partial\phi}{\partial t}; \quad z_0 = \frac{1}{2}\int dz(1 - \cos\theta). \tag{7.92}$$

With this definition[26] of z_0, to within a constant, the Zeeman energy is $z_0 \mathcal{A} P_z$ where $P_z = 2g\mu_{\mathrm{B}} B(m/v_c)$, is the classical expression for the pressure on the wall. Now in terms of the new collective coordinate ϕ_0, the conjugate momentum is $p_\phi = (\partial T/\partial\dot{\phi}_0) = 2(\hbar m\mathcal{A}/v_c)z_0$, i.e., to within a constant of proportionality, is the coordinate of the wall. Evidently it is more logical to take z_0 as the coordinate whence $p_z = -2(\hbar m\mathcal{A}/v_c)\phi_0$ is the momentum. In terms of this, the kinetic energy of the collective coordinate is $T_0 = -z_0\dot{p}_z$ and the Lagrangian $L_0 = -z_0(\dot{p}_z - \mathcal{A}P_z)$. This L_0 correctly, gives $(\dot{p}_z - \mathcal{A}P_z) = 0$ as an equation of motion.

When $K_\perp$ is finite it is necessary to account for the term $\hbar^2 m^2 K_\perp \sin^2\theta \sin^2\phi$. Interest is now on how this couples to the collective coordinates. Given that the solution in the rotating frame has $\phi = 0$, it is possible to replace $\phi \to \phi_0$, whence the term in the potential energy is $\hbar^2 m^2 K_\perp \sin^2\phi_0 \int dz \sin^2\theta$. Trivially

[26] Given the system extends from $-\ell$ to $+\ell$, then $\frac{1}{2}\int_{-\ell}^{\ell} dz(1 - \cos\theta(z)) = \frac{1}{2}\int_{-\ell}^{\ell} dz(1 - \cos\theta_0(z - z_0)) = \frac{1}{2}\int_{-\ell-z_0}^{\ell-z_0} dz(\cos\theta_0(z) - 1) = \frac{1}{2}\int_{-\ell-z_0}^{-\ell} 2 + \frac{1}{2}\int_{-\ell}^{\ell} dz(\cos\theta_0(z) - 1)z_0 + \int_{-\ell}^{\ell} dz(\cos\theta_0 - 1) \equiv z_0 + \text{const.}$, i.e., the two definitions of z_0 differ by a constant.

$\int dz \sin^2\theta = a\,(J/2A)^{1/2}\int dz \sin\theta\,(\partial\theta/\partial z) = a\,(J/2A)^{1/2}\int d(\cos\theta) = 2a\,(J/2A)^{1/2} = 2w$. So the new term[27] is $\hbar^2 2wm^2K_\perp \sin^2\phi_0 = \hbar^2 wm^2 K_\perp(1-\cos 2\phi_0)$. Now

$$L_0 = z_0\left(\frac{2\hbar m\mathcal{A}}{v_c}\frac{\partial\phi_0}{\partial t} + \mathcal{A}P_z\right) - \frac{\hbar^2 m^2 w\mathcal{A}K_\perp}{v_c}\left[1-\cos\left(2\phi_0\right)\right], \tag{7.93}$$

which gives,

$$\frac{\partial\phi_0}{\partial t} = \frac{v_c}{2\hbar m}P_z \tag{7.94}$$

and

$$\frac{\partial z_0}{\partial t} = \hbar m w K_\perp \sin 2\omega_{\rm L} t, \tag{7.95}$$

i.e., there is a yo-yo motion with twice the Larmor frequency. This motion corresponds to trajectories which map out constant energy surfaces. Thus when $K_\perp = 0$ this is just a circular motion, but when $K_\perp \neq 0$ the wall moves backward and forward in order to "borrow" the energy from the Zeeman term in order to go over the barrier to the rotational motion which represents the energy proportional to $K_\perp$. Finally it is to be observed that this formalism is valid independent of the nature of the pressure P_z applied to the wall.

7.7.1 *Relaxation in domain walls*

The Gilbert term has traditionally [54] been used to describe the relaxation of domain walls. In the current situation this leads not only to forward wall motion but also to a *pressure* on the wall. The simple adoption of Gilbert relaxation is wrong for two reasons (i) the existence of a domain wall implies a second broken symmetry, associated with the position of the wall, this in addition and beyond that associated with a single domain ferromagnet. The phenomenology of the relaxation of this new order parameter is quite different to that for a single domain. Even at a micro-magnetic level, relaxation is not simple Gilbert like since the equations are necessarily modified to reflect the stability of the topological defect that is a domain wall. (ii) When a spin current j_s is present, there is a divergence term which, at least in part, drives the motion of the wall and this implies that Gilbert relaxation should involve the total derivative, as described in Sec. (7.2.1). It will be shown below that the necessary modifications to Gilbert damping imply this total derivative becomes a "particle derivative".

In order to address (i) consider a domain wall driven by an applied longitudinal field alone, i.e., with $j_s = 0$. If in addition $K_\perp = 0$, assuming Gilbert relaxation and ignoring α^2, Eq. (7.94) becomes $(\partial\phi_0/\partial t) = g\mu_{\rm B}B/\hbar$,[28] and the

[27]For small p_z this generates the so-called Döring mass $m_{\rm D} = \hbar^2\mathcal{A}/(wv_cK_\perp)$.

[28]With Gilbert but *not* Landau–Lifshitz relaxation $(\partial\phi_0/\partial t) = g\mu_{\rm B}B/[(1+\alpha^2)\hbar]$ and implies Gilbert relaxation results in an energy shift. Since relaxation, without mode coupling, should not lead to a shift, it is implied that Landau–Lifshitz relaxation is more appropriate, see also immediately below.

wall acquires the velocity $v_r = \alpha w g\mu_B B/\hbar$. The fact that, to this order, there is no relaxation contribution to the pressure simply confirms the observation made in Sec. 7.5 that, with $K_\perp = 0$, Gilbert relaxation leads to a width without an energy shift.[29] As was also discussed in that section, when $K_\perp \neq 0$, shifts occur because of a coupling between clockwise and counter clockwise rotational precessional modes of the wall about the external field. It is therefore an important difference that here it remains the case that $(\partial\phi_0/\partial t) = g\mu_B B/\hbar$ even when $K_\perp$ is finite, showing that these modes remain uncoupled and therefore that the Gilbert term needs to be modified so that again unphysical energy shifts do *not* occur. Consider the magnetization $\boldsymbol{M}$ of the slice of the wall between z and $z + dz$, the appropriate Landau–Lifshitz–Gilbert, Eqs. (7.11), are,

$$\frac{D\boldsymbol{M}}{Dt} = -\frac{g\mu_B}{\hbar}\boldsymbol{M} \times \boldsymbol{B} + \frac{\alpha}{M}\boldsymbol{M} \times \frac{D\boldsymbol{M}}{Dt} \tag{7.96}$$

where $\boldsymbol{B} = -(\partial\mathcal{V}/\partial\boldsymbol{M})$, calculated with $B = K_\perp = 0$. Here the partial derivative replaced by the *particle derivative*:

$$\frac{D\boldsymbol{M}}{Dt} = \frac{\partial\boldsymbol{M}}{\partial t} + \dot{\phi}\frac{\partial\boldsymbol{M}}{\partial\phi} + \boldsymbol{v}_0 \cdot \boldsymbol{\nabla}\boldsymbol{M} = \frac{\partial\boldsymbol{M}}{\partial t} - \frac{g\mu_B}{\hbar}\boldsymbol{M} \times \boldsymbol{B} + \boldsymbol{v}_0 \cdot \boldsymbol{\nabla}\boldsymbol{M}, \tag{7.97}$$

and where $\boldsymbol{v}_0 = (\hbar m w K_\perp \sin 2\omega_L t + v_r)\hat{z}$[30]. This equation is written in the rotating frame discussed in the last section and for which ϕ_0 is given by Eq. (7.94). In the last form of this last equation $\dot{\phi}\frac{\partial\boldsymbol{M}}{\partial\phi}$ has been replaced by its explicit expression. Evidently the source of the pressure P_z is *not* to be included in the energy function $\mathcal{V}$ which determines $\boldsymbol{B}$ since its effect is included by the passage to the rotating frame. In *addition* that part of the $K_\perp$ which couples to the wall motion, rather than the internal degrees of freedom, must also be canceled from $\mathcal{V}$. The particle derivative is that taken at a point fixed relative to the center of the moving wall and is that frequently encountered in fluid dynamics. The usual static solution is then appropriate in the frame moving with velocity $v_0\hat{z}$ and rotating according to Eq. (7.94), i.e., Eq. (7.95) becomes

$$\frac{\partial z_0}{\partial t} = v_0 = v_r + \hbar m w K_\perp \sin 2\omega_L t. \tag{7.98}$$

7.7.2 *Angular momentum transfer in domain walls*

It is necessary to determine how a current is reflected in the Lagrangian. As with the pillar problem, the $J\boldsymbol{S}_i \cdot \boldsymbol{s}_i$ part of $\mathcal{H}$ is made diagonal by rotating the electron axis of quantization parallel to that of the local moment. The calculation is simpler than that for pillars since the angles between adjacent planes of the wall are always small. Thus in Eq. (7.45) it is possible to approximate $\sin\theta/d \approx$

[29] The nature of the coherent states which correspond to the motion of the wall are such that $\hbar(\partial\phi_0/\partial t)$ is the separation between energy levels connected by $M_i^{\pm}$.

[30] Including v_r here removes the $O(\alpha^2)$ contribution to the pressure mentioned above.

$(\partial\theta/\partial z)$ and multiply this by 2 since there are two neighboring planes with currents in the opposite sense and therefore which have the first derivative which add. Thus the effective vector potential is, $A_z = \hbar(\partial\theta/\partial z)(M_y M)$ and this again leads to a gauge field in the y-direction which is proportional to the current. This is being evaluated in the rotating frame within which the pressure P_z has been eliminated. The wall lies in a plane defined by $\phi_i = 0$ and A_z was generated by considering only rotations by the angle θ. This corresponds to a U(1) gauge theory with a single transverse field. The full theory with rotations by both angles θ and ϕ corresponds to the group SU(2) and there are necessarily two transverse and a longitudinal gauge field which in elementary particle theory would be associated with the $W^{\pm}$ and Z^0 particles. In order that the U(1) transverse field appears in the current Lagrangian formalism, i.e., a quantity linear in $(\partial\theta/\partial z)$ it is actually necessary to determine the terms linear in $(\partial\phi/\partial z)$. Calculation with the full SU(2) theory shows that the coupling to the conduction electron spin current arises from a term

$$\hbar v_s(\cos\theta - 1)\frac{\partial\phi}{\partial z}; \quad v_s = \frac{j_s v_c}{2eM} \tag{7.99}$$

in the Lagrangian density. It must be recognized that, *in general*, this is non-conservative effect. In order that energy be conserved, account must be taken of the smf/emf which appears in the electronic circuit. In the present context this new term leaves the equation of motion for ϕ *unchanged*, but modifies that for θ. With the coupling to the current included

$$\frac{\partial\theta}{\partial t} + (v_s - v_r)\frac{\partial\theta}{\partial z} + 2\hbar m^2 K_\perp \cos\phi \sin^2\theta = 0, \tag{7.100}$$

where the sign of v_r corresponds to the case when the pressure P_z is in opposition to the motion due to the current. Thus the equation of motion of the wall center becomes

$$\frac{\partial z_0}{\partial t} = (v_s - v_r) + \hbar m w K_\perp \sin 2\phi_0. \tag{7.101}$$

This would lead to a simple theory for the critical current if it was not for the fact that the velocities add when the pressure is not in opposition.

This derivation of the coupling of a current to a domain wall is very abstract. It is, in fact, not hard to show, by direct calculation of the equations of motion for the local magnetization, that this coupling simply accounts for the divergence term $\nabla\cdot\mathcal{J}$. While the magnitude of the spin current does not change, its direction does and this leads to a finite divergence. The change in angle of an electron as it advances by on lattice spacing is $\Delta\theta = (\partial\theta/\partial z)a$. The change in angular momentum is $\hbar\Delta M = \hbar\Delta\theta/2$ per electron. The number current $dn/dt = ja^2/e$ and so $dM/dt = (\Delta\theta/2)dn/dt = (ja^3/2e)(\partial\theta/\partial z)$. Finally that part of the derivative $d\theta/dt$ due to the current $(1/M)dM/dt = (jv_c/2eM)(\partial\theta/\partial z)$ as in Eq. (7.100).

The idea that the derivative in the Gilbert relaxation term needed to be modified due to the presence of spin currents, j_s, and the associated divergence term was already introduced in Sec. (7.2.1). Taking this into account the velocity which appears in the particle derivative is

$$\boldsymbol{v}_0 = \left(\frac{j_s v_c}{2eM} + \hbar m w K_\perp \sin 2\omega_{\mathrm{L}} t + v_r \right) \hat{z}. \tag{7.102}$$

The resulting quantity $\boldsymbol{v}_0 \cdot \boldsymbol{\nabla M}$ can be interpreted as a generalized divergence term which includes the spin currents due to the motion of the wall, i.e., Gilbert relaxation occurs in the frame in which the wall is a rest.

The modified Gilbert term as it appears in Eq. (7.100) can evidently be included by simply adding

$$\mathcal{V}_r = \hbar v_r (\cos\theta - 1) \frac{\partial \phi}{\partial z}, \tag{7.103}$$

to the Lagrangian density. Here v_r is a relaxation determined velocity. It is rather unusual that a non-conservative relaxation effect can be added to the Lagrangian in such a fashion.

Lastly in connection with wall relaxation, it is natural to ask under what conditions the relaxation reverts to the standard Gilbert form. In this connection the limit of zero field and pressure is problematic. The yo-yo motion of the wall is reflect by $\hbar m w K_\perp \sin 2\omega_{\mathrm{L}} t$ in $\boldsymbol{v}_0$. Since ω_{L} is proportional to P_z, the distance traveled diverges as $P_z \to 0$. When this length is greater than that of the wire the problem changes nature. The term proportional to $K_\perp$ no longer appears in $\boldsymbol{v}_0$ and must be re-introduced into the functional $\mathcal{V}$ which determines $\boldsymbol{B}$. Then if $j_s = 0$, and ignoring $\mathrm{O}(\alpha^2)$, traditional Gilbert relaxation is recovered.

7.7.3 *Pinning of domain walls*

The present theory does not produce any relaxation in the absence of a pressure P_z. It follows that there is no *intrinsic* pinning. Such a pinning only arises with standard Gilbert damping[55]. As already discussed, unmodified Gilbert relaxation violates the second law of thermodynamics.

The effect of *extrinsic* pinning depends very much on the details of the pinning potential $\mathcal{V}(z)$ and the value of $K_\perp$. It will be assumed that $K_\perp$ is large enough that the deviations in ϕ_0 due to the pressure $P_z = -(\partial\mathcal{V}/\partial z)/\mathcal{A}$ are small. This pinning pressure appears in Eq. (7.94). Now ϕ_0 is small and this equation is used to define a wall momentum $p_z = (\hbar m \mathcal{A}/v_c)\phi_0$. Again since ϕ_0 is small, and using the explicit expression for v_r, Eq. (7.101) can be written as:

$$\frac{\partial z_0}{\partial t} - v_s = \frac{1}{m_{\mathrm{D}}} p_z - \alpha' \frac{a^3}{2\hbar} P_z \tag{7.104}$$

where m_{D} is the Döring mass defined earlier and where α' is an uninteresting constant which depends on the detailed wall structure. Apart from the absence of a relaxation pressure, these are just the traditional equations which describe domain wall motion with the angular momentum transfer term added (see, for

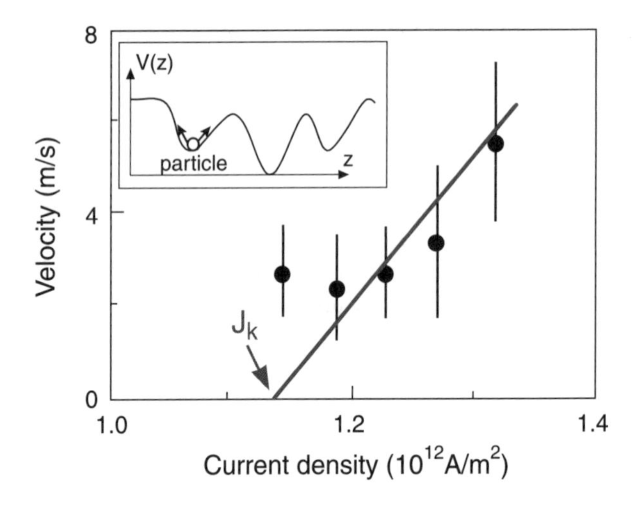

FIG. 7.11. Comparison of experimental data with the result Eq. (7.106). The experimental points are taken from Ref. [17]. The solid line corresponds to the value of C from Eq. (7.106) with $p \approx 0.7$ [53]. The insert shows the equivalent particle problem.

example, Ref. [54], Eqs. (10.1) and (11.2)). Without the current, v_s, term these are the homogeneous differential equations of motion for a "particle" moving in the potential $\mathcal{V}(z)$ as in the insert of Fig. (7.11). Assuming v_s and relaxation are small corrections, P_z might be replaced by its average $\langle P_z \rangle$ over the motion. With this

$$\frac{\partial z}{\partial t} - (v_s - v_r) = \frac{1}{m_{\rm D}} p \tag{7.105}$$

where $v_r = \alpha(a^3/2)\langle P_z \rangle$. The particular integral of this, i.e., the steady state solution, is obtained by eliminating the quantity $(v_s - v_r)$ by $z \to z - (v_s - v_r)t$. This causes the potential to become time dependent, i.e., the motion is that of a free particle driven by a time-dependent force with zero average and no relaxation. In the original frame, the resulting oscillations must be added to the constant velocity $(v_s - v_r)$. The non-conservative driving term has placed the particle at an energy above the top of the maxima in the pinning potential. The average $\langle P_z \rangle$ is non-zero since the "particle" spends more time in the regions where the retarding effects of this same term are the greatest, it depends strongly only on the velocity near the maxima in P_z and these are far from the top of the well so that the relevant velocities and therefore $\langle P_z \rangle$ are insensitive to small changes in the particle energy for the energies of interest. This justifies assuming v_r is a constant near to the critical current. The kinetic critical current j_k, is evidently given by $v_0 = v_r$, and, near to this threshold current, the average velocity,

$$v = pC(j - j_k); \qquad C \equiv \frac{a^3}{2eM}. \tag{7.106}$$

It is important that the velocity near threshold is greatly reduced but that C is independent of j_k, i.e., above the critical current, the angular momentum not destroyed by the pinning is 100% converted into motion of the wall.

In Fig. (7.11) this prediction, is compared with the experiments of Yamaguchi *et al.* [17]. Using the lattice constant for Permalloy, with $M = 1$, $C \approx 4.5 \times 10^{-11}\,\mathrm{m^3/C}$, and using $p \approx 0.7$ suggested in Ref. [17], $pC \approx 3.15 \times 10^{-11}\,\mathrm{m^3/C}$. This corresponds approximately to the gradient of the line shown in Fig. (7.11). Clearly, within a factor of 2 in either sense, this is consistent with the trend of the data points.

7.7.4 *The smf/emf produced by a domain wall*

When there is a current induced motion, and when there is a pressure exerted on a domain wall, it must be that the electrons which pass through the wall not only transfer angular momentum but also do work, or have work done on them, in order that energy be conserved. It is an interesting aspect of the domain wall problem that the wall can in a real sense be considered as a spin battery. This consists of a length of ferromagnetic wire in a static magnetic field. It might be "charged" by passing a current in the sense which increases the energy by moving the wall against the field pressure. The battery is "discharged" by connecting the wire to an external circuit and letting the wall relax by forcing a current around the circuit.

The smf/emf appropriate to the ground state is obtained by calculating the non-operator part of the time derivative of the vector potential

$$A_z = \sigma_z \hbar \frac{\partial \theta}{\partial z} \frac{M_y}{M}, \tag{7.107}$$

where $\sigma_z = \pm 1$ as the electron is of majority or minority spin. What is needed is the time derivative of M_y in the laboratory frame. Recall the wall is in a rotating frame and lies in the xz-plane so that $\phi = 0$. In the laboratory frame assuming $\phi = 0$ at $t = 0$, $M_y = M \sin\theta \sin\phi \approx M \sin\theta\phi$ so that $\dot{M}_y = M \sin\theta(d\phi/dt)$. Applying a force to the wall results in a force analogous to that in a Stern–Gerlach experiment given by:

$$\boldsymbol{f}_{is} = -\frac{\partial \boldsymbol{A}_i}{\partial t} = \sigma_z \frac{1}{2M} \frac{\partial \theta}{\partial z} \sin\theta_i \, M\hbar \frac{d\phi}{dt} \hat{z} = \sigma_z \frac{1}{2} \frac{\partial \cos\theta}{\partial z} \frac{P_z v_c}{2m} \hat{z}. \tag{7.108}$$

This Stern–Gerlach force couples to the spin degree of freedom and hence has a different sign for the majority and minority spins, reflected by the factor of σ_z. In a domain wall the current has an effective polarization p determined by the material. The force on an "average" electron therefore involves taking the average $\langle \sigma_z \rangle \equiv p$ which here is used to define p. The net force on such an average electron is therefore

$$\boldsymbol{f}_s = p\frac{1}{2}\frac{\partial \cos\theta}{\partial z}\frac{P_z v_c}{2m}\hat{z}. \tag{7.109}$$

Integrating this force through the wall gives,

$$\mathcal{E} = p\frac{v_c}{2em}P_z, \tag{7.110}$$

for the emf determined by the *total* pressure P_z. This is easily seen to reflect energy conservation. If the value of P_z appropriate to a magnetic field is used, this reduces to $\mathcal{E} = pg\mu_{\mathrm{B}}B$. Substituting Eq. (7.94) into Eq. (7.110) gives, rather generally,

$$\mathcal{E} = p\frac{\hbar}{e}\frac{d\phi}{dt}, \tag{7.111}$$

i.e., a particularly simple result for the emf generated by a domain wall.

The predicted current densities j are large. With $m = 1$ and $a \sim 3\,\text{Å}$ and a realistic $v \sim 1\text{m/s}$ the current density $j \sim 3 \times 10^{10}\text{A/m}^2$ or a current $i \sim 30\,\text{mA}$ for a micron square wire. The value of $\mathcal{E}$ reflects the value of B and for $1\,\text{T}$ corresponding to about $100\,\mu\text{V}$.

This result for $\mathcal{E}$ must *not* be interpreted as the spin weighted average of the potential shift of the conduction electrons. Such shifts can be a consequence of an smf/emf but are not the origin. What $\mathcal{E}$ reflects is the conversion of the magnetic energy stored in the wire into electrical energy. This is an important point. Consider an isolated wire with a pinned domain wall in a field. The field does result in opposite shifts of the energy of the majority electron on the two sides of the wall; however these shifts are compensated by a small charge transfer which bring the two Fermi levels back into line. Only when the pinning force is removed is there an smf/emf. If this is done suddenly, the force $\boldsymbol{f}_s$ would cause the electrons to be accelerated as they pass through the wall, but after a short time a potential difference equal to $\mathcal{E}$ develops so that the force due to the potential gradient exactly cancels $\boldsymbol{f}_s$ and a steady state current results. To emphasize the point, it might be noted that $\mathcal{E}$ exists if the conduction electron g-factor were zero since, within the s–d-exchange model at least, the energy stored in the wall is to a large part due to the local moment energy. The field-induced chemical potential shifts for a pinned wall would be absent and would only develop when the wall is in motion.

7.7.5 *Applications of the smf produced by a domain wall*

A basic memory device is illustrated in Fig. 7.12. The Bloch wall has two stable equilibrium positions 0 and 1 and might be switched from one to the other using an external current. In a current-only design, the device might be read by applying a current between a and b. If the wall lies in position 0 it will be dislodged by this reading pulse, and once it passes the half-way point, will induce an output in the circuit connecting b and c. Clearly, there will be no such current

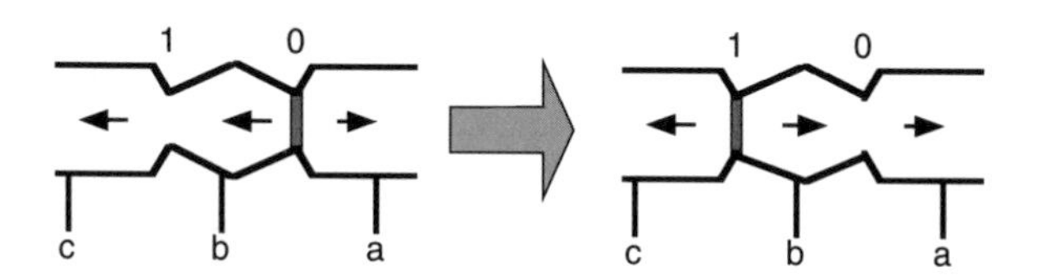

FIG. 7.12. A current-only read-write memory element. To the left the system is in the 0-state. A current between a and b will carry the wall past the unstable equilibrium point. As it moves towards the 1-state under the pressure P_z implicit in the device shape it will produce an output emf between b and c. No emf occurs if the system is in the 1-state. The system can be switched between 0 and 1 by passing a current from a to c.

pulse if the wall is initially in position 1. Here the pressure P_x is produced by the shape of the bridge.

Current and power amplification might be achieved by the device shown in Fig. 7.13. Here only one of the equilibrium situations 0 is stable. An initial pulse puts the system in the unstable equilibrium position 1. A current pulse between a and b causes the wall to leave 1 and the large P_z due to the narrowing of the wire produces a large induced smf/emf and hence a large current output in that part of the circuit which is connect to b and c. Isolation of input and output is afforded by the small distance 1 to b. The potential seen by the wall is also illustrated in Fig. 7.13. A large gain implies a small barrier, of height E_b,

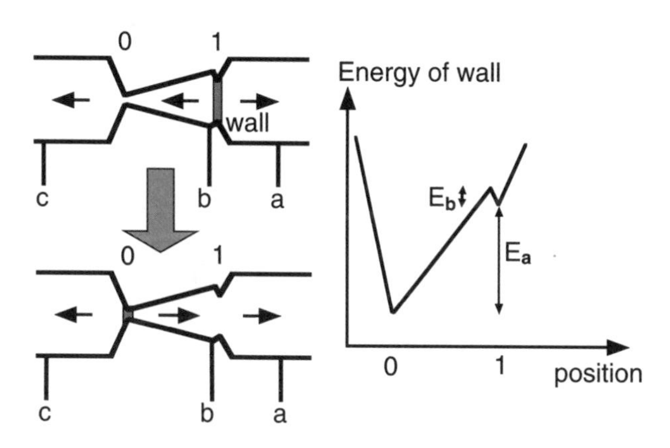

FIG. 7.13. An power amplifier. Starting with the initial state, top left, a pulse between a and b moves the wall from 1 to b, i.e., a point at which there is a P_z to the left. There is an emf between b and c as the wall moves between b and 0. The final state is shown at the bottom left. To the right is shown the energy profile of the device, see text.

at an energy E_a relative to 0, and which is large compared to E_b. In fact, the power gain g is limited by, and is approximately equal to, $g = (E_a/E_b)$, since the input pulse must raise the wall over the barrier, i.e., give it an energy E_a while the maximum energy which is given to the external output circuit is evidently E_a. In position 1 the wall might either tunnel or be thermally excited out of this unstable equilibrium, *however* the wall constitutes a macroscopic object and this affords an enormous reduction in both of these processes.

References

[1] S. I. Kiselev, J. C. Sankey, I. N. Krivorotov, N. C. Emley, R. J. Schoelkopf, R. A. Buhrmanand D. C. Ralph, Nature **425**, 380 (2003).

[2] S. Urazhdin, N. O. Birge, W. P. Pratt Jr., and J. Bass, Phys. Rev. Lett. **91**, 146803 (2003).

[3] H. W. Schumacher, C. Chappert, P. Crozat, R. C. Sousa, P. P. Freitas, J. Miltat, J. Fassbender, and B. Hillebrands, Phys. Rev. Lett. **90**, 017201 (2003).

[4] M. L. Polianski and P. W. Brouwer, Phys. Rev. Lett. **92**, 026602 (2004).

[5] J. C. Slonczewski, J. Magn. Magn. Mater. **159**, L1 (1996).

[6] L. Berger, Phys. Rev. B **54**, 9353 (1996).

[7] M. Tsoi, A. G. M. Jansen, J. Bass, W.-C. Chiang, M. Seck, V. Tsoi, and P. Wyder, Phys. Rev. Lett. **80**, 4281 (1998); erratum, *ibid.* **81**, 493 (1998).

[8] J. Z. Sun, J. Magn. Magn. Mater. **202**, 157 (1999).

[9] E. B. Myers, D. C. Ralph, J. A. Katine, R. N. Louie, R. A. Buhrman, Science **285**, 867 (1999).

[10] J. A. Katine, F. J. Albert, and R. A. Buhrman, E. B. Myers and D. C. Ralph Phys. Rev. Lett. **84**, 3149 (2000).

[11] M. AlHajDarwish, H. Kurt, S. Urazhdin, A. Fert, R. Loloee, W. P. Pratt, Jr., and J. Bass, Phys. Rev. Lett. **93**, 157203 (2004).

[12] J. Grollier, V. Cros, A. Hamzic, J. M. George, H. Jaffres, A. Fert, G. Faini, J. Ben Youssef, and H. Legall, Appl. Phys. Lett. **78**, 3663 (2001).

[13] J. Z. Sun, D. J. Monsma, T. S. Kuan, M. J. Rooks, D. W. Abraham, B. Özyilmaz, A. D. Kent, and R. H. Koch, J. Appl. Phys. **93**, 6859 (2003).

[14] H. W. Schumacher, C. Chappert, R. C. Sousa, P. P Freitas, and J. Miltat, Phys. Rev. Lett. **90**, 017204 (2003).

[15] M. Klaui, C. A. F. Vaz, J. A. C. Bland, W. Wernsdorfer, G. Faini, E. Cambril, and L. J. Heyderman, Appl. Phys. Lett. **83**, 105 (2003).

[16] D. A. Allwood, G. Xiong, M. D. Cooke, C. C. Faulkner, D. Atkinson, N. Vernier, R. P. Cowburn, Science **296**, 2003 (2002).

[17] A. Yamaguchi, T. Ono, S. Nasu, K. Miyake, K. Mibu, and T. Shinjo, Phys. Rev. Lett. **92**, 077205 (2004).

[18] M. Yamanouchi, D. Chiba, F. Matsukura, and H. Ohno, Nature **428**, 539 (2004).

[19] A. Yamaguchi, S. Nasu, H. Tanigawa, T. Ono, K. Miyake, K. Mibu, and T. Shinjo, Appl. Phys. Lett. **86**, 012511 (2005).

[20] E. Saitoh, H. Miyajima, T. Yamaoka, and G. Tatara, Nature **432**, 203 (2004).

[21] M. Klaui, C. A. F. Vaz, J. A. C. Bland, W. Wernsdorfer, G. Faini, E. Cambril, L. J. Heyderman, F. Nolting, and U. Rudiger, Phys. Rev. Lett. **94**, 106601 (2005).

[22] Y. Tserkovnyak, A. Brataas, G. E. W. Bauer, B. I. Halperin, cond-mat/0409242.

[23] I. Žutič, J. Fabian, S. D. Sarma, Rev. Mod. Phys. **76**, 323 (2004).

[24] Y. B. Bazaliy, B. A. Jones and S.-C. Zhang, Phys. Rev. B **57** R3213 (1998).

[25] S. Zhang, P. M. Levy, and A. Fert, Phys. Rev. Lett. **88**, 236601 (2002).

[26] J. C. Slonczewski, J. Magn. Magn. Mater. **195**, L261 (1999).

[27] L. Berger, J. Appl. Phys. **93**, 7693 (2003).

[28] G. E. W. Bauer, Y. Tserkovnyak, D. Huertas-Hernando, and A. Brataas, Phys. Rev. B **67**, 094421 (2003).

[29] Z. Li and S. Zhang, Phys. Rev. B **68**, 024404 (2003).

[30] See, for example, R. P. Feynman, R. B. Leighton, M. Sands, *The Feynman Lectures on Physics* (Addison Wesley Longman, 1970).

[31] S. E. Barnes and S. Maekawa, cond-mat/0410021.

[32] See review on the theory ESR of dilute magnetic alloys: S. E. Barnes, Adv. Phys. **30**, 801 (1981).

[33] P. Monod, H. Hurdequint, A. Janossy, J. Obert, and J. Chaumont, Phys. Rev. Lett. **29**, 1327, (1972).

[34] S. Mizukami,Y. Ando, and T. Miyazaki, J. Magn. Magn. Mater. **239**, 42, (2002).

[35] See, for example, W. Wernsdorfer, Adv. Chem. Phys. **118**, 99 (2001).

[36] M. Preda and S. E. Barnes, cond-mat/0112398.

[37] B. Özyilmaz, A. D. Kent, M. J. Rooks, and J. Z. Sun, Phys. Rev. B **71**, 140403 (2005).

[38] J. Z Sun, B. Özyilmaz, W. Chen, M. Tsoi, and A. D. Kent, J. Appl. Phys. **97**, 10C714 (2005).

[39] S. Ingvarsson, L. Ritchie, X. Y. Liu, Gang Xiao, J. C. Slonczewski, P. L. Trouilloud, and R. H. Koch, Phys. Rev. B **66**, 214416 (2002).

[40] P. Monod and F. Beuneu, Phys. Rev. B **19**, 911 (1979); F. Beuneu and P. Monod, B **18**, 2422 (1978).

[41] B. Heinrich, Y. Tserkovnyak, G. Woltersdorf, A. Brataas, R. Urban, and G. E. W. Bauer, Phys. Rev. Lett. **90**, 187601 (2003).

[42] K. Lenz, T. Tolinski, J. Lindner, E. Kosubek, and K. Baberschke, Phys. Rev. B **69**, 144422 (2004).

[43] J. Z. Sun, Phys. Rev B **62**, 570 (2000).

[44] S. E. Barnes, J. Phys. F: Met. Phys. **4**, 1535 (1974).

[45] Y. Ji, C. L. Chien, and M. D. Stiles, Phys. Rev. Lett. **90**, 106601 (2003).

[46] S. E. Barnes, J. Dupraz, and R. Orbach, J. Appl. Phys. **42** 1959 (1971).

[47] P. W. Anderson and P. R. Weiss, Rev. Mod. Phys. **25**, 269276 (1953).

[48] S. I. Kiselev, J. C. Sankey, I. N. Krivorotov, N. C. Emley, M. Rinkoski, C. Perez, R. A. Buhrman, and D. C. Ralph, Phys. Rev. Lett. **93**, 036601(2004).

[49] M. D. Stiles, Jiang Xiao, and A. Zangwill, Phys. Rev. B **69**, 054408 (2004).

[50] M. L. Polianski and P. W. Brouwer, Phys. Rev. Lett. **92**, 026602 (2004).

[51] B. Özyilmaz, A. D. Kent, J. Z. Sun, M. J. Rooks, and R. H. Koch, Phys. Rev. Lett. **93**, 176604 (2004).
[52] J. J. Versluijs, M. A. Bari, and J. M. D. Coey, Phys. Rev. Lett. **87**, 026601 (2001).
[53] S. E. Barnes and S. Maekawa, Phys. Rev. Lett. **95**, 107204 (2005).
[54] See for example, A. P. Malozemoff and J. C. Slonczewski, *Magnetic Domain Walls in Bubble Materials* (Academic, New York, 1979).
[55] G. Tatara and H. Kohno, Phys. Rev. Lett. **92**, 086601 (2004).

8 Spin injection and spin transport in hybrid nanostructures

S. Takahashi, H. Imamura, and S. Maekawa

8.1 Introduction

There is current interest in the emerging field of spin transport in magnetic nanostructures, because of their potential applications as spin-electronic devices [1–3]. Recent experimental and theoretical studies have demonstrated that the spin-polarized carriers injected from a ferromagnet (F) into a non-magnetic material (N), such as a normal metal, semiconductor, and superconductor, create a non-equilibrium spin accumulation in N. The efficient spin injection, spin accumulation, spin transfer, and spin detection are central issues for utilizing the spin degree of freedom as new functionalities in spin-electronic devices.

In this chapter, we describe the basic aspect of spin injection and spin transport in magnetic nanohybrid structures containing normal metals or superconductors. Particular emphasis is placed on the spin accumulation and spin current in a non-local spin device of F1/N/F2 structures, where F1 is a spin injector and F2 a spin detector. We solve the spin-dependent transport equations for the electrochemical potentials of up and down spins in the structure of arbitrary junction resistance ranging from a metallic contact to a tunneling regime, and examine optimal conditions for spin accumulation and spin current. The spin accumulation detected by F2 depends strongly on whether the junction interface is a metallic contact or a tunnel barrier; it is greatly improved when a tunnel barrier is used instead of a metallic contact, and therefore efficient spin injection and detection are achieved when both interfaces are tunnel junctions. On the other hand, a large spin-current injection from N, through the N/F interface, into F2 is realized when N is in metallic contact with F2 whose spin diffusion length is very short, like a Permalloy (Py), because F2 plays the role of strong sink for the spin current flowing in N. These findings indicate that tunnel junctions are favorable for a large spin accumulation, while metallic contacts are favorable for a large spin-current injection. Intriguing and useful devices are those containing a material with a low density of carriers that carry spins. In a superconductor device, the spin accumulation signal is greatly amplified by opening of the energy gap, because the same spin-injection necessitates a much larger spin-splitting of the Fermi energy. Semiconductor devices likewise exhibit a large spin accumulation signal. The spin-current induced anomalous Hall effect is also discussed.

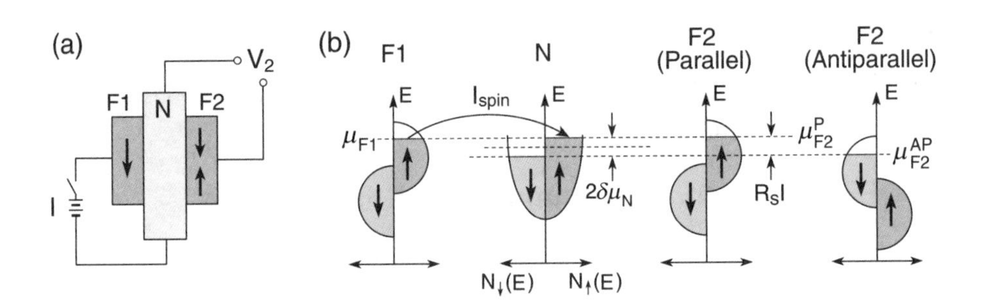

FIG. 8.1. (a) Three-terminal device of F1/N/F2 structure proposed by Johnson [5]. The arrows in F1 and F2 indicate the direction of the magnetizations. (b) Densities of states for the up and down spin subbands in each electrode in (a). The magnetization of F2 is either parallel or antiparallel to that of F1.

8.2 Spin injection, spin accumulation, and spin current

In 1985, Johnson and Silsbee [4] first reported that non-equilibrium spins injected from ferromagnets diffuse into Al films over the spin diffusion length of the order of 1 μm (or even several hundred μm for pure Al). In 1993, Johnson proposed a spin injection technique [5] in a F1/N/F2 structure (F1 is an injector and F2 a detector), which is represented conceptually in Fig. 8.1 in a pedagogical case where both F1 and F2 are half-metallic. When the current flows from F1 to N, up-spin electrons are injected into N, so that the populations of up-spin electrons increase by shifting the electrochemical potential (ECP) by $\delta\mu_{\rm N}$, while those of down-spin electrons decrease by shifting the ECP by $-\delta\mu_{\rm N}$, resulting in the spin splitting of ECP by $2\delta\mu_{\rm N}$, which corresponds to spin accumulation in N. When the magnetization of F2 is parallel to that of F1, the ECP of F2 coincides with the up-spin ECP of N, and when the magnetization is antiparallel, it coincides with the down-spin ECP. Therefore, the output voltage (V_2) depends on whether the magnetization of F2 is parallel or antiparallel to that of F1.

Recently, Jedema *et al.* have made spin injection and detection experiments in a permalloy/copper/permalloy (Py/Cu/Py) structure fabricated by advanced microfabrication techniques, and observed a clear spin accumulation signal at room temperature in the non-local geometry for measurement [6]. Subsequently, they have found that the efficiency of spin injection and accumulation is greatly improved in a cobalt/aluminum/cobalt (Co/I/Al/I/Co) structure when tunnel barriers ($\rm I = Al_2O_3$) are inserted between the Co and Al electrodes [7]. Figures 8.2(a) and (b) show a scanning electron microscope image of the Co/I/Al/I/Co structure and the experimental result of the spin accumulation signal. The bias current I is injected from Co 1 and taken out from the left end of N, and the spin accumulation at distance L from Co 1 is detected by measuring the potential difference (V) between Co 2 and N. Sweeping the magnetic field B from the negative side (or positive side), the non-local resistance V/I changes sign from positive to negative, and switches symmetrically around zero. The positive

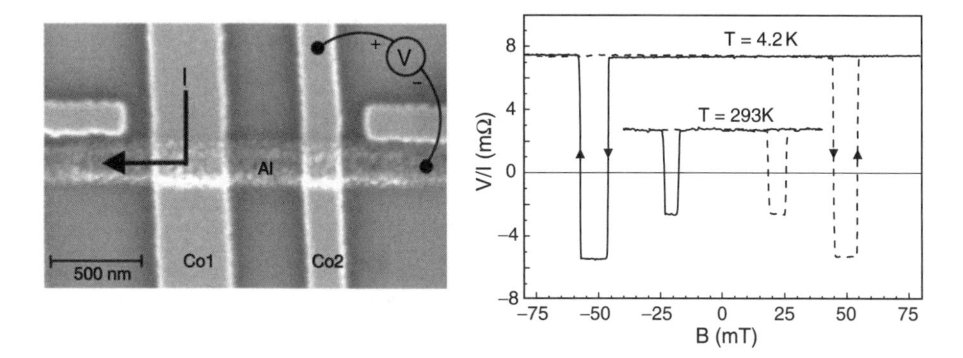

FIG. 8.2. Spin injection and detection device of Co/I/Al/I/Co (I = A_2O_3) in the non-local geometry for measurement by Jedema *et al.* [7]. (a) Scanning electron microscope image of the device and (b) spin accumulation resistance V/I as a function of the in-plane magnetic field B for a sample with a Co electrode spacing $L = 650\,\text{nm}$ at $T = 4.2\,\text{K}$ and room temperature.

(negative) value of the non-local resistance corresponds to the parallel (antiparallel) magnetizations. This result demonstrates that the experiment of non-local measurement can observe purely the spin degree of freedom (spin accumulation) in non-magnetic metals. Similar experiments have been made in other groups using the non-local measurement [8–16].

8.2.1 *Spin transport in non-local geometry*

We consider a spin injection and detection device which consists of a non-magnetic metal N connected to ferromagnets of injector F1 and detector F2 as shown in Fig. 8.3. F1 and F2 are ferromagnetic electrodes of width w_{F} and thickness d_{F}, and are separated by distance L, and N is a normal-metal electrode of width w_{N} and thickness d_{N}. The magnetizations of F1 and F2 are aligned either parallel or antiparallel. In this device, spin-polarized electrons are injected from F1 into N by flowing the current I from F1 to the left end of N. The spin accumulation at distance L from F1 is detected by F2, and the potential difference (V_2) between F2 and N is measured. Because of the absence of a voltage source on the right part of the device, there is no charge current in the electrodes that lie on the right side of F1. By contrast, the injected spins are diffused equally in both directions, creating spin accumulation not only on the left side but also on the right side. Accordingly, the spin and charge degrees of freedom are transported separately in the device; the advantage of the non-local measurement is that F2 probes only the spin degree of freedom.

The electrical current density $\mathbf{j}_\sigma$ for spin channel σ ($\uparrow$ or $\downarrow$) in each electrode is driven by the electric field $\mathbf{E}$ and the gradient of the carrier density deviation δn_σ from equilibrium:

$$\mathbf{j}_\sigma = \sigma_\sigma \mathbf{E} - eD_\sigma \nabla \delta n_\sigma,$$

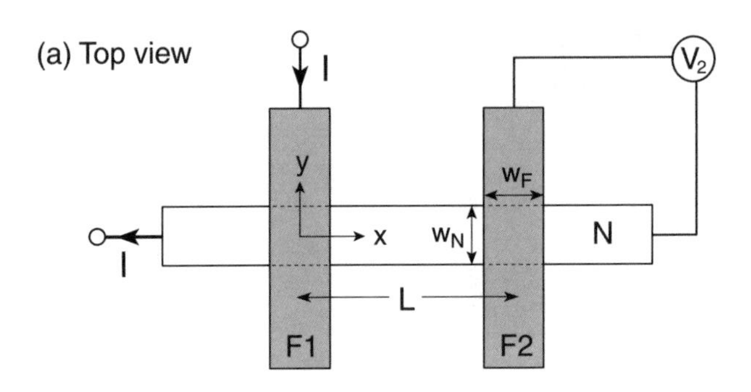

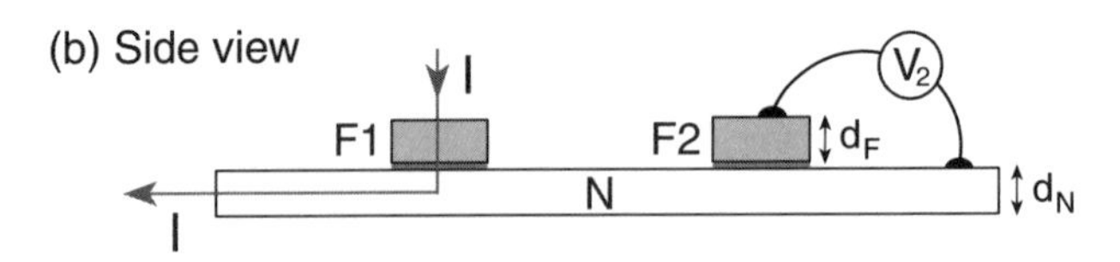

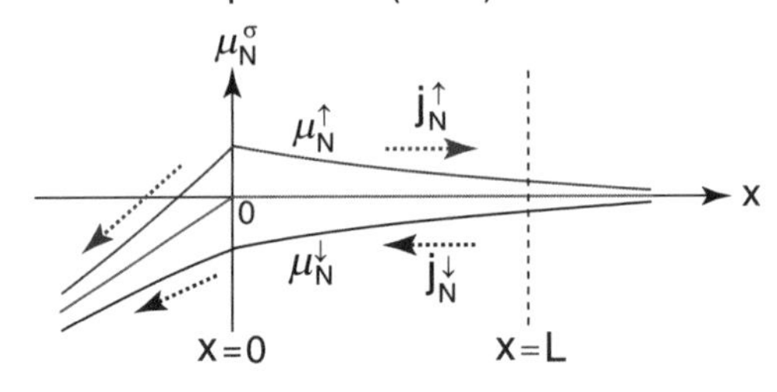

FIG. 8.3. Spin injection and detection device: (a) top view and (b) side view. The bias current I is applied from F1 to the left end of N. The spin accumulation at $x = L$ is detected by measuring the voltage V_2 between F2 and N. (c) Spatial variation of the electrochemical potential (ECP) for up and down spin electrons in N.

where σ_σ and D_σ are the electrical conductivity and the diffusion constant. Making use of $\delta n_\sigma = N_\sigma \delta\varepsilon_{\rm F}^\sigma$ (N_σ is the density of states in the spin subband and $\delta\varepsilon_{\rm F}^\sigma$ is the shift of the Fermi energy from equilibrium) and the Einstein relation $\sigma_\sigma = e^2 N_\sigma D_\sigma$, one has

$$\mathbf{j}_\uparrow = -(\sigma_\uparrow/e)\nabla\mu_\uparrow, \qquad \mathbf{j}_\downarrow = -(\sigma_\downarrow/e)\nabla\mu_\downarrow, \tag{8.1}$$

where $\mu_\sigma = \varepsilon_{\rm F}^\sigma + e\phi$ is the electrochemical potential (ECP) and ϕ the electric potential. The continuity equations for charge and spin in the steady state are

$$\nabla\cdot(\mathbf{j}_\uparrow + \mathbf{j}_\downarrow) = 0, \qquad \nabla\cdot(\mathbf{j}_\uparrow - \mathbf{j}_\downarrow) = -e\frac{\delta n_\uparrow}{\tau_{\uparrow\downarrow}} + e\frac{\delta n_\downarrow}{\tau_{\downarrow\uparrow}}, \tag{8.2}$$

where $\tau_{\sigma\sigma'}$ is the scattering time of an electron from spin state σ to σ'. Using these continuity equations and detailed balance $N_\uparrow/\tau_{\uparrow\downarrow} = N_\downarrow/\tau_{\downarrow\uparrow}$, one obtains [4, 17–20]

$$\nabla^2 \left(\sigma_\uparrow \mu_\uparrow + \sigma_\downarrow \mu_\downarrow\right) = 0, \tag{8.3a}$$

$$\nabla^2 \left(\mu_\uparrow - \mu_\downarrow\right) = \frac{1}{\lambda^2}\left(\mu_\uparrow - \mu_\downarrow\right), \tag{8.3b}$$

where λ is the spin-diffusion length

$$\lambda = \sqrt{D\tau_{\mathrm{sf}}},$$

with

$$\tau_{\mathrm{sf}}^{-1} = \frac{1}{2}(\tau_{\uparrow\downarrow}^{-1} + \tau_{\downarrow\uparrow}^{-1}), \quad D^{-1} = (N_\uparrow D_\downarrow^{-1} + N_\downarrow D_\uparrow^{-1})/(N_\uparrow + N_\downarrow).$$

In the following, the subscripts (or superscripts) "N" and "F" denote the quantities of N and F, respectively. The material parameters in N are spin-*independent*: $\sigma_{\mathrm{N}}^\uparrow = \sigma_{\mathrm{N}}^\downarrow = \frac{1}{2}\sigma_{\mathrm{N}}$, etc., and those in F spin-*dependent*: $\sigma_{\mathrm{F}}^\uparrow \neq \sigma_{\mathrm{F}}^\downarrow$ ($\sigma_{\mathrm{F}} = \sigma_{\mathrm{F}}^\uparrow + \sigma_{\mathrm{F}}^\downarrow$), etc. The spin-diffusion lengths of transition-metal ferromagnets are $\lambda_{\mathrm{F}} \sim 5\,\mathrm{nm}$ for Permalloy (Py), $\lambda_{\mathrm{F}} \sim 12\,\mathrm{nm}$ for CoFe, and $\lambda_{\mathrm{F}} \sim 50\,\mathrm{nm}$ for Co from studies of CPP-GMR (current-perpendicular-plane giant magnetoresistance) by Bass *et al.* [21–23], whereas those of non-magnetic metals are $\lambda_{\mathrm{N}} \sim 1\mu\mathrm{m}$ for Cu [6, 11], and $\lambda_{\mathrm{N}} \sim 0.65\,\mu\mathrm{m}$ for Al [7]. The fact that λ_{F} is extremely short compared with λ_{N} in the above conventional materials plays a crucial role for spin transport in devices made from those materials as shown below.

For the interfacial current across the junctions, we employ a model used in CPP-GMR by Valet and Fert [17]. Due to the spin-dependent interface resistance R_i^σ between different metals of the junction i ($i = 1, 2$), the ECP in each spin channel changes discontinuously at the interface when the current passes through the junction. The interfacial current I_i^σ from Fi (F1 or F2) to N is given in terms of the discontinuity $(\mu_{\mathrm{F}i}^\sigma - \mu_{\mathrm{N}}^\sigma)$ at the interface as [17–19]

$$I_i^\sigma = \frac{1}{eR_i^\sigma}\left(\mu_{\mathrm{F}i}^\sigma - \mu_{\mathrm{N}}^\sigma\right), \quad (i = 1, 2,\ \sigma = \uparrow\downarrow) \tag{8.4}$$

where the distribution of the current is assumed to be uniform over the interface. The total charge and spin currents across the ith interface are $I_i = I_i^\uparrow + I_i^\downarrow$ and $I_i^s = I_i^\uparrow - I_i^\downarrow$.

Equation (8.4) is applicable not only for a transparent contact but also for tunnel junctions. In the transparent case ($R_i^\sigma \to 0$), ECP in each spin channel is continuous at the interface, so that the spin splittings of ECPs on the F and N sides strongly influence each other. In the tunneling case, ECPs have large discontinuity at the interface, and their discontinuity is much larger than the splitting of the ECPs, so that the spin accumulation on the N side does not depend on the details of the spin splitting on the F side.

In real situations, the distribution of the current across the interface depends on the relative magnitude of the interface resistance to the electrode resistance [24]. When the interface resistance is much larger than the electrode resistance as in a tunnel junction, the current distribution is uniform in the contact area [25], which validates our assumption mentioned above. However, when the interface resistance is comparable to or smaller than the electrode resistance as in a metallic contact junction, the current distribution becomes inhomogeneous; the interface current has a large current density around a corner of the contact [14, 26]. In this case, the effective contact area through which the current flows is smaller than the contact area $A_{\rm J} = w_{\rm N} w_{\rm F}$ of the junctions.

When the bias current I flows from F1 to the left side of N ($I_1 = I$) and there is no charge current on the right side ($I_2 = 0$), the solution of Eqs (8.3a) and (8.3b) takes the form

$$\mu_{\rm N}^{\sigma} = \bar{\mu}_{\rm N} + \sigma \left(a_1 e^{-|x|/\lambda_{\rm N}} - a_2 e^{-|x-L|/\lambda_{\rm N}} \right), \tag{8.5}$$

where the first term describes the charge transport and takes $\bar{\mu}_{\rm N} = -(eI/\sigma_{\rm N})x$ for $x < 0$ and $\bar{\mu}_{\rm N} = 0$ (ground level of ECP) for $x > 0$, and the second term the shift in ECP of up ($\sigma = +$) and down ($\sigma = -$) spins; the a_1-term in the bracket represents the spin accumulation due to spin injection from F1 into N, while the a_2-term is the spin depletion due to spin extraction from N into F2. Note that, in the region of $x > 0$, the charge current ($j_{\rm N} = j_{\rm N}^{\uparrow} + j_{\rm N}^{\downarrow}$) is absent and only the spin current ($j_{\rm N}^{s} = j_{\rm N}^{\uparrow} - j_{\rm N}^{\downarrow}$) flows.

In the F1 and F2 electrodes in which the thickness is much larger than the spin diffusion length ($d_{\rm F} \gg \lambda_{\rm F}$) as in the case of Py or CoFe, the solutions close to the interfaces of junctions 1 and 2 are of the form

$$\mu_{\rm F1}^{\sigma} = \bar{\mu}_{\rm F1} + \sigma b_1 \left(\sigma_{\rm F1}/\sigma_{\rm F1}^{\sigma} \right) e^{-z/\lambda_{\rm F}}, \tag{8.6a}$$

$$\mu_{\rm F2}^{\sigma} = \bar{\mu}_{\rm F2} - \sigma b_2 \left(\sigma_{\rm F2}/\sigma_{\rm F2}^{\sigma} \right) e^{-z/\lambda_{\rm F}}, \tag{8.6b}$$

where $\bar{\mu}_{\rm F1} = -[eI/(\sigma_{\rm F} A_{\rm J})]z + eV_1$ represents the current flow of I in F1, $\bar{\mu}_{\rm F2} = eV_2$ is a constant electric potential in F2 with no charge current, and V_1 and V_2 are the voltage drops across junctions 1 and 2.

Using the matching condition that the charge and spin currents are continuous at the interfaces, we can determine a_i, b_i, and V_i as listed in Section 8.5. If the detected voltages V_2 in the parallel (P) and antiparallel (AP) alignments of magnetizations are denoted by $V_2^{\rm P}$ and $V_2^{\rm AP}$, respectively, then the spin accumulation signal detected by F2 is given by

$$R_s = (V_2^{\rm P} - V_2^{\rm AP})/I. \tag{8.7}$$

8.2.2 *Spin accumulation signal*

The spin accumulation signal R_s is calculated in the form [20]

$$R_s = R_{\rm N} \frac{\left(\frac{2P_1}{1-P_1^2} \frac{R_1}{R_{\rm N}} + \frac{2p_{\rm F}}{1-p_{\rm F}^2} \frac{R_{\rm F}}{R_{\rm N}} \right) \left(\frac{2P_2}{1-P_2^2} \frac{R_2}{R_{\rm N}} + \frac{2p_{\rm F}}{1-p_{\rm F}^2} \frac{R_{\rm F}}{R_{\rm N}} \right) e^{-L/\lambda_{\rm N}}}{\left(1 + \frac{2}{1-P_1^2} \frac{R_1}{R_{\rm N}} + \frac{2}{1-p_{\rm F}^2} \frac{R_{\rm F}}{R_{\rm N}} \right) \left(1 + \frac{2}{1-P_2^2} \frac{R_2}{R_{\rm N}} + \frac{2}{1-p_{\rm F}^2} \frac{R_{\rm F}}{R_{\rm N}} \right) - e^{-\frac{2L}{\lambda_{\rm N}}}}, \tag{8.8}$$

where R_i $(1/R_i = 1/R_i^{\uparrow} + 1/R_i^{\downarrow})$ is the interface resistance of junction i, $R_{\rm N}$ and $R_{\rm F}$ are the *spin accumulation resistances* of the N and F electrodes

$$R_{\rm N} = (\rho_{\rm N}\lambda_{\rm N})/A_{\rm N}, \qquad R_{\rm F} = (\rho_{\rm F}\lambda_{\rm F})/A_{\rm J}, \tag{8.9}$$

with the cross-sectional area $A_{\rm N} = w_{\rm N}d_{\rm N}$ of N and the contact area $A_{\rm J} = w_{\rm N}w_{\rm F}$ of the junctions, and P_i is the interfacial current spin-polarization of junction i and $p_{\rm F}$ the current spin-polarization of F1 and F2:

$$P_i = \left|R_i^{\uparrow} - R_i^{\downarrow}\right|/\left(R_i^{\uparrow} + R_i^{\downarrow}\right), \qquad p_{\rm F} = \left|\rho_{\rm F}^{\uparrow} - \rho_{\rm F}^{\downarrow}\right|/\left(\rho_{\rm F}^{\uparrow} + \rho_{\rm F}^{\downarrow}\right). \tag{8.10}$$

In metallic contact junctions, the spin polarization, P_i and $p_{\rm F}$, ranges around 50–70% from GMR experiments [21–23] and point-contact Andreev-reflection experiments [27], whereas in tunnel junctions, P_i ranges around 30–55% from superconducting tunneling spectroscopy experiments with alumina tunnel barriers [28–30], and $\sim$ 85% in MgO barriers [31, 32].

The spin accumulation signal R_s strongly depends on whether each junction is a metallic contact or a tunnel junction, and on the relative magnitude among R_1, R_2, $R_{\rm F}$, and $R_{\rm N}$. When $R_{\rm F}$ is much smaller than $R_{\rm N}$, e.g., $(R_{\rm F}/R_{\rm N}) \ll 1$, as in the case of Cu and Py, we have the following limiting cases. When both junctions are tunnel junctions, we have [4, 7]

$$R_s/R_{\rm N} = P_1P_2e^{-L/\lambda_{\rm N}}, \qquad (R_1, R_2 \gg R_{\rm N} \gg R_{\rm F}). \tag{8.11}$$

When one of the junctions is a transparent contact and the other is a tunnel junction, we have [20]

$$R_s/R_{\rm N} = \frac{2p_{\rm F}P_1}{1-p_{\rm F}^2}\left(\frac{R_{\rm F}}{R_{\rm N}}\right)e^{-L/\lambda_{\rm N}}, \qquad (R_1 \gg R_{\rm N} \gg R_{\rm F} \gg R_2) \tag{8.12a}$$

$$R_s/R_{\rm N} = \frac{2p_{\rm F}P_2}{1-p_{\rm F}^2}\left(\frac{R_{\rm F}}{R_{\rm N}}\right)e^{-L/\lambda_{\rm N}}, \qquad (R_2 \gg R_{\rm N} \gg R_{\rm F} \gg R_1). \tag{8.12b}$$

When both junctions are transparent contact, we have [6, 18, 19, 33]

$$R_s/R_{\rm N} = \frac{2p_{\rm F}^2}{(1-p_{\rm F}^2)^2}\left(\frac{R_{\rm F}}{R_{\rm N}}\right)^2 \sinh^{-1}(L/\lambda_{\rm N}), \quad (R_{\rm N} \gg R_{\rm F} \gg R_1, R_2). \tag{8.13}$$

Note that R_s in the above limiting cases is independent of R_i, except for the intermediate regime $(R_{\rm F} \ll R_i \ll R_{\rm N})$ where

$$R_s/R_{\rm N} = \frac{2P_1P_2}{(1-P_1^2)(1-P_2^2)}\left(\frac{R_1R_2}{R_{\rm N}^2}\right)\sinh^{-1}(L/\lambda_{\rm N}). \tag{8.14}$$

Figure 8.4 show the spin accumulation signal R_s for $R_{\rm F}/R_{\rm N} = 0.01$ [6], $p_{\rm F} = 0.7$, and $P_i = 0.4$. We see that R_s increases by one order of magnitude by

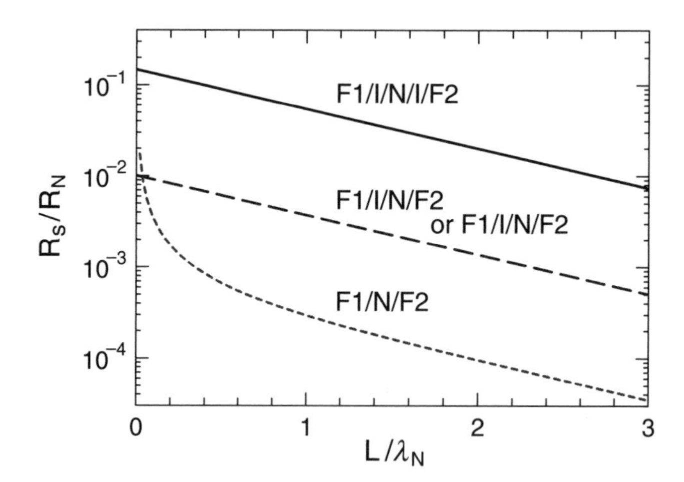

FIG. 8.4. Spin accumulation signal R_s as a function of distance L between F1 and F2.

replacing a transparent contact with a tunnel barrier, since the resistance mismatch, which is represented by the factor (R_F/R_N), is removed when a transparent contact is replaced with a tunnel junction. Note that the mismatch originates from a large difference in the spin diffusion lengths between N and F ($\lambda_N \gg \lambda_F$). Another example of resistance mismatch arises from the resistivity mismatch ($\rho_N \gg \rho_F$) as in the case of a non-magnetic semiconductor for N [33–35].

A question arises on whether the contacts of the metallic Py/Cu/Py structure [6] are transparent ($R_i/R_F \ll 1$) or tunneling-like ($R_i/R_N \gg 1$) [36]. If one uses the experimental values ($R_iA_J \sim 2 \times 10^{-12}\,\Omega\text{cm}^2$, $\lambda_F \sim 5$nm [21], $\rho_F \sim 10^{-5}\,\Omega$cm), one has $R_i/R_F \sim 0.4$, indicating that Py/Cu/Py lies in the transparent regime, so that Eq. (8.13) can be used to analyze the experimental data [6]. The value $R_N = 3\,\Omega$, which is estimated from the material parameters of Cu[31] in the Py/Cu/Py structure, yields $R_s = 1\,\text{m}\Omega$ at $L = \lambda_N$. If one takes into account the cross-shaped Cu of Ref. [6], one expects one-third of the above value [37], which is consistent with the experimental value of $R_s = 0.1\,\text{m}\Omega$ [6].

Figure 8.5 shows the experimental data of R_s as a function of distance L in Co/I/Al/I/Co [7], Py/I/Al/I/Py [13], Co/I/Cu/I/Co [16], and Py/Cu/Py [38, 39]. In the tunnel device of Co/I/Al/I/Co (I$=Al_2O_3$), fitting Eq. (8.11) to the experimental data yields λ_N = 650 nm at 4.2 K, λ_N = 350 nm at 293 K, $P_{1,2} = 0.1$, and $R_N = 3\,\Omega$.[32] The relation $\lambda_N^2 = D\tau_{sf}$ with $\lambda_N = 650$ nm and $D = 1/[2e^2N(0)\rho_N] \sim 40\,\text{cm}^2/\text{s}$ leads to $\tau_{sf} = 100$ ps at 4.2 K, which is the same order of magnitude as that evaluated from the Maki parameter $b = \hbar/(3\tau_{sf}\Delta_{Al}) \sim 0.01$ [31] in superconducting tunneling spectroscopy. In the metallic-contact device

[31] $\rho_N = 1.4\,\mu\Omega$cm, $\lambda_N = 1\,\mu$m, and $A_N = 100 \times 50\,\text{nm}^2$ for Cu [6].
[32] $\rho_N = 6\,\mu\Omega$cm, $\lambda_N = 0.65\,\mu$m, and $A_N = 250 \times 50\,\text{nm}^2$ for Al [7].

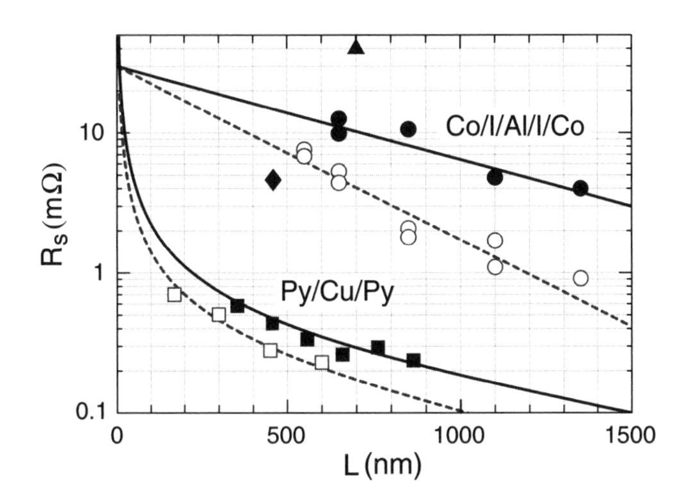

FIG. 8.5. Spin accumulation signal R_s as a function of distance L between two ferromagnetic electrodes in tunnel devices: (•, ∘) Co/I/Al/I/Co [7], (▲) Py/I/Al/I/Py [13], and (♦) Co/I/Cu/I/Co [16], and in metallic-contact devices: (□, ■) Py/Cu/Py [38, 39], where (•, ▲, ♦, ■) are the data at 4.2K and (∘, □) at room temperature.

of Py/Cu/Py, fitting Eq. (8.13) to the experimental data at 4.2 K yields $\lambda_{\rm N} = 920\,{\rm nm}$, $R_{\rm N} = 5\,\Omega$,[33] $[p_{\rm F}/(1-p_{\rm F}^2)](R_{\rm F}/R_{\rm N}) = 0.64\times10^{-3}$, and fitting Eq. (8.13) to those at 293 K yields $\lambda_{\rm N} = 700\,{\rm nm}$, $R_{\rm N} = 2.5\,\Omega$,[34] and $[p_{\rm F}/(1-p_{\rm F}^2)](R_{\rm F}/R_{\rm N}) = 10^{-3}$. In the tunneling regime, the spin splitting of ECP at position x in N is given by

$$2\delta\mu_{\rm N}(x) = P_1 e R_{\rm N} I e^{-|x|/\lambda_{\rm N}}, \tag{8.15}$$

whose maximum value is $\delta\mu_{\rm N}(0) \sim 15\,\mu{\rm V}$ for $P_1 \sim 0.1$, $R_{\rm N} = 3\,\Omega$, and $I = 100\,\mu{\rm A}$ [7], which is much smaller than the superconducting gap $\Delta \sim 200\mu{\rm eV}$ of an Al film. The effect of superconductivity on the spin accumulation is discussed in Section 8.3.

It is noteworthy that when F1 and F2 are half-metals ($p_{\rm F} \simeq 1$, $R_{\rm F} \gg R_{\rm N}$), we have the expression

$$R_s = p_{\rm F}^2 R_{\rm N} e^{-L/\lambda_{\rm N}}, \tag{8.16}$$

without tunnel barriers, which is the advantage of half-metals with high spin polarization.

[33] $\rho_{\rm N} = 3\,\mu\Omega{\rm cm}$, $\lambda_{\rm N} = 0.92\,\mu{\rm m}$, and $A_{\rm N} = 125\times45\,{\rm nm}^2$ for Cu [38].
[34] $\rho_{\rm N} = 2\,\mu\Omega{\rm cm}$, $\lambda_{\rm N} = 0.7\,\mu{\rm m}$, and $A_{\rm N} = 100\times80\,{\rm nm}^2$ for Cu [39].

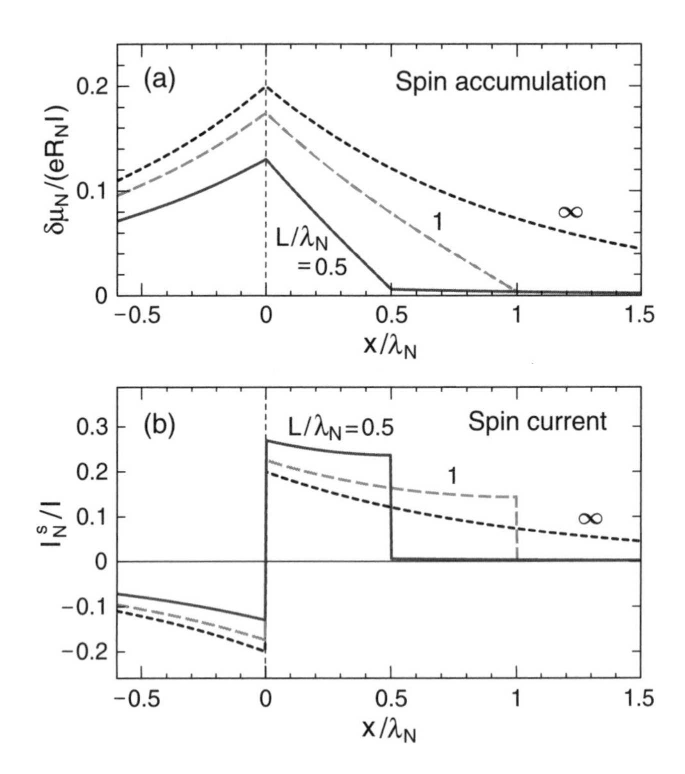

FIG. 8.6. Spatial variation of (a) spin accumulation $\delta\mu_{\rm N}$ and (b) spin current $I^s_{\rm N}$ flowing in the N electrode in a F1/I/N/F2 device, where the first junction is a tunnel junction and the second junction is a metallic-contact junction, for $L/\lambda_{\rm N} = 0.5$, 1.0, and ∞. The discontinuous change of spin current at $L/\lambda_{\rm N} = 0.5$ and 1.0 indicates that the spin current flows out of N into F2 through the N/F2 interface. The parameters are the same as those in Fig. 8.4. The inset shows the spin-current flow in the device.

8.2.3 *Non-local spin injection and manipulation*

We next investigate how the spin current flows through the structure, particularly the spin current across the N/F2 interface, because of the interest in spin-current induced magnetization switching [40–42] in non-local devices.

The distribution of spin accumulation and spin current is strongly influenced by the relative magnitude of the interface resistances (R_i) and the spin accumulation resistances ($R_{\rm F}, R_{\rm N}$). Figure 8.6(a) shows the spatial variation of spin accumulation $\delta\mu_{\rm N}$ in the N electrode of the F1/I/N/F2 structure. The short-dashed curve indicates $\delta\mu_{\rm N}$ in the absence of F2. When F2 is in contact with N at position of $L/\lambda_{\rm N} = 0.5$ and 1.0, the spin accumulation is strongly suppressed due to the metallic contact with F2 of very low $R_{\rm F}$, leaving very little accumulation on the right side of F2. This behavior has recently been observed in a spin injection and detection device with three Py electrodes [10, 11]. We notice that the slope of the curves between F1 and F2 ($0 < x < L$) becomes steeper than the short-dashed curve, indicating that the corresponding spin currents I_N^s between F1 and F2 become larger than that in the absence of F2 as shown in Fig. 8.6(b). In addition, the large discontinuous drop of I_N^s to nearly zero at $x = L$ is caused by spin current absorption by F2, indicating that most of the spin current flows out to F2 through the N/F2 interface. In the N region ($x > L$) on the right side of F2, both spin accumulation and spin current are very small. This implies that F2 with very small R_F like Py and CoFe and in metallic contact with N works as a strong spin absorber (an ideal spin sink).

For seeking the optimal conditions for spin current injection from N into F2, we calculate the spin current I_2^s across the N/F2 interface (*cf.* Eq. 8.81) and obtain

$$I_2^s = \frac{2\left(\frac{P_1}{1-P_1^2}\frac{R_1}{R_N} + \frac{p_{F1}}{1-p_{F1}^2}\frac{R_{F1}}{R_N}\right) e^{-L/\lambda_N} I}{\left(1 + \frac{2}{1-P_1^2}\frac{R_1}{R_N} + \frac{2}{1-p_{F1}^2}\frac{R_{F1}}{R_N}\right)\left(1 + \frac{2}{1-P_2^2}\frac{R_2}{R_N} + \frac{2}{1-p_{F2}^2}\frac{R_{F2}}{R_N}\right) - e^{-\frac{2L}{\lambda_N}}}. \quad (8.17)$$

A large spin-current injection occurs when junction 2 is a metallic contact ($R_2 \ll R_N$) and junction 1 is a tunnel junction ($R_1 \gg R_N$), yielding

$$I_2^s \approx P_1 I e^{-L/\lambda_N}, \quad (8.18)$$

for F2 like Py or CoFe with very short spin-diffusion length ($R_{F2} \ll R_N$). The spin current flowing in N on the left side of F2 is $I_N^s = P_1 I e^{-x/\lambda_N}$ ($0 < x < L$), which is two times larger than that in the absence of F2, while on the right side $I_N^s \approx 0$ ($x > L$). A similar situation results from a half-metallic F1 in metallic-contact with N, for which P_1 is replaced by $p_{F1} (\simeq 1)$ in Eq. (8.18).

When the F2 electrode is replaced with a nano-scale island comparable to or smaller than $w_N w_F$, the F2 island still works as a strong absorber (sink) for spin current; the spin-angular momentum is efficiently transferred from F1 to F2 by non-local spin injection. This provides a method for manipulating the orientation of magnetization in non-local devices [43].

8.3 Spin injection into superconductors

The spin accumulation signal in the spin injection and detection device containing a superconductor (S) such as Co/I/Al/I/Co is of great interest, because R_s is strongly influenced by opening the superconducting gap. We show that S becomes a low-carrier system for spin transport by opening the superconducting gap Δ and the resistivity of the spin current increases below the superconducting critical temperature T_c. In the F1/I/S/I/F2 tunneling device, the spin signal would increase due to the increase of R_N by opening of Δ below T_c (see Eq. 8.11). In the following, we consider the situation where the spin splitting of ECP is smaller than the gap Δ, for which the suppression of Δ due to spin accumulation [44–47] is neglected. While Andreev reflection plays an important role in metallic-contact junctions, we can neglect it in the tunneling device.

In the superconducting state, the equation for the spin splitting $(\mu_\uparrow - \mu_\downarrow)$ is the same as Eq. (8.3b) with $\lambda_{\rm N}$ in the normal state [50], which is intuitively understood as follows. Since the dispersion curve of the quasiparticle (QP) excitation energy is given by $E_k = \sqrt{\xi_k^2 + \Delta^2}$ with one-electron energy ξ_k [48], the QP's velocity

$$\tilde{v}_k = (1/\hbar)(\partial E_k/\partial k) = (|\xi_k|/E_k) v_k$$

is slower by the factor $|\xi_k|/E_k$ compared with the normal-state velocity $v_k (\approx v_{\rm F})$. By contrast, the impurity scattering time [49]

$$\tilde{\tau}_{\sigma\sigma'} = (E_k/|\xi_k|)\tau_{\sigma\sigma'},$$

is longer by the inverse of the factor. Then, the spin-diffusion length $\lambda_{\rm S} = (\tilde{D}\tilde{\tau}_{sf})^{1/2}$, where $\tilde{D} = \frac{1}{3}\tilde{v}_k^2\tilde{\tau}_{\rm tr} = (|\xi_k|/E_k)D$ and $\tilde{\tau}_{\rm tr}^{-1} = \sum_{\sigma'} \tilde{\tau}_{\sigma\sigma'}^{-1} = (E_k/|\xi_k|)\tau_{\rm tr}^{-1}$ [49], becomes

$$\lambda_{\rm S} = (\tilde{D}\tilde{\tau}_{sf})^{1/2} = (D\tau_{sf})^{1/2} = \lambda_{\rm N},$$

owing to the cancellation of the factor $|\xi_k|/E_k$. *The spin diffusion length in the superconducting state is the same as that in the normal state* [50, 51]. Consequently, we can use the same form of solutions for ECPs as in the normal-state, except for the coefficients which are modified by the onset of superconductivity.

The spin accumulation in S is determined by balancing the spin injection rate with the spin-relaxation rate:

$$\sum_i I_i^s + e\left(\frac{\partial S}{\partial t}\right)_{\rm sf} = 0, \tag{8.19}$$

where S is the total spins accumulated in S, and I_1^s and I_2^s are the spin injection and extraction rates through junction 1 and 2, respectively. At low temperatures the spin relaxation is dominated by spin-flip scattering via the spin-orbit interaction $V_{\rm so}$ at non-magnetic impurities or grain boundaries. The scattering matrix elements of $V_{\rm so}$ over quasiparticle states $|\mathbf{k}\sigma\rangle$ with momentum $\mathbf{k}$ and spin σ has the form:

$$\langle \mathbf{k}'\sigma'|V_{\rm so}|\mathbf{k}\sigma\rangle = i\eta_{\rm so}\left(u_{k'}u_k - v_{k'}v_k\right)\left[\boldsymbol{\sigma}_{\sigma'\sigma}\cdot(\mathbf{k}\times\mathbf{k}')\right]V_{\rm imp},$$

where $\eta_{\rm so}$ is the spin-orbit coupling parameter, $V_{\rm imp}$ is the impurity potential, $\boldsymbol{\sigma}$ is the Pauli spin matrix, and $u_k^2 = 1 - v_k^2 = \frac{1}{2}\left(1 + \xi_k/E_k\right)$ are the coherent factors [48]. Using the golden rule formula for the spin-flip scattering processes of $\langle \mathbf{k}' \downarrow |V_{\rm so}|\mathbf{k} \uparrow\rangle$ and $\langle \mathbf{k}' \uparrow |V_{\rm so}|\mathbf{k} \downarrow\rangle$, one obtains the spin-relaxation rate of the form [52, 53]

$$\left(\frac{\partial S}{\partial t}\right)_{\rm sf} = -\frac{N(0)}{\tau_{\rm sf}}\int_{\rm S} d\mathbf{r}\int_{\Delta}^{\infty}\left(f_{\mathbf{k}\uparrow} - f_{\mathbf{k}\downarrow}\right)dE_{\mathbf{k}}$$

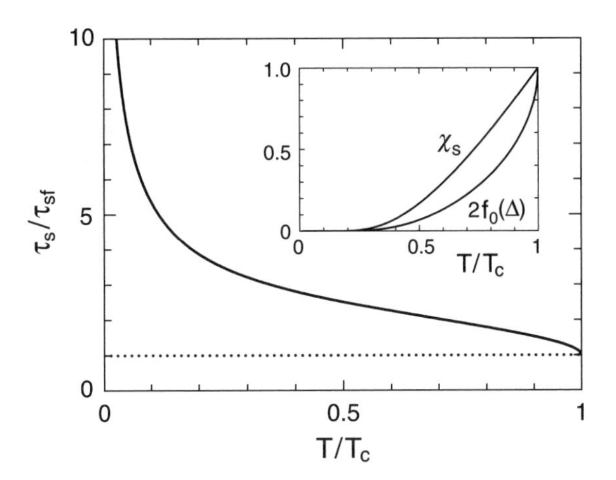

FIG. 8.7. Temperature dependence of the spin relaxation time τ_s in the superconducting state. The inset shows χ_s and $2f_0(\Delta)$ vs. T, which are used to calculate τ_s.

$$\approx -\frac{1}{\tau_{\rm sf}} 2f_0(\Delta) N(0) \int_{\rm S} d\mathbf{r} \delta\mu_{\rm S}(\mathbf{r}), \tag{8.20}$$

where $\tau_{\rm sf}$ is the spin-flip scattering time in the normal state and $f_{\mathbf{k}\sigma}$ the distribution function for a quasiparticle with momentum $\mathbf{k}$ and spin σ. In Eq. (8.20), we made use of the expansion with respect to the ECP shift $\delta\mu_{\rm S}$:

$$f_{\mathbf{k}\sigma} \sim f_0(E_{\mathbf{k}}) - \sigma[\partial f_0(E_{\mathbf{k}})/\partial E_{\mathbf{k}}]\delta\mu_{\rm S}.$$

On the other hand, the spin accumulation in S is given by

$$\mathcal{S} = \frac{1}{2} \int_{\rm S} d\mathbf{r} \sum_{\mathbf{k}} [f_{\mathbf{k}\uparrow} - f_{\mathbf{k}\downarrow}] \approx \chi_s(T) N(0) \int_{\rm S} d\mathbf{r} \delta\mu_{\rm S}(\mathbf{r}), \tag{8.21}$$

where $\chi_s(T)$ is the QP spin-susceptibility, called the Yosida function:

$$\chi_s(T) = 2 \int_{\Delta}^{\infty} \frac{E_{\mathbf{k}}}{\sqrt{E_{\mathbf{k}}^2 - \Delta^2}} \left(-\frac{\partial f_0}{\partial E_{\mathbf{k}}} \right) dE_{\mathbf{k}}, \tag{8.22}$$

whose asymptotic values are $\chi_s(T) \sim 1 - [7\zeta(3)/4\pi^2] (\Delta/k_{\rm B}T)^2$ near T_c and $\chi_s(T) \sim (\pi\Delta/2k_{\rm B}T)^{1/2} \exp[-\Delta/k_{\rm B}T]$ well below T_c.

The spin relaxation time τ_s of S in the superconducting state is determined from $(\partial\mathcal{S}/\partial t)_{\rm sf} = -S/\tau_s$, and given by

$$\tau_s(T) = \frac{\chi_s(T)}{2f_0(\Delta)} \tau_{\rm sf}, \tag{8.23}$$

which is the same as the result of Yafet [53] who studied the electron-spin resonance (ESR) in the superconducting state [54]. Figure 8.7 shows the temperature dependence of $\tau_s/\tau_{\rm sf}$. In the normal state above the superconducting critical temperature T_c, the spin relaxation time τ_s coincides with the spin-flip scattering time $\tau_{\rm sf}$. In the superconducting state below below T_c, τ_s becomes longer with decreasing T according to $\tau_s \simeq (\pi\Delta/2k_{\rm B}T)^{1/2}\tau_{\rm sf}$ at low temperatures.

Since the spin diffusion length in the superconducting state is the same as that in the normal state, the ECP shift in S has the form same as in Eq. (8.5), $\delta\mu_{\rm S} = \left(\tilde{a}_1 e^{-|x|/\lambda_{\rm N}} - \tilde{a}_2 e^{-|x-L|/\lambda_{\rm N}}\right)$, where the coefficients are calculated as follows. In the tunnel device, the tunnel spin currents are $I_1^s = P_1 I$ and $I_2^s \approx 0$, so that Eqs (8.19) and (8.20) give the coefficients $\tilde{a}_1 = P_1 R_{\rm N} eI/[2f_0(\Delta)]$ and $\tilde{a}_2 \approx 0$, leading to the spin splitting of ECP in the superconducting state

$$2\delta\mu_{\rm S}(x) = P_1 \frac{R_{\rm N} eI}{2f_0(\Delta)} e^{-|x|/\lambda_{\rm N}}, \tag{8.24}$$

indicating that the spin splitting in ECP is enhanced by the factor $1/[2f_0(\Delta)]$ compared with that in the normal state. The detected voltage V_2 by F2 at distance L is given by $V_2 = \pm P_2 \delta\mu_{\rm S}(L)$ for the P (+) and AP (−) alignments. Therefore, the spin signal R_s detected by F2 at distance L in the superconducting state becomes [20]

$$R_s = P_1 P_2 \frac{R_{\rm N}}{2f_0(\Delta)} e^{-L/\lambda_{\rm N}}. \tag{8.25}$$

The above result is directly obtained by the replacement $\rho_{\rm N} \to \rho_{\rm N}/[2f_0(\Delta)]$ in the normal-state result of Eq. (8.11), which results from the fact that the carrier density decreases in proportion to $2f_0(\Delta)$, and superconductors behave as a low carrier system for spin transport.

Figure 8.8 shows the T-dependence of spin signals $R_s = V_s/I$ below the superconducting critical temperature T_c. The values are normalized to those in the normal state. The rapid increase in R_s below T_c reflects the strong T-dependence in the resistivity of spin current. However, when the spin splitting $\delta\mu_{\rm S} \sim \frac{1}{2} eP_1 R_{\rm N} I/[2f(\Delta)]$ at $x = 0$ becomes comparable to or larger than Δ, the superconductivity is suppressed or destroyed by pair breaking due to spin accumulation [44, 45, 55–61], so that R_s deviates from the curve of Fig. 8.8 and decreases to the normal-state value. To test this prediction, it is highly desired to measure V_s in F1/I/S/I/F2 structures such as Co/I/Al/I/Co [7] or Py/I/Al/I/Py [13] in the superconducting state by lowering T below T_c.

A large enhancement of spin signal is also expected when a semiconducting material is used for N, because the carrier concentration is much lower and the resistivity is much larger than those of normal metals. A proper combination of a non-magnetic doped GaAs with ferromagnets such as CoFe, Py, and (Mn,Ga)As, as well as an appropriate choice of interface with or without a tunnel barrier, yield a large R_s proportional to either $\rho_{\rm GaAs}$ or $\rho_{\rm (Ga,Mn)As}$ as in Eqs (8.11), (8.12a), and (8.12b), and therefore the spin signal is expected to be larger by several orders of magnitude than that of the metallic case [8, 20]. This result is promising for applications for spin-electronic devices.

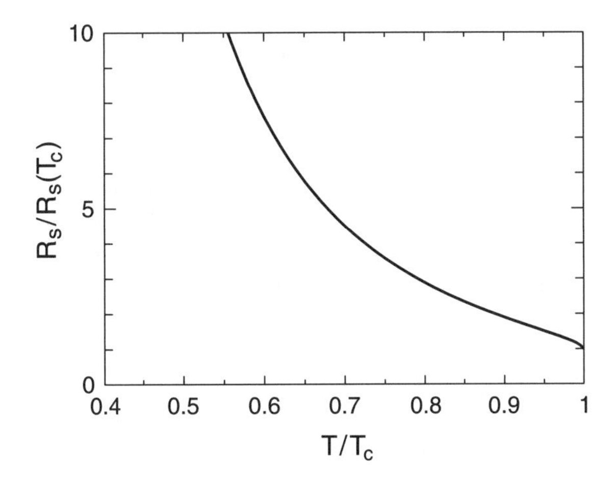

FIG. 8.8. Temperature dependence of the spin accumulation signal R_s in a F/I/S/I/F structure. The values of R_s are normalized to the value at the superconducting critical temperature T_c.

8.4 Spin Hall effect

A basic mechanism underlying the anomalous Hall effect (AHE) is the relativistic interaction between the spin and orbital motion of electrons (spin-orbit interaction) in metals or semiconductors. Conduction electrons moving in a crystal are scattered by local potentials created by defects or impurities in the crystal. The spin-orbit interaction at local potentials causes a spin-asymmetry in the scattering of conduction electrons [62]. In ferromagnetic materials, up-spin (majority) electrons are scattered preferentially in one direction and down-spin (minority) electrons are in the opposite direction, resulting in an anomalous Hall current that flows perpendicular to both the applied electric field and the magnetization directions. Another mechanism for AHE is the spin chirality in frustrated ferromagnets [63].

Spin injection techniques in nanohybrid structures makes it possible to cause AHE in *non-magnetic* conductors. When spin-polarized electrons are injected from a ferromagnet (F) to a non-magnetic electrode (N), these electrons moving in N are deflected by the spin-orbit interaction to induce spin and charge Hall currents in the transverse direction and accumulate spin and charge on the sides of N [64–66]. It is noteworthy to consider the following two cases. In the case that the charge current is absent and only the spin current flows in N, the spin-orbit interaction induces a charge Hall current and causes charge accumulation on the sides of N; the spin current in the longitudinal direction gives rise to the charge current in the transverse direction. In the case that an unpolarized charge current flows in N by an applied electric field, the spin-orbit interaction induces a spin Hall current and causes spin accumulation on the sides of N; the charge current in the longitudinal direction gives rise to the spin current in the transverse direction. In this way, the spin (charge) degree of freedom are converted to charge (spin) degree of freedom due to the spin-orbit interaction, which is important for spin-electronic applications. In addition to these *extrinsic* spin Hall effects, *intrinsic* spin Hall effects have been proposed in semiconductors which do not require impurities [67–71].

In the following, we consider the effect of the spin-orbit scattering on the spin and charge transport in a non-magnetic metal (N) such as Cu and Al, and discuss AHE in the presence of spin current (or charge current) in N, taking into account two mechanisms for AHE: *side jump* (SJ) and *skew scattering* (SS) [62,72–74], and derive formulas for the extrinsic spin and charge Hall effects [66].

8.4.1 *Basic formulation*

The spin-orbit coupling in the presence of non-magnetic impurities in a metal is derived as follows [75]. The impurity potentials $V(\mathbf{r})$ create an additional electric field $\mathbf{E}(\mathbf{r}) = -(1/e)\nabla V(\mathbf{r})$. When an electron passes through the field with velocity $\hat{\mathbf{p}}/m = (\hbar/i)\nabla/m$, the electron feels the effective magnetic field $\mathbf{B}_{\text{eff}}(\mathbf{r}) = -(1/mc)\hat{\mathbf{p}} \times \mathbf{E}(\mathbf{r})$. This yields the spin-orbit coupling

$$V_{\text{so}}(\mathbf{r}) = -\mu_B \boldsymbol{\sigma} \cdot \mathbf{B}_{\text{eff}}(\mathbf{r}) = \eta_{\text{so}} \boldsymbol{\sigma} \cdot \left[\nabla V(\mathbf{r}) \times \frac{1}{i}\nabla \right],$$

where $\boldsymbol{\sigma}$ is the Pauli spin operator and η_{so} is the spin-orbit coupling parameter. In the free-electron model, $\eta_{\text{so}} = \hbar^2/4m^2c^2$, which is corrected by the Thomas factor of one-half, and in real metals η_{so} is enhanced by several orders of magnitude for Bloch electrons [62]. The total impurity potential $U(\mathbf{r})$ is the sum of the ordinary impurity potential and the spin-orbit potential: $U(\mathbf{r}) = V(\mathbf{r}) + V_{\text{so}}(\mathbf{r})$.

The one-electron Hamiltonian H in the presence of the impurity potential $U(\mathbf{r})$ is given by

$$H = \sum_{\mathbf{k},\sigma} \xi_{\mathbf{k}} a^{\dagger}_{\mathbf{k}\sigma} a_{\mathbf{k}\sigma} + \sum_{\mathbf{k},\mathbf{k}'} \sum_{\sigma,\sigma'} U^{\sigma'\sigma}_{\mathbf{k}'\mathbf{k}} a^{\dagger}_{\mathbf{k}'\sigma'} a_{\mathbf{k}\sigma}. \tag{8.26}$$

Here, the first term is the kinetic energy of conduction electrons with energies $\xi_{\mathbf{k}} = (\hbar k)^2/2m - \varepsilon_{\text{F}}$, and the second term describes the scattering of conduction electrons whose scattering amplitude $U^{\sigma'\sigma}_{\mathbf{k}'\mathbf{k}}$ is given by

$$U^{\sigma'\sigma}_{\mathbf{k}'\mathbf{k}} = V_{\text{imp}} \left[\delta_{\sigma'\sigma} + i\eta_{\text{so}} \boldsymbol{\sigma}_{\sigma'\sigma} \cdot (\mathbf{k} \times \mathbf{k}') \right] \sum_i e^{i(\mathbf{k}-\mathbf{k}')\cdot\mathbf{r}_i}, \tag{8.27}$$

where $V_{\text{imp}} \sum_i e^{i(\mathbf{k}-\mathbf{k}')\cdot\mathbf{r}_i}$ are the matrix elements of the weak δ-function potential $V(\mathbf{r}) \approx V_{\text{imp}} \sum_i \delta(\mathbf{r} - \mathbf{r}_i)$ for impurity potentials at position $\mathbf{r}_i$.

The velocity of electrons is calculated from the velocity operator $\hat{\mathbf{v}} = d\mathbf{r}/dt = (1/i\hbar)[\mathbf{r}, H]$ by taking the matrix element between the scattering state $|\mathbf{k}^+\sigma\rangle$ in the presence of impurities [76]:

$$\mathbf{v}^{\sigma}_{\mathbf{k}} = \langle \mathbf{k}^+\sigma | \hat{\mathbf{v}} | \mathbf{k}^+\sigma \rangle = \frac{\hbar\mathbf{k}}{m} + \boldsymbol{\omega}^{\sigma}_{\mathbf{k}}, \tag{8.28}$$

where the first term is the usual velocity and the second term is the *anomalous velocity* arising from the spin-orbit scattering

$$\boldsymbol{\omega}_{\mathbf{k}}^{\sigma} = \frac{\eta_{\text{so}}}{\hbar} \sum_{i} \langle \mathbf{k}^{+}\sigma | \boldsymbol{\sigma} \times \nabla V(\mathbf{r} - \mathbf{r}_i) | \mathbf{k}^{+}\sigma \rangle, \tag{8.29}$$

with the scattering state $|\mathbf{k}^{+}\sigma\rangle$ in the Born approximation

$$|\mathbf{k}^{+}\sigma\rangle = |\mathbf{k}\sigma\rangle + \sum_{\mathbf{k}'} |\mathbf{k}'\sigma\rangle \frac{V_{\text{imp}} \sum_i e^{i(\mathbf{k}-\mathbf{k}')\cdot\mathbf{r}_i}}{\xi_{\mathbf{k}} - \xi_{\mathbf{k}'} + i\delta}, \tag{8.30}$$

where $|\mathbf{k}\sigma\rangle$ is the one-electron state of conduction electrons.

Substituting Eq. (8.30) into Eq. (8.29), the anomalous velocity is calculated up to first order in η_{so}:

$$\boldsymbol{\omega}_{\mathbf{k}}^{\sigma} = \alpha_{\text{H}}^{\text{SJ}} \left[\boldsymbol{\sigma}_{\sigma\sigma} \times \frac{\hbar \mathbf{k}}{m} \right], \tag{8.31}$$

with the dimensionless coupling parameter of the side jump

$$\alpha_{\text{H}}^{\text{SJ}} = \frac{m\eta_{\text{so}}}{\hbar \tau_{\text{tr}}^{0}} = \frac{\hbar \bar{\eta}_{\text{so}}}{3mD} = \frac{\hbar}{2\varepsilon_{\text{F}} \tau_{\text{tr}}^{0}} \bar{\eta}_{\text{so}}, \tag{8.32}$$

where $\tau_{\text{tr}}^{0} = 1/[(2\pi/\hbar) n_{\text{imp}} N(0) V_{\text{imp}}^{2}]$ is the transport relaxation time due to impurity scattering without the correction of spin orbit scattering, $\bar{\eta}_{\text{so}} = k_{\text{F}}^{2} \eta_{\text{so}}$ the dimensionless spin-orbit coupling parameter, and $D = (1/3) \tau_{\text{tr}}^{0} v_{\text{F}}^{2}$ the diffusion constant.

If one introduces the current operator for electrons with spin σ

$$\hat{\mathbf{J}}_{\sigma} = e \sum_{\mathbf{k}} \left[\frac{\hbar \mathbf{k}}{m} + \boldsymbol{\omega}_{\mathbf{k}}^{\sigma} \right] a_{\mathbf{k}\sigma}^{\dagger} a_{\mathbf{k}\sigma}, \tag{8.33}$$

($e = -|e|$ is the electronic charge), the "total" charge current $\mathbf{J}_q = \mathbf{J}_{\uparrow} + \mathbf{J}_{\downarrow}$ and the "total" spin current $\mathbf{J}_s = \mathbf{J}_{\uparrow} - \mathbf{J}_{\downarrow}$ are expressed as

$$\mathbf{J}_q = \mathbf{J}_q' + \alpha_{\text{H}}^{\text{SJ}} \left[\hat{\mathbf{z}} \times \mathbf{J}_s' \right], \tag{8.34}$$

$$\mathbf{J}_s = \mathbf{J}_s' + \alpha_{\text{H}}^{\text{SJ}} \left[\hat{\mathbf{z}} \times \mathbf{J}_q' \right], \tag{8.35}$$

where $\mathbf{J}_q'$ and $\mathbf{J}_s'$ are the charge and spin currents which do not include the side -jump contribution:

$$\mathbf{J}_q' = e \sum_{\mathbf{k}} \frac{\hbar \mathbf{k}}{m} \left[f_{\mathbf{k}\uparrow} + f_{\mathbf{k}\downarrow} \right], \qquad \mathbf{J}_s' = e \sum_{\mathbf{k}} \frac{\hbar \mathbf{k}}{m} \left[f_{\mathbf{k}\uparrow} - f_{\mathbf{k}\downarrow} \right], \tag{8.36}$$

and $f_{\mathbf{k}\sigma} = \langle a_{\mathbf{k}\sigma}^{\dagger} a_{\mathbf{k}\sigma} \rangle$ is the distribution function of an electron with energy $\xi_{\mathbf{k}}$ and spin σ. The second terms in Eqs (8.34) and (8.35) are the charge and spin Hall currents due to the *side jump* arising from the anomalous velocity. In addition to the side jump contribution, there is the *skew scattering* contribution which originates from the anisotropic scattering due to the spin-orbit interaction, and appears as a modification of the distribution function caused by the spin-orbit scattering. We calculate the distribution function $f_{\mathbf{k}\sigma}$ based on the Boltzmann transport equation in the next section.

The spin density S is given by

$$\mathcal{S} = \sum_{\mathbf{k}} \left[\langle a^{\dagger}_{\mathbf{k}\uparrow} a_{\mathbf{k}\uparrow} \rangle - \langle a^{\dagger}_{\mathbf{k}\downarrow} a_{\mathbf{k}\downarrow} \rangle \right] = \sum_{\mathbf{k}} \left(f_{\mathbf{k}\uparrow} - f_{\mathbf{k}\downarrow} \right). \tag{8.37}$$

8.4.2 *Scattering probability and Boltzmann equation*

The Boltzmann transport equation in steady state has the form

$$\boldsymbol{v}_{\mathbf{k}} \cdot \nabla f_{\mathbf{k}\sigma} + \frac{e\mathbf{E}}{\hbar} \cdot \nabla_{\mathbf{k}} f_{\mathbf{k}\sigma} = \left(\frac{\partial f_{\mathbf{k}\sigma}}{\partial t} \right)_{scatt}, \tag{8.38}$$

where $\boldsymbol{v}_{\mathbf{k}} = \hbar\mathbf{k}/m$, $\mathbf{E}$ is the external electric field, $f_{\mathbf{k}\sigma}$ is the distribution function, and the right-hand term is the collision term due to impurity scattering. The impurity scattering process for carriers can be described by the $\hat{T}$ matrix, which has the form within the second-order Born approximation:

$$\langle \mathbf{k}'\sigma' | \hat{T} | \mathbf{k}\sigma \rangle = \left[V_{\mathbf{k}'\mathbf{k}} + \sum_{\mathbf{q}} \frac{V_{\mathbf{k}'\mathbf{q}} V_{\mathbf{q}\mathbf{k}}}{\xi_{\mathbf{k}} - \xi_{\mathbf{q}} + i\delta} \right] \delta_{\sigma'\sigma} + i\eta_{\mathrm{so}} V_{\mathbf{k}'\mathbf{k}} (\mathbf{k} \times \mathbf{k}') \cdot \boldsymbol{\sigma}_{\sigma'\sigma}.$$

where $V_{\mathbf{k}'\mathbf{k}} = \langle \mathbf{k}' | V | \mathbf{k} \rangle$. The scattering probability $P^{\sigma'\sigma}_{\mathbf{k}'\mathbf{k}}$ from the state $|\mathbf{k}\sigma\rangle$ to the state $|\mathbf{k}'\sigma'\rangle$ is calculated by

$$P^{\sigma'\sigma}_{\mathbf{k}'\mathbf{k}} = \frac{2\pi}{\hbar} n_{\mathrm{imp}} |\langle \mathbf{k}'\sigma' | \hat{T} | \mathbf{k}\sigma \rangle|^2 \delta(\xi_{\mathbf{k}} - \xi_{\mathbf{k}'}) = P^{\mathbf{k}'\mathbf{k}(1)}_{\sigma'\sigma} + P^{\mathbf{k}'\mathbf{k}(2)}_{\sigma'\sigma},$$

where $P^{\sigma'\sigma(1)}_{\mathbf{k}'\mathbf{k}}$ and $P^{\sigma'\sigma(2)}_{\mathbf{k}'\mathbf{k}}$ are the first-order and second-order scattering probabilities:

$$P^{\sigma'\sigma(1)}_{\mathbf{k}'\mathbf{k}} = \frac{2\pi}{\hbar} n_{\mathrm{imp}} V^2_{\mathrm{imp}} \left(\delta_{\sigma\sigma'} + |\eta_{\mathrm{so}} (\mathbf{k}' \times \mathbf{k}) \cdot \boldsymbol{\sigma}_{\sigma\sigma'}|^2 \right) \delta(\xi_{\mathbf{k}'} - \xi_{\mathbf{k}}), \tag{8.39a}$$

$$P^{\sigma'\sigma(2)}_{\mathbf{k}'\mathbf{k}} = \frac{(2\pi)^2}{\hbar} n_{\mathrm{imp}} V^3_{\mathrm{imp}} N(0) \left[\eta_{\mathrm{so}} (\mathbf{k}' \times \mathbf{k}) \cdot \boldsymbol{\sigma}_{\sigma\sigma} \right] \delta_{\sigma\sigma'} \delta(\xi_{\mathbf{k}'} - \xi_{\mathbf{k}}). \tag{8.39b}$$

The change of the distribution function $f_{\mathbf{k}\sigma}(\mathbf{r})$ due to impurity scattering is related to the scattering probability $P^{\sigma'\sigma}_{\mathbf{k}'\mathbf{k}}$ by

$$\left(\frac{\partial f_{\mathbf{k}\sigma}}{\partial t} \right)_{scatt} = \sum_{\mathbf{k}'\sigma'} \left[P^{\sigma\sigma'}_{\mathbf{k}\mathbf{k}'} f_{\mathbf{k}'\sigma'} - P^{\sigma'\sigma}_{\mathbf{k}'\mathbf{k}} f_{\mathbf{k}\sigma} \right]$$

$$= \sum_{\mathbf{k}'\sigma'} P_{\mathbf{k}'\mathbf{k}}^{\sigma'\sigma(1)} \left[f_{\mathbf{k}'\sigma'} - f_{\mathbf{k}\sigma} \right] + \sum_{\mathbf{k}'\sigma'} P_{\mathbf{k}'\mathbf{k}}^{\sigma'\sigma(2)} \left[f_{\mathbf{k}'\sigma} + f_{\mathbf{k}\sigma} \right]. \quad (8.40)$$

The first term in the brackets is the scattering-in term ($\mathbf{k}'\sigma' \to \mathbf{k}\sigma$) while the second term is the scattering-out term ($\mathbf{k}\sigma \to \mathbf{k}'\sigma'$). It is convenient to separate $f_{\mathbf{k}\sigma}$ into three parts [77]

$$f_{\mathbf{k}\sigma} = f_{k\sigma}^0 + g_{\mathbf{k}\sigma}^{(1)} + g_{\mathbf{k}\sigma}^{(2)}, \quad (8.41)$$

where $f_{k\sigma}^0$ is an *undirectional* distribution function defined by the average of $f_{\mathbf{k}\sigma}$ with respect to the solid angle $\Omega_{\mathbf{k}}$ of $\mathbf{k}$:

$$f_{k\sigma}^0 = \int f_{\mathbf{k}\sigma} \frac{d\Omega_{\mathbf{k}}}{(4\pi)},$$

and $g_{\mathbf{k}\sigma}^{(1)}$ and $g_{\mathbf{k}\sigma}^{(2)}$ are *directional* distribution functions, i.e., $\int g_{\mathbf{k}\sigma}^{(i)} d\Omega_{\mathbf{k}} = 0$, and are associated with the first-order and the second-order transitions, respectively, which are calculated in the following way.

8.4.2.1 *First order solution* The first term in Eq. (8.40) is calculated as

$$\begin{aligned} \sum_{\mathbf{k}'\sigma'} \left[P_{\mathbf{k}\mathbf{k}'}^{\sigma\sigma'(1)} f_{\mathbf{k}'}^{\sigma'} - P_{\mathbf{k}'\mathbf{k}}^{\sigma'\sigma(1)} f_{\mathbf{k}}^{\sigma} \right] &= -\frac{g_{\mathbf{k}\sigma}^{(1)}}{\tau_0(\theta)} - \frac{f_{\mathbf{k}\sigma} - f_{k-\sigma}^0}{\tau_{\text{sf}}(\theta)} \\ &= -\frac{g_{\mathbf{k}\sigma}^{(1)}}{\tau_{\text{tr}}(\theta)} - \frac{f_{k\sigma}^0 - f_{k-\sigma}^0}{\tau_{\text{sf}}(\theta)}, \end{aligned} \quad (8.42)$$

where $\tau_{\text{tr}}(\theta)$ is the transport relaxation time

$$1/\tau_{\text{tr}}(\theta) = 1/\tau_0(\theta) + 1/\tau_{\text{sf}}(\theta),$$

and $\tau_0(\theta)$ and $\tau_{\text{sf}}(\theta)$ are the spin-conserving and spin-flip relaxation times

$$\frac{1}{\tau_0(\theta)} = \sum_{\mathbf{k}'} P_{\mathbf{k}\mathbf{k}'}^{\uparrow\uparrow(1)} = \sum_{\mathbf{k}'} P_{\mathbf{k}\mathbf{k}'}^{\downarrow\downarrow(1)} = \frac{1}{\tau_{\text{tr}}^0} \left(1 + \frac{1}{3} \bar{\eta}_{\text{so}}^2 \sin^2\theta \right), \quad (8.43\text{a})$$

$$\frac{1}{\tau_{\text{sf}}(\theta)} = \sum_{\mathbf{k}'} P_{\mathbf{k}\mathbf{k}'}^{\uparrow\downarrow(1)} = \sum_{\mathbf{k}'} P_{\mathbf{k}\mathbf{k}'}^{\downarrow\uparrow(1)} = \frac{\bar{\eta}_{\text{so}}^2}{3\tau_{\text{tr}}^0} \left(1 + \cos^2\theta \right), \quad (8.43\text{b})$$

with θ the angle between $\mathbf{k}$ and the x axis. Then, the Boltzmann equation (8.38) with the collision term (8.42) is [65, 78]

$$\boldsymbol{v}_{\mathbf{k}} \cdot \frac{\partial f_{\mathbf{k}\sigma}}{\partial \mathbf{r}} + \frac{e\mathbf{E}}{\hbar} \cdot \frac{\partial f_{\mathbf{k}\sigma}}{\partial \mathbf{k}} = -\frac{g_{\mathbf{k}\sigma}^{(1)}}{\tau_{\text{tr}}(\theta)} - \frac{f_{k\sigma}^0 - f_{k-\sigma}^0}{\tau_{\text{sf}}(\theta)}, \quad (8.44)$$

where the first term in Eq. (8.44) in the r.h.s. describes momentum relaxation by spin-conserving scattering and the second term spin relaxation by spin-flip scattering. Since $\tau_{\rm tr}(\theta) \ll \tau_{\rm sf}(\theta)$, momentum relaxation occurs first and then is followed by the *slow* spin relaxation.

The first-order solution due to momentum relaxation is obtained as

$$g^{(1)}_{\mathbf{k}\sigma} \approx -\tau_{\rm tr}(\theta)\left(\boldsymbol{v}_{\mathbf{k}}\cdot\nabla + \frac{e\mathbf{E}}{\hbar}\cdot\nabla_{\mathbf{k}}\right) f^0_{k\sigma}, \tag{8.45}$$

The distribution function $f^0_{k\sigma}$ is the local equilibrium one with Fermi energy $\varepsilon^\sigma_{\rm F}(\mathbf{r}) = \varepsilon_{\rm F} + \sigma\delta\mu_{\rm N}(\mathbf{r})$ shifted by $\pm\delta\mu_{\rm N}(\mathbf{r})$ for the up and down spin bands from equilibrium, $f^0_{k\sigma} = f_0(\xi_{\mathbf{k}} - \sigma\delta\mu_{\rm N})$, which is expanded as

$$f^0_{k\sigma} \approx f_0(\xi_{\mathbf{k}}) - \sigma\frac{\partial f_0(\xi_{\mathbf{k}})}{\partial\xi_{\mathbf{k}}}\delta\mu_{\rm N}(\mathbf{r}), \tag{8.46}$$

where $f_0(\xi_{\mathbf{k}})$ is the Fermi distribution function. Therefore, Eq. (8.45) becomes

$$g^{(1)}_{\mathbf{k}\sigma} \approx \tau_{\rm tr}(\theta)\frac{\partial f_0(\xi_{\mathbf{k}})}{\partial\xi_{\mathbf{k}}}\boldsymbol{v}_{\mathbf{k}}\cdot\nabla\mu^\sigma_{\rm N}(\mathbf{r}), \tag{8.47}$$

with the electrochemical potential (ECP)

$$\mu^\uparrow_{\rm N}(\mathbf{r}) = \varepsilon_{\rm F} + e\phi + \delta\mu_{\rm N}(\mathbf{r}), \qquad \mu^\downarrow_{\rm N}(\mathbf{r}) = \varepsilon_{\rm F} + e\phi - \delta\mu_{\rm N}(\mathbf{r}), \tag{8.48}$$

and the electric potential ϕ of $\mathbf{E} = -\nabla\phi$.

The spin-flip scattering by the spin-orbit coupling causes a slow relaxation for spin accumulation $(\mu^\uparrow_{\rm N} - \mu^\downarrow_{\rm N}) = 2\delta\mu_{\rm N}$. Substituting the above solution in Eq. (8.44) and summing over $\mathbf{k}$, one obtains the spin diffusion equation:

$$\nabla^2\left(\mu^\uparrow_{\rm N} - \mu^\downarrow_{\rm N}\right) = \frac{1}{\lambda^2_{\rm N}}\left(\mu^\uparrow_{\rm N} - \mu^\downarrow_{\rm N}\right), \tag{8.49}$$

with $\lambda_{\rm N} = \sqrt{D\tau_{\rm sf}}$, $D = (1/3)\tau_{\rm tr}v^2_{\rm F}$, and

$$1/\tau_{\rm tr} = \langle 1/\tau_{\rm tr}(\theta)\rangle_{\rm av} = (1 + \bar\eta^2_{\rm so})/\tau^0_{\rm tr}, \tag{8.50}$$

$$1/\tau_{\rm sf} = \langle 1/\tau_{\rm sf}(\theta)\rangle_{\rm av} = (4\bar\eta^2_{\rm so}/9)/\tau^0_{\rm tr}. \tag{8.51}$$

8.4.2.2 *Second-order solution* The second-order term in the Boltzmann equation is given by

$$\sum_{\mathbf{k}'\sigma'}\left[P^{\sigma'\sigma(1)}_{\mathbf{k}'\mathbf{k}}\left(g^{(2)}_{\mathbf{k}\sigma} - g^{(2)}_{\mathbf{k}'\sigma'}\right) - P^{\sigma'\sigma(2)}_{\mathbf{k}'\mathbf{k}}\left(g^{(1)}_{\mathbf{k}\sigma} + g^{(1)}_{\mathbf{k}'\sigma'}\right)\right] = 0. \tag{8.52}$$

Making use of Eqs (8.39a), (8.39b), and (8.47), the solution of the second-order (*skew scattering*) term becomes

$$g^{(2)}_{\mathbf{k}\sigma} = -\alpha^{\rm SS}_{\rm H}\tau_{\rm tr}(\theta)\frac{\partial f_0(\xi_{\mathbf{k}})}{\xi_{\mathbf{k}}}\left(\boldsymbol{\sigma}_{\sigma\sigma}\times\boldsymbol{v}_{\mathbf{k}}\right)\cdot\nabla\mu^\sigma_{\rm N}(\mathbf{r}), \tag{8.53}$$

where $\alpha^{\rm SS}_{\rm H}$ is the dimensionless parameter of skew scattering

$$\alpha^{\rm SS}_{\rm H} = (2\pi/3)\bar\eta_{\rm so}N(0)V_{\rm imp}. \tag{8.54}$$

The first and second solutions are used to calculate the spin and charge currents, $\mathbf{J}'_s$ and $\mathbf{J}'_q$, in Eq. (8.36).

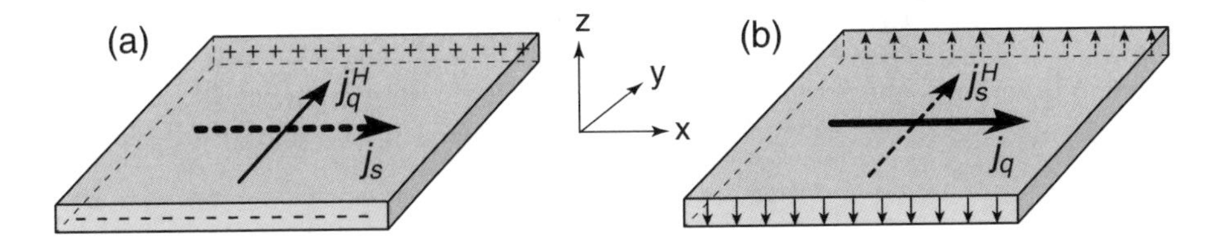

FIG. 8.9. (a) Spin-current induced Hall effect in which the spin current $\mathbf{j}_s$ along the x direction induces the charge Hall current $\mathbf{j}_q^{\mathrm{H}}$ in the y direction, creating charge accumulation on the sides of the sample. (b) Charge-current induced Hall effect in which the charge current $\mathbf{j}_q$ along the x direction induces the spin Hall current $\mathbf{j}_s^{\mathrm{H}}$ in the y direction, creating spin accumulation over the spin diffusion length from the sides of the sample.

8.4.3 *Spin and charge Hall currents*

The solutions of the Boltzmann equation yield the distribution function

$$f_{\mathbf{k}\sigma} \approx f_0(\xi_{\mathbf{k}}) - \sigma \frac{\partial f_0(\xi_{\mathbf{k}})}{\partial \xi_{\mathbf{k}}} \delta\mu_{\mathrm{N}}(\mathbf{r}) + \tau_{\mathrm{tr}}(\theta) \frac{\partial f_0(\xi_{\mathbf{k}})}{\partial \xi_{\mathbf{k}}} \left[\boldsymbol{v}_{\mathbf{k}} - \alpha_{\mathrm{H}}^{\mathrm{SS}} \boldsymbol{\sigma}_{\sigma\sigma} \times \boldsymbol{v}_{\mathbf{k}} \right] \cdot \nabla \mu_{\mathrm{N}}^{\sigma}(\mathbf{r}), \tag{8.55}$$

from which the spin and charge currents in Eq. (8.36) are calculated as

$$\mathbf{J}_s' = \mathbf{j}_s + \alpha_{\mathrm{H}}^{\mathrm{SS}} \left[\hat{\mathbf{z}} \times \mathbf{j}_q \right], \tag{8.56}$$

$$\mathbf{J}_q' = \mathbf{j}_q + \alpha_{\mathrm{H}}^{\mathrm{SS}} \left[\hat{\mathbf{z}} \times \mathbf{j}_s \right], \tag{8.57}$$

where $\mathbf{j}_s$ and $\mathbf{j}_q$ are the *longitudinal* spin and Ohmic currents:

$$\mathbf{j}_s = -\frac{\sigma_{\mathrm{N}}}{e} \nabla \delta\mu_{\mathrm{N}}, \tag{8.58}$$

$$\mathbf{j}_q = \sigma_{\mathrm{N}} \mathbf{E}, \tag{8.59}$$

where $\sigma_{\mathrm{N}} = 2e^2 N(0) D$ is the electrical conductivity, $\delta\mu_{\mathrm{N}} = \frac{1}{2}(\mu_{\mathrm{N}}^{\uparrow} - \mu_{\mathrm{N}}^{\downarrow})$ is the chemical potential shift, and $D = (1/3)\tau_{\mathrm{tr}} v_{\mathrm{F}}^2$ the diffusion constant. The second terms in Eqs (8.56) and (8.57) are the transverse Hall currents due to the skew scattering induced by the charge and spin currents. Therefore, the "total" spin and charge currents in Eqs (8.34) and (8.35) are written as

$$\mathbf{J}_q = \mathbf{j}_q + \mathbf{j}_q^{\mathrm{H}}, \tag{8.60}$$

$$\mathbf{J}_s = \mathbf{j}_s + \mathbf{j}_s^{\mathrm{H}}, \tag{8.61}$$

where

$$\mathbf{j}_q^{\mathrm{H}} = \alpha_{\mathrm{H}} \left[\hat{\mathbf{z}} \times \mathbf{j}_s \right] = -\frac{\alpha_{\mathrm{H}} \sigma_{\mathrm{N}}}{e} \left(\hat{\mathbf{z}} \times \nabla \delta\mu_{\mathrm{N}} \right), \tag{8.62}$$

$$\mathbf{j}_s^{\mathrm{H}} = \alpha_{\mathrm{H}} \left[\hat{\mathbf{z}} \times \mathbf{j}_q \right] = \alpha_{\mathrm{H}} \sigma_{\mathrm{N}} \left(\hat{\mathbf{z}} \times \mathbf{E} \right), \tag{8.63}$$

with $\alpha_{\mathrm{H}} = \alpha_{\mathrm{H}}^{\mathrm{SJ}} + \alpha_{\mathrm{H}}^{\mathrm{SS}}$.

Equation (8.62) indicates that the spin current $\mathbf{j}_s$ induces the transverse *charge* current (charge Hall current) $\mathbf{j}_q^{\rm H}$, while Eq. (8.63) indicates the charge current $\mathbf{j}_q$ induces the transverse *spin* current (spin Hall current) $\mathbf{j}_s^{\rm H}$, as shown in Fig. 8.9. Equations (8.60) and (8.61) are expressed in the matrix forms

$$\begin{bmatrix} J_{q,x} \\ J_{s,y} \end{bmatrix} = \begin{bmatrix} \sigma_{xx} & -\sigma_{xy} \\ \sigma_{xy} & \sigma_{xx} \end{bmatrix} \begin{bmatrix} E_x \\ -\nabla_y \delta\mu_{\rm N}/e \end{bmatrix}, \tag{8.64}$$

$$\begin{bmatrix} J_{s,x} \\ J_{q,y} \end{bmatrix} = \begin{bmatrix} \sigma_{xx} & -\sigma_{xy} \\ \sigma_{xy} & \sigma_{xx} \end{bmatrix} \begin{bmatrix} -\nabla_x \delta\mu_{\rm N}/e \\ E_y \end{bmatrix}. \tag{8.65}$$

Here, $\sigma_{xx} = \sigma_{\rm N}$ is the longitudinal conductivity and $\sigma_{xy} = \sigma_{\rm N}\alpha_{\rm H}$ is the Hall conductivity contributed from the side-jump and skew-scattering:

$$\sigma_{xy} = \sigma_{xy}^{\rm SJ} + \sigma_{xy}^{\rm SS}, \tag{8.66}$$

where

$$\sigma_{xy}^{\rm SJ} = \alpha_{\rm H}^{\rm SJ}\sigma_{\rm N} = \frac{e^2}{\hbar}\eta_{\rm so} n_e, \tag{8.67}$$

$$\sigma_{xy}^{\rm SS} = \alpha_{\rm H}^{\rm SS}\sigma_{\rm N} = \frac{e^2}{2\hbar}\eta_{\rm so} n_e \frac{n_e}{n_{\rm imp}} \frac{1}{N(0)V_{\rm imp}}, \tag{8.68}$$

with $n_e = (4/3)N(0)\varepsilon_{\rm F}$ the carrier (electron) density. It is noteworthy that the side-jump conductivity $\sigma_{xy}^{\rm SJ}$ is *independent* of impurity concentration.

The ratio of the SJ and SS Hall contributions is

$$\frac{\sigma_{xy}^{\rm SJ}}{\sigma_{xy}^{\rm SS}} = 2\frac{n_{\rm imp}}{n_e} N(0)V_{\rm imp} = \frac{3}{4\pi}\frac{\hbar}{\varepsilon_{\rm F}\tau_{\rm tr}} \frac{1}{N(0)V_{\rm imp}}. \tag{8.69}$$

In ordinary non-magnetic metals, the ratio is very small because of $n_{\rm imp} \ll n_e$ and $N(0)V_{\rm imp} \sim 1$, so that the SS contribution dominates the SJ contribution. However, in dirty metals or in low-carrier materials such as doped semiconductors (p- and n-type GaAs) where $n_{\rm imp} \sim n_e$, the SJ conductivity is comparable to or even larger than the SS conductivity in the anomalous Hall effect [70, 79].

8.4.4 *Spin-orbit coupling parameter*

It is interesting to note that the product $\rho_{\rm N}\lambda_{\rm N}$ is expressed in terms of the spin-orbit coupling parameter $\bar\eta_{\rm so}$ as

$$\rho_{\rm N}\lambda_{\rm N} = \frac{\sqrt{3}\pi}{2}\frac{R_{\rm K}}{k_{\rm F}^2}\sqrt{\frac{\tau_{\rm sf}}{\tau_{\rm tr}}} = \frac{3\sqrt{3}\pi}{4}\frac{R_{\rm K}}{k_{\rm F}^2}\frac{1}{\bar\eta_{\rm so}}, \tag{8.70}$$

where $R_{\rm K} = h/e^2 \sim 25.8\,{\rm k\Omega}$ is the quantum resistance. The relation (8.70) provides a new method to obtain information on the spin-orbit coupling in non-magnetic metals. Using the experimental data of $\rho_{\rm N}$ and $\lambda_{\rm N}$ and the Fermi momentums of $k_{\rm F}^{\rm Cu} = 1.36\times10^8\,{\rm cm}^{-1}$ and $k_{\rm F}^{\rm Al} = 1.75\times10^8\,{\rm cm}^{-1}$ [81] in Eq. (8.70),

Table 8.1 Spin-orbit coupling parameter of Cu and Al.

	$\lambda_{\rm N}$ (nm)	$\rho_{\rm N}$ ($\mu\Omega$cm)	$\tau_{\rm tr}/\tau_{\rm sf}$	$\bar{\eta}_{\rm so}$
Cu	1000[a]	1.43[a]	0.70×10^{-3}	0.040
Cu	546[b]	3.44[b]	0.41×10^{-3}	0.030
Al	650[c]	5.90[c]	0.36×10^{-4}	0.009
Al	1200[d]	1.25[d]	0.23×10^{-3}	0.023

[a]From Ref. [6], [b]from Ref. [16], [c]from Ref. [7], [d]from Ref. [80].

we obtain the value of the spin-orbit coupling parameter $\bar{\eta}_{\rm so}$ in Cu and Al, which is listed in Table 8.1. We note that the values of $\bar{\eta}_{\rm so}$ estimated from the spin injection and detection method are 10^2–10^3 times as large as the free-electron values.

8.4.5 *Non-local spin Hall effect*

We consider a spin-injection Hall device shown in Fig. 8.10. The magnetization of F at $x = 0$ points in the z direction. The spin injection is made by flowing the current I from F to the left end of N (the negative direction of x), while the Hall voltage ($V_{\rm H} = V_{\rm H}^+ - V_{\rm H}^-$) is measured by the Hall probe in the right side of N ($x > 0$). In the positive region of x, only the spin current $\mathbf{j}_s$ flows in the x-direction and the charge current $\mathbf{j}_q$ is absent. Therefore, from Eqs (8.60) and (8.61), we obtain the total spin current $\mathbf{J}_s \cong \mathbf{j}_s$ and the total charge current

$$\mathbf{J}_q = \alpha_{\rm H}\left[\hat{\mathbf{z}} \times \mathbf{j}_s\right] + \sigma_{\rm N}\mathbf{E}, \tag{8.71}$$

where the first term is the Hall current induced by the spin current, and the second term is the Ohmic current which builds up in the transverse direction as opposed to the Hall current. In the open circuit condition in the transverse direction, the y component of Eq. (8.71) vanishes: $\mathbf{J}_q^y = 0$, which yields the relation between the transverse electric field E_y and the transport spin current $\mathbf{j}_s = (j_s, 0, 0)$:

$$E_y = -\alpha_{\rm H}\rho_{\rm N} j_s. \tag{8.72}$$

Integrating Eq. (8.72) with respect to y, we obtain the Hall voltage at $x = L$

$$V_{\rm H} = \alpha_{\rm H} w_{\rm N} \rho_{\rm N} j_s, \tag{8.73}$$

where $w_{\rm N}$ is the width of N.

For the tunneling case, the spin current is given by

$$j_s \approx \frac{1}{2} P_1 (I/A_{\rm N}) e^{-L/\lambda_{\rm N}}, \tag{8.74}$$

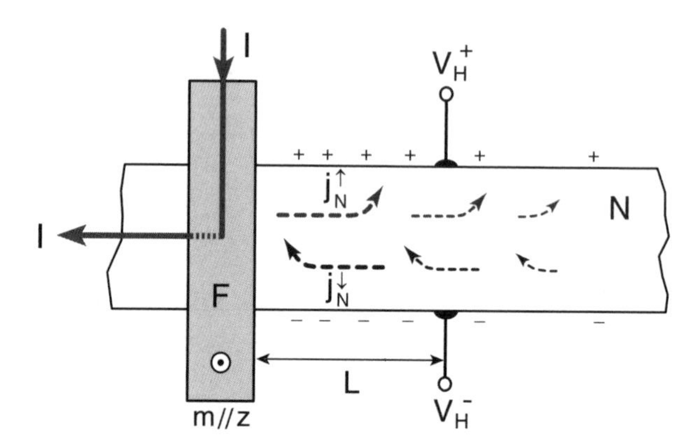

FIG. 8.10. Non-local spin-injection Hall device. The magnetic moment of F is aligned perpendicular to the plane. The non-local Hall voltage $V_{\rm H}$ is induced in the transverse direction by injection of spin-polarized current.

so that the *non-local* Hall resistance $R_{\rm H} = V_{\rm H}/I$ is

$$R_{\rm H} = \frac{1}{2}\alpha_{\rm H} P_1 \frac{\rho_{\rm N}}{d_{\rm N}} e^{-L/\lambda_{\rm N}}. \tag{8.75}$$

Typical values of $P_1 \sim 0.3$, $\alpha_{\rm H} \approx \alpha_{\rm H}^{\rm SS} \sim \bar{\eta}_{\rm so} \sim 0.04$ ($N(0)V_{\rm imp} \sim 1$), $\rho_{\rm N} \sim 1\,\mu\Omega\,{\rm cm}^2$, and $d_{\rm N} \sim 100\,{\rm nm}$ gives a rather large value for $R_{\rm H}$ of the order of $1\,{\rm m}\Omega$, indicating that the spin-current induced Hall effect is measurable in the non-local geometry. Recently, the spin-current induced Hall effect has been reported in local and non-local geometries of Co/Al [82] and Py/Cu [83].

It is interesting to see whether the anomalous Hall effect appears in the superconducting state. The spin current carried by QPs in a superconductor is deflected by spin-orbit impurity scattering to accumulate the QP charge (charge imbalance) in the transverse direction. The QP charge accumulation is compensated by the Cooper pair charge due to overall charge neutrality, thereby creating the electric potential to maintain the ECP of pairs to be constant in space (otherwise the pairs are accelerated). This spin and charge coupling leads to a novel anomalous Hall effect in superconductors [66].

8.5 Appendix: Electrochemical potentials in F1/N/F2

From matching conditions that the charge and spin currents are continuous at the interfaces of F1/N/F2, the coefficients a_1 and a_2 in the electrochemical potential (ECP) of Eq. (8.5) in the N electrode are determined as

$$a_1 = \frac{(P_1 r_1 + p_{\rm F1} r_{\rm F1})(1 + 2r_2 + 2r_{\rm F2})}{(1 + 2r_1 + 2r_{\rm F1})(1 + 2r_2 + 2r_{\rm F2}) - e^{-2L/\lambda_{\rm N}}} R_{\rm N} eI, \tag{8.76}$$

$$a_2 = -\frac{(P_1 r_1 + p_{\rm F1} r_{\rm F1})\, e^{-L/\lambda_{\rm N}}}{(1 + 2r_1 + 2r_{\rm F1})(1 + 2r_2 + 2r_{\rm F2}) - e^{-2L/\lambda_{\rm N}}} R_{\rm N} eI, \tag{8.77}$$

where r_i and r_{Fi} $(i = 1, 2)$ are the normalized resistances:

$$r_i = \frac{1}{(1 - P_i^2)} \frac{R_i}{R_N}, \qquad r_{Fi} = \frac{1}{(1 - p_{Fi}^2)} \frac{R_{Fi}}{R_N}. \tag{8.78}$$

The spin accumulation voltage V_2 detected by F2 is given by

$$V_2 = \pm 2 \frac{(P_1 r_1 + p_{F1} r_{F1})(P_2 r_2 + p_{F2} r_{F2})}{(1 + 2r_1 + 2r_{F1})(1 + 2r_2 + 2r_{F2}) - e^{-2L/\lambda_N}} R_N I, \tag{8.79}$$

where " + " and " − " correspond to the parallel and antiparallel alignments of magnetizations, respectively. The coefficients, b_1 and b_2, in ECP of Eqs (8.6a) and (8.6b) in F1 and F2 are related to a_1 and a_2 through

$$b_1 = \frac{1}{2} p_{F1} R_{F1} e I - (R_{F1}/R_N) a_1, \qquad b_2 = (R_{F2}/R_N) a_2. \tag{8.80}$$

The interfacial spin currents across junctions 1 and 2 are given by

$$I_1^s = (2/eR_N) a_1, \qquad I_2^s = -(2/eR_N) a_2. \tag{8.81}$$

Acknowledgements

The authors would like to thank M. Ichimura, J. Martinek, K. Tanikawa, T. Yamashita for helpful discussions and their collaboration. This work was supported by CREST, MEXT, JSPS, and the NAREGI Nanoscience Project, Japan.

References

[1] *Spin Dependent Transport in Magnetic Nanostructures*, edited by S. Maekawa and T. Shinjo (Taylor and Francis, 2002).
[2] *Semiconductor Spintronics and Quantum Computation*, edited by D. D. Awschalom, D. Loss, and N. Samarth (Springer-Verlag, 2002).
[3] I. Žutić, J. Fabian, and S. Das Sarma, *Spintronics: Fundamentals and applications*, Rev. Mod. Phys. **76**, 323 (2004).
[4] M. Johnson and R. H. Silsbee, Phys. Rev. Lett. **55**, 1790 (1985); *ibid.* **60**, 377 (1988); Phys. Rev. B **37**, 5326 (1988).
[5] M. Johnson, Phys. Rev. Lett. **70**, 2142 (1993).
[6] F. J. Jedema, A. T. Filip and B. J. van Wees, Nature (London) **410**, 345 (2001)
[7] F. J. Jedema, H. B. Heersche, A. T. Filip, J. J. A. Baselmans, and B. J. van Wees, Nature (London) **416**, 713 (2002).
[8] J.-M. George, A. Fert, and G. Faini, Phys. Rev. B **67**, 12410 (2003).
[9] M. Urech, J. Johansson, V. Korenivski and D. B. Haviland, J. Magn. Magn. Mater. **272-276**, E1469 (2004).
[10] T. Kimura, J. Hamrle, Y. Otani, K. Tsukagoshi, and Y. Aoyagi, Appl. Phys. Lett. **85**, 3795 (2004);

[11] T. Kimura, J. Hamrle, Y. Otani, Phys. Rev. B **72**, 14461 (2005).
[12] Y. Ji, A. Hoffmann, J. S. Jiang, and S. D. Bader, Appl. Phys. Lett. **85**, 6218 (2004).
[13] K. Miura, T. Ono, S. Nasu, T. Okuno, K. Mibu, and T. Shinjo, J. Magn. Magn. Mater. **286**, 142 (2005).
[14] T. Kimura, J. Hamrle, Y. Otani, K. Tsukagoshi, and Y. Aoyagi, J. Magn. Magn. Mater. **286**, 88 (2005).
[15] D. Beckmann, H. B. Weber, and H. v. Lohneysen, Phys. Rev. Lett. **93**, 197003 (2004).
[16] S. Garzon, I. Žutić, and R. A. Webb, Phys. Rev. Lett. **94**, 176601 (2005).
[17] T. Valet and A. Fert, Phys. Rev. B **48**, 7099 (1993).
[18] A. Fert and S. F. Lee, Phys. Rev. B **53**, 6554 (1996).
[19] S. Hershfield and H. L. Zhao, Phys. Rev. B **56**, 3296 (1997).
[20] S. Takahashi and S. Maekawa, Phys. Rev. B **67**, 052409 (2003).
[21] J. Bass and W. P. Pratt Jr., J. Magn. and Magn. Mater. **200**, 274 (1999).
[22] S. D. Steenwyk, S. Y. Hsu, R. Loloee, J. Bass, and W. P. Pratt Jr., J. Magn. Magn. Mater. **170**, L1 (1997).
[23] S. Dubois, L. Piraux, J.-M. George, K. Ounadjela, J. L. Duvail, and A. Fert, Phys. Rev. B **60**, 477 (1999).
[24] R. J. M. van de Veerdonk, J. Nowak, R. Meservey, J. S. Moodera, and W. J. M. de Jonge, Appl. Phys. Lett. **71**, 2839 (1997).
[25] M. Ichimura, S. Takahashi, K. Ito, and S. Maekawa, J. Appl. Phys. **95**, 7225 (2004).
[26] J. Hamrle, T. Kimura, T. Yang, and Y. Otani, Phys. Rev. B **71**, 094434 (2005).
[27] R. J. Soulen Jr., J. M. Byers, M. S. Osofsky, B. Nadgorny, T. Ambrose, S. F. Cheng, P. R. Broussard, C. T. Tanaka, J. Nowak, J. S. Moodera, A. Barry, J. M. D. Coey, Science **282**, 85 (1998).
[28] R. Meservey and P. M. Tedrow, Phys. Rep. **238**, 173 (1994).
[29] J. S. Moodera and G. Mathon, J. Magn. Magn. Mater. **200**, 248 (1999).
[30] D. J. Monsma and S. S. P. Parkin, Appl. Phys. Lett. **77**, 720 (2000).
[31] S. S. P. Parkin, C. Kaiser, A. Panchula, P. M. Rice, B. Hughes, M. Samant, and S.-H. Yang, Nature Materials **3**, 862 (2004).
[32] S. Yuasa, T. Nagahama, A. Fukushima, Y. Suzuki, and K. Ando, Nature Materials **3**, 868 (2004).
[33] E. I. Rashba, Phys. Rev. B **62**, R16267 (2000).
[34] A. Fert and H. Jaffrès, Phys. Rev. B **64**, 184420 (2001).
[35] G. Schmidt, G. Richter, P. Grabs, C. Gould, D. Ferrand, and L. W. Molenkamp, Phys. Rev. B **62**, R4790 (2000).
[36] M. Johnson, Nature **416**, 809 (2002); F. J. Jedema, A. T. Filip, and B. J. van Wees, *ibid.*, 810 (2002).
[37] M. Johnson and J. Byers, Phys. Rev. B **67**, 125112 (2003).
[38] S. Garzon, Ph. D. Thesis (Univ. Maryland, 2005).
[39] T. Kimura, J. Hamrle, and Y. Otani, J. Magn. Soc. Jpn. **29**, 192 (2005).
[40] J. C. Slonczewski, J. Magn. Magn. Mater. **159**, L1 (1996).

[41] L. Berger, Phys. Rev. B **54**, 9353 (1996).
[42] J. A. Katine, F. J. Albert, R. A. Buhrman, E. B. Myers, and D. C. Ralph, Phys. Rev. Lett. **84**, 3149 (2000).
[43] T. Kimura, Y. Otani, and J. Hamrle, cond-mat/0508599.
[44] S. Takahashi, H. Imamura, and S. Maekawa, Phys. Rev. Lett. **82**, 3911 (1999).
[45] S. Takahashi, H. Imamura, and S. Maekawa, J. Appl. Phys. **85**, 5227 (2000).
[46] S. Takahashi, H. Imamura, and S. Maekawa, Physica C **341-348**, 1515 (2000).
[47] J. Johansson, V. Korenivski, D. B. Haviland, and A. Brataas, Phys. Rev. Lett. **93**, 216805 (2004)
[48] M. Tinkham, *Introduction to Superconductivity* (McGraw-Hill, New York, 1996).
[49] J. Bardeen, G. Rickayzen, and L. Tewordt, Phys. Rev. **113**, 982 (1959).
[50] T. Yamashita, S. Takahashi, H. Imamura, and S. Maekawa, Phys. Rev. B **65**, 172509 (2002).
[51] J. P. Morten, A. Brataas, and W. Belzig [Phys. Rev. B **70**, 212508 (2004)] have pointed out that, in the *elastic* transport regime, the spin diffusion length is renormalized in the superconducting state.
[52] S. Takahashi, T. Yamashita, H. Imamura, S. Maekawa, J. Magn. Magn. Mater. **240**, 100 (2002).
[53] Y. Yafet, Phys. Lett. **98**, 287 (1983).
[54] D. Lubzens and S. Schultz, Phys. Rev. Lett. **36**, 1104 (1976).
[55] V. A. Vas'ko, V.A. Larkin, P. A. Kraus, K. R. Nikolaev, D. E. Grupp, C. A. Nordman, and A. M. Goldman, Phys. Rev. Lett. **78**, 1134 (1997).
[56] Z. W. Dong, R. Ramesh, T. Venkatesan, M. Johnson, Z. Y. Chen, S. P. Pai, V. Talyansky, R. P. Sharma, R. Shreekala, C. J. Lobb, and R. L. Greene, Appl. Phys. Lett. **71**, 1718 (1997).
[57] T. Nojima, M. Iwata, T. Hyodo, S. Nakamura, N. Kobayashi, Physica C **412**, 147 (2004).
[58] C. D. Chen, Watson Kuo, D. S. Chung, J. H. Shyu, and C. S. Wu, Phys. Rev. Lett. **88**, 047004 (2002).
[59] J. Johansson, M. Urech, D. Haviland, and V. Korenivski, J. Appl. Phys. **93**, 8650 (2003).
[60] D. Wang and J. G. Lu, J. Appl. Phys. **97**, 10A708 (2005); G. Bergmann, J. Lu, and D. Wang, Phys. Rev. B **71**, 134521 (2005).
[61] T. Daibou, M. Oogane, Y. Ando, and T. Miyazaki, (unpublished).
[62] *The Hall Effect and its Applications*, edited by C. L. Chien and C. R. Westgate (Plenum, New York, 1980).
[63] Y. Taguchi, Y. Oohara, H. Yoshizawa, N. Nagaosa, and Y. Tokura, Science **291**, 2573 (2001).
[64] J. E. Hirsch, Phys. Rev. Lett. **83**, 1834 (1999).
[65] S. Zhang, Phys. Rev. Lett. **85**, 393 (2001).
[66] S. Takahashi and S. Maekawa, Phys. Rev. Lett. **88**, 116601 (2002).
[67] R. Karplus and J. M. Luttinger, Phys. Rev. **95**, 1154 (1954).
[68] S. Murakami, N. Nagaosa, and S.-C. Zhang, Science **301**, 1348 (2003).

[69] J. Sinova, D. Culcer, Q. Niu, N. A. Sinitsyn, T. Jungwirth, and A. H. MacDonald, Phys. Rev. Lett. **92**, 126603 (2002).
[70] Y. K. Kato, R. C. Myers, A. C. Gossard, D. D. Awschalom, Science **306**, 1910 (2004).
[71] J. Wunderlich, B. Kaestner, J. Sinova, and T. Jungwirth Phys. Rev. Lett. **94**, 047204 (2005).
[72] J. Smit, Physica C **24**, 39 (1958).
[73] L. Berger, Phys. Rev. B **2**, 4559 (1970).
[74] A. Crépieux and P. Bruno, Phys. Rev. B **64**, 14416 (2001).
[75] J. J. Sakurai and San Fu Tuan, *Modern Quantum Mechanics* (Addison-Wesley, 1985).
[76] S. K. Lyo and T. Holstein, Phys. Rev. Lett. **29**, 423 (1972).
[77] J. Kondo, Prog. Theor. Phys. **27**, 772 (1964).
[78] J-Ph Ansermet, J. Phys. C **10**, 6027 (1998).
[79] W.-L. Lee, S. Watauchi, V. L. Miller, R. J. Cava, N. P. Ong, Science **303** (2004) 1647.
[80] F. J. Jedema, M. S. Nijboer, A. T. Filip, and B. J. van Wees, Phys. Rev. B **67**, 085319 (2003).
[81] N. W. Ashcroft and D. Mermin, *Solid State Physics* (Saunders College, 1976).
[82] Y. Otani, T. Ishiyama, S.G. Kim, K. Fukamichi, M. Giroud, and B. Pannetier, J. Magn. Magn. Mater. **239**, 135 (2002).
[83] T. Kimura, Y. Otani, K. Tsukagoshi and Y. Aoyagi, J. Magn. Magn. Mater. **272-276**, e1333 (2004).

9 Andreev reflection at ferromagnet/ superconductor interfaces

H. Imamura, S. Takahashi, and S. Maekawa

Andreev reflection (AR) is a scattering process where electrical current is converted to supercurrent at an interface between a normal metal and a superconductor [1]. When an electron near the Fermi surface is incident on a superconductor from a normal metal, the electron is reflected as a hole with opposite group velocity. AR in non-magnetic-metal/superconductor (NM/SC) junctions has been extensively studied both theoretically and experimentally in the last two decades. Comprehensive reviews on AR in hybrid superconducting nanostructures was published by Lambert and Raimondi [2] and by Beenakker [3].

Recently much attention has been focused on AR in nanohybrids of ferromagnet and superconductor for a variety of different reasons. AR in a ferromagnet/superconductor (FM/SC) point contact provides a powerful tool for determining the spin polarization of conduction electrons [4–6]. The coherence length of a superconductor can be estimated by measuring the tunnel magnetoresistance of a FM/SC/FM double junction system [7, 8]. A novel scattering process called "crossed AR" is predicted for a SC with two ferromagnetic leads, where AR occurs even if the ferromagnetic leads are half metallic.

In this chapter recent progress in AR in magnetic nanohybrids is reviewed. In Section 9.1 we give an introduction of the Blonder, Tinkham, and Klapwijk (BTK) theory [9] for a FM/SC junction. In Section 9.2 the experimental and theoretical work on FM/SC point contacts are presented. In Section 9.3 we show that the coherence length of a superconductor can be estimated by analyzing the tunnel magnetoresistance of a FM/SC/FM double junction system. Crossed AR in a superconductor connected with two ferromagnetic leads is discussed in Section 9.4.

9.1 Basic theory of Andreev reflection

We first introduce the basic theory of AR for a FM/SC junction [9]. We assume that the current flows along the z-axis and the system has translational symmetry in the transverse (x and y) direction and therefore the wave vector parallel to the interface $\boldsymbol{k}_{\parallel} = (k_x, k_y)$ is conserved. Electrons in the FM, which is the region $z < 0$, are described by the Stoner model with exchange field h_0. Quasiparticle excitations in the SC, which is the region $z > 0$, are given by the usual Bogoliubov transformation [10]. The current flowing through the junction is obtained by taking the difference of the distributions between the right-going electrons, $f_\sigma^{\rightarrow}(E)$, and left-going ones, $f_\sigma^{\leftarrow}(E)$, as

$$I = \frac{e}{h} \sum_{\boldsymbol{k}_\parallel, \sigma} \int_{-\infty}^{\infty} \left[f_{\boldsymbol{k}_\parallel,\sigma}^{\rightarrow}(E) - f_{\boldsymbol{k}_\parallel,\sigma}^{\leftarrow}(E) \right] dE, \tag{9.1}$$

where h is the Plank constant, e is the charge of the electron, E is the energy measured from the Fermi energy μ_{F}, and $\sigma = +(-)$ for the spin up (down) band. To avoid cumbersome notation, the subscript $\boldsymbol{k}_\parallel$ is dropped from now on.

The distribution functions $f_\sigma^{\rightarrow}(E)$ and $f_\sigma^{\leftarrow}(E)$ are obtained by solving the scattering problem described by the following Bogoliubov–de Gennes (BdG) equation [11]:

$$\begin{pmatrix} H_0 - h_{\mathrm{ex}}(z)\sigma & \Delta(z) \\ \Delta(z) & -H_0 - h_{\mathrm{ex}}(z)\sigma \end{pmatrix} \psi_\sigma(z) = E\psi_\sigma(z). \tag{9.2}$$

The single-particle Hamiltonian H_0 is defined as

$$H_0 \equiv -\frac{\hbar^2}{2m}\frac{\partial^2}{\partial z^2} + \frac{\hbar^2}{2m}k_\parallel^2 - \mu_{\mathrm{F}} + \frac{\hbar^2 k_{\mathrm{F}}}{m} Z\delta(z), \tag{9.3}$$

where $\hbar \equiv h/2\pi$, m is the effective mass of the electron, and $k_{\mathrm{F}} \equiv \sqrt{2m\mu_{\mathrm{F}}}/\hbar$ is the Fermi wavenumber. The last term in Eq. (9.3) represents a δ-function type scattering potential at the interface characterized by the dimensionless parameter Z. The exchange field $h_{\mathrm{ex}}(z)$ is given by

$$h_{\mathrm{ex}}(z) = \begin{cases} h_0 & (z < 0) \\ 0 & (0 < z) \end{cases} \tag{9.4}$$

and the superconducting gap is

$$\Delta(z) = \begin{cases} 0 & (z < 0) \\ \Delta & (0 < z). \end{cases} \tag{9.5}$$

The solution of the BdG equation takes the form

$$\psi_\sigma = \begin{pmatrix} f_\sigma(z) \\ g_\sigma(z) \end{pmatrix}, \tag{9.6}$$

where $f_\sigma(z)$ and $g_\sigma(z)$ are the electron and hole components of the wavefunction, respectively. Substituting Eq. (9.6) into Eq. (9.2) the general solutions of the BdG equation in the SC are obtained as

$$\psi_{\pm k^+}(z) = \begin{pmatrix} u_0 \\ v_0 \end{pmatrix} e^{\pm ik^+ z}, \quad \psi_{\pm k^-}(z) = \begin{pmatrix} v_0 \\ u_0 \end{pmatrix} e^{\pm ik^- z}, \tag{9.7}$$

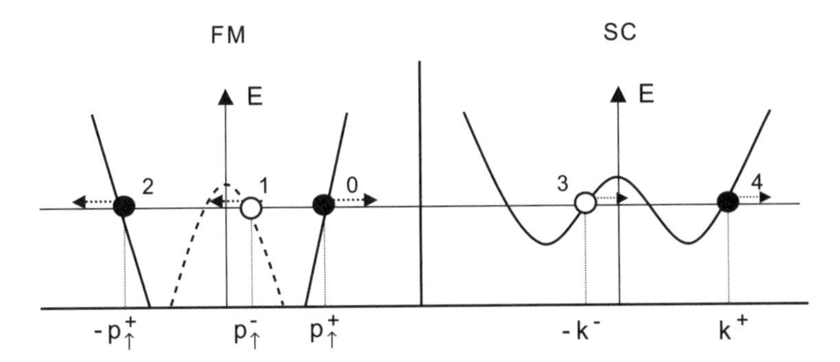

FIG. 9.1. Schematic diagram of excitation energy vs. momentum at the FM/SC interface. Open circles denote holes, filled circles represent electrons, and arrows point in the direction of the group velocity. This figure describes an incident electron at (0), along with resulting reflected (1,2) and transmitted (3,4) particles.

where u_0 and v_0 are the coherence factors defined as

$$u_0^2 = 1 - v_0^2 = \frac{1}{2}\left[1 + \frac{\sqrt{E^2 - \Delta^2}}{E}\right], \tag{9.8}$$

and $k^{+(-)}$ is the z component of the wavenumber of an electron-(hole-)like quasiparticle:

$$k^{\pm} = \frac{\sqrt{2m}}{\hbar}\sqrt{\mu_{\mathrm{F}} - \hbar^2 k_{\parallel}^2/2m \pm \sqrt{E^2 - \Delta^2}}. \tag{9.9}$$

In the FM, the general solutions can be written in the form

$$\psi_{\pm p_\sigma^+}(z) = \begin{pmatrix} 1 \\ 0 \end{pmatrix} e^{\pm i p_\sigma^+ z}, \quad \psi_{\pm p_\sigma^-}(z) = \begin{pmatrix} 0 \\ 1 \end{pmatrix} e^{\pm i p_\sigma^- z}, \tag{9.10}$$

where $p_\sigma^{+(-)}$ is the z component of the wavenumber of an electron (hole) with spin σ:

$$p_\sigma^{\pm} = \frac{\sqrt{2m}}{\hbar}\sqrt{\mu_{\mathrm{F}} - \hbar^2 k_{\parallel}^2/2m \pm E \pm \sigma h_{\mathrm{ex}}}. \tag{9.11}$$

Let us consider the scattering of an electron with spin up injected into the SC from the FM. We have four scattering processes as shown in Fig. 9.1 in which the incident electron is indicated by 0. The processes 1, 2, 3, and 4 correspond to AR, normal reflection, transmission as a hole-like quasiparticle, and transmission as an electron-like quasiparticle, respectively . The wavefunction in the FM is expressed as the sum of an incident wave, Andreev reflected wave, and normal reflected wave:

$$\psi_\sigma^{\rm FM}(z) = \begin{pmatrix}1\\0\end{pmatrix} e^{ip_\sigma^+ z} + a_\sigma \begin{pmatrix}0\\1\end{pmatrix} e^{ip_\sigma^- z} + b_{\sigma,nl} \begin{pmatrix}1\\0\end{pmatrix} e^{-ip_\sigma^+ z}, \tag{9.12}$$

where a_σ and b_σ are coefficients for AR and normal reflection, respectively. In the SC we have

$$\psi_\sigma^{\rm SC}(z) = c_\sigma \begin{pmatrix}u_0\\v_0\end{pmatrix} e^{ik^+ z} + d_\sigma \begin{pmatrix}v_0\\u_0\end{pmatrix} e^{-ik^- z}, \tag{9.13}$$

where c_σ and d_σ are the coefficients for transmission waves as a hole-like quasiparticle and as an electron-like quasiparticle, respectively. The coefficients $a_\sigma, b_\sigma, c_\sigma$, and d_σ are determined by matching the wavefunctions $\psi_\sigma^{\rm FM}$ and $\psi_\sigma^{\rm SC}$ at the interface:

$$\begin{cases} \psi_\sigma^{\rm FM}(0) = \psi_\sigma^{\rm SC}(0) \\ \left.\dfrac{d\psi_\sigma^{\rm SC}}{dz}\right|_{z=0} - \left.\dfrac{d\psi_\sigma^{\rm FM}}{dz}\right|_{z=0} = 2k_{\rm F} Z \psi_\sigma^{\rm FM}(0) . \end{cases} \tag{9.14}$$

Once we obtain the wavefunction by substituting the solution of Eq. (9.14) into Eqs(9.12) and (9.13), we can evaluate the probabilities for each scattering process. The probability density $P(z)$ for finding either an electron or a hole at a particular time and place is expressed as $P_\sigma(z) = |f_\sigma(z)|^2 + |g_\sigma(z)|^2$, where $f_\sigma(z)$ and $g_\sigma(z)$ are the electron and hole components of the wavefunction in Eq. (9.6). The probability density satisfies the following conservation relation

$$\frac{d}{dt}P + \frac{d}{dx}J_P = 0, \tag{9.15}$$

where the probability current density is defined as

$$J_P = \frac{\hbar}{m}{\rm Im}\left(f^* \nabla f - g^* \nabla g\right). \tag{9.16}$$

The probability current density for the incident electron is given by

$$J_0 = \frac{\hbar}{m} p_\sigma^+, \tag{9.17}$$

and those for the four scattering processes are

$$J_1 = -\frac{\hbar}{m} p_\sigma^- \left|a_\sigma\right|^2, \tag{9.18}$$

$$J_2 = -\frac{\hbar}{m} p_\sigma^+ \left|b_\sigma\right|^2, \tag{9.19}$$

$$J_3 = \frac{\hbar}{m}\left({\rm Re}\, k^+\right) \left|c_\sigma\right|^2 \left(u_0^2 - v_0^2\right), \tag{9.20}$$

$$J_4 = \frac{\hbar}{m}\left({\rm Re}\, k^-\right) \left|c_\sigma\right|^2 \left(u_0^2 - v_0^2\right) \left|a_\sigma\right|^2. \tag{9.21}$$

Following the BTK model [9] we introduce the probabilities $A_\sigma(E)$, $B_\sigma(E)$, $C_\sigma(E)$, and $D_\sigma(E)$ for the processes 1, 2, 3, and 4, respectively. From Eqs(9.18)–(9.21) we have

$$A_\sigma \equiv \left|\frac{J_1}{J_0}\right| = \frac{p_\sigma^-}{p_\sigma^+}\left|a_\sigma\right|^2, \tag{9.22}$$

$$B_\sigma \equiv \left|\frac{J_2}{J_0}\right| = \left|b_\sigma\right|^2, \tag{9.23}$$

$$C_\sigma \equiv \left|\frac{J_3}{J_0}\right| = \frac{\mathrm{Re}\,k^+}{p_\sigma^+}\left|c_\sigma\right|^2\left(u_0^2 - v_0^2\right), \tag{9.24}$$

$$D_\sigma \equiv \left|\frac{J_4}{J_0}\right| = \frac{\mathrm{Re}\,k^-}{p_\sigma^+}\left|d_\sigma\right|^2\left(u_0^2 - v_0^2\right). \tag{9.25}$$

The probabilities for $E < 0$ can be obtained in a similar way.

Suppose that the bias voltage, V, is applied and the Fermi level of the ferromagnetic lead is shifted by eV. The distribution function of right-going electrons is given by

$$f_\sigma^{\rightarrow}(E) = f_0(E - eV), \tag{9.26}$$

where $f_0(E)$ is the Fermi–Dirac distribution function. The distribution function of the left-going electrons can be described as

$$\begin{aligned} f_\sigma^{\leftarrow}(E) = {} & A_\sigma(E)[1 - f_0(-E - eV)] + B_\sigma(E) f_0(E - eV) \\ & + [C_\sigma(E) + D_\sigma(E)] f_0(E). \end{aligned} \tag{9.27}$$

Substituting Eqs (9.26) and (9.27) into Eq. (9.1) we have

$$\begin{aligned} I = \frac{e}{h}\sum_{\boldsymbol{k}_\parallel,\sigma}\int_{-\infty}^{\infty} & [f_0(E - eV) - A_\sigma(E)[1 - f_0(-E - eV)] \\ & - B_\sigma(E) f_0(E - eV) - [C_\sigma(E) + D_\sigma(E)]\, f_0(E)\; dE. \end{aligned} \tag{9.28}$$

The probabilities $C_\sigma(E)$ and $D_\sigma(E)$ can be eliminated by using the relation $A_\sigma(E) + B_\sigma(E) + C_\sigma(E) + D_\sigma(E) = 1$. Then, the conductance is given by

$$G = \frac{dI}{dV} = \frac{e}{h}\sum_{\boldsymbol{k}_\parallel,\sigma}\int_{-\infty}^{\infty}\frac{d\,f_0(E - eV)}{dV}\left[1 + A_\sigma(-E) - B_\sigma(E)\right] dE. \tag{9.29}$$

At zero temperature the derivative of the Fermi distribution function can be replaced by the δ-function and the conductance is expressed as

$$G = \frac{e^2}{h}\sum_{\boldsymbol{k}_\parallel,\sigma}\left[1 + A_\sigma(-eV) - B_\sigma(eV)\right]. \tag{9.30}$$

9.2 Point-contact Andreev reflection

In this section we review recent work on AR in a ballistic FM/SC point contact, which is known as a powerful technique for measuring the spin polarization of conduction electrons. Usually the spin polarization of the FM is determined by the spin-dependent tunneling technique [12]. However, this technique is restricted to magnetic materials which can be incorporated into a tunnel junction with a superconductor and requires a complex fabrication process. In 1998, a new technique called point-contact Andreev reflection (PCAR) for measuring the spin polarization of a ferromagnet was proposed by Soulen *et al.* [5, 13] and independently by Upadhyay *et al.* [6]. The PCAR technique is much easier to put into practice than the tunneling technique. With no restrictions on the sample geometry, one can avoid the complex fabrication steps. They put a superconducting tip on a ferromagnet and measure the bias voltage dependence of the normalized conductance. By analyzing the observed normalized conductance curves they can estimate the spin polarization of the FM [14–17].

The first intuitive and simple description of the conductance through a ballistic FM/SC point contact was presented by de Jong and Beenakker [4]. They consider a FM/SC point contact where the FM is contacted through a small area with a SC. In the contact region the number of spin-up transmitting channels ($N_\uparrow$) is larger than that of spin-down transmitting channels ($N_\downarrow$), i.e., $N_\uparrow \geq N_\downarrow$. They suppose that there is no partially transmitting channels and neglect mixing of channels for simplicity. When the SC is in the normal conducting state, all scattering channels (transverse modes in the point contact at the Fermi level) are fully transmitted, yielding the conductance

$$G_{\rm FN} = \frac{e^2}{\hbar}\left(N_\uparrow + N_\downarrow\right). \tag{9.31}$$

When the SC is in the superconducting state, all the spin-down incident electrons in $N_\downarrow$ channels are Andreev reflected and give a double contribution to the conductance, $2(e^2/\hbar)N_\downarrow$. However, spin-up incident electrons in some channels cannot be Andreev reflected since the density of states for spin-down electrons is smaller than that for spin-up electrons. Then only a fraction ($N_\downarrow/N_\uparrow$) of the $N_\uparrow$ channels can be Andreev reflected and the resulting conductance is $2(e^2/\hbar)(N_\downarrow/N_\uparrow)N_\uparrow$. The total conductance at zero bias voltage ($V = 0$) is given by the sum of these two contributions:

$$G_{\rm FS} = \frac{e^2}{\hbar}2\left(N_\downarrow + \frac{N_\downarrow}{N_\uparrow}N_\uparrow\right) = 4N_\downarrow. \tag{9.32}$$

Taking the ratio of Eq. (9.32) to Eq. (9.31), we obtain the normalized conductance

$$\frac{G_{\rm FS}}{G_{\rm FN}} = 2\left(1 - P\right), \tag{9.33}$$

where P is the spin polarization of transmitting channels defined as

$$P \equiv \frac{N_\uparrow - N_\downarrow}{N_\uparrow + N_\downarrow}. \tag{9.34}$$

Equation (9.33) shows that the normalized conductance is a monotonic decreasing function of P and it vanishes if the FM is half-metallic ($P = 1$).

In PCAR experiments, however, the unknown parameters to be determined are not only the spin polarization but the strength of the interfacial scattering. It is necessary to analyze the full bias voltage dependence of the normalized conductance on the basis of the BTK theory [5, 6, 13–16] to obtain the spin polarization. The conduction across the FM/SC interface can be divided into an unpolarized part (non-magnetic channel) and a completely polarized part (half-metallic channel). By applying the BTK theory the conductance for the non-magnetic channel is obtained as

$$G_N(V, Z) \propto \int_{-\infty}^{\infty} \frac{d\, f_0(E - V, T)}{dV} [1 + A_N(E, Z) - B_N(E, Z)]\, dE, \tag{9.35}$$

where $f_0(E - V, T)$ is the Fermi distribution function and A_N and B_N are probabilities for the Andreev reflection and normal reflection, respectively. At zero temperature the derivative of the Fermi distribution function can be replaced by the delta function $\delta(E - V)$ and we have

$$G_N(V, Z) \propto 1 + A_N(V, Z) - B_N(V, Z). \tag{9.36}$$

The r.h.s. of Eq. (9.36) is given by

$$1 + A_N(V, Z) - B_N(V, Z) = \begin{cases} \dfrac{2(1 + \beta^2)}{\beta^2 + (1 + 2Z^2)^2}, & |eV| < \Delta, \\[2ex] \dfrac{2\beta}{1 + \beta + 2Z^2}, & |eV| > \Delta, \end{cases} \tag{9.37}$$

where $\beta \equiv E/(\sqrt{|\Delta^2 - E^2|})$ and Z is the dimensionless height of the interfacial scattering potential. As for the half-metallic channels the conductance can be obtained in a similar manner as

$$G_H(V, Z) \propto 1 + A_H(V, Z) - B_H(V, Z), \tag{9.38}$$

where

$$1 + A_H(V, Z) - B_H(V, Z) = \begin{cases} 0, & |eV| < \Delta, \\[2ex] \dfrac{4\beta}{(1 + \beta)^2 + 4Z^2}, & |eV| > \Delta. \end{cases} \tag{9.39}$$

Introducing the spin polarization for PCAR defined by the spin-resolved current as $P \equiv (I_\uparrow - I_\downarrow)/(I_\uparrow + I_\downarrow)$, the total conductance of the FM/SC point contact can be expressed as

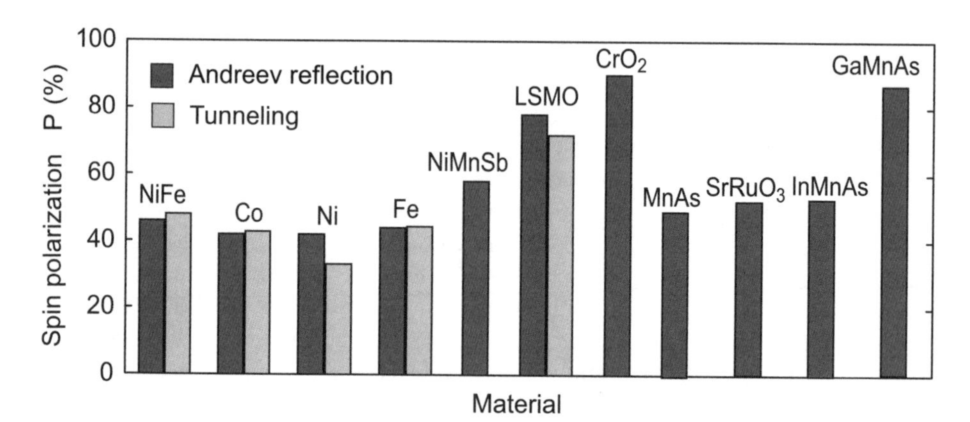

FIG. 9.2. Measured values of the spin polarization P for various materials. Results of NiFe, Co, Ni, Fe, NiMnSb, LSMO, and CrO_2 are adapted from Ref. [5]. Those for MnAs, $SrRu0_3$, InMnAs, and GaMnAs are from Refs. [18], [19], [20], and [21], respectively.

$$G(V, Z, P) = (1 - P)G_N(V, Z) + PG_H(V, Z). \tag{9.40}$$

The spin polarization P and the barrier height Z are estimated by fitting the experimentally observed conductance curve using Eqs(9.36)–(9.40). The experimental results of P by performing the PCAR measurement are shown in Fig. 9.2. For some materials we also plot the spin polarization obtained by tunneling experiments. The agreement between the two methods is generally good.

It is important to note that since the system we consider is a point contact, the spin polarization P measured by PCAR is not the spin polarization of the ferromagnetic material but is determined by the spin-up and spin-down current flowing through the contact region. If the contact region is superconducting, the number of transmitting channels is the same for both spin-up and spin-down quasiparticles and the estimated spin polarization should be different from that for the ferromagnetic contact where the contact region is ferromagnetic. The material of the contact region depends on the details of each experiment. Therefore, we need to clarify the difference between the superconducting and ferromagnetic contact [22, 23]. To see the difference, we numerically solve the scattering problem of the FM/SC point contact consisting of three co-axial cylinders as shown in Figs 9.3(a) and (b). The left and right electrodes with diameter $W_{\rm E}$ are connected by a contact with diameter $W_{\rm C}(< W_{\rm E})$ and length D. We employ the Stoner model with exchange field h_0 for the ferromagnet. Since the system has cylindrical symmetry, the wavefunctions Ψ_σ can be expressed as

$$\Psi_\sigma(\boldsymbol{r}) = \sum_{n,l} N^l_{\sigma n} \psi^n_{\sigma l}(z) J_n \left(\frac{2\gamma_{nl}}{W_{\rm E(C)}} r \right) e^{in\phi}, \tag{9.41}$$

where γ_{nl} is the lth zero of the Bessel function $J_n(r)$, and $N^l_{\sigma n}$ is the normalization constant [24]. The set of quantum numbers (n, l, σ) defines the channel. For convenience we introduce the abbreviations F/S/S and F/F/S to refer to the superconducting and ferromagnetic contacts, respectively. Assuming that an electron in channel (n, l, σ) is incident from the left electrode, the wavefunction in the left electrode in the F/S/S system is given by

$$\Psi^{\mathrm{L}}(z) = e^{iq^{n+}_{\sigma l} z} \begin{pmatrix} 1 \\ 0 \end{pmatrix} J_n\left(\frac{2\gamma_{nl}}{W_{\mathrm{E}}} r\right) e^{in\phi} + \sum_{s=1}^{M_{\mathrm{E}}} \left[a^n_{\sigma ls} e^{iq^{n-}_{\sigma s} z} \begin{pmatrix} 0 \\ 1 \end{pmatrix} + b^n_{\sigma ls} e^{-iq^{n+}_{\sigma s} z} \begin{pmatrix} 1 \\ 0 \end{pmatrix} \right] J_n\left(\frac{2\gamma_{ns}}{W_{\mathrm{E}}} r\right) e^{in\phi}. \tag{9.42}$$

In the contact region $(-D/2 < z < D/2)$,

$$\Psi^{\mathrm{C}}(z) = \sum_{s=1}^{M_{\mathrm{C}}} \left[\left(\alpha^n_{\sigma ls} e^{ip^{n+}_s z} + \beta^n_{\sigma ls} e^{-ip^{n+}_s z} \right) \begin{pmatrix} u_0 \\ v_0 \end{pmatrix} + \left(\xi^n_{\sigma ls} e^{ip^{n-}_s z} + \eta^n_{\sigma ls} e^{-ip^{n-}_s z} \right) \begin{pmatrix} v_0 \\ u_0 \end{pmatrix} \right] J_n\left(\frac{2\gamma_{ns}}{W_{\mathrm{C}}} r\right) e^{in\phi}. \tag{9.43}$$

In the right electrode $(z > D/2)$,

$$\Psi^{R}(z) = \sum_{s=1}^{M_{\mathrm{E}}} \left[c^n_{\sigma ls} e^{ik^{n+}_s z} \begin{pmatrix} u_0 \\ v_0 \end{pmatrix} + d^n_{\sigma ls} e^{-ik^{n-}_s z} \begin{pmatrix} v_0 \\ u_0 \end{pmatrix} \right] J_n\left(\frac{2\gamma_{ns}}{W_{\mathrm{E}}} r\right) e^{in\phi}. \tag{9.44}$$

Here, the wavevectors below the superconducting energy gap $(E \leq \Delta)$ are given by

$$q^{n\pm}_{\sigma l} = \sqrt{\frac{2m}{\hbar^2} (\mu_{\mathrm{F}} \pm h_0 \sigma \pm E) - \left(\frac{2\gamma_{nl}}{W_{\mathrm{E}}}\right)^2}, \tag{9.45}$$

$$p^{n\pm}_l = \sqrt{\frac{2m}{\hbar^2} \left(\mu_{\mathrm{F}} \pm \sqrt{E^2 - \Delta^2}\right) - \left(\frac{2\gamma_{nl}}{W_{\mathrm{C}}}\right)^2}, \tag{9.46}$$

$$k^{n\pm}_l = \sqrt{\frac{2m}{\hbar^2} \left(\mu_{\mathrm{F}} \pm \sqrt{E^2 - \Delta^2}\right) - \left(\frac{2\gamma_{nl}}{W_{\mathrm{E}}}\right)^2}. \tag{9.47}$$

The wavevectors above the superconducting energy gap $(E > \Delta)$ are the same as those given by Eq. (9.47) except for the wavevector k^{n-}_l in the right electrode. In order to deal with the evanescent waves, we write the wavevector k^{n-}_l as

$$k^{n-}_l = \left(\sqrt{\frac{2m}{\hbar^2} \left(\mu_{\mathrm{F}} - \sqrt{E^2 - \Delta^2}\right) - \left(\frac{2\gamma_{nl}}{W_{\mathrm{E}}}\right)^2} \right)^*. \tag{9.48}$$

The number of channels in the electrode and contact is truncated by the cutoff constant M_{E} and M_{C}, respectively. The cutoff constants are taken to be large enough to express the stationary scattering state.

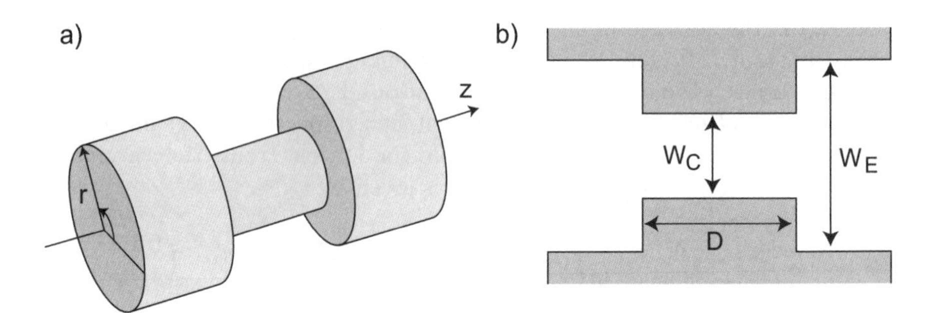

FIG. 9.3. (a) Geometry of the point contact which is represented by the coaxial cylinders. (b) Cross-section along the z-direction. The length and width of the contact region are D and W_{C}, respectively. The width of the electrodes is W_{E}[22].

For the F/F/S system, the wavefunction inside the contact is rewritten as

$$\Psi^{\mathrm{C}} = \sum_{s=1}^{M_{\mathrm{C}}} \left[\left(\alpha^n_{\sigma ls} e^{ip^{n+}_{\sigma s} z} + \beta^n_{\sigma ls} e^{-ip^{n+}_{\sigma s} z} \right) \begin{pmatrix} 1 \\ 0 \end{pmatrix} + \left(\xi^n_{\sigma ls} e^{ip^{n-}_{\bar{\sigma} s} z} + \eta^n_{\sigma ls} e^{-ip^{n-}_{\bar{\sigma} s} z} \right) \begin{pmatrix} 0 \\ 1 \end{pmatrix} \right] \times J_n \left(\frac{2\gamma_{ns}}{W_{\mathrm{C}}} r \right) e^{in\phi},$$

where the wavevectors $p^{n\pm}_{\sigma l}$ are given by

$$p^{n\pm}_{\sigma l} = \sqrt{\frac{2m}{\hbar^2} (\mu_{\mathrm{F}} \pm h_0 \sigma \pm E) - \left(\frac{2\gamma_{nl}}{W_{\mathrm{C}}} \right)^2 }. \tag{9.49}$$

The coefficients $a^n_{\sigma ls}$, $b^n_{\sigma lm}$, $c^n_{\sigma ls}$, $d^n_{\sigma ls}$, $\alpha^n_{\sigma ls}$, $\beta^n_{\sigma ls}$, $\xi^n_{\sigma ls}$ and $\eta^n_{\sigma ls}$ are obtained by matching the slope and value of the wavefunction at the boundary of the contact $z = \pm\frac{D}{2}$ [25,26]. The probabilities for the AR, $A^n_{\sigma ls}$, and the ordinary reflection, $B^n_{\sigma ls}$, can be obtained in a similar manner as in Section 9.1. The conductance is expressed by using the probabilities $A^n_{\sigma ls}$ and $B^n_{\sigma ls}$ as

$$G = \frac{e}{h} \sum_{\sigma=\pm} \sum_{n,l} \int_{-\infty}^{\infty} \frac{d}{dV} f_0(E - eV, T) \times \left[1 + \sum_s A^n_{\sigma ls}(-E) - \sum_s B^n_{\sigma ls}(E) \right] dE. \tag{9.50}$$

In numerical calculation, the cutoff constants M_{C} and M_{E} are taken to be three times as many as the number of open channels which are determined by the condition that $\mathrm{Re}[(2m/\hbar^2)(\mu_{\mathrm{F}} - \sqrt{E^2 - \Delta^2}) - (2\gamma_{nl}/W_{\mathrm{C(E)}})^2] > 0$. The diameter of the electrodes and the length of the contact are taken to be $W_{\mathrm{E}} = 60.8/k_{\mathrm{F}}$ and $D = 5.0/k_{\mathrm{F}}$, respectively. The superconducting energy gap is assumed to be

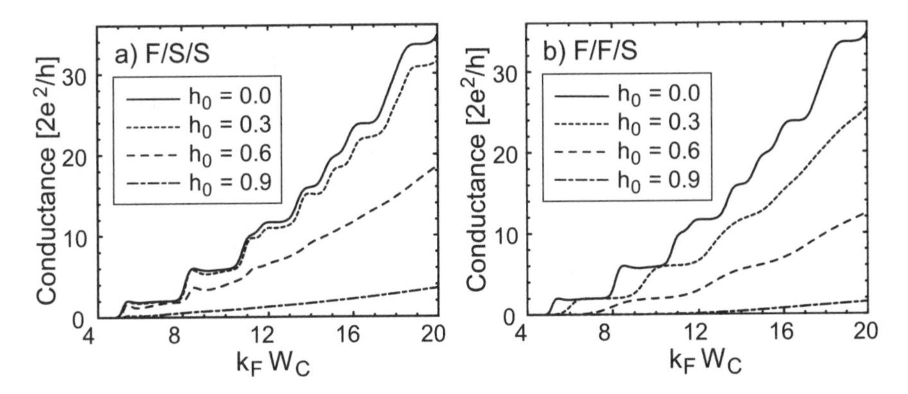

FIG. 9.4. (a) Zero-bias conductance for the F/S/S point contact at $T = 0$. No interfacial scattering is considered. The horizontal axis represents the width of the contact $W_{\rm C}$ multiplied by the Fermi wavenumber $k_{\rm F}$. The solid, dotted, dashed and dot-dashed lines represent the conductance curves for $h_0 = 0, 0.3, 0.6$ and $0.9\mu_{\rm F}$, respectively at $T = 0$. The superconducting gap of SC is taken to be $\Delta = 1.5 \times 10^{-5}\mu_{\rm F}$. (b) The same plot for the F/F/S point contact[22].

$\Delta/\mu_{\rm F} = 1.5 \times 10^{-5}$, which is of the same order of that for Al [27]. The position of the interfacial scattering is located at $z_0 = -D/2(+D/2)$ for the F/S/S (F/F/S) system. The temperature is assumed to be zero and the superconducting gap of SC is taken to be $\Delta = 1.5 \times 10^{-5}\mu_{\rm F}$.

In Fig. 9.4(a) the zero-bias conductance of the F/S/S system is plotted against the width of contact $W_{\rm C}$. The plateau of quantized conductance disappear as the exchange fields h_0 increase due to the mismatch of the Fermi wavelength. In the adiabatic picture, the number of transmitting channels is determined by the condition that $\mathrm{Re}[(2m/\hbar^2)(\mu_{\rm F} - \sqrt{E^2 - \Delta^2}) - (2\gamma_{nl}/W_{\rm C})^2] > 0$ and does not depend on the strength of the exchange field in the ferromagnetic electrode. Therefore, the positions of the conductance steps do not shift if the exchange field h_0 increases. The conductance for the F/F/S system also decreases as the exchange field increases as shown in Fig. 9.4(b). Note that the width $W_{\rm C}$ at which the new transmitting channel opens increases with increasing exchange field, h_0. The shift of the conductance steps can be explained as follows. In the ferromagnetic contact, electrons in the spin-up and spin-down bands feel the different exchange field $-h_0$ and h_0, respectively. Therefore, the number of transmitting spin-down channels $N_\downarrow$ is smaller than that of transmitting spin-up channels $N_\uparrow$. As pointed out by de Jong and Beenakker [3], the number of channels contributing to AR is restricted by $N_\downarrow$. In the adiabatic picture [28], $N_\downarrow$ is determined by the condition that $(2m/\hbar^2)(\mu_{\rm F} - E - h_0) - (2\gamma_{nl}/W_{\rm C})^2 > 0$. Therefore, the number of transmitting channels for the same contact width decreases with increasing the exchange field. In the narrow F/F/S system, the suppression of AR discussed by de Jong and Beenakker appears as the shift of the width of

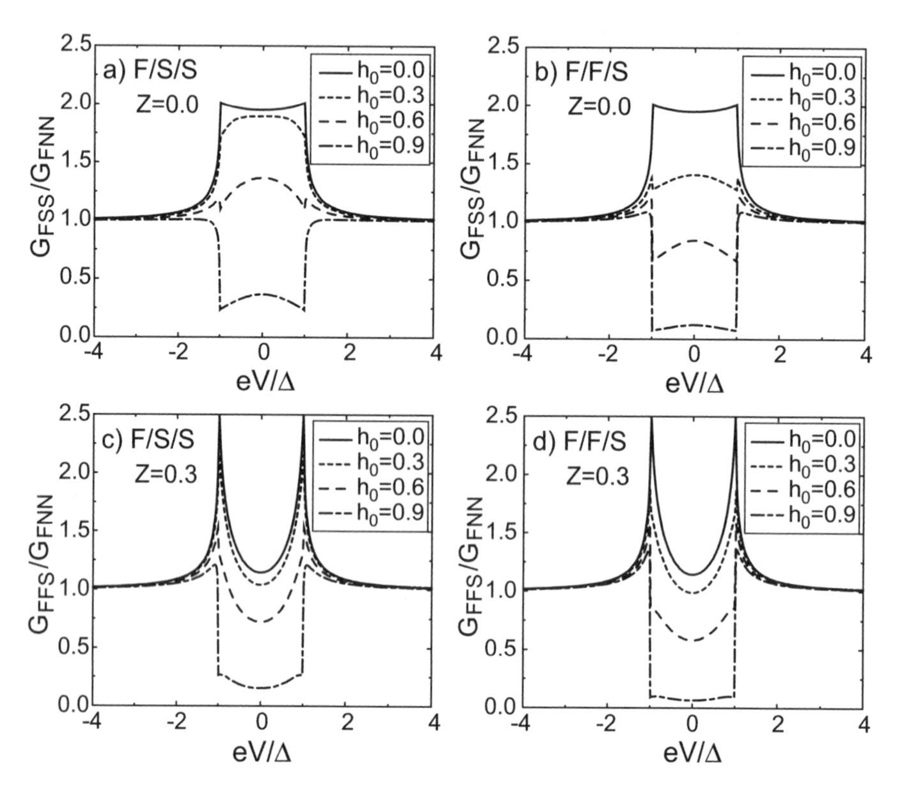

FIG. 9.5. Conductance for the F/S/S and F/F/S point contacts is plotted against the bias voltage at $T = 0$. The width of the contact is taken to be $k_F W_C = 18.5$. The strength of the interfacial scattering Z is assumed to be zero for panels (a) and (b), and $Z = 0.3$ for panels (c) and (d). The horizontal axis represents the applied bias voltage normalized by the superconducting gap Δ. The solid, dotted, dashed and dot-dashed lines represent the conductance curves for $h_0 = 0, 0.3, 0.6$ and $0.9\mu_F$, respectively. The temperature is assumed to be zero and the superconducting gap of the SC is taken to be $\Delta = 1.5 \times 10^{-5}\mu_F$[22].

the contact W_C at which the new transmitting channel opens as shown in Fig. 9.4(b). Even when the contact is superconducting (F/S/S), the conductance is suppressed due to the mismatch of the Fermi wavelength as shown in Fig. 9.4(a). However, the magnitude of suppression is smaller than that for the F/F/S system.

Figures 9.5(a) and (c) show the conductance (G)-voltage (V) curves for the F/S/S system, and Figs 9.5(b) and (d) show those for the F/F/S system. The conductance is normalized by that for the FM/NC point contact. One can see that the normalized conductance for the F/F/S system is smaller than that for the F/S/S system since the number of transmitting channels is limited by AR.

This implies that the spin-polarization obtained by analyzing the PCAR experiment depends on whether the contact region is ferromagnetic or superconducting. If the contact region is superconducting, we would underestimate the spin polarization of the FM. The G-V curves for the contact with an interfacial scattering potential of $Z = 0.3$ are also plotted in Figs 9.5(c) and (d). Similar to the case of $Z = 0$, AR for the F/F/S system is more suppressed than for the F/S/S system. Without the interfacial scattering potential, $Z = 0$, the G-V curve has a bump at zero bias voltage as shown in Figs 9.5(a) and (b). However, the G-V curve for $Z = 0.3$ shows a dip as shown in Figs 9.5(c) and (d). These dips and bumps in the G-V curve are similar to the energy dependence of conductance for F/S tunnel junctions discussed by Žutić and Valls [29, 30].

9.3 Ferromagnet/superconductor/ferromagnet double junctions

Andreev reflection includes a conversion process of the quasiparticle (QP) current to the supercurrent carried by Cooper pairs in the range of the penetration depth, which is approximately equal to the Ginzburg–Landau (GL) coherence length [10], from the FM/SC interface [9]. In a FM/SC single-junction system, injected quasiparticles are completely converted into Cooper pairs whose spin wavefunction is singlet. However, if the SC is sandwiched by two ferromagnetic leads and the thickness of the SC layer is smaller than the GL coherence length the spin information carried by quasiparticles can be transmitted to the other ferromagnet. Recently, Gu *et al.* observed the magnetoresistance due to the overlap of the QP penetration in the SC by AR [7]. Their results are well explained by the BKT theory [31].

In this section we review the theory of AR in the FM/SC/FM double junction system and derive an expression for the current through the junction. We show that the recent experimental results by Gu *et al.* [7] agree very well with the BTK theory and, thus, the information about the coherence length of SC is obtained by analyzing the excess resistance. We consider the FM1/SC/FM2 double junction system consisting of three rectangular blocks as shown in Figs 9.6(a) and (b) [31]. The cross-section of the system is a square with side W and the thickness of the SC is L. The current flows along the z-axis and the interfaces between FM1/SC and SC/FM2 are located at $z = -L/2$ and $z = L/2$, respectively. For simplicity, we assume that the system is symmetric: FM1 and FM2 are made of the same ferromagnetic materials and the potentials for the left and right interfaces are the same. The system is described by the Bogoliubov–de Gennes (BdG) equation(9.2). The exchange field $h_{\rm ex}(z)$ is given by

$$h_{\rm ex}(z) = \begin{cases} h_0 & (z < -L/2), \\ 0 & (-L/2 < z < L/2), \\ \pm h_0 & (L/2 < z), \end{cases} \tag{9.51}$$

where $+h_0$ and $-h_0$ represent the exchange fields for the parallel and antiparallel alignments, respectively. The superconducting gap is expressed as

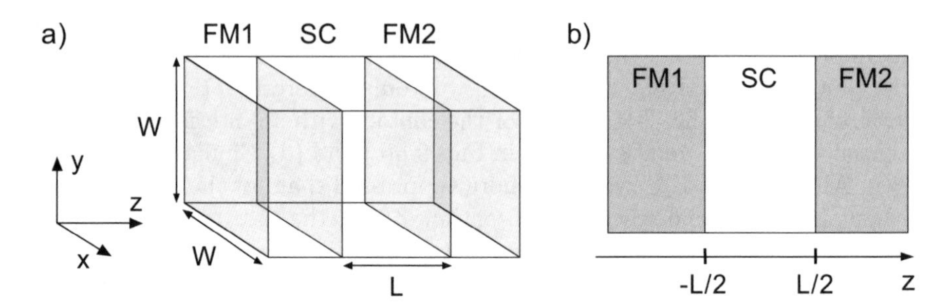

FIG. 9.6. (a) Schematic diagram of a ferromagnet/superconductor/ferromagnet (FM1/SC/FM2) double junction system. A superconductor with a thickness of L is sandwiched by two semi-infinite ferromagnetic electrodes. The system is rectangular and the cross-section is a square with side W. (b) The current flows along the z-axis. The interfaces between FM1/SC and SC/FM2 are located at $z = -L/2$ and $z = L/2$, respectively [31].

$$\Delta(z) = \begin{cases} 0 \ (z < -L/2, L/2 < z), \\ \Delta \ (-L/2 < z < L/2). \end{cases} \tag{9.52}$$

We assume that the temperature dependence of the superconducting gap is given by $\Delta = \tanh(1.74\sqrt{T_\mathrm{c}/T - 1})$, where Δ_0 is the superconducting gap at $T = 0$ and T_c is the superconducting critical temperature. The interfacial scattering potential is defined as

$$V(z) = \frac{\hbar^2 k_\mathrm{F}}{m} Z \left[\delta(z + L/2) + \delta(z - L/2)\right]. \tag{9.53}$$

Since the system is rectangular, the wavefunction in the transverse (x and y) directions is given by

$$\mathrm{S}_{nl}(x, y) \equiv \sin(n\pi x/W) \sin(l\pi y/W), \tag{9.54}$$

where n and l are the quantum numbers which define the channel. The eigenvalue of the transverse mode for the channel (n,l) is

$$E_{nl} = \frac{\hbar^2}{2m}\left[\left(\frac{n\pi}{W}\right)^2 + \left(\frac{l\pi}{W}\right)^2\right]. \tag{9.55}$$

Let us consider the scattering of an electron with spin up in the channel (n, l) injected into the SC from the FM1 (0 in Fig. 9.7). There are the following eight scattering processes: Andreev reflection (1 in Fig. 9.7) and normal reflection (2 in Fig. 9.7) at the interface of FM1/SC, transmission to the SC (3, 4 in Fig. 9.7) and reflection at the interface of SC/FM2 (5, 6 in Fig. 9.7), transmission as an

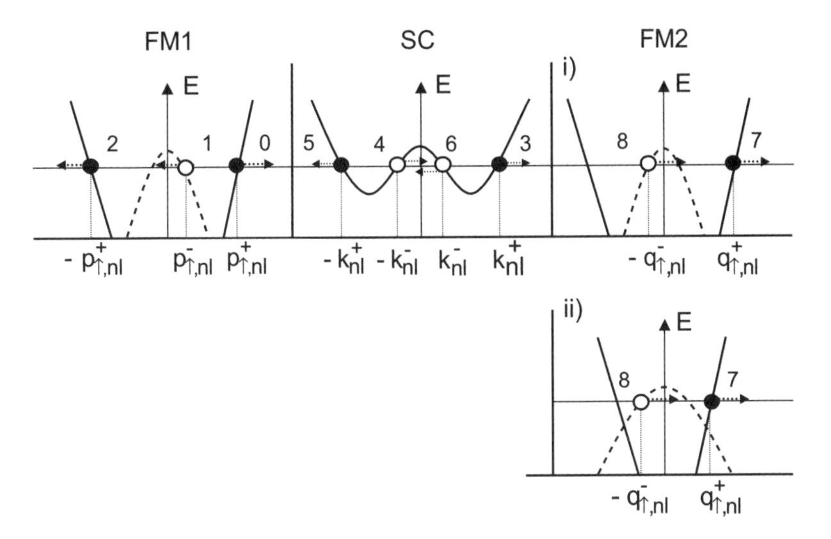

FIG. 9.7. Schematic diagrams of energy vs. momentum for the FM1/SC/FM2 double junction system with the parallel and antiparallel alignments of the magnetizations shown in panels (i) and (ii), respectively. The open circles denote holes, the closed circles electrons, and the arrows point in the direction of the group velocity. The incident electron with spin up in the channel (n, l) is denoted by 0, along with the resulting scattering processes: Andreev reflection (1), normal reflection (2) at the interface of FM1/SC, transmission to the SC (3, 4) and reflection at the interface of SC/FM2 (5, 6), transmission as an electron to the FM2 (7) and that as a hole (8) [31].

electron to the FM2 (7 in Fig. 9.7) and that as a hole (8 in Fig. 9.7). Therefore, the wavefunction in the FM1 ($z < -L/2$) is given by

$$\Psi^{\mathrm{FM1}}_{\sigma,nl}(\boldsymbol{r}) = \Bigg[\begin{pmatrix} 1 \\ 0 \end{pmatrix} e^{ip^{+}_{\sigma,nl}\left(z+\frac{L}{2}\right)} + a_{\sigma,nl} \begin{pmatrix} 0 \\ 1 \end{pmatrix} e^{ip^{-}_{\sigma,nl}\left(z+\frac{L}{2}\right)}$$

$$+ b_{\sigma,nl} \begin{pmatrix} 1 \\ 0 \end{pmatrix} e^{-ip^{+}_{\sigma,nl}\left(z+\frac{L}{2}\right)} \Bigg] \mathrm{S}_{nl}(x,y). \tag{9.56}$$

In the SC ($-L/2 < z < L/2$), we have

$$\Psi^{\mathrm{SC}}_{\sigma,nl}(\boldsymbol{r}) = \Bigg[\alpha_{\sigma,nl} \begin{pmatrix} u_0 \\ v_0 \end{pmatrix} e^{ik^{+}_{nl}\left(z+\frac{L}{2}\right)} + \beta_{\sigma,nl} \begin{pmatrix} v_0 \\ u_0 \end{pmatrix} e^{-ik^{-}_{nl}\left(z+\frac{L}{2}\right)}$$

$$+ \xi_{\sigma,nl} \begin{pmatrix} u_0 \\ v_0 \end{pmatrix} e^{-ik^{+}_{nl}\left(z-\frac{L}{2}\right)} + \eta_{\sigma,nl} \begin{pmatrix} v_0 \\ u_0 \end{pmatrix} e^{ik^{-}_{nl}\left(z-\frac{L}{2}\right)} \Bigg] \mathrm{S}_{nl}(x,y), \tag{9.57}$$

and in the FM2 ($L/2 < z$),

$$\Psi_{\sigma,nl}^{\mathrm{FM2}}(\boldsymbol{r}) = \left[c_{\sigma,nl} \begin{pmatrix} 1 \\ 0 \end{pmatrix} e^{iq_{\sigma,nl}^{+}\left(z-\frac{L}{2}\right)} + d_{\sigma,nl} \begin{pmatrix} 0 \\ 1 \end{pmatrix} e^{-iq_{\sigma,nl}^{-}\left(z-\frac{L}{2}\right)} \right] \mathrm{S}_{nl}(x,y). \quad (9.58)$$

Here $p_{\sigma,nl}^{\pm}$, $k_{nl}^{\pm}$ and $q_{\sigma,nl}^{\pm}$ are the wavenumbers in the FM1, SC and FM2, respectively:

$$p_{\sigma,nl}^{\pm} = \frac{\sqrt{2m}}{\hbar}\sqrt{\mu_{\mathrm{F}} \pm E \pm \sigma h_{\mathrm{ex}} - E_{nl}}, \quad (9.59)$$

$$k_{nl}^{\pm} = \frac{\sqrt{2m}}{\hbar}\sqrt{\mu_{\mathrm{F}} \pm \sqrt{E^2 - \Delta^2} - E_{nl}}. \quad (9.60)$$

For the parallel alignment, the wavenumber $q_{\sigma,nl}^{\pm}$ in FM2 is defined by the same formula as $p_{\sigma,nl}^{\pm}$:

$$q_{\sigma,nl}^{\pm} = \frac{\sqrt{2m}}{\hbar}\sqrt{\mu_{\mathrm{F}} \pm E \pm \sigma h_0 - E_{nl}}. \quad (9.61)$$

On the contrary, $q_{\sigma,nl}^{\pm}$ for the antiparallel alignment is defined as

$$q_{\sigma,nl}^{\pm} = \frac{\sqrt{2m}}{\hbar}\sqrt{\mu_{\mathrm{F}} \pm E \mp \sigma h_0 - E_{nl}}. \quad (9.62)$$

The coefficients $a_{\sigma,nl}, b_{\sigma,nl}, c_{\sigma,nl}, d_{\sigma,nl}, \alpha_{\sigma,nl}, \beta_{\sigma,nl}, \xi_{\sigma,nl}$, and $\eta_{\sigma,nl}$ are determined by matching the wavefunctions at the left and right interfaces. Solving Eq. (9.14), the probabilities of transmission and reflection are calculated following the BTK theory [9]. When an electron with σ-spin is injected from the FM1, the probability of the Andreev reflection $R_{\sigma,nl}^{he}$, normal reflection $R_{\sigma,nl}^{ee}$, and transmission as an electron and as a hole to the FM2, $T_{\sigma,nl}^{e'e}$ and $T_{\sigma,nl}^{h'e}$, are given by

$$\begin{cases} R_{\sigma,nl}^{he}(E) = \dfrac{p_{\sigma,nl}^{-}}{p_{\sigma,nl}^{+}} a_{\sigma,nl}^{*} a_{\sigma,nl}, \\ R_{\sigma,nl}^{ee}(E) = b_{\sigma,nl}^{*} b_{\sigma,nl}, \\ T_{\sigma,nl}^{e'e}(E) = \dfrac{q_{\sigma,nl}^{+}}{p_{\sigma,nl}^{+}} c_{\sigma,nl}^{*} c_{\sigma,nl}, \\ T_{\sigma,nl}^{h'e}(E) = \dfrac{q_{\sigma,nl}^{-}}{p_{\sigma,nl}^{+}} d_{\sigma,nl}^{*} d_{\sigma,nl}, \end{cases} \quad (9.63)$$

where the prime $e'(h')$ in Eq. (9.63) indicates the electron (hole) in the FM2. The probabilities for an incident electron with energy $E < 0$ can be calculated in a similar way. Using the fact that the BdG equation describing the scattering

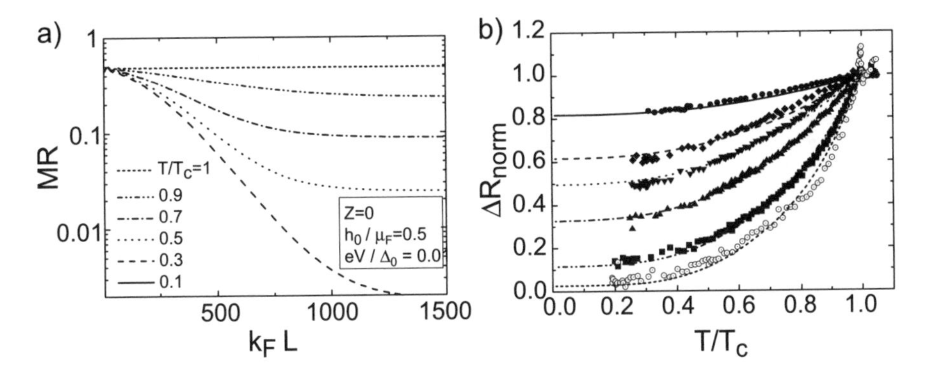

FIG. 9.8. (a) MR as a function of the thickness of the SC, $k_F L$. From top to bottom, T/T_c is 1, 0.9, 0.7, 0.5, 0.3, and 0.1. We assume $\xi_Q(E = T = 0) = 200/k_F$. (b) Excess resistance, $\Delta R = R_{AP} - R_P$, normalized by the value at T_c is plotted as a function of T/T_c. The solid curves show theoretical results for the thickness of the SC, L = 30, 40, 50, 60, 80, and 100 nm from top to bottom, where k_F is taken to be 1 Å^{-1} for Nb. The symbols show the experimental results by Gu *et al.* [7] for the thickness of Nb, t_{Nb} = 30, 40, 50, 60, 80, and 100 nm from top to bottom [31].

process of an incident electron with energy E and spin σ coincides with that of a hole with energy $-E$ and spin $-\sigma$, we can express the current as

$$I = \frac{e}{h} \sum_{nl,\sigma} \int_0^\infty \left(R^{eh}_{nl,\sigma} + R^{he}_{nl,\sigma} + T^{ee'}_{nl,\sigma} + T^{hh'}_{nl,\sigma} \right) \times \left[f_0 \left(E - \tfrac{eV}{2} \right) - f_0 \left(E + \tfrac{eV}{2} \right) \right] dE. \tag{9.64}$$

It should be noted that this expression for the current reduces to that derived by Lambert [32] for the NM/SC/NM double junction system when $h_0 = 0$. The magnetoresistance (MR) is defined as

$$\mathrm{MR} \equiv \frac{R_{AP} - R_P}{R_P}, \tag{9.65}$$

where $R_{P(AP)} = V/I_{P(AP)}$ is the resistance in the parallel (antiparallel) alignment.

In Fig. 9.8(a), MR is plotted as a function of the thickness of the SC, L, multiplied by the Fermi wavenumber k_F. We assume that the strength of the interfacial barrier is $Z = 0$ and the exchange field is $h_0 = 0.5\mu_F$. The side length of the cross-section is taken to be $W = 1000/k_F$. When the SC is in the normal conducting state ($T/T_c = 1$), MR is constant since we neglect the spin-flip scattering in the SC. When the SC is in the superconducting state (T/T_c = 0.1, 0.3, 0.5, 0.7, and 0.9), MR decreases with increasing thickness of the SC. MR at low temperatures $T/T_c \ll 1$ shows an exponential decrease in a wide range of

$k_F L$. The decrease of MR due to superconductivity is explained by considering the decay of the quasiparticle current in the SC. In the energy region below the superconducting gap ($|E| < \Delta$) where the energy of the transverse mode E_{nl} is smaller than the Fermi energy μ_F, the wavenumber $k_{nl}^{\pm}$ is expanded as

$$k_{nl}^{\pm} \sim \frac{\sqrt{2m}}{\hbar}\left(\mu_F \pm i\sqrt{\Delta^2 - E^2}\right)^{\frac{1}{2}} \sim k_F \pm i\frac{1}{2\xi_Q}. \tag{9.66}$$

The imaginary part in Eq. (9.66) gives the exponential decay term $\exp(-z/\xi_Q)$ in j_Q, where ξ_Q is the penetration depth given by

$$\xi_Q = \frac{\hbar v_F}{2\sqrt{\Delta^2 - E^2}}, \tag{9.67}$$

where v_F is the Fermi velocity. In the SC, the QP current with spin decreases exponentially and changes to the supercurrent carried by Cooper pairs with no spin in the range of ξ_Q from the interfaces. As a result, it becomes difficult to transfer spins from FM1 to FM2 and MR decreases with increasing thickness of the SC. The finite MR in the region of large L is due to QPs with energy above the superconducting gap ($E > \Delta$). The diffusive effect on the Andreev reflection is incorporated into our theory by replacing the penetration depth ξ_Q in the ballistic theory with the penetration depth in the dirty-limit ξ_Q^D [33].

Figure 9.8(b) shows the excess resistance $\Delta R = R_{AP} - R_P$ normalized by the value at (in the experiment, T slightly above) T_c (ΔR_{norm}) as a function of temperature. The solid curves indicate the calculated results and the symbols the experimental ones [7]. From Fig. 9.8(b), we obtain $\xi_Q^D(E = T = 0) = 46$, 36, 36, 33, 30, and 27 nm for the curves of $L = 30$, 40, 50, 60, 80, and 100 nm, respectively, where k_F is taken to be $1\,\text{Å}^{-1}$ for Nb [27]. These results indicate that Δ in the Nb film is reduced compared to that in a bulk Nb due to the proximity effect. Actually, the magnitude of the superconducting gap depends on the position z in the Nb film by the proximity effect. Here, we interpret the value of Δ as the averaged value of the superconducting gap with respect to z in the Nb film [31].

9.4 Crossed Andreev reflection

In the last section, we showed that in a FM/SC/FM double junction system the spin information is carried by evanescent quasiparticles in the SC as long as the thickness of the SC is less than the GL coherence length. In this section, we consider a system where the same bias voltage is applied to both of two ferromagnetic leads and the SC is earthed so that the current flows from two ferromagnetic leads to the SC (see Fig. 9.9 (a)). In such a system there is a novel quantum phenomenon called the crossed AR where an electron injected from FM1 into the SC captures another one in FM2 to form a Cooper pair in the SC [31, 32, 34–43]: i.e., an Andreev-reflected hole is not created in FM1 but in FM2. Deutscher and Feinberg [35] have discussed crossed Andreev reflection and MR

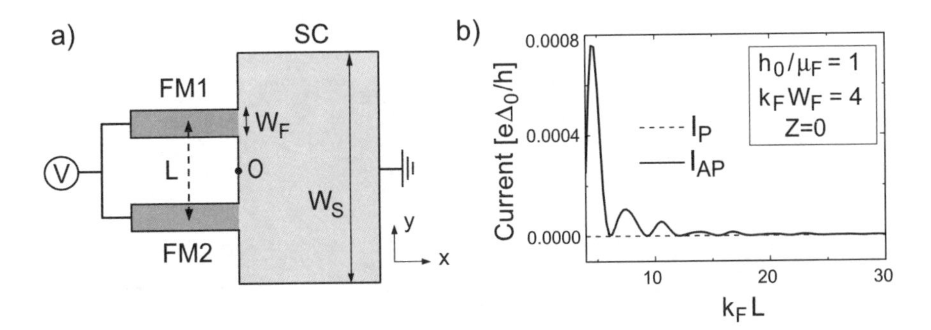

FIG. 9.9. (a) Schematic diagram of a superconductor (SC) with two ferromagnetic leads (FM1 and FM2). FM1 and FM2 with width $W_{\rm F}$ are connected to the SC with width $W_{\rm S}$ at $x = 0$. The distance between FM1 and FM2 is L. (b) The current as a function of L. FM1 and FM2 are half-metals ($h_0/\mu_{\rm F} = 1$). The solid and dashed lines are for the currents in the antiparallel and parallel alignments of the magnetizations, respectively [31].

by using the BTK theory [9]. They argued that crossed Andreev reflection should occur when the distance between FM1 and FM2 is of the order of or less than the size of the Cooper pairs (the coherence length), and calculated the probability of crossed Andreev reflection in the case that both ferromagnetic leads are half-metals and the spatial separation of FM1 and FM2 is neglected (one-dimensional model), i.e., the effect of the distance between two ferromagnetic leads on crossed AR is not incorporated. Subsequently, Falci *et al.* [37] have discussed crossed AR and the elastic cotunneling in the tunneling limit by using the lowest order perturbation of the tunneling Hamiltonian. However, to elucidate the effect of crossed AR on the spin transport more precisely, it is important to explore how crossed AR depends on the distance between two ferromagnetic leads as well as on the exchange field of FM1 and FM2, for arbitrary transparency of the interface from the metallic to the tunneling limit. Here, we present a theory for crossed AR [31]. The total current of the system can be expressed as

$$I = \sum_{\sigma,m,i} \left[I^{i,e}_{\sigma,m} + I^{i,h}_{\sigma,m} \right], \tag{9.68}$$

where $I^{i,e}_{\sigma,m}$ and $I^{i,h}_{\sigma,m}$ are the currents carried by electrons with spin σ in channel m in the i-th ferromagnetic electrode FMi defined as

$$\begin{aligned} I^{i,e}_{\sigma,m} = \frac{e}{h} \sum_{l=1}^{\infty} \int_0^{\infty} \Big\{ & \left(R^{i,he}_{\sigma,lm} + \tilde{R}^{i,he}_{\sigma,lm} \right) \Big[f_0(E) - f_0(E+eV) \Big] \\ & + \left(1 - R^{i,ee}_{\sigma,lm} - \tilde{R}^{i,ee}_{\sigma,lm} \right) \Big[f_0(E-eV) - f_0(E) \Big] \Big\} dE, \end{aligned} \tag{9.69}$$

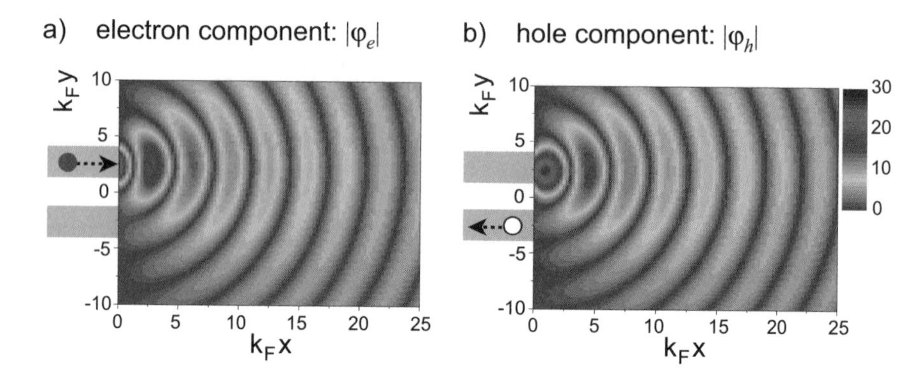

FIG. 9.10. Spacial variation of the absolute values of the (a) electron component $|\varphi_e|$ and (b) hole component $|\varphi_h|$ in the wave function of the SC when a spin up electron is injected from FM1 to SC. The distance between FM1 and FM2 is taken to be $k_{\rm F}L = 5$.

$$I^{i,e}_{\sigma,m} = \frac{e}{h}\sum_{l=1}^{\infty}\int_0^{\infty}\Big\{\Big(R^{i,eh}_{\sigma,lm} + \tilde{R}^{i,eh}_{\sigma,lm}\Big)\Big[f_0(E-eV) - f_0(E)\Big] + \Big(1 - R^{i,hh}_{\sigma,lm} - \tilde{R}^{i,hh}_{\sigma,lm}\Big)\Big[f_0(E) - f_0(E+eV)\Big]\Big\}dE. \quad (9.70)$$

Here $R^{i,he}_{\sigma,m}$, $R^{i,ee}_{\sigma,m}$, $\tilde{R}^{i,he}_{\sigma,m}$, and $\tilde{R}^{i,he}_{\sigma,m}$ are the probabilities of AR, normal reflection, crossed AR, and crossed normal reflection in FMi, respectively. These probabilities are obtained by solving the scattering problems as shown in the previous section. The magnetoresistance is defined as in Eq. (9.65). In the numerical calculation, the temperature, the applied bias voltage, the width of the SC, and the superconducting order parameter are $T/T_{\rm c} = 0.01$, $eV/\Delta_0 = 0.01$, $W_{\rm S} = 1000/k_{\rm F}$, and $\Delta_0/\mu_{\rm F} = 200$, respectively, where $k_{\rm F}$ is the Fermi wave number.

First, we consider the case that FM1 and FM2 are half-metals ($h_0/\mu_{\rm F} = 1$) and neglect the interfacial barrier, i.e., $Z = mH/\hbar^2 k_{\rm F} = 0$. The width of FM1 and FM2 is taken to be $W_{\rm F} = 4/k_{\rm F}$, where only one propagating mode exists in the system. We obtain the maximum possible value of MR, i.e., MR $= -1$ independently of L. In order to understand this behavior, we consider the L dependence of the currents in the parallel and antiparallel alignments as shown in Fig. 9.9(a). When an electron with spin up in FM1 is injected into the SC, ordinary AR does not occur because electrons with down spin are absent in FM1. In the parallel alignment, crossed AR does not occur either because there are no electrons with down spin in FM2. Therefore, no current flows in the system as shown in Fig. 9.9(a). On the other hand, in the antiparallel alignment, while the ordinary AR is absent, crossed AR occurs because there are electrons with down spin in FM2, which is a member of a Cooper pair, for an incident electron with spin-up from FM1, and, therefore, a finite current flows in the system as shown in Fig 9.9(a). As a result, we find MR $= -1$ irrespective of L in the case of

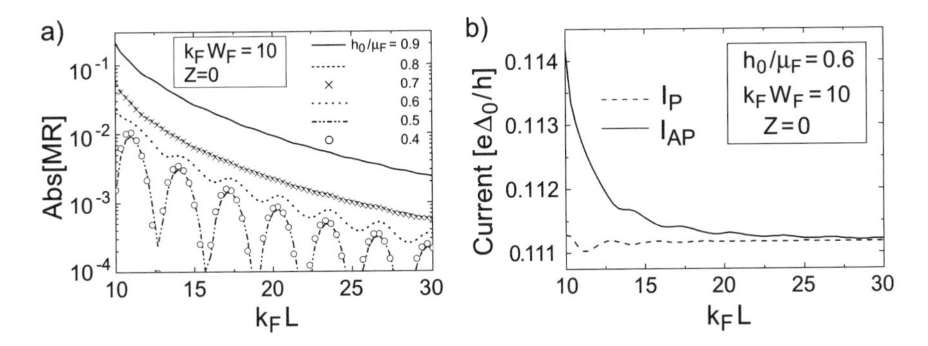

FIG. 9.11. (a) The absolute value of MR as a function of L in the case that the exchange fields h_0/μ_F are 0.4, 0.5, 0.6, 0.7, 0.8, and 0.9. (b) The current as a function of L in the case of $h_0/\mu_F = 0.6$. The solid and dashed lines are for the currents in the antiparallel and parallel alignments, respectively [31].

half-metallic FM1 and FM2. The current in the antiparallel alignment decreases oscillating with increasing L due to interference between the wave functions in FM1 and FM2. The probability of crossed Andreev reflection decreases rapidly as $k_F L^{-3}$ [31].

In order to understand the mechanism of crossed Andreev reflection more clearly, we consider the spacial variation of the wave function in the SC. Figures 9.10(a) and (b) show the absolute value of the electron and hole components, φ_e and φ_h, respectively, which are defined as $\Psi_{SC} = {}^t(\varphi_e, \varphi_h)$. Here, we consider the case that a spin-up electron is injected to the SC from FM1. As shown in Fig. 9.10(a), the injected electron penetrates into the SC as an evanescent quasiparticle and the electron component of the wave function is diffracted in the SC, oscillating with period π/k_F and decaying in the range of ξ. Because the quasiparticle is viewed as a composite particle between the electron and hole components, the hole component of the wave function emerges from the interior of the SC towards the contacts as shown in Fig. 9.10(b). If the hole component connects the wave function of the hole in FM2, the hole is reflected back to FM2 (crossed Andreev reflection). As seen from Fig. 9.10(b), the absolute value of the hole component at the interface has a peak at $y = 2.5/k_F$, and decays oscillatory along the y-direction. The L dependence of the probability of crossed Andreev reflection originates from the wave nature of the hole component in the SC, and is characterized by the Fermi wave number k_F. Therefore, the wave nature of the evanescent quasiparticle in SC is essential for crossed Andreev reflection.

We next consider the L dependence of MR for several values of the exchange field in the case that $W_F = 10/k_F$, and $Z = 0$ (Fig. 9.11a). In this case, there are several propagating modes in FM1 and FM2. The magnitude of the MR decreases with increasing L for each value of the exchange field. This behavior of the MR is understood by considering the L dependence of the current in the parallel and antiparallel alignments. As shown in Fig. 9.11(b), in the case that $h_0 =$

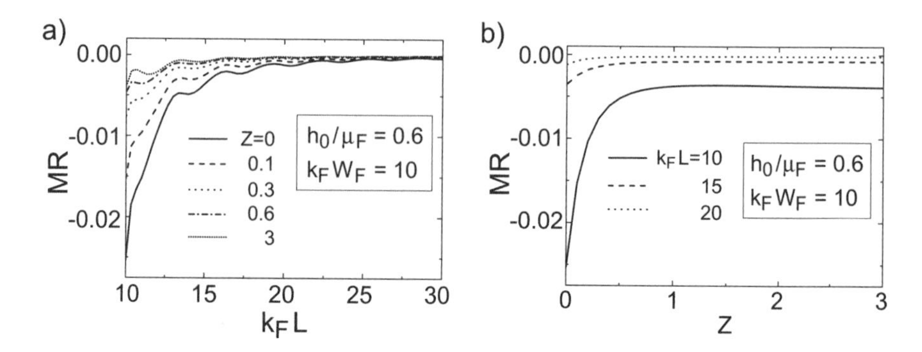

FIG. 9.12. (a) The absolute value of the MR as a function of L in the case that the exchange fields h_0/μ_F are 0.4, 0.5, 0.6, 0.7, 0.8, and 0.9. (b) The current as a function of L in the case of $h_0/\mu_\mathrm{F} = 0.6$. The solid and dashed lines are for the currents in the antiparallel and parallel alignments, respectively [31].

$0.6\mu_\mathrm{F}$, the finite current in the parallel alignment flows because ordinary Andreev reflection occurs, and is almost independent of L. On the other hand, the current in the antiparallel alignment decreases with increasing L since the contributions of crossed Andreev reflection process to the current decreases with increasing L, and therefore the magnitude of the MR decreases with increasing L. In this case, the oscillation of the current in the antiparallel alignment is suppressed because electrons and holes in the several propagating modes l contribute to the current and wash out the oscillation.

Figure 9.12(a) shows the L dependence of the MR for $h_0 = 0.6\mu_\mathrm{F}$ and several values of interfacial barrier parameter Z. The MR approaches zero with increasing L and shows a strong dependence on the height of the interfacial barrier Z. The fact that the MR decreases with increasing L is explained in the same way as in the case of no interfacial barriers as shown in Fig. 9.12(a). To investigate the Z dependence of the MR in detail, we calculate the Z dependence of the MR for $k_\mathrm{F}L = 10, 15$, and 20 as shown in Fig. 9.12(a). The magnitude of the MR decreases with increasing Z in the range of $Z \lesssim 0.5$ and is almost constant for L in the range of $Z \gtrsim 0.5$. This dependence is understood as follows. The MR consists of the denominator I_AP and the numerator $I_\mathrm{P} - I_\mathrm{AP}$, which mainly come from the process of ordinary AR and crossed AR, respectively. Crossed AR is more sensitive to scattering at the interfacial barriers than ordinary Andreev reflection, and therefore the value of $I_\mathrm{P} - I_\mathrm{AP}$ decreases more rapidly than that of I_AP in the range of $Z \lesssim 0.5$, and therefore the magnitude of the MR decreases with increasing Z for $k_\mathrm{F}L = 10, 15$, and 20 as shown in Fig. 9.12(b).

Acknowledgements

The authors thank K. Kikuchi and T. Yamashita for valuable contributions to the works described in this chapter. This work was supported by CREST, MEXT.KAKENHI, the NAREGI Nanoscience Project, and the NEDO Grant.

References

[1] A. F. Andreev, Sov. Phys-jetp. Engl. Trans. **19**, 1228 (1964).
[2] C. J. Lambert and R. Raimondi, J. Phys.-Condens. Matter **10**, 901 (1998).
[3] C. W. J. Beenakker, Rev. Mod. Phys. **69**, 731 (1997).
[4] M. J. M. de Jong and C. W. J. Beenakker, Phys. Rev. Lett. **74**, 1657 (1995).
[5] R. J. Soulen, Jr., J. M. Byers, M. S. Osofsky, B. Nadgorny, T. Ambrose, S. F. Cheng, P. R. Broussard, C. T. Tanaka, J. Nowak, J. S. Moodera, A. Barry, and J. M. D. Coey, Science **282**, 85 (1998).
[6] S. K. Upadhyay, A. Palanisami, R. N. Louie, and R. A. Buhrman, Phys. Rev. Lett. **81**, 3247 (1998).
[7] J. Y. Gu, J. A. Caballero, R. D. Slater, R. Loloee, and W. P. Pratt, Jr., Phys. Rev. B **66**, 140507 (2002).
[8] T. Yamashita, S. Takahashi, H. Imamura, and S. Maekawa, Phys. Rev. B **65**, 172509 (2002).
[9] G. E. Blonder, M. Tinkham, and T. M. Klapwijk, Phys. Rev. B **25**, 4515 (1982).
[10] M. Tinkham, *Introduction to Superconductivity* (McGraw-Hill, New York, 1976).
[11] P. G. de Gennes, *Superconductivity of Metals and Alloys* (Benjamin, New York, 1966).
[12] R. Meservey and P. M. Tedrow, Phys. Rep. **238**, 173 (1994).
[13] R. J. Soulen, Jr., M. S. Osofsky, B. Nadgorny, T. Ambrose, P. Broussard, S. F. Cheng, J. Byers, C. T. Tanaka, J. Nowack, J. S. Moodera, G. Laprade, A. Barry, M. D. Coey, J. Appl. Phys. **85**, 4589 (1999).
[14] Y. Ji, G. J. Strijkers, F. Y. Yang, C. L. Chien, J. M. Byers, A. Anguelouch, G. Xiao, and A. Gupta, Phys. Rev. Lett. **86**, 5585 (2001a).
[15] I. I. Mazin, A. A. Golubov, and B. Nadgorny, J. Appl. Phys. **89**, 7576 (2001).
[16] Y. Ji, G. J. Strijkers, F. Y. Yang, and C. L. Chien, Phys. Rev. B **64**, 224425 (2001b).
[17] J. S. Parker, S. M. Watts, P. G. Ivanov, Phys. Rev. Lett. **88**, 196601 (2002).
[18] R. P. Panguluri, G. Tsoi, B. Nadgorny, S. H. Chun, N. Samarth, and I. I. Mazin, Phys. Rev. B **68**, 201307(R) (2003).
[19] P. Raychaudhuri, A. P. Mackenzie, J. W. Reiner, and M. R. Beasley, Phys. Rev. B **67**, 020411(R) (2003).
[20] R. P. Panguluri, B. Nadgorny, T. Wojtowicz, W. L. Lim, X. Liu, and J. K. Furdyna, Appl. Phys. Lett. **84**, 4947 (2004).
[21] J. G. Barden, J. S. Parker, P. Xiong, S. H. Chun, and N. Samarth, Phys. Rev. Lett. **91**, 056602 (2003).
[22] K. Kikuchi, H. Imamura, S. Takahashi, and S. Maekawa, Phys. Rev. B **65**, 020508(R) (2001).
[23] H. Imamura, K. Kikuchi, S. Takahashi, and S. Maekawa, J. Appl. Phys. **91**, 7032 (2002).
[24] E. N. Bogachek, A. N. Zagoskin, and I. O. Kulik, Sov. J. Low Temp. Phys. **16**, 796 (1991).
[25] A. Furusaki, H. Takayanagi, and M. Tsukada, Phys. Rev. Lett. **67**, 132 (1991).
[26] A. Furusaki, H. Takayanagi, and M. Tsukada, Phys. Rev. B **45**, 10563 (1992).

[27] N. W. Ashcroft and N. D. Mermin, *Solid State Physics* (Saunders College, Philadelphia, 1976).
[28] A. Yacoby and Y. Imry, Phys. Rev. B **41**, 5341 (1990).
[29] I. Žutić and O. Valls, Phys. Rev. B **61**, 1555 (2000).
[30] I. Žutić, J. Fabian, and S. Das Sarma, Rev. Mod. Phys. **76**, 323 (2004).
[31] T. Yamashita, H. Imamura, S. Takahashi, and S. Maekawa, Phys. Rev. B **67**, 094515 (2003).
[32] C. J. Lambert, J. Phys.-Condens. Matter **3**, 6579 (1991).
[33] A. A. Abrikosov, L. P. Gorokov, and I. E. Dzyaloshinski, *Methods of Quantum Field Theory in Statistical Physics* (Dover, New York, 1975).
[34] J. M. Byers and M. E. Flatté, Phys. Rev. Lett. **74**, 306 (1995).
[35] G. Deutscher and D. Feinberg, Appl. Phys. Lett. **76**, 487 (2000).
[36] G. Deutscher, J. Supercond. **15**, 43 (2002).
[37] G. Falci, D. Feinberg, and F. W. J. Hekking, Europhys. Lett. **54**, 255 (2001).
[38] R. Mélin, J. Phys.-Condens. Matter **13**, 6445 (2001).
[39] R. Mélin and D. Feinberg, Eur. Phys. J. B **26**, 101 (2002).
[40] R. Mélin, H. Jirari, and S. Peysson, J. Phys.-Condens. Matter **15**, 5591 (2003).
[41] V. Apinyan and R. Mélin, Eur. Phys. J. B **25**, 373 (2002).
[42] Y. Zhu, Q. Sun, and T. Lin, Phys. Rev. B **65**, 024516 (2002).
[43] D. Beckmann, H. B. Weber, and H. v. Löhneysen, Phys. Rev. Lett. **93**, 197003 (2004).

Index